Boris N. Khoromskij

Tensor Numerical Methods in Scientific Computing

Author
DrSci. Boris N. Khoromskij
Max Planck Institute for
Mathematics in the Sciences
Inselstr. 22–26
04103 Leipzig
Germany
bokh@mis.mpg.de

ISBN 978-3-11-037013-3
e-ISBN (PDF) 978-3-11-036591-7
e-ISBN (EPUB) 978-3-11-039139-8
ISSN 1865-3707

Library of Congress Control Number: 2018934808

Bibliographic information published by the Deutsche Nationalbibliothek
The Deutsche Nationalbibliothek lists this publication in the Deutsche Nationalbibliografie; detailed bibliographic data are available on the Internet at http://dnb.dnb.de.

Typesetting: le-tex publishing services GmbH, Leipzig
Printing and binding: CPI books GmbH, Leck

www.degruyter.com

Contents

1 Introduction

Everything should be made as simple
as possible, but not simpler.
A. Einstein

This book is motivated by several lecture courses on tensor numerical methods in scientific computing given by the author at the Max Planck Institute for Mathematics in the Sciences and the University of Leipzig in 2005/2007, at the University of Linz/RICAM in 2009, the University and ETH Zürich in 2010, the University of Rome Tor Vergata in 2011, as well as at summer schools at the University of Jyväskylä in 2015, and at the Shanghai Jiao Tong University in 2017. These lecture courses aimed to demonstrate that the newly developed tensor numerical methods provide powerful tools for multidimensional scientific computing.

The most difficult computational problems nowadays arise from *high dimensionality* encountered in numerical modeling of problems in quantum chemistry and material science, in molecular dynamics, in computer simulations of stochastic processes, or in machine learning. Specifically, the solution of multidimensional partial differential equations (PDEs), which model complicated physical phenomena, requires the approximation of multivariate functions and integrodifferential operators defined in $\mathbb{R}^d$ with $d \geq 3$.

The *challenge of many dimensions* can easily be seen in the following examples. The many-particle Schrödinger equation is posed in $\mathbb{R}^d$ with $d = 3N$, where N is the number of electrons in a molecular system. Even in simplified quantum chemical models, like the Hartree–Fock equation or density functional theory, where the dimensionality of the problem is reduced to $d = 3$, the singular kernels of the integral operators involved remain functions in six spatial dimensions. Recent numerical methods for stochastic PDEs reduce the problem with stochastic coefficients to a solution of the deterministic equation in physical space variables, where coefficients depend on the high dimensional parameter with the number of dimensions up to several hundred. In stochastic modeling of multiparticle chemical reactions by the Fokker–Planck or chemical master equations the dimensionality of dynamical problems is related to the number of interacting particles and may vary from several tens to hundreds.

These difficulties were recognized a long time ago in both the computational physics/chemistry and scientific computing communities. We refer to Paul Dirac (1929): "The underlying physical laws necessary for the mathematical theory of a large part of physics and the whole of chemistry are thus completely known, and the difficulty is only that the exact application of these laws leads to equations much too complicated to be soluble." Richard Bellman [25] (1961) described the problem of the exponential increase in the amount of data associated with adding extra dimensions to a mathematical space: "In view of all that we have said in the foregoing sections,

https://doi.org/10.1515/9783110365917-001

the many obstacles we appear to have surmounted, what casts the pall over our victory celebration? It is the curse of dimensionality, a malediction that has plagued the scientist from earliest days."

The point is that most of the traditional numerical approaches exhibit computational complexity that grows exponentially in the physical dimension d as n^d, and hence, for large d they become infeasible because of "the curse of dimensionality." Commonly used approaches, like fast multipole [126], sparse grids and hyperbolic cross approximations [55, 330], hierarchical matrices with low rank blocks [133, 140, 149], and wavelet multiresolution methods [265], already rely on a certain separation of variables, where the separation principle usually applies to the pair of variables, originating from the so called Schmidt decomposition. These methods still inherit the exponential complexity scaling in dimension d in a slightly relaxed form so that the numerical modeling may become challenging already for 3D problems.

The first attempt at a low parametric representation of multivariate functions traces back to Kolmogorov's paradigm (1957) [234], which proves the existence of a univariate implicit parametrization for the class of Lipshitz continuous d dimensional functions. However, the constructive implementation of such a decomposition remains an open problem.

Methods that allow linear scaling in the dimension parameter d rely on the idea of tensor product constructions at all computationally extensive stages of the solution process. Tensor product approaches were initially investigated in connection with the analytical low rank approximation of functions and operators in $\mathbb{R}^d$. A mathematical discipline that these belong to is called a nonlinear approximation theory. We may summarize that the novel tensor numerical methods for the solution of the multidimensional PDEs emerged as a *bridging* of the nonlinear approximation theory for multivariate functions and operators, with the traditional and recently developed multilinear algebra.

In multilinear algebra the so called polyadic or canonical tensor decomposition, the expression of a multidimensional tensor by a sum of products, was invented by F. Hitchcock in 1927 [165]. Later on L. Tucker in 1966 introduced the Tucker tensor decomposition, which allows an efficient algorithmic implementation [350]. These rank structured representations of multidimensional data arrays have been extensively discussed in the literature on principal component analysis (PCA) methods mainly focusing on purely algebraic techniques. Traditional applications of the so called multiway decomposition have been regarded for a long time in chemometrics, signal processing, and statistical data modeling, where a low accuracy fitting of small size data arrays is usually required.

However, in modern scientific computing the amount of data is on a much larger scale. Indeed, in the numerical solution of multidimensional PDEs the higher order tensors represent multivariate functions and operators in $\mathbb{R}^d$ discretized on large d-fold spatial grids of size $n \times \cdots \times n$, presupposing the storage size $O(n^d)$, where the univariate grid parameter n may vary from many hundreds to many thousands. Even in low dimensions, say for $d = 3$, severe limitations may arise in computations on large spatial grids of size n^3 required for accurate representation of functions with multiple singularities and highly inhomogeneous or oscillating data. This is the case in 3D electronic structure calculations for large molecules or lattice structured systems, in numerical modeling of multiparticle interaction potentials, and in many other problems requiring high accuracy. For such problems the main prerequisite for application of the canonical and Tucker tensor formats would be a fast decay of the approximation error in the rank parameter.

In 2006 it was discovered that for a class of function related tensors the error in their Tucker tensor approximation decays exponentially fast with respect to the Tucker rank parameter [206]. It was proven that the rank structured approximation in this basic tensor format ensures the accuracy of order $O(n^{-p})$, which can be achieved with the rank bound $O(\log n)$ uniformly in the dimension, thus reducing the numerical costs to $O(dn \log^q n)$. Previously, the exponentially fast convergence of the canonical approximation in tensor rank was proven by I. Gavrilyuk, W. Hackbusch, and B. Khoromskij for the class of analytic radial basis functions (see [110, 112, 141]), where the sinc approximation techniques were gainfully applied. Numerical multilinear canonical tensor decomposition remains a difficult problem.

Furthermore, the favorable feature that for function related tensors the Tucker ranks typically depend only logarithmically on the grid size has led to the idea of the multigrid Tucker decomposition [212]. This approach allows us to reduce the cost $O(n^{d+1})$ of the most computationally consuming part in the standard Tucker algorithm, that is the higher order singular value decomposition (HOSVD) [241, 247, 248], thus enabling decomposition of function related tensors in $O(n^d)$ computational cost. However, HOSVD requires the full size tensors, which is not feasible for numerical modeling in physics, quantum chemistry, and in multidimensional scientific computing. Thus the HOSVD approach does not break the curse of dimensionality and actually has a limited significance in computational practice.

To that end, a significant advance was brought about by the so called reduced higher order singular value decomposition (RHOSVD) as part of the canonical-to-Tucker (C2T) transform introduced by B. Khoromskij and V. Khoromskaia in 2008, [212]. It was demonstrated that for the Tucker decomposition of function related tensors given in the canonical form (say, resulting from certain algebraic transforms or analytic approximations) there is no need to build a full tensor. Indeed, it is enough to find the orthogonal basis only for directional matrices of the canonical tensor, which consist of skeleton vectors in every single dimension. The C2T decomposition proved to be an efficient tool for reducing the redundant rank parameter in the large canonical tensors.

This progress in tensor approximation methods led up to the efficient grid based solver for the nonlinear 3D integrodifferential Hartree–Fock equation [183, 195, 214]. The rank structured tensor numerical approach enables calculation of the convolution integral operators with the Coulomb potential in $\mathbb{R}^3$ in $O(n \log n)$ complexity, so that the univariate grid size of the order of $n \approx 10^5$ could be employed. Thus, for all integral operators in the Hartree–Fock equation the 3D analytical integration is completely avoided, namely, it is substituted by the grid based tensor algorithms in 1D complexity [184, 185, 212].

In the recent years the results on low rank canonical and Tucker tensor decomposition of function related tensors have been gainfully applied to the problems arising in the classical potential theory, in ab initio models for electronic and molecular structure calculations, to stochastic/parametric PDEs, for some multidimensional dynamical models, in geometric homogenization theory, and in isogeometric analysis, as well as for approximation of the complicated 6D integrals in the Boltzmann equation modeling the dilute gas.

The matrix product states (MPS) or tensor train (TT) product type tensor formats are aimed at multilinear algebra in higher dimensions. The MPS type representations have long been recognized as the suitable tools in the numerical simulations of the spin type systems, FCI equations in computational quantum chemistry, and in the quantum information theory; see for example [356, 365] and the survey papers [321, 322, 354]. A major impact on further developments of the tensor numerical methods in scientific computing was due to the introduction of the tensor train (TT) multiplicative format by I. Oseledets and E. Tyrtyshnikov in 2009 [289, 292]. This tensor format is well suited to the function related multilinear algebra in higher dimensions. The TT tensor format became popular mostly due to the development of the advanced TT Toolbox, [290]. The closely related hierarchical Tucker (HT) tensor representation was introduced and analyzed in 2009 by W. Hackbusch and S. Kühn, [146]. Both the TT and HT tensor formats were established on the basis of an earlier concept of hierarchical dimension splitting [206]. Note that although the multiplicative TT and HT parametrizations formally apply to any full format tensor in higher dimensions, they become computationally feasible only when using the RHOSVD like procedures applied either to the canonical format input, or to tensors already given in the TT form. The HOSVD in MPS type formats was discussed in [123, 289, 356].

The numerical analysis of problems related to the grid based representation of highly nonregular multivariate functions has led to the development of new classes of tensor formats that can be considered as a further generalization of the traditional canonical, Tucker, and MPS/TT type parametrizations, which were initially established for approximation of rather general numerical data.

The quantized tensor train (QTT) approximation of function generated vectors, introduced by B. Khoromskij in 2009, [196, 197], was proven to provide the logarithmic data compression $O(d \log n)$ on the wide class of functions in $\mathbb{R}^d$ sampled on a tensor grid of size n^d. It was proven in [196, 197] that for function generated vector of size $n = q^L$, its reshaping into a $q \times \dots \times q$ hypercube allows a small TT rank decomposition of the resultant L dimensional tensor. For example, if one samples an exponential function on a uniform grid in the form of a long vector of size $n = 2^{20}$, the quantized image can be represented in the QTT tensor format by using as few as $2 \cdot 20$ real numbers. The low rank TT representation of the reshaped $2^L \times 2^L$ matrices, observed in numerical experiments, was reported in 2009 by I. Oseledets [286, 287]. This can be viewed as the prototype of the QTT representation to the discretized operators.

In recent years, the QTT tensor approximation method has been proven to provide an efficient tensor tool to tackle a wide class of real life problems, where the traditional tensor approaches are not efficient; see the detailed discussion in Chapter 4. In particular, the efficient TT and QTT based solvers have been developed for the solution of the challenging dynamical Fokker–Planck [84] and chemical master [86] equations capable of treating the d dimensional problems with d equal to several tens; see Section 5.3. The QTT format proved to be efficient for accurate integration of highly oscillating functions [226] and for solving the elliptic equations in geometric homogenization [224].

The tensor structured approach appeared to be efficient in summation of the long range interaction potentials on large finite lattices in the nonperiodic setting. The recent method of summing the electrostatic potentials on 3D $L \times L \times L$ lattices using assembled canonical or Tucker vectors reduces summation to amazing $O(L)$ computational work, instead of $O(L^3)$ in the Ewald type approaches [188, 189].

The novel range separated (RS) tensor format was recently proposed and analyzed by P. Benner, V. Khoromskaia and B. Khoromskij in [30] in relation with the rank structured tensor approximation of highly nonregular functions with multiple singularities in $\mathbb{R}^3$, sampled on the fine $n \times n \times n$ grid. These may be the electrostatic potentials of a large atomic system like a biomolecule or the radial basis functions in the modeling of scattered data. The multiple calculation of electrostatic potentials is the main computational bottleneck in the numerical simulation of many-particle dynamics. Clearly, the traditional tensor formats are infeasible for representation of such nonstructured data. It was proven [30] that the sums of long range contributions from all particles can be represented in a form of the low rank canonical or Tucker tensor at the $O(dn)$ storage cost almost independently of the number of particles N. The basic computational tool here is the RHOSVD algorithm; see details in Section 5.5. The representation complexity of the short range part is $O(N)$. The main advantage of the RS tensor format is that the separation of the long and short range parts is performed just by sorting skeleton vectors in the canonical tensor representation of the single generating kernel.

The new QTT and RS tensor representations in tensor numerical analysis differ from the traditional formats of the multilinear algebra because they originate from approximation theory applied to specific classes of functions, and therefore they are intrinsically ‘function related tensor formats’. They may be useful in a wide range of challenging computational problems in the electronic structure calculations for finite lattices, in geometric homogenization theory, in calculation of interaction potentials for many-particle systems, in multiparticle dynamics, in multidimensional scattered data modeling, for accurate integration of highly oscillating functions, and other problems. In particular, the new approach to the solution of elliptic equations with highly oscillating coefficients based on the QTT approximation was proposed in [224, 225]. The numerical primer in the stochastic homogenization theory for elliptic problems was recently reported in [193].

There is an extensive literature on different aspects of multilinear algebra, tensor calculus and related issues. To that end, we refer to the following monographs [1, 50, 127, 138, 176, 192, 246, 275, 370], surveys [55, 68, 69, 72, 125, 153, 191, 209, 210, 232, 321], and lecture notes [198, 200, 203], among others. Examples of the low rank approximation techniques in multivariate data analysis can be found in [30, 39, 80, 85, 188, 219, 291, 306].

In spite of the considerable promises, there is still a demand for understanding the deep mathematics behind the tensor methodology, for practical realization, and for construction of the efficient and reliable algorithms for real life problems. We note that the progress in theoretical understanding of tensor techniques appears in the process of numerical solution of the challenging problems. The purpose of this research monograph is to describe and analyze the recent tensor numerical methods based on the concept of *low rank tensor approximation to operators and functions*, and to demonstrate their efficiency on various examples of multidimensional PDEs. These methods are designed to meet the claims of modern applications, thus opening prospects for efficient numerical simulation of demanding problems stemming from high dimensional elliptic differential, integral, or temporal parabolic equations, which may depend on many parameters.

In general, there are still limitations for the application of the tensor numerical approaches to problems with complicated geometries in low dimensions, $d \leq 3$, where the standard FEM and BEM techniques combined with high performance computing remain appropriate.

Note that the existence of low rank tensor approximations to discretized functions and operators can be explained by highly redundant data representations, which are typical for the traditional grid based numerical techniques. In this way, the rank structured tensor parametrization provides the means to find the 'hidden structure' in the functions solving the governing PDEs, indicating that the essential information can be represented on low parametric manifolds such that complexity in the relevant physical data typically scales only linearly in the dimensionality parameters.

This research monograph offers an introduction to tensor numerical methods and suggests constructive recipes and algorithmic schemes for the numerical treatment of multidimensional problems in scientific computing. In particular, it presents the efficient tensor structured algorithms for solving the Hartree–Fock spectral problems, the dynamical Fokker–Planck and chemical master equations, and the stochastic/parametric PDEs, as well as for the elliptic equations with highly oscillating coefficients arising in geometric and stochastic homogenization theory. It can be viewed as a short tour through a number of papers on tensor numerical methods published by the author in the recent decade of his work at the Max Planck Institute for Mathematics in the Sciences in Leipzig, Germany.

This book is addressed to a broad audience of readers including undergraduate and postgraduate students as well as researchers in numerical analysis, engineering, and natural sciences.

Leipzig, October 2017 Boris N. Khoromskij

Historical remarks and acknowledgments

I would like to thank Wolfgang Hackbusch for inviting me to the newly organized Institute in 1999, and for our productive collaboration during the time when he was the director at the Max Planck Institute (MPI) for Mathematics in the Sciences (MIS) in Leipzig. This resulted in our 25 joint papers on $\mathcal{H}$-matrix and rank-structured approximation techniques. I also thank Ivan Gavrilyuk for our fruitful collaboration on sinc approximation of multidimensional functions and operators. I appreciate our starting collaborative work in 2007–2011 with Heinz-Juergen Flad, Venera Khoromskaia, and Reinhold Schneider, that made a breakthrough in the development and first application of the tensor numerical methods in ab initio quantum chemistry.

When I first invited Eugene Tyrtyshnikov to the MPI in Leipzig in 2003 nobody could predict the results of this meeting. I am grateful to Eugene for introducing me to the world of multilinear algebra and for our successful collaboration in the starting period. In 2009 he came to the MPI in Leipzig together with his group of young researchers Ivan Oseledets, Sergey Dolgov, Dmitry Savostyanov, and Vladimir Kazeev. Over the following years, during their intensive visits to the MPI in Leipzig, we made a number of interesting research projects on tensor numerical and multilinear algebra.

I am grateful to Christoph Schwab for our inspiring joint work in 2009–2010 on the novel tensor numerical approaches for stochastic/parametric PDEs. I am appreciative to Peter Benner, the director at the Max Planck Institute for Dynamics of Complex Technical Systems in Magdeburg, for our successful collaboration, which has led to a number of promising results. I would like to thank Felix Otto, the director at the MPI MIS in Leipzig, for our collaboration on the unexplored numerical techniques for stochastic homogenization problems. I thank Ivan Oseledets, Ulrich Langer, Stefan Sauter, and Sergey Repin for interesting joint works on investigation of tensor numerical methods. I thank my former colleagues from the MPI in Leipzig, Heinz-Juergen Flad, Lars Grasedyck, Jan Schneider, and Alexander Litvinenko for our joint works. I thank my former PhD student Sergey Dolgov for our productive collaboration on multiparticle dynamics, QTT tensor approximation, and parametric PDEs.

I greatly appreciate my wife and colleague at the MPI in Leipzig, Venera Khoromskaia, for her initiative and valuable work on the development of tensor numerical methods in computational quantum chemistry. I am grateful to my daughter, Diana Khoromskaia, for motivating me to write this book.

2 Theory on separable approximation of multivariate functions

2.1 Representing multivariate functions via separation of variables

In the following we consider a function f of several variables (continuous or discrete) as a mapping of the domain of definition $\Omega \in \mathbb{R}^d$ to $\mathbb{R}$. The 'structural complexity' of this mapping can be measured by the amount of correlations between different variables, which leads to the concept of separation of variables. In this section we discuss different functional formats that allow the quantitative characterization of the amount of separability by estimation of the specific rank parameters, which can be interpreted as the result of application of a certain approximation scheme to the function of interest.

From the computational point of view the most interesting special case in scientific computing describes the functions of discrete variable, i.e., multidimensional tensors

$$\mathbf{F} : \mathbb{R}^{n_1} \times \cdots \times \mathbb{R}^{n_d} \mapsto \mathbb{R},$$

which normally represent the computable entities in numerical simulations of physical problems, such as potentials, velocity fields, forces, wave functions, etc. sampled over $n_1 \times n_2 \times \cdots \times n_d$ tensor grid in $\mathbb{R}^d$. The direct explicit representation of a function generated tensor $f \mapsto \mathbf{F}$ requires $n_1 n_2 \ldots n_d$ storage size that increases exponentially in dimension d. This phenomenon, usually called the 'curse of dimensionality', is the main source of 'big data' in numerical analysis and scientific computations.

There are a number of traditional methods like polynomial or h–p approximation, sparse grids, Fourier and wavelet transforms, which are based on the representation of the initial discretized function in a certain problem independent basis set. This allows us to somewhat diminish the curse of dimensionality to a weaker form $C(\log n)^d$, which slightly extends the range of tractability of traditional numerical methods to the dimension parameter of the order or several ones.

The concept of separation of variables attempts to use the intrinsic properties of a function of interest, which allows us to construct the nonlinear parametrization capable of representing the discretized function in the nonredundant form with almost linear storage scaling in d. Separation of variables has long appeared to be one of the commonly used principles in approximation theory, computational physics and chemistry, data analysis, etc. The classical examples of how such an approach works are given by the Schmidt decomposition of bivariate functions, numerical schemes based on Gaussian basis sets in computational quantum chemistry, or principal component analysis in data processing.

https://doi.org/10.1515/9783110365917-002

2.1.1 Schmidt decomposition of bivariate functions

The problem on approximation of a function $f(x, y) : L^2([0,1]^2) \mapsto \mathbb{R}$ by bilinear forms

$$f(x, y) \approx \sum_{k=1}^{R} u_k(x) v_k(y) \quad \text{in} \quad L^2([0,1]^2) \tag{2.1}$$

is a continuous analogue to the rank decomposition of a rectangular matrix. The problem of best R-term approximation of a bivariate function in form (2.1) was solved thanks to E. Schmidt [327] (celebrated theorem on Schmidt decomposition, 1907).

Let $\{\sigma_k(J_f)\}$, $\sigma_1 \geq \sigma_2 \geq \cdots \geq 0$, be a nonincreasing sequence of singular values of the integral operator,

$$J_f(g) := \int_0^1 f(x, y) g(y) dy ,$$

$$\sigma_k(J_f) := \lambda_k[(A)^{1/2}] , \quad A = J_f^* J_f , \quad J_f^* \text{ is adjoint to } J_f ,$$

with orthonormal sequences $\{\varphi_k(x)\}$, $\{\psi_k(y)\}$, i.e.,

$$A\psi_k(y) = \lambda_k \psi_k(y) ; \quad A^* \varphi_k(x) = \lambda_k \varphi_k(x) , \quad k = 1, 2, \ldots$$

and $\lambda_k(A)$ is a sequence of eigenvalues of operator A. The iterated kernel function of A is represented by

$$f_A(x, y) := \int_0^1 f(x, z) f(z, y) dz .$$

Now, the *Schmidt decomposition* is given by

$$f(x, y) = \sum_{k=1}^{\infty} \sigma_k(J_f) \varphi_k(x) \psi_k(y) .$$

The best bilinear R-term approximation property reads as

$$\left\| f(x, y) - \sum_{k=1}^{R} \sigma_k \varphi_k(x) \psi_k(y) \right\|_{L^2} = \inf_{u_k, v_k \in L^2,\, k=1,\ldots,R} \left\| f(x, y) - \sum_{k=1}^{R} u_k(x) v_k(y) \right\|_{L^2} .$$

Schmidt decomposition ensures that for $d = 2$ the best bilinear approximation can be realized by successive application of one-term approximation, i.e., by the so called pure greedy algorithm; see Section 2.1.7. In the case of symmetric and positive definite kernels the corresponding result is known as Mercer's decomposition (1909), [276].

The Schmidt decomposition is a continuous analogue to singular value decomposition (SVD) of a matrix. It is interesting to observe that for Nyström's approximation to the integral operator J_f the Schmidt decomposition of the discrete problem is reduced to matrix SVD.

2.1.2 Tensor product of Hilbert spaces

For a multivariate function $f(x_1, \ldots, x_d), f: L^2([0,1]^d) \mapsto \mathbb{R}$, the formal analogy to the representation (2.1) reads as

$$f(x_1, \ldots, x_d) \approx \sum_{k=1}^{R} \prod_{\ell=1}^{d} f_k^{(\ell)}(x_\ell) \quad \text{in} \quad L^2([0,1]^d)\,, \tag{2.2}$$

where each term represents a *separable function* (rank-1 element). Clearly, for small R, the representation (2.2) has a very simple structure completely described by a few univariate functions $f_k^{(\ell)}(x_\ell)$ depending on f. However, the generalization of the constructive Schmidt decomposition for best R-term approximation to the case $d \geq 3$ is not practically possible due to several reasons. In general, the construction of nearly best R-term ansatz in (2.2) is the challenging nonlinear approximation problem, which can be solved in analytic form only in some special cases, to be discussed in the following.

The rigorous consideration of problem (2.2) requires a specific functional space to treat the target function f and univariate functions in the right hand side of (2.2). To this end, in what follows we recall the main definitions and basis properties of *tensor product of Hilbert spaces*, which provide the convenient and most commonly used functional setting for theoretical and numerical analysis of separable approximations.

Let H_ℓ ($\ell = 1, \ldots, d$) be a real, separable Hilbert space of functions. Following the construction in [308] we introduce a tensor product of Hilbert spaces H_ℓ.

Definition 2.1. A tensor product of Hilbert spaces H_ℓ, further denoted by

$$\mathbb{H} = H_1 \otimes \cdots \otimes H_d \equiv \bigotimes_{\ell=1}^{d} H_\ell\,,$$

is introduced as the closure of a set of finite sums, $\sum_k \bigotimes_{\ell=1}^{d} w_k^{(\ell)}$, of dual multilinear forms (linear functionals) on $H_1 \times \ldots \times H_d$. Given a d-tuple $(v^{(1)}, \ldots, v^{(d)}) \in H_1 \times \ldots \times H_d$, a single form is defined by

$$\bigotimes_{\ell=1}^{d} w^{(\ell)}\,(v^{(1)}, \ldots, v^{(d)}) := \prod_{\ell=1}^{d} \langle w^{(\ell)}, v^{(\ell)} \rangle_{H_\ell}\,.$$

The scalar product of rank-1 (separable) elements in $\mathbb{H}$ is determined via

$$\langle w^{(1)} \otimes \cdots \otimes w^{(d)}, v^{(1)} \otimes \cdots \otimes v^{(d)} \rangle_{\mathbb{H}} = \prod_{\ell=1}^{d} \langle w^{(\ell)}, v^{(\ell)} \rangle_{H_\ell}\,,$$

and is extended to the finite sums of rank-1 elements by linearity.

The bilinear form $\langle \cdot, \cdot \rangle_{\mathbb{H}}$ is also called the *induced scalar product*.

The construction of tensor product of Hilbert spaces can be applied to the complex spaces H_ℓ with respect to the induced scalar product over the field $\mathbb{C}$. Moreover, the more general construction of a tensor product of Banach spaces may be considered as well (see [131, 138, 318] for the discussion). However, in the following we remain in the framework of tensor product of Hilbert spaces setting because:

(a) the majority of applications in scientific computing are formulated in Hilbert spaces,
(b) the basic multilinear algebra algorithms are designed for scalar products, and
(c) the numerical methods for PDEs are based on Hilbert space setting.

Basic properties of tensor product of Hilbert spaces are formulated in the following statements [308]:

Lemma 2.2. *The bilinear form* $\langle\cdot,\cdot\rangle_{\mathbb{H}}$ *is well defined and is positive definite.*

Proof. Let us consider the case $d = 2$, and verify that $\langle\lambda,\lambda'\rangle$ does not depend on the particular tensor representation of λ and λ'. To that end it is enough to show that if μ is a finite sum representing a zero form, then $\langle\eta,\mu\rangle = 0$ for all $\eta \in \mathbb{H}$. Let $\eta = \sum_{i=1}^{N} c_i(\varphi_i \otimes \varphi_i)$, then

$$\begin{aligned}\langle\eta,\mu\rangle &= \left\langle \sum_{i=1}^{N} c_i(\varphi_i \otimes \varphi_i), \mu \right\rangle \\ &= \sum_{i=1}^{N} c_i\langle\varphi_i \otimes \psi_i, \mu\rangle = \sum_{i=1}^{N} c_i\mu(\varphi_i, \psi_i) = 0\,,\end{aligned} \tag{2.3}$$

since μ is a zero form. Hence, $\langle\cdot,\cdot\rangle$ is well defined.

Now, suppose $\lambda = \sum_{k=1}^{M} d_k(\eta_k \otimes \mu_k)$. Then $\{\eta_k\}_{k=1}^{M}$ and $\{\mu_k\}_{k=1}^{M}$ generate subspaces $\mathcal{M}_1 \subset H_1$ and $\mathcal{M}_2 \subset H_2$ respectively. Let $\{\varphi_j\}_{j=1}^{N_1}$ and $\{\psi_\ell\}_{\ell=1}^{N_1}$ be orthonormalized basis sets for M_1 and M_2; then any η_k can be represented by φ_j, and any μ_k by ψ_ℓ, hence we obtain

$$\lambda = \sum_{j=1,\,\ell=1}^{N_1,N_2} c_{j\ell}(\varphi_j \otimes \psi_\ell)\,.$$

But the orthogonality of basis sets leads to

$$\begin{aligned}\langle\lambda,\lambda\rangle &= \left\langle \sum_{j,\ell} c_{j\ell}(\varphi_j \otimes \psi_\ell), \sum_{i,m} c_{im}(\varphi_i \otimes \psi_m) \right\rangle \\ &= \sum c_{j\ell}c_{im}\langle\varphi_j,\varphi_i\rangle\langle\psi_\ell,\psi_m\rangle = \sum_{j,\ell} |c_{j\ell}|^2\,,\end{aligned} \tag{2.4}$$

hence it is clear that $\langle\lambda,\lambda\rangle = 0$ implies that all $c_{j\ell} = 0$ and λ is a zero form. Therefore, the form $\langle\cdot,\cdot\rangle$ is positively defined. Extension to the case of $d > 2$ is straightforward. □

The next lemma describes the construction of orthogonal tensor product basis in a tensor product of Hilbert spaces setting.

Lemma 2.3. *If* $\{\phi^{(\ell)}_{k_\ell}\}$ *is an orthonormal basis in* H_ℓ, *then* $\{\Phi_{\mathbf{k}}\} = \{\bigotimes_{\ell=1}^{d} \phi^{(\ell)}_{k_\ell}\}$, $\mathbf{k} = (k_1, \dots, k_d) \in \mathbb{N}^d$, *is the orthonormal basis in* $\mathbb{H}$.

Proof. Again, without loss of generality, consider the case $d = 2$. For simplicity of notation let us consider the case when both H_1 and H_2 are infinite and separable (other cases can be considered in a similar way). Clearly, the set $\{\varphi_k \otimes \psi_\ell\}$ is orthonormal, hence it is enough to show that $\mathbb{H}$ belongs to a closed space S spanned by $\{\varphi_k \otimes \psi_\ell\}$.

Let $\varphi \otimes \psi \in \mathbb{H}$. Since both $\{\varphi_k\}$ and $\{\psi_\ell\}$ are basis sets, then the representations $\varphi = \sum c_k \varphi_k$ and $\psi = \sum d_\ell \psi_\ell$, with $\sum |c_k|^2 \leq \infty$ and $\sum |d_\ell|^2 \leq \infty$ hold. Hence, $\sum_{k,\ell} |c_k d_\ell|^2 \leq \infty$. Therefore, the properties of Hilbert spaces ensure that in S there exists a vector $\mu = \sum_{k,\ell} c_k d_\ell (\varphi_k \otimes \psi_\ell)$.

Straightforward calculations show that

$$\left\| \varphi \otimes \psi - \sum_{k<M,\, \ell<N} c_k d_\ell (\varphi_k \otimes \psi_\ell) \right\| \to 0$$

if $M, N \to \infty$, which completes the proof. □

The tensor product of univariate (real valued) continuous functions $f^{(\ell)} : I_\ell \to \mathbb{R}$, $I_\ell = [a_\ell, b_\ell]$, is a d-variate function, referred to as separable or rank-1, and defined pointwise on $\Pi = I_1 \times \cdots \times I_d$ as

$$f := \bigotimes_{\ell=1}^{d} f^{(\ell)} : \Pi \to \mathbb{R}\,, \quad \text{where} \quad f(x_1, \dots, x_d) = \prod_{\ell=1}^{d} f^{(\ell)}(x_\ell)\,.$$

Example 2.4. Prove that $L^2(\Pi) = \bigotimes_{\ell=1}^{d} L^2(I_\ell)$. (Hint: Apply Lemma 2.3).

Remark 2.5. Note that $H_0^1(\Pi) \neq \bigotimes_{\ell=1}^{d} H_0^1(I_\ell)$, since $H_0^1(I_\ell) \subset C(I_\ell)$, but $H_0^1(\Pi)$ is not a subset of $C(\Pi)$. Hence, we only have $\bigotimes_{\ell=1}^{d} H_0^1(I_\ell) \subset H_0^1(\Pi)$.

Example 2.6. Denote by $H^{\otimes n}$ the n-fold tensor product of spaces H. If $H = L^2(\mathbb{R})$, then an element $\psi \in \mathcal{F}(H) := \oplus_{n=0}^{\infty} H^{\otimes n}$, of the so called Fock space over H, $\mathcal{F}(H)$, is a sequence of functions

$$\psi = \{\psi_0,\ \psi_1(x_1),\ \psi_2(x_1, x_2),\ \psi_3(x_1, x_2, x_3), \dots\}\,,$$

such that

$$|\psi_0|^2 + \sum_{n=1}^{\infty} \int_{\mathbb{R}^n} |\psi_n(x_1, \dots, x_n)|^2 dx_1 \dots dx_n < \infty\,.$$

The finite expansion in $\mathcal{F}(H)$ as above is also known as ANOVA (analysis of variance) representation.

In the physical literature, the subspaces of $\mathcal{F}(H)$ consisting of symmetric or antisymmetric functions with respect to permutation of two arguments are called the *boson or fermion Fock spaces*, respectively.

Example 2.7 ([308]). Let $H = L^2(\mathbb{R})$ and $H^{\otimes n} = L^2(\mathbb{R}) \otimes \cdots \otimes L^2(\mathbb{R}) = L^2(\mathbb{R}^{\otimes n})$, then define $S_n H^{\otimes n}$ as a subspace in $L^2(\mathbb{R}^{\otimes n})$, consisting of all functions invariant with respect to any permutation of their arguments. A *symmetric or boson Fock space over H* is defined by

$$\mathcal{F}_s(H) = \bigoplus_{n=0}^{\infty} S_n H^{\otimes n} \subset \mathcal{F}(H) .$$

Let $\mathcal{P}_n$ be a group of permutations of n elements, and let $\varepsilon(\cdot)$ be a function from $\mathcal{P}_n$ in $\{1, -1\}$, which is equal to 1 on symmetric and -1 on antisymmetric permutations respectively. Introduce $A_n = \frac{1}{n!}\sum_{\sigma\in\mathcal{P}_n} \varepsilon(\sigma)\sigma$, $\sigma \in \mathcal{P}_n$, then A_n is an orthogonal projector in $H^{\otimes n}$. Its range $A_n H^{\otimes n}$ is called an n-fold antisymmetric tensor product of H. If $H = L^2(\mathbb{R})$, then $A_n H^{\otimes n}$ is a subspace in $L^2(\mathbb{R}^{\otimes n})$, consisting of all functions that are antisymmetric with respect to permutations of two coordinates. The subspace

$$\mathcal{F}_a(H) = \bigoplus_{n=0}^{\infty} A_n H^{\otimes n} \subset \mathcal{F}(H) ,$$

is called the *antisymmetric or fermion Fock space over H.*

Note that the construction of tensor product of Hilbert spaces can be applied to the particular case of multi-index data arrays, i.e., to the functions of discrete arguments (dth order tensors).

Definition 2.8. (dth order tensor). A function of d discrete arguments,

$$f\colon I_1 \times \cdots \times I_d \to \mathbb{R} ,$$

i.e., a multidimensional array over $I_1 \times \cdots \times I_d$ with $I_\ell = \{1, \ldots, n_\ell\}$, is called the dth order tensor. The respective tensor product of Hilbert spaces $\mathbb{H} = \bigotimes_{\ell=1}^{d} \mathbb{R}^{I_\ell} = \mathbb{R}^{I_1\times\cdots\times I_d}$ is equipped with Euclidean scalar product and Frobenius norm.

High order tensors are the main ingredients for representation of functions and operators in numerical analysis of multidimensional PDEs. All the related issues will be discussed in Chapters 3–5.

2.1.3 Additive representation in the canonical and Tucker type form

This section discusses the traditional low parametric additive type separable representations of functions based on sums of rank-1 elements. For definiteness, in the following we use L^2 setting as in Example 2.4. The more detailed discussion of these representations in the case of multi-index data arrays (tensors) will be addressed in Chapter 3. The functional rank structured decompositions can be viewed as prototypes of their discrete analogues, since they can be easily transformed from the functional to the discrete form of dth order tensors by certain sampling in the finite product basis set. The low rank separable representations are attractive in numerical analysis since they can be specified by only a small number $O(d)$ of univariate functions.

Definition 2.9. (Canonical functional format). Define a set $\mathcal{C}_R$ as a subset of elements in $\mathbb{H}$, requiring at most R terms (rank-R functions),

$$\mathcal{C}_R = \left\{ w \in \mathbb{H}: \; w = \sum_{k=1}^{R} w_k^{(1)} \otimes w_k^{(2)} \otimes \cdots \otimes w_k^{(d)}, \;\; w_k^{(\ell)} \in H_\ell \right\}.$$

A function $w \in \mathcal{C}_R$ can be represented by the description of $d\,R$ elements $w_k^{(\ell)} \in H_\ell$.

The *advantage* of the canonical functional format is the tremendous storage reduction of the respective discrete representation in form of $n^{\otimes d}$ arrays, removing d from the exponential, $n^d \to d\,R\,n$ (linear scaling in d).

However, the *limitations* are critical: It applies efficiently only to special classes of functions, for example given in analytic form by the Laplace transform (Section 2.3). In the finite dimensional setting the robust algebraic R-term decomposition methods are not known.

Consider the orthogonal subspace representation, which is the functional counterpart of the so called Tucker format of multidimensional tensors. Given a tuple of dimension (also interpreted as a rank parameter) parameters, $\mathbf{r} = (r_1, \ldots, r_d) \in \mathbb{N}^d$, choose

$$V_\ell = \operatorname{span} \left\{ \phi_k^{(\ell)} \right\}_{k=1}^{r_\ell} \subset H_\ell\,, \quad r_\ell := \dim V_\ell \le \dim H_\ell\,, \quad 1 \le \ell \le d\,,$$

with orthogonal basis, and build the tensor subspace, $\mathbb{V} = V_1 \otimes V_2 \otimes \cdots \otimes V_d \subset \mathbb{H}$. Each $v \in \mathbb{V}$ can be represented by a sum of rank-1 functions (set $\mathbf{k} = (k_1, \ldots, k_d)$)

$$v = \sum_{\mathbf{k}=\mathbf{1}}^{\mathbf{r}} b_{\mathbf{k}} \phi_{k_1}^{(1)} \otimes \phi_{k_2}^{(2)} \otimes \cdots \otimes \phi_{k_d}^{(d)}\,, \qquad b_{\mathbf{k}} \in \mathbb{R}\,. \tag{2.5}$$

The set of Tucker type functions includes all elements in $\mathbb{H}$ that can be represented in form (2.5) over all possible subspaces V_ℓ with fixed dimensions r_ℓ.

Definition 2.10 (Tucker functional format). Given the dimension parameters, $\mathbf{r} = (r_1, \ldots, r_d)$, define the set of Tucker type functions in $\mathbb{H}$,

$$\mathcal{T}_{\mathbf{r}} := \left\{ v \in \mathbb{V} := \bigotimes_{\ell=1}^{d} V_\ell \subset \mathbb{H}: \;\; \forall\, V_\ell \text{ s.t. } \dim V_\ell \le r_\ell \right\}.$$

Fixed $\mathbb{V}$, then the representation (2.5) with $b_{\mathbf{k}} \in \mathbb{R}$ holds for each element in $\mathbb{V}$.

Every $v \in \mathcal{T}_{\mathbf{r}}$ can be represented by $\prod_{\ell=1}^{d} r_\ell$ reals, and the description of $\sum_{\ell=1}^{d} r_\ell$ functions $\phi_k^{(\ell)} \in V_\ell$.

The orthogonality of Tucker representation normally ensures robustness of numerical algorithms. However, there is the limitation in the case of higher dimensions: the storage size for the core coefficients $[b_{\mathbf{k}}]$ still scales exponentially in d, r^d, where $r = \max r_\ell$.

2.1.4 Functional matrix product states decomposition

Along with additive type separation models we also discuss the product type decomposition, which is the functional counterpart of the so called matrix product states (MPS) representation of the multidimensional tensors, known in physics and quantum information theory since [322, 355, 365]. Consequently, we call it the *matrix product functions* (MPF) representation. Alternative notations like FTT (functional tensor train [288]) inherit the conventional abbreviations in multilinear algebra. The more detailed discussion of MPS type formats will be postponed until Chapter 3.

Definition 2.11. Given the product index set $\mathcal{J} := \times_{\ell=1}^{d} J_\ell$, $J_\ell = \{1, \ldots, r_\ell\}$ and set $J_0 = J_d = \{1\}$. The rank-$\mathbf{r}$ matrix product function format includes all elements in $\mathbb{H}$, which are representable as products of functional tritensors over $\mathcal{J}$,

$$\mathrm{MPF}[\mathbf{r}] := \left\{ f \in \mathbb{H} : f = \sum_{\mathbf{j}\in\mathcal{J}} \bigotimes_{\ell=1}^{d} G^{(\ell)}_{j_{\ell-1}j_\ell}, \quad \text{with} \quad G^{(\ell)} \in \mathbb{R}^{J_{\ell-1}\times H_\ell \times J_\ell} \right\}. \tag{2.6}$$

In the case of continuous function in $\mathbb{H}$, we have pointwise representation by the product of matrices, i.e., MPF,

$$f(x_1, \ldots, x_d) = \sum_{\mathbf{j}\in\mathcal{J}} G^{(1)}_{j_d j_1}(x_1) G^{(2)}_{j_1 j_2}(x_2) \ldots G^{(d)}_{j_{d-1} j_d}(x_d) \equiv G^{(1)}(x_1) \ldots G^{(d)}(x_d).$$

Here $G^{(1)}(x_1)$ is a row $1 \times r_1$ vector function depending on x_1, $G^{(\ell)}(x_\ell)$ is a matrix of size $r_{\ell-1} \times r_\ell$ with functional elements depending on x_ℓ, $\ell = 2, \ldots, d-1$, and $G^{(d)}(x_d)$ is a column vector of size $r_{d-1} \times 1$, depending on x_d.

A function $f \in \mathrm{MPF}[\mathbf{r}] \subset \mathbb{H}$ is represented by a product of matrices, each depending on a single variable. Sampling of matrix product functions representation on a n^d tensor grid requires $O(dr^2 n)$ storage size.

Remark 2.12. Note that a R-term canonical functional representation belongs to the class of matrix product functions factorizations with ranks $r_\ell = R$ and diagonal tritensors $G^{(\ell)}$, $(\ell = 1, \ldots, d-1)$ such that $G^{(\ell)}_{j_{\ell-1}j_\ell}(x_\ell) = 0$ if $j_{\ell-1} \neq j_\ell$, i.e., $\mathcal{C}_R \subset \mathrm{MPF}[\mathbf{r}]$. Likewise, $\mathcal{C}_R$ belongs to the class of rank $\mathbf{r} = (R, \ldots, R)$ Tucker tensors with the diagonal core, i.e., $\mathcal{C}_R \subset \mathcal{T}_{\mathbf{r}}$. For $d = 2$ all three formats coincide.

The particular distinctions between representations provided by these three models will be demonstrated by examples in Section 2.1.5. The common attractive features in numerical treatment are due to (a) small number of representation parameters associated with storage of $O(d)$ univariate functions and some coefficients, and (b) perfect multilinear data structure that allows us to conduct function operator calculus. For example, the latter is not the case for superposition type representations.

Note that the canonical, Tucker, and MPF ranks of a given real valued function may differ when considered in the fields $\mathbb{R}$ and $\mathbb{C}$. In some cases the construction of such decompositions can be done much easier, leading to smaller separation ranks if considered over a complex field.

In what follows, we will explain how to select real and imaginary parts of the complex valued function given in the MPF format. The result is formulated by the following theorem (continuous counterpart of the algebraic representation in [85]; see Section 3.6). If there is no ambiguity, we use simplified notations, $A^{(p)}(x_p) = A^{(p)}_{x_p}$.

Theorem 2.13. *The complex valued MPF* $f(x_1, \ldots, x_d) = A^{(1)}_{x_1} \ldots A^{(d)}_{x_d}$, *with ranks* $r_0 = r_d, r_1, \ldots, r_{d-1}, r_d$ *can be represented in a form* $f(x_1, \ldots, x_d) = \hat{A}^{(1)}_{x_1} \ldots \hat{A}^{(d)}_{x_d}$, *with rank parameters* $r_0 = r_d, 2r_1, \ldots, 2r_{d-1}, r_d$, *and with*

$$\hat{A}^{(1)}_{x_1} = \begin{bmatrix} \operatorname{Re} A^{(1)}_{x_1} & \operatorname{Im} A^{(1)}_{x_1} \end{bmatrix}, \quad \hat{A}^{(p)}_{x_p} = \begin{bmatrix} \operatorname{Re} A^{(p)}_{x_p} & \operatorname{Im} A^{(p)}_{x_p} \\ -\operatorname{Im} A^{(p)}_{x_p} & \operatorname{Re} A^{(p)}_{x_p}. \end{bmatrix}, \quad p = 2, \ldots, d-1,$$
$$\hat{A}^{(d)}_{x_d} = \begin{bmatrix} \operatorname{Re} A^{(d)}_{x_d} \\ -\operatorname{Im} A^{(d)}_{x_d} \end{bmatrix} + \mathrm{i} \begin{bmatrix} \operatorname{Im} A^{(d)}_{x_d} \\ \operatorname{Re} A^{(d)}_{x_d} \end{bmatrix}, \tag{2.7}$$

where the functional cores $\hat{A}^{(p)}_{x_p}$, $p = 1, \ldots, d-1$ *are all real valued.*

Proof. The proof is constructive and follows the corresponding arguments in the algebraic case [85]. Let $A^{(p)}_{x_p} = B^{(p)}_{x_p} + \mathrm{i}C^{(p)}_{x_p}$, with real valued $B^{(p)}_{x_p} = \operatorname{Re} A^{(p)}_{x_p}$ and $C^{(p)}_{x_p} = \operatorname{Im} A^{(p)}_{x_p}$. Then

$$A^{(1)}_{x_1} = \begin{bmatrix} B^{(1)}_{x_1} & C^{(1)}_{x_1} \end{bmatrix} \begin{bmatrix} I \\ \mathrm{i}I \end{bmatrix},$$

where I is an $r_1 \times r_1$ identity matrix. Define real valued core $\hat{A}^{(1)}_{x_1} = [B^{(1)}_{x_1} \; C^{(1)}_{x_1}]$ and multiply $[I \;\; \mathrm{i}I]^{\mathrm{T}}$ to the right,

$$\begin{bmatrix} I \\ \mathrm{i}I \end{bmatrix} A^{(2)}_{x_2} = \begin{bmatrix} I \\ \mathrm{i}I \end{bmatrix} \left(B^{(2)}_{x_2} + \mathrm{i}C^{(2)}_{x_2} \right) = \begin{bmatrix} B^{(2)}_{x_2} + \mathrm{i}C^{(2)}_{x_2} \\ -C^{(2)}_{x_2} + \mathrm{i}B^{(2)}_{x_2} \end{bmatrix} = \begin{bmatrix} B^{(2)}_{x_2} & C^{(2)}_{x_2} \\ -C^{(2)}_{j_2} & B^{(2)}_{x_2} \end{bmatrix} \begin{bmatrix} I \\ \mathrm{i}I \end{bmatrix}.$$

Hence, we can define

$$\begin{bmatrix} B^{(2)}_{x_2} & C^{(2)}_{x_2} \\ -C^{(2)}_{x_2} & B^{(2)}_{x_2} \end{bmatrix} \begin{bmatrix} I \\ \mathrm{i}I \end{bmatrix} =: \hat{A}^{(2)}_{x_2} \begin{bmatrix} I \\ \mathrm{i}I \end{bmatrix},$$

where in the right hand side $\hat{A}^{(2)}_{x_2}$ is a new real valued functional core and I is $r_2 \times r_2$ identity matrix. We continue the process and establish (2.7). The last functional core reads

$$\hat{A}^{(d)}_{x_d} = \begin{bmatrix} I \\ \mathrm{i}I \end{bmatrix} A^{(d)}_{x_d} = \begin{bmatrix} I \\ \mathrm{i}I \end{bmatrix} \left(B^{(d)}_{x_d} + \mathrm{i}C^{(d)}_{x_d} \right) = \begin{bmatrix} B^{(d)}_{x_d} + \mathrm{i}C^{(d)}_{x_d} \\ -C^{(d)}_{x_d} + \mathrm{i}B^{(d)}_{x_d} \end{bmatrix} = \begin{bmatrix} B^{(d)}_{x_d} \\ -C^{(d)}_{x_d} \end{bmatrix} + \mathrm{i} \begin{bmatrix} C^{(d)}_{x_d} \\ B^{(d)}_{x_d} \end{bmatrix},$$

which completes the proof. □

The relation (2.7) allows us to represent explicitly real and imaginary parts of a matrix product function since all factors but one in the corresponding factorization are real.

Corollary 2.14. *Real and imaginary parts in a MPF of* $f(x_1, \dots, x_d)$ *in (2.7) are given by*

$$\operatorname{Re} f(x_1, \dots, x_d) = \left(\prod_{p=1}^{d-1} \hat{A}_{x_p}^{(p)} \right) \operatorname{Re} \hat{A}_{x_d}^{(d)} , \quad \operatorname{Im} f(x_1, \dots, x_d) = \left(\prod_{p=1}^{d-1} \hat{A}_{x_p}^{(p)} \right) \operatorname{Im} \hat{A}_{x_d}^{(d)} .$$

This corollary provides a powerful tool to derive explicit MPF real valued representations of a function with minimal rank having at hand the complex valued factorization, as will be shown in the following paragraph.

2.1.5 Examples on the explicit canonical, Tucker, and MPF design

Examples in this section illustrate how the explicit canonical, Tucker, and MPF representations for multivariate functions may look for some simple but instructive instances. In particular, we will demonstrate when the exact canonical rank of a function may be an order of magnitude larger than the corresponding one for the Tucker and MPF decompositions.

Example 2.15. The Tucker approximation of the continuous function in $\mathbb{H} = L^2(I^d)$, $I = [-1, 1]$, can be constructed by the tensor product polynomial interpolation of order $\mathbf{r}$,

$$f(x_1, \dots, x_d) \approx \sum_{\mathbf{j}=1}^{\mathbf{r}} f(v_{j_1}, \dots, v_{j_d}) \prod_{\ell=1}^{d} L_{j_\ell}(x_\ell) , \tag{2.8}$$

where L_{j_ℓ} is a set of the Lagrange polynomials on $[-1, 1]$ at, say, Chebyshev–Gauss–Lobatto grid, $\{v_{j_\ell}\}, j_\ell = 1, \dots, r_\ell$. Here both the core coefficients tensor and orthogonal basis sets are given explicitly.

The approximation error of the Tucker type tensor interpolant (2.8) decays exponentially in the Tucker rank-$\mathbf{r}$ for smooth enough functions; see Section 2.2.

Example 2.16. Let $\mathbb{H} = L^2(I^d)$.

1. Rank-d function $f(x) = f_1(x_1) + f_2(x_2) + \cdots + f_d(x_d)$ can be approximated by a rank–2 expansion at any prescribed accuracy,
$$f(x) \approx \frac{\prod_{\ell=1}^{d}(1 + \varepsilon f_\ell(x_\ell)) - 1}{\varepsilon} + O(\varepsilon) , \quad \text{as } \varepsilon \to 0 .$$
2. Exponential transform leads to the exact rank-1 decomposition,
$$g(x) = \exp(f_1(x_1) + \cdots + f_d(x_d)) = \prod_{\ell=1}^{d} \exp(f_\ell(x_\ell)) .$$
3. Introduce the rank–2 functional vectors $U^{(\ell)} = (1_\ell, x_\ell)^{\mathrm{T}}$, then the rank-$\mathbf{r}$, $\mathbf{r} = (2, 2, \dots, 2)$, orthogonal Tucker representation of $f(x) = \sum_{\ell=1}^{d} x_\ell$ can be constructed,
$$f = \sum_{|\mathbf{k}|=d+1} U_{k_1}^{(1)} \otimes U_{k_2}^{(2)} \otimes \cdots \otimes U_{k_d}^{(d)} , \quad k_\ell \in \{1, 2\} , \tag{2.9}$$

implying the explicit representation to the Tucker core in (2.5) of size $2^{\otimes d}$,

$$b_{\mathbf{k}} = 1\,, \quad \text{if} \quad |\mathbf{k}| \equiv k_1 + \cdots + k_d = d+1\,, \quad \text{and} \quad b_{\mathbf{k}} = 0 \quad \text{otherwise}\,.$$

Here the Tucker core is sparse and the number of nonzero elements in $[b_{\mathbf{k}}]$ is exactly $d \ll 2^d$, i.e., the number of active summands in (2.9) is exactly d.

4. The function $f(x) = \sin(\sum_{\ell=1}^d x_\ell)$ exhibits the canonical rank–2 representation over the field $\mathbb{C}$,

$$\sin\left(\sum_{\ell=1}^d x_\ell\right) = \frac{e^{i\sum_{\ell=1}^d x_\ell} - e^{-i\sum_{\ell=1}^d x_\ell}}{2i}\,.$$

Consequently, the Tucker ranks of $f(x)$ do not exceed 2 (prove this statement).

In some cases the matrix product function decomposition can be constructed explicitly as in the following examples, where we assume for ease of presentation that f_ℓ are continuous functions on $H_\ell = H$ for all $\ell = 1, \ldots, d$.

Example 2.17. FTT rank of $f(x) = f_1(x_1) + f_2(x_2) + \cdots + f_d(x_d)$ is 2, due to the explicit decomposition

$$f(x) = \begin{pmatrix} f_1(x_1) & 1 \end{pmatrix} \begin{pmatrix} 1 & 0 \\ f_2(x_2) & 1 \end{pmatrix} \cdots \begin{pmatrix} 1 & 0 \\ f_{d-1}(x_{d-1}) & 1 \end{pmatrix} \begin{pmatrix} 1 \\ f_d(x_d) \end{pmatrix},$$

which can be easily verified by induction in view of the identities $(a, b, c \in \mathbb{C})$,

$$a + b = \begin{pmatrix} a & 1 \end{pmatrix} \begin{pmatrix} 1 \\ b \end{pmatrix}, \quad a + b + c = \begin{pmatrix} a & 1 \end{pmatrix} \begin{pmatrix} 1 & 0 \\ b & 1 \end{pmatrix} \begin{pmatrix} 1 \\ c \end{pmatrix}.$$

In Section 2.1.8, the explicit MPF decomposition of the function $f(x) := \sum_{j=1}^d x_j$ will be used as a prototype construction to design the matrix product representation of multivariate elliptic operators.

It is possible to obtain the MPF representation for functions $\sin(\sum_{j=1}^d x_j)$ and $\cos(\sum_{j=1}^d x_j)$ by simply extracting the real and imaginary parts of the rank-1 complex function $\exp(-i\sum_{j=1}^d x_j)$ as in Corollary 2.14.

Lemma 2.18. *Rank–2 MPF decomposition of $f(x) := \sin(\sum_{j=1}^d x_j)$, $x \in \mathbb{R}^d$, reads*

$$f(x) = \begin{pmatrix} \sin x_1 & \cos x_1 \end{pmatrix} \begin{pmatrix} \cos x_2 & -\sin x_2 \\ \sin x_2 & \cos x_2 \end{pmatrix} \cdots \begin{pmatrix} \cos x_{d-1} & -\sin x_{d-1} \\ \sin x_{d-1} & \cos x_{d-1} \end{pmatrix} \begin{pmatrix} \cos x_d \\ \sin x_d \end{pmatrix}.$$

Likewise, for the function $g(x) := \cos(\sum_{j=1}^d x_j)$, $x \in \mathbb{R}^d$, we have

$$g(x) = \begin{pmatrix} \sin x_1 & \cos x_1 \end{pmatrix} \begin{pmatrix} \cos x_2 & -\sin x_2 \\ \sin x_2 & \cos x_2 \end{pmatrix} \cdots \begin{pmatrix} \cos x_{d-1} & -\sin x_{d-1} \\ \sin x_{d-1} & \cos x_{d-1} \end{pmatrix} \begin{pmatrix} \sin x_d \\ -\cos x_d \end{pmatrix}.$$

For both functions the Tucker ranks do not exceed 2.

Proof. The result on MPF decomposition follows by Corollary 2.14 applied to the rank-1 representation of the complex exponential function $\exp(-i\sum_{j=1}^{d} x_j)$. The bound on Tucker ranks is justified by the observation that each mode in the MPF decomposition spans at most two linear independent functions. □

Note that the MPF representations of sine and cosine functions as above differ only in the last matrix factor. The direct proof of Lemma 2.18 can based on induction argument as follows [288]:

$$\begin{aligned} f(x) &= \sin x_1 \cos(x_2 + \cdots + x_d) + \cos x_1 \sin(x_2 + \cdots + x_d) \\ &= \begin{pmatrix} \sin x_1 & \cos x_1 \end{pmatrix} \begin{pmatrix} \cos(x_2 + \cdots + x_d) \\ \sin(x_2 + \cdots + x_d) \end{pmatrix} \\ &= \begin{pmatrix} \sin x_1 & \cos x_1 \end{pmatrix} \begin{pmatrix} \cos x_2 & -\sin x_2 \\ \sin x_2 & \cos x_2 \end{pmatrix} \begin{pmatrix} \cos(x_3 + \cdots + x_d) \\ \sin(x_3 + \cdots + x_d) \end{pmatrix}. \end{aligned}$$

A similar argument allows us to prove the rank–2 MPF decompositions for functions $f(x) := \sin(\sum_{j=1}^{d} f_j(x_j))$ and $f(x) := \cos(\sum_{j=1}^{d} f_j(x_j))$, $x \in \mathbb{R}^d$.

The discrete analogues of the above presented separable functional decompositions remain valid; see Chapter 3.

2.1.6 Nonlinear approximation of functions in separable form

One may think about a constructive approximation of a given multivariate function $f = f(x_1, \ldots, x_d) \in \mathbb{H}$, $d \geq 2$, in one of the rank structured forms described above.

Since $\mathcal{T}_{\mathbf{r}}$, $\mathcal{C}_R$ and MPF[**r**] are not linear spaces (kind of nonlinear manifolds), we arrive at a challenging *nonlinear approximation* problem on estimation

$$f \in \mathbb{H}: \quad \sigma(f, \mathcal{S}) := \inf_{s \in \mathcal{S}} \|f - s\|_{\mathbb{H}}, \tag{2.10}$$

where $\mathcal{S} = \{\mathcal{T}_{\mathbf{r}}, \mathcal{C}_R, FTT[\mathbf{r}]\}$. For $d = 2$ the problem can be solved by Schmidt decomposition. The advantage is that all possible candidates in the chosen approximation set $\mathcal{S}$ are determined by only a small number of univariate functions.

Why might the problem (2.10) be difficult for $d \geq 3$? One of the difficulties is illustrated by the following instructive example [278]:

Proposition 2.19. *The trigonometric identity* ($d \geq 2$)

$$f(x) := \sin\left(\sum_{j=1}^{d} x_j\right) = \sum_{j=1}^{d} \sin(x_j) \prod_{k \in \{1,\ldots,d\} \setminus \{j\}} \frac{\sin(x_k + \alpha_k - \alpha_j)}{\sin(\alpha_k - \alpha_j)} \tag{2.11}$$

holds for any $\alpha_k \in \mathbb{R}$, *such that* $\sin(\alpha_k - \alpha_j) \neq 0$ *for all* $j \neq k$.

In the case $d = 2$, the assertion (2.11) is trivial. For $d \geq 3$ it can be proven by induction. The complete proof can be found in [278].

Expansion (2.11) shows the lack of uniqueness (ambiguity) of the best rank-d tensor representation. In this case there exists a continuum of minimizers in (2.10) specified by the choice of parameters $\{\alpha_k\}$, $k = 1, \dots, d$. Hence, the minimization process might be nonrobust due to the presence of multiple local minima.

Remark 2.20. Both the Tucker and MPF ranks of f in (2.11) are equal to 2 (Lemma 2.18).

Another problem is illustrated by Example 2.16 indicating that $\mathcal{C}_R$ is a nonclosed set (a sequence of rank–2 elements converges to a d-term representation).

The development of tensor methods based on separation of variables states several principal questions (no ultimate answers are known):
- For which classes of functions can the 'curse of dimensionality' be relaxed?
- How can one solve (2.10) efficiently? (i.e., generalize the truncated Schmidt decomposition).
- When can one expect fast (exponential) convergence in the rank parameters?
- How can one solve the basic physical equations on the 'nonlinear tensor manifold' $\mathcal{S}$ getting rid of the 'curse of dimensionality'?

The systematic use of the approximation theory and modern multilinear algebra provides constructive answers to the above questions in the case of many real life problems, leading to developments of tensor structured numerical methods in high dimensional scientific computing.

At this point it might be interesting to take a look from the historical perspective ([52]) on how the problem of many dimensions was tackled in the classical approximation theory. We discuss this issue with the example of *Kolmogorov's paradigm* related to the Hilbert 13th problem: "A solution of the algebraic equation of degree 7 cannot be written as superposition of continuous bivariate functions."

This problem was solved as a consequence of the celebrated theorem by A. N. Kolmogorov, 1957 on the superposition of univariate functions [234]; see also [235].

Algorithm AK. (Kolmogorov's superposition theorem, [234]). For $d \geq 2$, any function $f \in \mathbb{C}([0,1]^d)$ can be represented in the form

$$f(x_1, \dots, x_d) = \sum_{i=1}^{2d+1} g_i \left(\sum_{\ell=1}^{d} \phi_{i\ell}(x_\ell) \right),$$

where functions $\phi_{i\ell}\colon [0,1] \to \mathbb{R}$ do not depend on f and belong to the class $Lip1$ (i.e., Lipschitz continuous with exponent one), while $g_i\colon \mathbb{R} \to \mathbb{R}$ are continuous functions.

From the point of view of low parametric approximation of functions of many variables *Theorem AK* is not constructive. But in our context it says that in the discrete setting, any function $f \in \mathbb{C}([0,1]^d)$ can be represented by $O(2dN + (2d+1)dN)$ reals, where N corresponds to the size of the interpolating table for g_i; see [52].

Kolmogorov's result on the superposition of univariate functions indicates that the only existence of low parametric representation of a multivariate function (which may be interesting in the complexity theory) does not solve the problem from a computational point of view unless there are no constructive algorithms to compute the particular low parametric representation. In the case of Kolmogorov's parametrization, the additional difficulty is due to the fact that construction of algorithms for numerical approximation of a continuous function with controllable accuracy is not a well defined problem.

2.1.7 On canonical decomposition by greedy algorithm

For $\mathcal{S} = \mathcal{C}_R$, the canonical representation can be considered in the framework of best R-term approximation with regard to a redundant dictionary of rank-1 functions [343].

Definition 2.21. A system $\mathcal{D}$ of functions in $\mathbb{H}$ is called a dictionary if each $g \in \mathcal{D}$ has norm one and its linear span is dense in $\mathbb{H}$.

Given $R \in \mathbb{N}$, denote by $\Sigma_R(\mathcal{D})$ the collection of $s \in \mathbb{H}$, which can be written in the form

$$s = \sum_{g \in \Lambda} c_g g\,, \quad \Lambda \subset \mathcal{D}\colon \#\Lambda \le R \quad \text{with } c_g \in \mathbb{R}\,.$$

For $f \in \mathbb{H}$, the *best R-term approximation* error is defined by

$$\sigma_R(f, \mathcal{D}) := \inf_{s \in \Sigma_R(\mathcal{D})} \|f - s\|\,.$$

Since our particular goal is the low rank separable decomposition of a function, this problem can be viewed as the construction of best rank-R canonical approximation in the dictionary

$$\mathcal{D} := \bigcup_{r=1}^{\infty} \mathcal{C}_r\,.$$

The simple approach to tackle the above approximation problem numerically is the so called *pure greedy algorithm* (PGA); see [343] for the detailed disposition.

The PGA inductively computes an estimate to the best R-term approximation. For given $f \in \mathbb{H}$ let $g = g(f) \in \mathcal{D}$ be an element maximizing $|\langle f, g\rangle|$ (in our special rank-R setting, it is best rank-1 approximation by nonlinear maximization). Define

$$G(f) := \langle f, g\rangle g\,, \quad R(f) := f - G(f)\,.$$

The PGA reads as: Given $f \in \mathbb{H}$, introduce

$$R_0(f) := f \quad \text{and} \quad G_0(f) := 0\,.$$

Then, for all $1 \le m \le R$, we inductively define

$$G_m(f) := G_{m-1}(f) + G(R_{m-1}(f))\,,$$
$$R_m(f) := f - G_m(f) = R(R_{m-1}(f))\,.$$

The practical efficiency of the PGA algorithm is determined by the convergence rate in R. Applying PGA to functions characterized via the approximation property

$$\sigma_R(f, \mathcal{D}) \leq R^{-q}, \quad R = 1, 2, \ldots,$$

with some $q \in (0, 1/2]$ (low order approximation), leads to the error bound [343]

$$\|f - G_R(f, \mathcal{D})\| \leq C(q, \mathcal{D}) C R^{-q}, \quad R = 1, 2, \ldots,$$

which is 'too pessimistic' in applications. Moreover, the convergence may be very slow even under the existence of good rank-R approximation. We refer to [12, 45], where the improved convergence rate for PGA was investigated for some special problem classes.

We are interested in the constructive R-term separable approximation on a class of analytic functions (possibly with point singularities), providing exponential convergence in the number of terms $R = 1, 2, \ldots,$

$$\sigma_R(f, \mathcal{D}) \leq C \exp(-\alpha R^q), \quad q = 1 \text{ or } q = 1/2 .$$

In the next sections, we show that the asymptotically optimal R-term approximation can be constructed explicitly via the quadrature and interpolation based *sinc* approximation, which can also be combined with the direct fitting by exponential sums.

Finally, we consider the special version of PGA called *greedy completely orthogonal decomposition*, which can be useful in numerical analysis since it leads to the simple and robust approximation scheme. The decomposition in $\mathcal{C}_R$,

$$f = \sum_{k=1}^{R} a_k v_k, \quad v_k = \phi_k^{(1)}(x_1) \otimes \cdots \otimes \phi_k^{(d)}(x_d) \in \mathcal{C}_1, \tag{2.12}$$

is called *completely orthogonal* if

$$\langle \phi_k^{(\ell)}, \phi_m^{(\ell)} \rangle = \delta_{k,m} \quad \forall \ell = 1, \ldots, d .$$

In other words greedy completely orthogonal decomposition is defined as PGA under the orthogonality constraint on $\Phi^{(\ell)}$:

$$\Phi^{(\ell)} = [\phi_1^{(\ell)}, \ldots, \phi_R^{(\ell)}] \quad \text{is the orthogonal set for} \quad \ell = 1, \ldots d .$$

Similar to the Schmidt decomposition, it is implemented by successive application of a 1-term algorithm, which is simple for implementation.

In the case of discrete d-variate functions (tensors) the following statement holds:

Lemma 2.22 ([373] Orthogonal Tucker with a diagonal core). *Let $f \in \mathbb{H}$ allow a rank-R completely orthogonal decomposition. Then the greedy completely orthogonal decomposition algorithm correctly computes it. If $a_1 > a_2 > \cdots > a_R > 0$, then the completely orthogonal decomposition is unique.*

Proof. Let f have the representation (2.12). Then the greedy completely orthogonal decomposition in (2.12) reduces to solving (for G_m, $m = 1, \dots, R$)

$$\max_{u \in \mathcal{C}_1,\, \|u\|=1} \langle f, u \rangle \quad \text{with} \quad \Phi^{(\ell)} = [\phi_1^{(\ell)}, \dots, \phi_R^{(\ell)}] - \text{orthogonal} \quad \text{(simple problem)},$$

and letting $a_m = \langle f, u_m \rangle$. Set $m = 1$, then for

$$G_1 = \prod_{\ell=1}^{d} \left(\sum_{k=1}^{R} c_{k,\ell} \phi_k^{(\ell)}(x_\ell) \right),$$

we have

$$\sum_{k=1}^{R} a_k \prod_{\ell=1}^{d} c_{k,\ell} \to \max \quad \text{with} \quad \sum_{k=1}^{R} c_{k,\ell}^2 = 1, \; \ell = 1, \dots, d.$$

Assuming without loss of generality that $a_1 \geq a_2 \geq \cdots \geq a_R > 0$, we obtain the solution of optimization problem: $c_{1,\ell} = 1, c_{2,\ell} = \cdots = c_{r,\ell} = 0$, implying $G_1 = v_1$ (here we the use symmetry in ℓ). This ensures $\langle f, G_1 \rangle = a_1$, hence we obtain inductively

$$G_m = \sum_{k=1}^{m} a_k v_k, \quad \phi_k^{(1)}(x_1) \otimes \cdots \otimes \phi_k^{(d)}(x_d) \in \mathcal{C}_1,$$

finalizing the proof of the first statement. Uniqueness is deduced by the orthogonality assumption. □

The greedy algorithm attracts attention in numerical analysis of multidimensional problems due to its simplicity. However, in general, it is not stable and the convergence may be very slow. The essential limitation of its stable version, i.e., the greedy completely orthogonal decomposition method, is the poor approximation properties of the representation in form (2.12).

In recent years several attempts have been made in the direction of enhancement of PGA algorithms in application to the solution of differential equations; we refer to [59] for the detailed exposition.

2.1.8 Tensor structured representation of operators

The formatted low parametric separable representations can be applied also to linear operators acting between a pair of tensor product Hilbert spaces, leading to the concept of canonical, Tucker, and *matrix product* operators. The detailed discussion of this issue will be presented in Chapter 3 for the case of multifold matrices representing the discrete elliptic operator in $\mathbb{R}^d$ and its inverse, the elliptic Green function, convolution and Fourier transforms, matrix exponential, and some other examples.

Let A_ℓ be some linear continuous operators on H_ℓ, $\ell = 1, \dots, d$. The tensor product of these operators, $A_{(d)} = A_1 \otimes A_2 \otimes \cdots \otimes A_d \colon \mathbb{H} \to \mathbb{H}$, is defined by the action on rank-1 element $f = \prod_{\ell=1}^{d} f_\ell$, $f_\ell \in H_\ell$ as follows:

$$A_{(d)} f := A_1 f_1 \otimes A_2 f_2 \otimes \cdots \otimes A_d f_d \in \mathbb{H}.$$

Likewise, this definition can be extended to the case of unbounded operators $A_\ell : H_\ell \to H'_\ell$, via their action on a single rank-1 function in $\mathbb{H}$ [308]. In this case the problem illustrated in Remark 2.5 makes the construction more involved. However, for our applications, we are mainly interested in the case of bounded operators describing, for example, the discretized multidimensional PDEs.

To fix the idea, we only consider the simple example of d dimensional Laplacian type operator $A_{(d)} : \mathbb{H} \to \mathbb{H}$ defined by the rank-d canonical tensor representation,

$$A_{(d)} := A_1 \otimes I_2 \otimes \cdots \otimes I_d + \cdots + I_1 \otimes \cdots \otimes I_{d-1} \otimes A_d \,,$$

where I_ℓ is the identity operator and A_ℓ is the bounded linear operator on H_ℓ. The action of $A_{(d)}$ on rank-1 separable element $f = \prod_{\ell=1}^{d} f_\ell$ is defined by function

$$A_{(d)} f := A_1 f_1 \otimes f_2 \otimes \cdots \otimes f_d + \cdots + f_1 \otimes \cdots \otimes f_{d-1} \otimes A_d f_d \,,$$

whose canonical rank does not exceed d.

Along the line of Example 2.16 (c), we introduce the 2×1 operator valued vectors $U^{(\ell)} = (I_\ell, A_\ell)^T$, then the rank-$\mathbf{r}$ Tucker representation of $A_{(d)}$, with $\mathbf{r} = (2, 2, \ldots, 2)$, takes a form

$$A_{(d)} = \sum_{|\mathbf{k}|=d+1} U^{(1)}_{k_1} \otimes U^{(2)}_{k_2} \otimes \cdots \otimes U^{(d)}_{k_d} \,, \quad k_\ell \in \{1, 2\} \,.$$

Now the action of $A_{(d)}$ on rank-1 element is defined by the rank-**2** functional Tucker tensor

$$A_{(d)} f := \sum_{|\mathbf{k}|=d+1} \begin{pmatrix} f_1 \\ A_1 f_1 \end{pmatrix}_{k_1} \otimes \begin{pmatrix} f_2 \\ A_2 f_2 \end{pmatrix}_{k_2} \otimes \cdots \otimes \begin{pmatrix} f_d \\ A_d f_d \end{pmatrix}_{k_d} \,.$$

Similar to Example 2.17, we derive the rank–2 matrix product operators decomposition of $A_{(d)}$,

$$A_{(d)} = \begin{pmatrix} A_1 & 1_1 \end{pmatrix} \bowtie \begin{pmatrix} 1_2 & 0 \\ A_2 & 1_2 \end{pmatrix} \bowtie \cdots \bowtie \begin{pmatrix} 1_{d-1} & 0 \\ A_{d-1} & 1_{d-1} \end{pmatrix} \bowtie \begin{pmatrix} 1_d \\ A_d \end{pmatrix} ,$$

where the rank product operation $\bowtie$ is defined as a matrix product of the two corresponding operator core matrices, their blocks being multiplied by means of tensor product operation $\otimes$ in a tensor product Hilbert space. Application to a rank-1 element is defined similar to the Tucker case, resulting in the single rank–2 MPF term,

$$A_{(d)} f = \begin{pmatrix} A_1 f_1 & f_1 \end{pmatrix} \cdot \begin{pmatrix} f_2 & 0 \\ A_2 f_2 & f_2 \end{pmatrix} \cdots \begin{pmatrix} f_{d-1} & 0 \\ A_{d-1} f_{d-1} & f_{d-1} \end{pmatrix} \cdot \begin{pmatrix} f_d \\ A_d f_d \end{pmatrix} .$$

A more detailed discussion of matrix product operator representations in the case of finite dimensional operators (matrices) acting on multidimensional vectors will be addressed in Chapter 3.

2.2 Analytic methods of separable approximation

2.2.1 The problem setting

In this section, we discuss the analytic methods of separable approximation in $\mathbb{R}^d$, in particular, polynomial and trigonometric interpolation as well as approximation by exponential sums. Tensor product interpolation naturally leads to the Tucker decomposition while expansions by exponential sums can be treated as the example of a canonical model.

First, we sketch the classical tools for best polynomial approximation with a focus on the asymptotic error bounds on the class of analytic functions. Then we proceed with a separable approximation by the tensor product polynomial interpolation. The closely related technique is based on a trigonometric interpolation.

As an example, we present the nontrivial application to the oscillating Helmholtz kernel and provide the upper bounds for the canonical and Tucker ranks in terms of the frequency parameter.

We then consider the approach based on a separation by the exponential fitting, which is well suited for the rank structured approximation of functions in the form $f(x_1 + \cdots + x_d)$, arising in various applications.

Analytic methods of the canonical tensor product decomposition to the nonlocal operators and a separable approximation to the multivariate functions can be based on the sinc interpolation or sinc quadratures. Sinc methods will be considered in Sections 2.3 and 2.4.

The main *approximation problem* can be formulated as follows: Given a multivariate function $F\colon \Omega^d \to \mathbb{R}$, $\Omega \in \{\mathbb{R}, \mathbb{R}_+, (a,b)\}$, $(d \geq 2)$, approximate it by a separable expansion

$$F_r(x_1, \dots, x_d) := \sum_{k=1}^{r} c_k \Phi_k^{(1)}(x_1) \dots \Phi_k^{(d)}(x_d) \approx F, \quad c_k \in \mathbb{R},$$

where the set of univariate functions $\{\Phi_k^{(\ell)}\colon \Omega \to \mathbb{R}\}$, $(1 \leq \ell \leq d,\ 1 \leq k \leq r)$ may be fixed (linear approximation) or chosen adaptively to the approximating function F (nonlinear approximation). For numerical efficiency the *separation rank* $r \in \mathbb{N}$ is supposed to be reasonably small.

2.2.2 Best polynomial approximation

We begin with a discussion of approximations based on the Chebyshev polynomials. The Chebyshev polynomials, $T_n(w)$, $w \in \mathbb{C}$, are defined recursively by

$$T_0(w) = 1, \qquad T_1(w) = w,$$
$$T_{n+1}(w) = 2wT_n(w) - T_{n-1}(w), \quad n = 1, 2, \dots .$$

The trigonometric representation $T_n(x) = \cos(n \arccos x)$, $x \in B := [-1, 1]$, implies the boundary relations $T_n(1) = 1$, $T_n(-1) = (-1)^n$. Applying the conformal mapping

$$w = \frac{1}{2}\left(z + \frac{1}{z}\right)$$

leads to the elegant representation

$$T_n(w) = \frac{1}{2}(z^n + z^{-n}). \tag{2.13}$$

The favorable approximation properties of Chebyshev polynomials $\{T_n\}$ on the reference interval B will be formulated for the class of analytic functions. In the complex plane $\mathbb{C}$, we introduce the circular ring

$$\mathcal{R}_\rho := \{z \in \mathbb{C}:\ 1/\rho < |z| < \rho\} \quad \text{with} \quad \rho > 1\,,$$

and denote by $\mathcal{E}_\rho = \mathcal{E}_\rho(B)$ the *Bernstein's regularity ellipse* [36] (with foci at $w = \pm 1$ and the sum of semiaxes equal to $\rho > 1$),

$$\mathcal{E}_\rho := \{w \in \mathbb{C}:\ |w-1| + |w+1| \le \rho + \rho^{-1}\}\,.$$

Note that w is a conformal transform of $\{\xi \in \mathcal{R}_\rho:\ |\xi| > 1\}$ onto $\mathcal{E}_\rho$ as well as of $\{\xi \in \mathcal{R}_\rho:\ |\xi| < 1\}$ onto $\mathcal{E}_\rho$ (but not $\mathcal{R}_\rho$ onto $\mathcal{E}_\rho$). Besides, there holds

$$w(1/z) = w(z)\,.$$

Now we consider the best polynomial approximation by the Chebyshev series. The result is a consequence of Laurent's Theorem.

Theorem 2.23 (Laurent's Theorem). *Let $f: \mathbb{C} \to \mathbb{C}$ be analytic and bounded by $M > 0$ in $\mathcal{R}_\rho$ with $\rho > 1$ (in the following we say $f \in \mathcal{A}_\rho$), and set*

$$C_n := \frac{1}{2\pi}\int_0^{2\pi} f(e^{i\theta})e^{in\theta}d\theta\,, \quad n = 0,\ \pm 1,\ \pm 2, \ldots\,.$$

Then for all $z \in \mathcal{R}_\rho$, there holds $f(z) = \sum_{n=-\infty}^{\infty} C_n z^n$, where the series converges to $f(z)$ for all $z \in \mathcal{R}_\rho$. Moreover, we have $|C_n| \le M/\rho^{|n|}$, and for all $\theta \in [0, 2\pi]$ and an arbitrary integer m,

$$\left| f(e^{i\theta}) - \sum_{n=-m}^{m} C_n e^{in\theta} \right| \le \frac{2M}{\rho - 1}\rho^{-m}\,.$$

The constructive result on the best polynomial approximation can be formulated as follows [64]:

Theorem 2.24 (Chebyshev series). *Let F be analytic and bounded by M in $\mathcal{E}_\rho$ (with $\rho > 1$). Then the expansion*

$$F(w) = C_0 + 2\sum_{n=1}^{\infty} C_n T_n(w)\,, \tag{2.14}$$

holds for all $w \in \mathcal{E}_\rho$, *where*

$$C_n = \frac{1}{\pi} \int_{-1}^{1} \frac{F(x) T_n(x)}{\sqrt{1 - x^2}} dx .$$

Moreover, there holds $|C_n| \le M/\rho^n$. *For* $m = 1, 2, 3, \dots$, *the m-term truncation error is bounded by*

$$\left| F(x) - C_0 - 2 \sum_{n=1}^{m} C_n T_n(x) \right| \le \frac{2M}{\rho - 1} \rho^{-m} , \quad x \in B . \tag{2.15}$$

Proof. Each $f \in \mathcal{A}_{\rho,s} := \{f \in \mathcal{A}_\rho \colon C_{-n} = C_n\}$ has a representation (Theorem 2.23)

$$f(z) = C_0 + \sum_{n=1}^{\infty} C_n (z^n + z^{-n}) , \quad z \in \mathcal{R}_\rho . \tag{2.16}$$

(2.16) implies that $f(1/z) = f(z), \quad z \in \mathcal{R}_\rho$. Note that $w = \frac{1}{2}(z + \frac{1}{z})$ provides a one to one correspondence of functions F that are analytic and bounded by M in $\mathcal{E}_\rho$ with functions f in $\mathcal{A}_{\rho,s}$. Since under this mapping we have (2.13), it follows that if f defined by (2.16) is in $\mathcal{A}_{\rho,s}$, then the corresponding transformed function $F(w) = f(z(w))$, that is analytic and bounded by M in $\mathcal{E}_\rho$, is represented by (2.14). Now the result follows directly by Theorem 2.23. □

It is worth noting that Theorem 2.24 provides the same approximation error as for the best polynomial approximation as proven by S. N. Bernstein, 1912 [36]; see also [340] for theorems on the exponential approximation error of the Chebyshev interpolant. The detailed and comprehensive discussion of related issues may be found in [344].

2.2.3 Chebyshev interpolation

In general, the best polynomial approximation is not computable. However, for continuous functions, almost the best polynomial approximation can be constructed by a simple interpolation procedure. First, we consider the Lagrangian polynomial interpolation.

Let $\mathcal{P}_N(B)$ be a set of polynomials of degree $\le N$ on B. Define by $[\mathcal{I}_N F](x) \in \mathcal{P}_N(B)$ the interpolation polynomial of $F \in C(B)$ with respect to the Chebyshev–Gauss–Lobatto [64] (CGL) nodes

$$\xi_j = \cos \frac{\pi j}{N} \in B , \quad j = 0, 1, \dots, N , \quad \text{with } \xi_0 = 1, \ \xi_N = -1 ,$$

where ξ_j are zeros of the polynomials $(1 - x^2) T_N'(x)$, $x \in B$. The Lagrangian interpolant $\mathcal{I}_N$ of F has the form

$$\mathcal{I}_N F := \sum_{j=0}^{N} F(\xi_j) l_j(x) \in \mathcal{P}_N(B) , \tag{2.17}$$

where $l_j(x)$ denote the set of interpolation polynomials

$$l_j := \prod_{k=0, j\neq k}^{N} \frac{x-\xi_k}{\xi_j-\xi_k} \in \mathcal{P}_N(B)\,, \quad j=0,\ldots,N\,.$$

Clearly, the interpolation property $\mathcal{I}_N(\xi_j) = F(\xi_j)$ holds since $l_j(\xi_j) = 1$ and $l_j(\xi_k) = 0 \;\forall\; k \neq j$.

Stability of the Chebyshev and other interpolation procedures can be characterized by the asymptotic behavior of the *Lebesque constant*. Given the set $\{\xi_j\}_{j=0}^N$ of interpolation points on $B = [-1,1]$ and the associated Lagrangian interpolation operator $\mathcal{I}_N$, the so called Lebesque constant is defined as the minimal number $\Lambda_N \in \mathbb{R}_{>1}$, such that

$$\|\mathcal{I}_N u\|_{\infty,B} \le \Lambda_N \|u\|_{\infty,B} \quad \forall\, u \in C(B)\,. \tag{2.18}$$

In the case of *Chebyshev interpolation* it can be shown that Λ_N grows at most logarithmically in N, e.g.,

$$\Lambda_N \le \frac{2}{\pi}\log N + 1\,.$$

The interpolation points that produce the smallest value Λ_N^* of all Λ_N are not known, but Bernstein [37] proved that

$$\Lambda_N^* = \frac{2}{\pi}\log N + O(1)\,.$$

Now the error bound for polynomial interpolation can be derived using the results for the best approximation.

Theorem 2.25. *Let $u \in C^\infty[-1,1]$ have an analytic extension to $\mathcal{E}_\rho$ that is bounded by $M > 0$ in $\mathcal{E}_\rho$ (with $\rho > 1$). Then the error estimate holds*

$$\|u - \mathcal{I}_N u\|_{\infty,I} \le (1+\Lambda_N)\frac{2M}{\rho-1}\rho^{-N}\,, \quad N \in \mathbb{N}_0\,. \tag{2.19}$$

Proof. Estimate (2.15) implies for the best polynomial approximations to u on $[-1,1]$,

$$\min_{v\in\mathcal{P}_N} \|u - v\|_{\infty,B} \le \frac{2M}{\rho-1}\rho^{-N}\,.$$

The interpolation operator $\mathcal{I}_N$ is a projection, that is for all $v \in \mathcal{P}_N$ there holds $\mathcal{I}_N v = v$. Now, we apply the triangle inequality (for arbitrary $v \in \mathcal{P}_N$) to obtain

$$\|u - \mathcal{I}_N u\|_{\infty,B} = \|u - v - \mathcal{I}_N(u-v)\|_{\infty,B} \le (1+\Lambda_N)\|u-v\|_{\infty,B}\,,$$

which concludes the proof. □

Theorem 2.25 indicates that the Chebyshev interpolant provides asymptotically the same convergence rate as for the best polynomial approximation (say by Chebychev series). However, the polynomial interpolant is easily computable in practice.

2.2.4 Tensor product polynomial interpolation

Tensor product polynomial interpolation can be constructed by the product of univariate interpolants; see e.g., [145].

Given $N \in \mathbb{N}$, the set of interpolating functions $\{\varphi_j(x)\}$, $x \in B$, and sampling points $\xi_j \in B$ $(j = 0, 1, \dots, N)$, such that $\varphi_j(\xi_i) = \Delta_{ij}$, the *Lagrangian interpolant* of a continuous function $F: B \to \mathbb{R}$ with respect to the system $\{\varphi_j(x)\}$, $\mathcal{I}_N: C(B) \mapsto \operatorname{span}\{\varphi_j(x)\}_{j=1}^N$, has a form

$$\mathcal{I}_N F := \sum_{j=0}^{N} F(\xi_j)\varphi_j(x)\,, \qquad F \in C(B)\,, \tag{2.20}$$

providing the interpolation property, $\mathcal{I}_N(\xi_j) = F(\xi_j)$ $(j = 0, 1, \dots, N)$. Here the polynomial, trigonometric polynomial, or sinc interpolation can be applied.

Consider a multivariate function defined in the box $B^d = B_1 \times B_2 \times \cdots \times B_d$, with $B_k = B$,

$$f = f(x_1, \dots, x_d)\,, \quad f: B^d \to \mathbb{R}\,, \quad d \geq 2\,.$$

Let $\{\varphi_{j_k}^{(k)}(x_k)\}$ $(k = 1, \dots, d)$ be a set of interpolating functions on B_k. Define the Nth order *tensor product interpolation* operator

$$\mathbf{I}_N: C(B^d) \to C(B^d)\,, \quad \mathbf{I}_N f = \mathcal{I}_N^1 \times \mathcal{I}_N^2 \times \cdots \times \mathcal{I}_N^d f\,,$$

where $\mathcal{I}_N^k f$ is the Lagrangian interpolation with respect to the variable x_k, at nodes $\{\xi_{j_k}\} \in B_k$, $k = 1, \dots, d$, given by

$$\mathcal{I}_N^k f(x_1, \dots, x_k, \dots, x_d) = \sum_{j_k=0}^{N} f(x_1, \dots, \xi_{j_k}, \dots, x_d)\varphi_{j_k}^{(k)}(x_k)\,.$$

The *tensor product interpolant* $\mathbf{I}_N$ in d variables reads

$$\mathbf{I}_N f := \sum_{\mathbf{j}_1=0}^{N} \cdots \sum_{\mathbf{j}_d=0}^{N} f(\xi_{j_1}, \dots, \xi_{j_d})\varphi_{j_1}^{(1)}(x_1) \dots \varphi_{j_d}^{(d)}(x_d)\,.$$

In the case of CGL nodes, the interpolation points $\overline{\xi}_\alpha \in B^d$, $\alpha = (j_1, \dots, j_d) \in \mathbb{N}_0^d$, are obtained by the Cartesian product of 1D nodes,

$$\overline{\xi}_\alpha := \left(\cos\frac{\pi j_1}{N}, \dots, \cos\frac{\pi j_d}{N}\right)\,.$$

In the following we shall use either polynomial or sinc interpolation. The error bound for the tensor product polynomial interpolant is characterized by the univariate stability constant Λ_N defined in (2.18). First, we observe that $\mathbf{I}_N$ is the projection map

$$\mathbf{I}_N: C(B^d) \to \mathbb{P}_N := \{p_1 \times \cdots \times p_d : p_j \in \mathcal{P}_N,\ j = 1, \dots d\}\,,$$

implying stability of $\mathbf{I}_N$ in the multidimensional case (2.18),

$$\|\mathbf{I}_N f\|_{\infty,B^d} \leq \Lambda_N^d \|f\|_{\infty,B^d} \quad \forall f \in C(B^d) .$$

To derive an analogue of Theorem 2.25, let us introduce the product domain

$$\mathcal{E}_\rho^{(j)} := B_1 \times \cdots \times B_{j-1} \times \mathcal{E}_\rho(I_j) \times B_{j+1} \times \cdots \times B_d ,$$

and denote by X_{-j} the $(d-1)$ dimensional (single hole) subset of variables

$$X_{-j} := \{x_1, \ldots, x_{j-1}, x_{j+1}, \ldots, x_d\} \quad \text{with } x_j \in B_j ,\ j = 1, \ldots, d .$$

Assumption 2.26. *Given $f \in C^\infty(B^d)$, assume there is $\rho > 1$ such that for all $j = 1, \ldots, d$, and each fixed $\xi \in X_{-j}$, there exists an analytic extension of $f(x_j, \xi)$ to $\mathcal{E}_\rho(B_j) \subset \mathbb{C}$ with respect to x_j, $\hat{f}_j(\omega, \xi)$, bounded in $\mathcal{E}_\rho(B_j)$ by certain constant $M_j > 0$, independent on ξ.*

Next theorem estimates the error of multivariate polynomial interpolation [145].

Theorem 2.27. *For $f \in C^\infty(B^d)$, let Assumption 2.26 be satisfied. Then the interpolation error can be estimated by*

$$\|f - \mathbf{I}_N f\|_{\infty,B^d} \leq \Lambda_N^d \frac{2M_\rho(f)}{\rho - 1} \rho^{-N} , \tag{2.21}$$

where Λ_N is the maximal Lebesque constant for the 1D interpolants $\mathcal{I}_N^k$, $k = 1, \ldots, d$, and

$$M_\rho(f) := \max_{1\leq j\leq d} \left\{ \max_{\omega \in \mathcal{E}_\rho^{(j)}} |\hat{f}_j(\omega, \xi)| \right\} .$$

Proof. Multiple use of (2.18), (2.19) combined with the triangle inequality lead to

$$\begin{aligned}
|f - \mathbf{I}_N f| &\leq |f - \mathcal{I}_N^1 f| + |\mathcal{I}_N^1 (f - \mathcal{I}_N^2 \times \cdots \times \mathcal{I}_N^d f)| \\
&\leq |f - \mathcal{I}_N^1 f| + |\mathcal{I}_N^1 (f - \mathcal{I}_N^2 f)| \\
&\quad + |\mathcal{I}_N^1 \mathcal{I}_N^2 (f - \mathcal{I}_N^3 f)| + \cdots + |\mathcal{I}_N^1 \times \cdots \times \mathcal{I}_N^{d-1} (f - \mathcal{I}_N^d f)| \\
&\leq \Bigg[(1 + \Lambda_N) \max_{x \in \mathcal{E}_\rho^{(1)}} |\hat{f}_1(x, \xi)| + \Lambda_N (1 + \Lambda_N) \max_{x \in \mathcal{E}_\rho^{(2)}} |\hat{f}_2(x, \xi)| \\
&\qquad + \cdots + \Lambda_N^{d-1} (1 + \Lambda_N) \max_{x \in \mathcal{E}_\rho^{(d)}} |\hat{f}_d(x, \xi)| \Bigg] \frac{2}{\rho - 1} \rho^{-N} \\
&\leq \frac{(1 + \Lambda_N)(\Lambda_N^d - 1)}{\Lambda_N - 1} \frac{2M_\rho}{\rho - 1} \rho^{-N} .
\end{aligned}$$

Hence (2.21) follows since for $\lambda = \Lambda_N > 1$ we have $\frac{(1+\lambda)(\lambda^n - 1)}{\lambda - 1} \leq \lambda^n$. ☐

Note that the polynomial approximation can be easily transformed to the interpolation by trigonometric polynomials on the interval $t \in [0, 2\pi]$, by the change of variables, $x = \cos(t)$, $x \in [-1, 1]$.

In the following section we consider the separable polynomial/trigonometric approximation to some oscillating functions associated with the Helmholtz kernel arising in various applications.

2.2.5 Separable approximation of the Helmholtz kernel

In this section, we construct the exponentially convergent tensor decompositions to the classical Helmholtz kernel

$$G(x, y) = \frac{e^{i\kappa\|x-y\|}}{\|x-y\|}, \qquad \kappa \in \mathbb{R},$$

such that its real and imaginary parts

$$\frac{\cos(\kappa\|x-y\|)}{\|x-y\|} \quad \text{and} \quad \frac{\sin(\kappa\|x-y\|)}{\|x-y\|}, \qquad x, y \in \mathbb{R}^d$$

are treated separately. The kernel function $G(x, y)$ arises, for example, in finite element method/boundary element method (FEM/BEM) simulations for 3D scattering problems. Tensor product representations can be gainfully applied to calculation of BEM and volume integrals in the case of step type domains; see examples in Figure 2.1. In the case of polyhedral like or 'smooth' geometries the separability requirements can be fulfilled for patchwise approximations (for example isogeometric analysis (IGA) method in BEM).

Fig. 2.1: Examples of step type geometries in 3D.

Our goal is the constructive separable approximation to the real valued oscillatory potentials

$$f_{1,\kappa}(\|x\|) := \frac{\sin(\kappa\|x\|)}{\|x\|}; \quad f_{2,\kappa}(\|x\|) := \frac{1}{\|x\|} - \frac{\cos(\kappa\|x\|)}{\|x\|} = \frac{2\sin^2(\frac{\kappa}{2}\|x\|)}{\|x\|},$$

and to the related kernel functions (see examples in Figure 2.2)

$$f_{1,\kappa}(\|x-y\|), \quad f_{2,\kappa}(\|x-y\|), \quad \frac{1}{\|x-y\|}, \quad x, y \in \mathbb{R}^d.$$

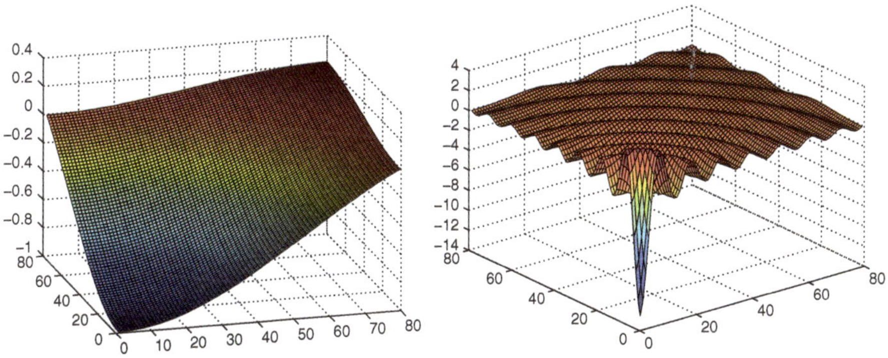

Fig. 2.2: $f_{1,\kappa}(\|x\|)$, slice for $d = 3$, $\|x\| \leq \pi$, $\kappa = 1$ (left), $\kappa = 15$ (right).

The main question thus arises: Are the Tucker, canonical and MPF models robust in κ?

First, we estimate the Tucker and canonical ε ranks of $f_{1,\kappa}$. The main result is due to the following theorem ([206]):

Theorem 2.28. *Given a threshold $\varepsilon > 0$, there is a rank-R canonical respectively rank-$\mathbf{r}$ Tucker ε-approximation to the function $f_{1,\kappa}\colon [0, \frac{2\pi}{\sqrt{d}}]^d \to \mathbb{R}$ in max norm, such that the following rank estimates hold (assuming that $\mathbf{r} = (r, \dots, r)$):*

$$R \leq Cd(|\log \varepsilon| + \kappa)\,, \quad r \leq C(|\log \varepsilon| + \kappa)\,.$$

Proof. Set $t = \|x\|^2$ and approximate the entire function

$$g(t) = \frac{\sin(\kappa\sqrt{t})}{\sqrt{t}}\,,$$

$t \in [0, 2\pi]$ by the trigonometric polynomials in t up to the accuracy ε in the max norm. Using the change of variables consider the *entire function*

$$f(z) = g(\arccos(z))\,, \quad z = \cos(t)\,, \quad z \in [-1, 1]$$

that has the maximum value $O(e^{\kappa})$ on the Bernstein's regularity ellipse of size $O(1)$.

Applying the Chebyshev series to $f(z)$,

$$f(z) \approx C_0 + \sum_{m=1}^{M} C_m T_m(z)\,, \quad z \in [-1, 1]\,,$$

leads to approximation by trigonometric polynomials with $O(|\log \varepsilon| + \kappa)$ terms, where each trigonometric term has the form $\cos(mt) = \cos(m\|x\|^2)$.

Now we recall that the multivariate function $h(x) := \cos(m(x_1^2 + \cdots + x_d^2))$ has a separation rank $R \leq d$, i.e., $h \in \mathfrak{C}_d$, in view of the rank-d representation for the function $s(x) := \sin(x_1^2 + \cdots + x_d^2)$, (Proposition 2.19),

$$\cos\left(\sum_{i=1}^{d} x_j\right) = \sum_{i=1}^{d} \sin\left(tx_j + \frac{\pi}{2d}\right) \prod_{k\in\{1,\dots,d\}\setminus\{j\}} \frac{\sin(x_k + \frac{\pi}{2d} + \alpha_k - \alpha_j)}{\sin(\alpha_k - \alpha_j)}\,,$$

for any $\alpha_k \in \mathbb{R}$, such that $\sin(\alpha_k - \alpha_j) \neq 0$, for all $j \neq k$. Finally, the result for the canonical rank follows by applying the Chebyshev trigonometric approximation to $f(z)$ taking into account that $h \in \mathfrak{C}_d$.

The bound on Tucker ranks follows from the property $h \in \mathfrak{T}_2$; see Lemma 2.18. □

The low rank canonical approximation to the nonoscillatory Newton kernel $\frac{1}{\|x\|}$ was developed in [38, 141]. This result is used in the treatment of the Helmholtz kernel.

Remark 2.29. The Newton kernel $\frac{1}{\|x\|}$, for $\|x\| \geq \varepsilon$, is proven to have a low rank Tucker/canonical ε-approximation with $r \leq R = O(|\log \varepsilon|^2)$ (see Chapters 3, 4 on sinc methods).

Next, we estimates the Tucker and canonical ranks of $f_{2,\kappa}$. To that end, let us introduce the function

$$f_0(t) := \frac{\sin^2(\kappa/2\sqrt{t})}{t} \equiv (f_{1,\kappa/2}(t))^2 .$$

Theorem 2.30. *Given $\varepsilon > 0$, there is the canonical/Tucker ε-approximation in max norm to the function $f_{2,\kappa}\colon [\varepsilon, \frac{2\pi}{\sqrt{d}}]^d \to \mathbb{R}$, such that the following rank estimates hold:*

$$r \leq R \leq Cd|\log \varepsilon|^2(|\log \varepsilon| + \kappa) .$$

Proof. Factorize the function

$$f_{2,\kappa}(t) = 2\frac{\sin^2(\kappa/2\sqrt{t})}{\sqrt{t}}, \qquad t = \|x\|^2 \in [0, 2\pi] ,$$

to obtain

$$f_{2,\kappa}(t) = 2\sqrt{t} f_0(t) = 2\|x\| f_0(t) \quad \text{with} \quad f_0(t) = (f_{1,\kappa/2}(t))^2 .$$

Applying to the entire function $f_0\colon [0, 2\pi] \to \mathbb{R}$ the same argument as in Theorem 2.28, we prove its separable approximation in the class $\mathfrak{C}_R$ (on the continuous level) that allows the κ-dependent rank estimate

$$R \leq Cd(|\log \varepsilon| + \kappa) .$$

At this point, we note that the multivariate function $\|x\| \equiv \frac{\|x\|^2}{\|x\|}$ allows the exponentially convergent sinc approximation with ε rank of the order ([204, 206]),

$$\text{rank}_C(\|x\|) = O(d|\log \varepsilon|^2) , \tag{2.22}$$

which proves the bound on the canonical rank R. □

Figure 2.3 shows the convergence history for the Tucker approximation applied to functions $f_{1,\kappa}$ and $f_{2,\kappa}$ for $d = 3$, depending on $\kappa \in [1, 15]$. It clearly indicates the relation $r \sim C + \kappa$ for different values of $\varepsilon_1 = 10^{-3}$ and $\varepsilon_2 = 10^{-4}$.

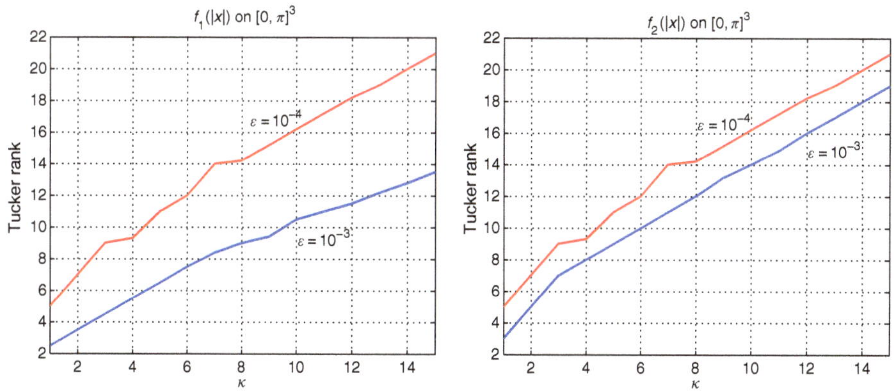

Fig. 2.3: Convergence history for the Tucker approximation applied to $f_{1,\kappa}$, $f_{2,\kappa}$ vs. $\kappa \in [1, 15]$.

Let us derive the rank estimates for matrix product decomposition of functions $f_{1,\kappa}$ and $f_{2,\kappa}$.

Theorem 2.31. *For any $d \geq 3$, we have for ε rank,*

$$\mathrm{rank}_{\mathrm{MPF}}(f_{1,\kappa}(\|x\|)) \leq C(|\log \varepsilon| + \kappa)\,,$$

$$\mathrm{rank}_{\mathrm{MPF}}(f_{2,\kappa}(\|x\|)) \leq C|\log \varepsilon|^2(|\log \varepsilon| + \kappa)\,, \quad \textit{for} \quad \|x\| \geq \varepsilon\,.$$

Proof. Every term in the rank-($|\log \varepsilon| + \kappa$) trigonometric approximation to functions $f_{1,\kappa}(t)$, respectively $f_0(t)$, has the explicit rank–2 matrix product functional representation due to Lemma 2.18. Now estimate (2.22) completes the proof. □

Theorems 2.28, 2.30, and 2.31 indicate linear scaling of the tensor rank in the frequency parameter κ, which can lead to the remarkable reduction of the numerical cost in the case of moderate and high frequencies.

We conclude this section with a brief discussion on the complexity issues of the constructed low rank tensor approximations. To that end, assume that the continuous potentials $f_{1,\kappa}$ and $f_{2,\kappa}$ are represented on a tensor grid of size $n \times n \times n$. For simplicity, we may choose $\varepsilon = O(1/n)$. Recall that the approximation condition in high frequency domains reads as $\kappa \leq Cn$.

Comparing the storage demands for the canonical and Tucker decompositions, we find that the first one is prior under the relation $dRn \leq r^d + rdn$, which implies the condition that ensures the priority of the canonical representation

$$d\left(\frac{R}{r} - 1\right) n \leq r^{d-1}\,.$$

Let $d = 3$, and assume that $\frac{R}{r} = 2$, then the canonical representation is preferable in the range of parameters

$$\sqrt{3n} \leq r\,.$$

The approximation condition, $\kappa \leq Cn$, implies that the Tucker model scales linear in $N_{\text{vol}} = n^3$. However, the relaxed approximation condition $\kappa \leq Cn^{2/3}$ leads to linear scaling in $N_{\text{BEM}} = n^2$ for the Tucker model.

In turn, the canonical decomposition scales at most as $O(n^2)$ for any d. On the other hand, the matrix product function representation is not competitive in high frequency regimes because of the quadratic complexity scaling in κ.

We conclude that the canonical tensor decomposition of the oscillating Helmholtz potential outperforms by an order of magnitude the 'best' wavelet based method of complexity $O(n^3 \log n + \kappa^3 \log \kappa)$; see [40].

2.2.6 Separation by exponential fitting

Separable approximation by exponential sums can be based either on the interpolation and quadratures method (sinc quadratures) or on direct exponential fitting (Proni type methods, nonlinear minimization by Remez type algorithms of rational approximation). These separation methods can be applied, in particular, to multivariate functions depending on a sum of single variables, say the radial function $f(x) = f(\|x\|)$, $x \in \mathbb{R}^d$, providing its explicit canonical decomposition.

The approximation by exponential sums can be also applied to matrix valued functions, $f(A)$, where $A = \sum_{i=1}^{d} A_i$ is a sum of pairwise commutable matrices A_i. The typical examples are given by elliptic operator inverse and matrix exponential.

To fix the idea, we show how the canonical approximation can be derived via separation by integral representations like the Laplace transform. Assume that a function of $\rho = \sum_{i=1}^{d} x_i$ is given by the integral

$$f(\rho) = \int_{\Omega} G(t) e^{\rho F(t)} dt\,, \quad \Omega \in \{\mathbb{R}, \mathbb{R}_+, (a,b)\}\,,$$

and suppose that the above integral can be approximated by an accurate R-term quadrature, then one obtains the separable rank-R approximation

$$f(x_1 + \cdots + x_d) \approx \sum_{\nu=1}^{R} c_\nu G(t_\nu) e^{\rho F(t_\nu)} = \sum_{\nu=1}^{R} c_\nu G(t_\nu) \prod_{i=1}^{d} e^{x_i F(t_\nu)}\,,$$

with weights $c_\nu G(t_\nu)$.

The construction of efficient quadratures for a class of analytic functions can be based on the sinc methods. In Section 2.3, we consider the sinc quadrature techniques applied to the Laplace transform of the target function $f(\rho)$. This approach applies, in particular, to the functions $f(\rho)$ defined by the Green's kernels and other classical potentials, for example

$$f_1(x) = \frac{1}{x_1 + \cdots + x_d}\,, \quad x_i \geq 0\,; \quad \text{the Laplace transform} \quad \frac{1}{\rho} = \int_0^\infty e^{-\rho t} dt,\ \rho > 0\,,$$

$$f_2(x) = 1/\|x\|, \quad x \in \mathbb{R}^d, \quad \|x\| > 0; \quad \text{the Laplace transform} \quad \frac{1}{\sqrt{\rho}} = \frac{2}{\pi}\int_0^\infty e^{-\rho t^2}\,dt\,.$$

Moreover, a separable representation for the function $f_1(x)$ directly applies to the canonical decomposition of the elliptic operator inverse, while the sinc quadrature for the function $f_2(x)$ provides the low rank approximation to the classical Newton kernel (see Section 2.4 and Chapter 3 for the detailed discussion).

In what follows, we briefly discuss the theoretical background behind the direct exponential fitting. The best n-term approximation of $f(\rho)$ by the exponential sums,

$$f(\rho) \approx \sum_{\nu=1}^{n} \omega_\nu e^{-t_\nu \rho}, \quad t_\nu \in \mathbb{C} \tag{2.23}$$

say, with respect to the L^∞ or L^2 norm, leads to an approximation whose separation rank $r = n$ is close to optimal. This quasioptimal n-term approximation can then be optimized by the algebraic methods.

Following [50], for $n \geq 1$ consider the set E_n^0 of exponential sums on $[0, \mathbb{R}_+)$,

$$E_n^0 := \left\{ u = \sum_{\nu=1}^{n} \omega_\nu e^{-t_\nu x} : \; \omega_\nu, \, t_\nu \in \mathbb{R} \right\}.$$

One can address the problem of finding the best approximation to f over the set E_n^0, characterized by the best approximation error,

$$d(f, E_n^0) := \inf_{v \in E_n^0} \|f - v\|_\infty\,. \tag{2.24}$$

The existence of an approximation by exponentials is due to the classical *big Bernstein theorem*: If f is completely monotone for $x \geq 0$, i.e.,

$$(-1)^n f^{(n)}(x) \geq 0 \quad \text{for all} \;\; n \geq 0\,, \;\; x \geq 0\,,$$

then it is the restriction of the Laplace transform of a measure to $\mathbb{R}_+$:

$$f(z) = \int_{\mathbb{R}_+} e^{-tz}\,d\mu(t)\,.$$

Exponential decay of the approximation error on $[a, b]$, $0 < a < b$ can be proven for a class of analytic functions. For technical reasons, we recall the complete elliptic integral of the first kind with modulus κ,

$$\mathbf{K}(\kappa) = \int_0^1 \frac{dt}{\sqrt{(1-t^2)(1-\kappa^2 t^2)}} \qquad (0 < \kappa < 1)$$

and define $\mathbf{K}'(\kappa) := \mathbf{K}(\kappa')$ by $\kappa^2 + (\kappa')^2 = 1$.

Proposition 2.32. *[50] Assume that f is completely monotone and analytic for* $\mathrm{Re}\, z > 0$, *and let* $0 < a < b$. *Then for the uniform approximation on the interval* $[a, b]$,

$$\lim_{n\to\infty} d(f, E_n^0)^{1/n} \le \frac{1}{\omega^2}\,, \qquad \omega = \exp\frac{\pi\mathbf{K}(\kappa)}{\mathbf{K}'(\kappa)} < 1\,, \qquad \textit{with } \kappa = \frac{a}{b}\,. \tag{2.25}$$

In the case of function $f(\rho)$ of the type f_1 and f_2 as mentioned above, we may assume $\rho \in [1, R]$, i.e., $\kappa = 1/R$ for $1 \ll R$. Now we plug into (2.25) the asymptotic of the complete elliptic integrals,

$$\mathbf{K}(\kappa') = \ln\frac{4}{\kappa} + C_1\kappa + \dots \qquad \text{for } \kappa' \to 1\,,$$

$$\mathbf{K}(\kappa) = \frac{\pi}{2}\left\{1 + \frac{1}{4}\kappa^2 + C_1\kappa^4 + \dots\right\} \qquad \text{for } \kappa \to 0\,,$$

to obtain

$$\frac{1}{\omega^2} = \exp\left(-\frac{2\pi\mathbf{K}(\kappa)}{\mathbf{K}(\kappa')}\right) \approx \exp\left(-\frac{\pi^2}{\ln(4R)}\right) \approx 1 - \frac{\pi^2}{\ln(4R)}\,.$$

The latter expression indicates that the number n of different terms to achieve a tolerance $\varepsilon > 0$ is estimated by

$$n \approx \frac{|\log\varepsilon|}{|\log\omega^{-2}|} \approx \frac{|\log\varepsilon|\ln(4R)}{\pi^2}\,.$$

It it worth noting that this result shows nearly the same asymptotic convergence rate in n, the number of terms, as that for the *constructive sinc quadrature* approximation to be presented in Sections 2.3 and 2.4 below. The latter leads to a simple approximation scheme based only on the interpolation of the target function at n explicitly given sampling points. The sinc quadrature approximation error is usually estimated in L^∞ norm.

In the rest of this paragraph we consider in more detail the exponential approximation in L^2 norm. The best approximation to $f(\rho)$, $\rho \in [1, R]$ with respect to a weighted L^2 norm is reduced to the minimization of an explicitly given differentiable functional: Given $R > 1$, $n \ge 1$, find the $2n$ parameters $\alpha_1, \omega_1, \dots, \alpha_n, \omega_n \in \mathbb{R}$, such that

$$F_W(R; \alpha_1, \omega_1, \dots, \alpha_n, \omega_n) := \int_1^R W(x)\left(f(x) - \sum_{i=1}^n \omega_i e^{-\alpha_i x}\right)^2 dx = \min\,.$$

In the important particular case of $f(x) = 1/x$ and $W(x) = 1$, the integral can be calculated in a closed form

$$F_1(R; \alpha_1, \omega_1, \dots, \alpha_n, \omega_n) = 1 - \frac{1}{R} - 2\sum_{i=1}^n \omega_i\,[Ei(-\alpha_i) - Ei(-\alpha_i R)]$$
$$+ \frac{1}{2}\sum_{i=1}^n \frac{\omega_i^2}{\alpha_i}\left[e^{-2\alpha_i} - e^{-2\alpha_i R}\right] + 2\sum_{1\le i<j\le n} \frac{\omega_i\omega_j}{\alpha_i + \alpha_j}\left[e^{-(\alpha_i+\alpha_j)} - e^{-(\alpha_i+\alpha_j)R}\right],$$

Tab. 2.1: Best approximation to $1/\sqrt{\rho}$ in weighted $L^2([1,R])$ norm.

R	10	50	100	200	$\|\cdot\|_{L^\infty}$	$W(\rho) = 1/\sqrt{\rho}$
$r = 4$	$3.7 \cdot 10^{-4}$	$9.6 \cdot 10^{-4}$	$1.5 \cdot 10^{-3}$	$2.2 \cdot 10^{-3}$	$1.9 \cdot 10^{-3}$	$4.8 \cdot 10^{-3}$
$r = 5$	$2.8 \cdot 10^{-4}$	$2.8 \cdot 10^{-4}$	$3.7 \cdot 10^{-4}$	$5.8 \cdot 10^{-4}$	$4.2 \cdot 10^{-4}$	$1.2 \cdot 10^{-3}$
$r = 6$	$8.0 \cdot 10^{-5}$	$9.8 \cdot 10^{-5}$	$1.1 \cdot 10^{-4}$	$1.6 \cdot 10^{-4}$	$9.5 \cdot 10^{-5}$	$3.3 \cdot 10^{-4}$
$r = 7$	$3.5 \cdot 10^{-5}$	$3.8 \cdot 10^{-5}$	$3.9 \cdot 10^{-5}$	$4.7 \cdot 10^{-5}$	$2.2 \cdot 10^{-5}$	$8.1 \cdot 10^{-5}$

where $Ei(x) = -\int_{-\infty}^{x} \frac{e^t}{t} dt$ is the integral exponential function. In the case $R = \infty$, the expression for $F_1(\infty; \dots)$ can be further simplified.

Gradient or Newton type methods with a proper choice of the initial guess can be used to obtain the minimizer of F_1. However, the convergence of nonlinear iterations with large n might be very slow. Table 2.1 presents the results of calculations for $f(\rho) = 1/\sqrt{\rho}$ using the weighted $L^2([1,R])$ norm, which were performed by the MATLAB subroutine FMINS based on the global minimization by direct search.

In general, the integral in F_W may be approximated by a certain quadrature. Optimization with respect to the maximum norm leads to the nonlinear minimization problem (2.24) involving $2n$ parameters $\{\omega_\nu, t_\nu\}_{\nu=1}^{n}$. The numerical scheme can be based on the *Remez algorithm* of rational approximation, see e.g., [138].

To conclude this section we note that in computational practice the approximation of multivariate functions by exponential sums usually applies to functions depending on a sum of variables (say, spherically symmetric or radial functions). In this case the approaches based on the direct approximation by sinc methods with a subsequent rank optimization using multilinear algebra tools are proved to be in preference to the methods based on the direct minimization of the cost functional of type F_W. Hence, in the following section we briefly discuss the main ingredients of the sinc approximation methods.

2.3 Introduction to sinc approximation methods

In this section we discuss sinc approximation methods, which play an important role in the construction of efficient numerical algorithms for low rank approximation of Green's kernels and other classes of spherically symmetric functions. First, we recall essentials of the Fourier transform (FT). Then we consider the following issues: sampling theorem, sinc quadratures and interpolation on $(-\infty, \infty)$, exponential convergence rate for functions in the Hardy space $H^1(D_\delta)$, and improved sinc quadratures. Finally, sinc methods on an arc (a, b) are addressed with a focus on the special case of polynomial and exponential decay of a function on $[0, \infty)$.

Our presentation is mainly based on the results in monographs [263, 334] and also on the recent papers [108–111].

2.3.1 Fourier transform in $L^1(\mathbb{R})$ and in $L^2(\mathbb{R})$

In the current discussion of the Fourier transform we follow [265], where a lot of practically interesting details have been addressed.

Continuous *Fourier transform* is defined by

$$\hat{f}(\omega) := \int_{\mathbb{R}} f(t)e^{-i\omega t}dt\,.$$

If $f \in L^1(\mathbb{R})$ then $\hat{f} \in C^0(\mathbb{R})$ and $|\hat{f}(\omega)| \leq \int_{\mathbb{R}} |f(t)|dt < +\infty$. Suppose that $f, \hat{f} \in L^1(\mathbb{R})$, then the *inverse Fourier transform* is given by

$$f(t) := \frac{1}{2\pi}\int_{\mathbb{R}} \hat{f}(\omega)e^{i\omega t}d\omega\,.$$

For f, $h \in L^1(\mathbb{R})$ the *convolution* of two functions given by

$$g(t) = f * h := \int_{\mathbb{R}} f(t-u)h(u)du$$

satisfies

$$g = \frac{1}{2\pi}\int_{\mathbb{R}} \hat{g}(\omega)e^{i\omega t}d\omega \in L^1(\mathbb{R}) \quad \text{with} \quad \hat{g}(\omega) = \hat{h}(\omega)\hat{f}(\omega)\,.$$

The Fourier transform in $L^2(\mathbb{R})$ can be treated in the Hilbert space setting, such that the inner products of f, $h \in L^2(\mathbb{R})$ and $L^2(\mathbb{R})$ norm are defined, respectively, by

$$\langle f, h\rangle = \int_{\mathbb{R}} f(t)h^*(t)dt\,, \quad \|f\|^2 = \langle f, f\rangle = \int_{\mathbb{R}} |f(t)|^2\,dt\,.$$

Let f, $h \in L^1(\mathbb{R}) \cap L^2(\mathbb{R})$. The *Parseval and Plancherel formulas* read, respectively, as

$$\langle f, h\rangle = \frac{1}{2\pi}\int_{\mathbb{R}} \hat{f}(\omega)\hat{h}^*(\omega)d\omega\,, \quad \|f\|^2 = \frac{1}{2\pi}\int_{\mathbb{R}} |\hat{f}(\omega)|^2 d\omega\,.$$

The global regularity of $f(t)$ can be controlled by the decay rate of $|\hat{f}(\omega)|$, i.e.,

$$|f^{(k)}(t)| \leq \frac{1}{2\pi}\int_{\mathbb{R}} |\hat{f}(\omega)||\omega|^k d\omega\,, \quad k = 0, 1, \ldots$$

and $f^{(k)}$ is continuous, if the corresponding weighted integrals converge.

We collect the important relations between $f(t)$ and its Fourier transform, $\hat{f}(\omega)$ (see [265]):

(A) Inverse: $\hat{f}(t) \Longleftrightarrow 2\pi f(-\omega)$

(B) Convolution: $(h * f)(t) \Longleftrightarrow \hat{h}(\omega)\hat{f}(\omega)$

(C) Multiplication: $h(t)f(t) \Longleftrightarrow \frac{1}{2\pi}(\hat{h} * \hat{f})(\omega)$

(D) Translation: $f(t-u) \Longleftrightarrow e^{-iu\omega}\widehat{f}(\omega)$
(E) Modulation: $e^{ivt}f(t) \Longleftrightarrow \widehat{f}(\omega - v)$
(F) Scaling: $f(t/s) \Longleftrightarrow |s|\widehat{f}(s\omega)$
(G) Time derivatives: $f^{(p)}(t) \Longleftrightarrow (i\omega)^p\widehat{f}(\omega)$
(H) Frequency derivatives: $(-it)^p f(t) \Longleftrightarrow \widehat{f}^{(p)}(\omega)$
(I) Complex conjugate: $f^*(t) \Longleftrightarrow \widehat{f}^*(-\omega)$
(J) Hermitian symmetry: $f(t) \in \mathbb{R} \Longleftrightarrow \widehat{f}(-\omega) = \widehat{f}^*(\omega)$

Since each frequency $e^{i\omega t}$ is amplified by a factor $\widehat{h}$, in signal processing a convolution is called a frequency filtering with a transfer function of a filter $\widehat{h}$ [265].

In what follows, we present some examples on the Fourier transform to be used later on.

Example 2.33. For a Dirac δ (tempered distribution) concentrated at the origin $t = 0$, i.e., $\int_{\mathbb{R}} \delta(t)f(t)dt = f(0)$,

$$\widehat{\delta}(\omega) = \int_{\mathbb{R}} \delta(t)e^{-i\omega t}dt = 1 \quad \text{(formal representation)}.$$

The so called characteristic function plays an important role in many applications.

Example 2.34. The FT of the *characteristic (indicator, step) function*

$$f(t) = \chi_{[-T,T]}(t) = \begin{cases} 1 & \text{if } t \in [-T, T], \\ 0 & \text{otherwise} \end{cases}$$

is given by

$$\widehat{f}(\omega) = \frac{1}{2\pi}\int_{-T}^{T} e^{-i\omega t}dt = \frac{2\sin(T\omega)}{\omega} \notin L^1(\mathbb{R}) \quad \text{(not integrable)}.$$

The basic relation between the characteristic and sinc functions is described as follows.

Corollary 2.35. *An ideal low pass filter has a transfer function* $\widehat{h} = \chi_{[-\xi,\xi]}(\omega)$, *thus its inverse FT (impulse response) is*

$$h(t) = \frac{1}{2\pi}\int_{-\xi}^{\xi} e^{i\omega t}d\omega = \frac{\sin(\xi t)}{\pi t}.$$

With $\xi = \pi$, *we obtain the classical* sinc *function, which provides the important example of the so called band limited signals.*

Functions $\chi_{[-\pi,\pi]}(t)$ (Haar scaling function) and $\mathrm{sinc}(t)$ have the complementary (in fact, the opposite) features in the *time* and *frequency* (Fourier) domains. Numerous *wavelet families* suggest certain compromises between these two 'extreme cases'.

The next examples describe FT of the translated Dirac and Gaussian functions.

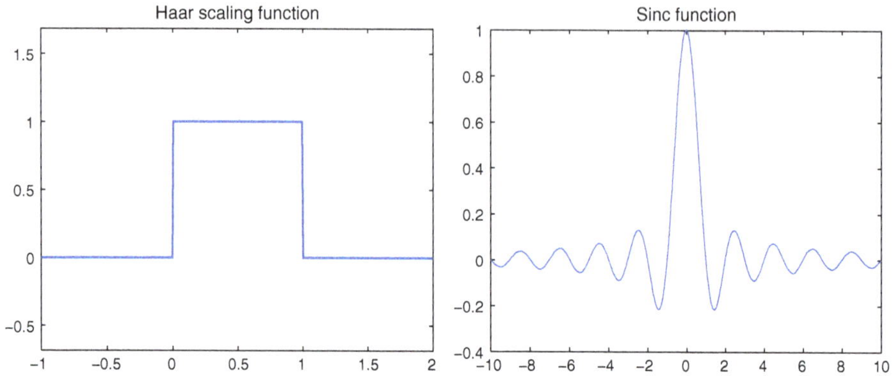

Fig. 2.4: Haar (indicator) and sinc scaling functions.

Example 2.36. A FT for a translated Dirac $\delta_\tau(t) = \delta(t-\tau)$ is calculated by evaluating $\mathrm{e}^{-i\omega t}$ at $t = \tau$:

$$\widehat{\delta}_\tau(\omega) = \int_{\mathbb{R}} \delta(t-\tau)\mathrm{e}^{-i\omega t}dt = \mathrm{e}^{-i\omega\tau} .$$

Consequently, for the Dirac comb $c(t) = \sum_{n=-\infty}^{\infty} \Delta(t-nT)$ we have $\widehat{c} = \sum_{n=-\infty}^{\infty} \mathrm{e}^{-inT\omega}$.

Example 2.37. A FT of a Gaussian $f(t) = \exp(-t^2) \in C^\infty$ is also a Gaussian:

$$\widehat{f}(\omega) = \sqrt{\pi}\exp(-\omega^2/4) .$$

Hint: it is easy to check using properties (G), (H) that $2\widehat{f}'(\omega)+\omega\widehat{f}(\omega) = 0$, which proves the statement.

2.3.2 Sampling theorem. Sinc interpolation

In this section we address the classical question of how to discretize analog signals on $\mathbb{R}$. The sinc function, already introduced as the Fourier transform of the characteristic function $\chi_{[-\pi,\pi]}(t)$, allows us to describe a class of signals $f(t)$, $t \in \mathbb{R}$, which can be discretized by recording their sample values $\{f(n\mathfrak{h})\}_{n\in\mathbb{Z}}$ at intervals $\mathfrak{h} > 0$.

The sinc function (also called Cardinal function) is given as

$$\mathrm{sinc}(x) := \frac{\sin(\pi x)}{\pi x} \quad \text{with convention } \mathrm{sinc}(0) = 1 .$$

V. A. Kotelnikov (1933), [236] and J. Whittaker (1935), [366] proved a celebrated theorem: *band limited signals*, which can be exactly reconstructed via their sampling values.

Theorem 2.38 (Sampling Theorem). *If the support of $\widehat{f}$ is included in $[-\pi/\mathfrak{h}, \pi/\mathfrak{h}]$ then for $t \in \mathbb{R}$*

$$f(t) = \sum_{n=-\infty}^{\infty} f(n\mathfrak{h}) S_{n,\mathfrak{h}}(t) , \quad \textit{with } S_{n,\mathfrak{h}}(t) = \operatorname{sinc}(t/\mathfrak{h} - n) .$$

As preliminaries to the proof, we make remarks (a)–(c) to justify the auxiliary Lemma; see [265] for more details.

(a) The *Poisson formula* is (in the sense of distributions)

$$\widehat{c} = \sum_{n=-\infty}^{\infty} \mathrm{e}^{-in\mathfrak{h}\omega} = \frac{2\pi}{\mathfrak{h}} \sum_{k=-\infty}^{\infty} \delta\left(\omega - \frac{2k\pi}{\mathfrak{h}}\right) . \tag{2.26}$$

Recall that $\widehat{c} = \sum_{n=-\infty}^{\infty} \mathrm{e}^{-in\mathfrak{h}\omega}$ is the FT of the Dirac comb $c(t) = \sum_{n=-\infty}^{\infty} \delta(t - n\mathfrak{h})$ (Example 2.36). Since $\widehat{c}$ is $\frac{2\pi}{\mathfrak{h}}$ -periodic, it suffices to prove that $\widehat{c}_{[-\pi/\mathfrak{h},\pi/\mathfrak{h}]} = \frac{2\pi}{\mathfrak{h}}\delta$.

(b) To any sample $f(n\mathfrak{h})$, we associate a Dirac and introduce the weighted Dirac sum

$$f_d(t) := \sum_{n=-\infty}^{\infty} f(n\mathfrak{h})\delta(t - n\mathfrak{h}) .$$

Since the FT of $\delta(t - n\mathfrak{h})$ is $\mathrm{e}^{-in\mathfrak{h}\omega}$, one obtains $\widehat{f}_d = \sum_{n=-\infty}^{\infty} f(n\mathfrak{h})\mathrm{e}^{-in\mathfrak{h}\omega}$.

(c) Now $f(t)$ can be computed from the sample values $f(n\mathfrak{h})$ due to the simple relation between Fourier transforms $\widehat{f}_d$ and $\widehat{f}$ as in the following Lemma:

Lemma 2.39. *The FT of f_d is given by*

$$\widehat{f}_d(\omega) = \frac{1}{\mathfrak{h}} \sum_{k=-\infty}^{\infty} \widehat{f}\left(\omega - \frac{2k\pi}{\mathfrak{h}}\right) .$$

Proof. $f(n\mathfrak{h})\delta(t - n\mathfrak{h}) = f(t)\delta(t - n\mathfrak{h})$ implies

$$f_d(t) := f(t) \sum_{n=-\infty}^{\infty} \delta(t - n\mathfrak{h}) \equiv f(t)c(t) .$$

Computing the FTs

$$\widehat{f}_d = \frac{1}{2\pi}\widehat{f} * \widehat{c}(\omega) , \tag{2.27}$$

we apply the *Poisson formula* (2.26) to represent $\widehat{c}(\omega)$. Since $\widehat{f} * \delta(\omega - \xi) = \widehat{f}(\omega - \xi)$, inserting the above formula into (2.27) proves the assertion. □

Now we are in a position to prove the sampling theorem; see [265].

Proof. If $n \neq 0$, the support of $\widehat{f}(\omega - n\pi/\mathfrak{h})$ does not intersect the support of $\widehat{f}(\omega)$ since $\widehat{f}(\omega) = 0$ for $|\omega| > \pi/\mathfrak{h}$. Thus Lemma 2.39 implies

$$\widehat{f}_d(\omega) = \frac{\widehat{f}(\omega)}{\mathfrak{h}} \quad \text{if} \quad |\omega| \le \frac{\pi}{\mathfrak{h}} .$$

Recall that the FT of $S_{0,\mathfrak{h}} = \operatorname{sinc}(t/\mathfrak{h})$ is $\widehat{S}_{0,\mathfrak{h}} = \mathfrak{h}\chi_{[-\pi/\mathfrak{h},\pi/\mathfrak{h}]}$.

Since $\mathrm{supp}(\widehat{f}) \in [-\pi/\mathfrak{h}, \pi/\mathfrak{h}]$, the previous relation results in $\widehat{f}(\omega) = \widehat{S}_{0,\mathfrak{h}}(\omega)\widehat{f}_d(\omega)$. The inverse FT of this equation, that is $f(t) = S_{0,\mathfrak{h}} * f_d(t)$, leads to the required result, taking into account that $S_{n,\mathfrak{h}}(t) = S_{0,\mathfrak{h}}(t - n\mathfrak{h})$,

$$f(t) = S_{0,\mathfrak{h}} * \sum_{n=-\infty}^{\infty} f(n\mathfrak{h})\delta(t - n\mathfrak{h}) = \sum_{n=-\infty}^{\infty} f(n\mathfrak{h})S_{0,\mathfrak{h}}(t - n\mathfrak{h}),$$

which proves the sampling equation. □

Sampling theorem plays an important role in tele/radio communications, signal processing, stochastic models, etc.

Sampling theorem can be viewed as a decomposition in orthogonal basis. Define the space $\mathbf{U}_\mathfrak{h}$ as a set of functions whose FTs have a support included in $[-\pi/\mathfrak{h}, \pi/\mathfrak{h}]$.

Lemma 2.40 ([334]). *A set of functions $\{S_{n,\mathfrak{h}}(t)\}_{n\in\mathbb{Z}}$ is an orthogonal basis of the space $\mathbf{U}_\mathfrak{h}$. If $f \in \mathbf{U}_\mathfrak{h}$ then*

$$f(n\mathfrak{h}) = \frac{1}{\mathfrak{h}} \langle f(t), S_{n,\mathfrak{h}}(t)\rangle .$$

Corollary 2.41. *The* sinc *interpolation formula of Theorem 2.38 can be interpreted as a decomposition of $f \in \mathbf{U}_\mathfrak{h}$ in an orthogonal basis of $\mathbf{U}_\mathfrak{h}$:*

$$f(t) = \frac{1}{\mathfrak{h}} \sum_{n=-\infty}^{\infty} \langle f(\cdot), S_{n,\mathfrak{h}}(\cdot)\rangle\, S_{n,\mathfrak{h}}(t) .$$

If $f \notin \mathbf{U}_\mathfrak{h}$, one finds the orthogonal projection of f in $\mathbf{U}_\mathfrak{h}$.

The interesting question arises: When does the sinc interpolant

$$C(f, \mathfrak{h})(x) = \sum_{k=-\infty}^{\infty} f(k\mathfrak{h})S_{k,\mathfrak{h}}(x) \tag{2.28}$$

represent a function exactly? The answer is known for a class of entire functions (Paley–Wiener space $\mathbf{W}(\pi/\mathfrak{h})$).

Definition 2.42. Let $\mathfrak{h} > 0$, and let $\mathbf{W}(\pi/\mathfrak{h})$ denote the family of entire functions, such that $\int_{\mathbb{R}} |f(t)|^2\, dt < \infty$, and for all $z \in \mathbb{C}$

$$|f(z)| \le Ce^{\pi|z|/\mathfrak{h}} \quad \text{with constant } C > 0 .$$

Proof of the following theorem is a consequence of the classical Paley–Wiener theorem (see [334] for details).

Theorem 2.43. *$\{\mathfrak{h}^{-1/2}S_{k,\mathfrak{h}}(x)\}_{k\in\mathbb{Z}}$ is a complete $L^2(\mathbb{R})$-orthonormal sequence in $\mathbf{W}(\pi/\mathfrak{h})$. Every $f \in \mathbf{W}(\pi/\mathfrak{h})$ has the cardinal series representation*

$$f(x) = C(f, \mathfrak{h})(x) , \quad x \in \mathbb{R} .$$

In the following sections we discuss the approximation error of the truncated sinc interpolant $C(f, \mathfrak{h})$ on certain classes of analytic functions by selecting a small number of terms in (2.28). This approach provides a powerful tool for the constructive low rank approximation of commonly used Green's kernels and interaction potentials.

2.3.3 Sinc approximation of analytic functions

It turns out that the sinc interpolant $C(f, \mathfrak{h})$ provides an incredibly accurate approximation on $\mathbb{R}$ for functions that are analytic and uniformly bounded on the strip

$$D_\delta := \{z \in \mathbb{C} : |\operatorname{Im} m\, z| \leq \delta\}\,, \quad 0 < \delta < \tfrac{\pi}{2}\,,$$

such that

$$N(f, D_\delta) := \int_{\mathbb{R}} (|f(x + i\delta)| + |f(x - i\delta)|)\, dx < \infty\,.$$

This defines the Hardy space further defined by $H^1(D_\delta)$.

First, we recall the standard sinc approximation results for analytic functions (see [334] and for the detailed considerations there). For $f \in H^1(D_\delta)$, we have exponential convergence in $1/\mathfrak{h}$

$$\sup_{x\in\mathbb{R}} |f(x) - C(f, \mathfrak{h})(x)| = O(\mathrm{e}^{-\pi\delta/\mathfrak{h}})\,, \quad \mathfrak{h} \to 0\,. \tag{2.29}$$

Using the sinc interpolant, it is possible to derive the sinc quadratures for analytic integrands. Likewise, if $f \in H^1(D_\delta)$, the integral

$$I(f) = \int_{\Omega} f(x)dx \quad (\Omega = \mathbb{R} \text{ or } \Omega = \mathbb{R}_+)$$

can be approximated with exponential convergence rate by the sinc quadrature (trapezoidal rule)

$$T(f, \mathfrak{h}) := \mathfrak{h} \sum_{k=-\infty}^{\infty} f(k\mathfrak{h}) \quad \left(\equiv \int_{\mathbb{R}} C(f, \mathfrak{h})(x)dx \approx I(f) \right)\,,$$

$$|I(f) - T(f, \mathfrak{h})| = O(\mathrm{e}^{-\pi\delta/\mathfrak{h}})\,, \quad \mathfrak{h} \to 0\,. \tag{2.30}$$

It is worth noting that analogues estimates hold for (computable) truncated sums

$$C_M(f, \mathfrak{h}) := \sum_{k=-M}^{M} f(k\mathfrak{h}) S_{k,\mathfrak{h}}(x)\,, \quad T_M(f, \mathfrak{h}) := \mathfrak{h} \sum_{k=-M}^{M} f(k\mathfrak{h})\,. \tag{2.31}$$

For further applications, we recall standard error estimates on $\mathbb{R}$.

Theorem 2.44 ([334]). *If $f \in H^1(D_\delta)$ and $|f(x)| \leq C \exp(-b|x|)$ for all $x \in \mathbb{R}$ $b, C > 0$, then*

$$\|f - C_M(f, \mathfrak{h})\|_\infty \leq C \left[\frac{\mathrm{e}^{-\pi\delta/\mathfrak{h}}}{2\pi\delta} N(f, D_\delta) + \frac{1}{b\mathfrak{h}} \mathrm{e}^{-b\mathfrak{h}M} \right]\,, \tag{2.32}$$

$$|I(f) - T_M(f, \mathfrak{h})| \leq C \left[\frac{\mathrm{e}^{-2\pi\delta/\mathfrak{h}}}{1 - \mathrm{e}^{-2\pi\delta/\mathfrak{h}}} N(f, D_\delta) + \frac{1}{b} \mathrm{e}^{-b\mathfrak{h}M} \right]\,. \tag{2.33}$$

Proof. We only sketch the main steps of the proof. The first term in the sum in the right hand side of (2.32) represents the *approximation error* (2.29),

$$\|f(x) - C(f, \mathfrak{h})(x)\|_\infty \leq \frac{N(f, D_\delta)}{2\pi\delta \sinh(\pi\delta/\mathfrak{h})},$$

while the second one gives the *truncation error*

$$\|C(f, \mathfrak{h})(x) - C_M(f, \mathfrak{h})(x)\|_\infty \leq \sum_{|k|\geq M+1} |f(k\mathfrak{h})|$$
$$\leq 2C \sum_{k=M+1}^{\infty} \mathrm{e}^{-bk\mathfrak{h}} \leq \frac{2C}{b\mathfrak{h}} \mathrm{e}^{-b\mathfrak{h}M},$$

which proves (2.32). Similar arguments apply to (2.33). □

Finally, we specify the particular exponential convergence rate in M. For *interpolation error* (2.32), the choice

$$\mathfrak{h} = \sqrt{\pi\delta/bM}$$

implies the exponential convergence rate

$$\|f - C_M(f, \mathfrak{h})\|_\infty \leq C M^{1/2} \mathrm{e}^{-\sqrt{\pi\delta bM}}. \tag{2.34}$$

In fact, for the chosen $\mathfrak{h}$, the first term in the right hand side in (2.32) dominates, hence (2.34) follows. For the *quadrature error* (2.33), the 'quasioptimal' choice

$$\mathfrak{h} = \sqrt{2\pi\delta/bM}$$

yields

$$|I(f) - T_M(f, \mathfrak{h})| \leq C\mathrm{e}^{-\sqrt{2\pi\delta bM}}. \tag{2.35}$$

Practically convenient choice of the strip size Δ is given by $\delta = \pi/2$. Hence, we proved the following corollary:

Corollary 2.45. *Set $\delta = \pi/2$ and choose $\mathfrak{h} = \frac{\pi}{\sqrt{bM}}$, then the sinc quadrature approximation error is bounded by*

$$|I(f) - T_M(f, \mathfrak{h})| \leq C\mathrm{e}^{-\pi\sqrt{bM}}. \tag{2.36}$$

Choose $\mathfrak{h} = \sqrt{\pi\delta/bM}$, then the sinc interpolation error is estimated by

$$\|f - C_M(f, \mathfrak{h})\|_\infty \leq C M^{1/2} \mathrm{e}^{-\pi\sqrt{\frac{1}{2}bM}}. \tag{2.37}$$

We observe that in both cases the leading exponential in $\mathrm{e}^{-\pi\sqrt{bM}}$ scales as $O(\sqrt{M})$. For functions with faster than exponential decay as $|x| \to \infty$ it is possible to improve this scaling in M.

2.3.4 Improved error bound in the case of double exponential decay

The standard error bound can be improved to almost linear exponential behavior. If f has a double exponential decay as $|x| \to \infty$, i.e.,

$$|f(x)| \le C \exp(-b e^{a|x|}) \quad \text{for all } x \in \mathbb{R} \text{ with } a, b, C > 0, \tag{2.38}$$

the convergence rate of sinc interpolation and quadrature can be improved up to $O(e^{-cM/\log M})$ (Theorem 2.44). The main result is given by the following theorem:

Theorem 2.46 ([109]). *Let $f \in H^1(D_\delta)$ with some $\delta < \frac{\pi}{2}$, and let (2.38) hold. Then the choice of the step size $\mathfrak{h} = \log(\frac{2\pi a M}{b})/(aM)$ leads to the quadrature error*

$$|I - T_M(f, \mathfrak{h})| \le C\, N(f, D_\delta) e^{-2\pi\delta a M/\log(2\pi a M/b)} . \tag{2.39}$$

The choice $\mathfrak{h} = \log(\frac{\pi a M}{b})/(aM)$ ensures the interpolation error

$$\|f - C_M(f, \mathfrak{h})\|_\infty \le C \frac{N(f, D_\delta)}{2\pi\delta} e^{-\pi\delta a M/\log(\pi a M/b)} . \tag{2.40}$$

Proof. We improve the error bound for $|I - T(f, \mathfrak{h})|$ as in Theorem 2.44. The first term in the right hand side in (2.33), representing the approximation error, remains the same. For the truncation error, the double exponential decay of the integrand leads to the simple estimates

$$\begin{aligned}
\sum_{k:\ |k|>M} \exp(-b e^{a|k\mathfrak{h}|}) &= 2 \sum_{k=M+1}^{\infty} \exp(-b e^{a|k\mathfrak{h}|}) \\
&\le 2 \int_M^\infty \exp(-b e^{a|x\mathfrak{h}|}) dx \\
&\le \frac{2 e^{-a\mathfrak{h}M}}{ab\mathfrak{h}} \exp(-b e^{a\mathfrak{h}M}) .
\end{aligned}$$

Hence, the improved *quadrature error* has a bound

$$|I - T_M(f, \mathfrak{h})| \le C \left[\frac{e^{-2\pi\delta/\mathfrak{h}}}{1 - e^{-2\pi\delta/\mathfrak{h}}} N(f, D_\delta) + \frac{e^{-a\mathfrak{h}M}}{ab} \exp(-b e^{a\mathfrak{h}M}) \right] .$$

Now (2.62) follows by substitution of $\mathfrak{h}$ since the first term in the previous estimate dominates.

The approximation error of $C_M(f, \mathfrak{h})$ allows the same estimate as in the standard case in (2.32). To prove (2.40), we notie that the *truncation error* bound is determined by the decay rate of f as $|x| \to \infty$,

$$\begin{aligned}
\|C(f, \mathfrak{h})(x) - C_M(f, \mathfrak{h})(x)\|_\infty &\le \sum_{|k|\ge M+1} |f(k\mathfrak{h})| \\
&\le 2C \sum_{k=M+1}^{\infty} e^{-b e^{ak\mathfrak{h}}} \le \frac{2C}{ba\mathfrak{h} e^{a\mathfrak{h}M}} e^{-b e^{a\mathfrak{h}M}} .
\end{aligned}$$

Tab. 2.2: Sinc interpolation to the Mexican hat, separation rank $r = M + 1$.

$\alpha\backslash M$	4	9	16	25	36	49	64	81	100
1	0.05	$6 \cdot 10^{-4}$	$7 \cdot 10^{-7}$	$1 \cdot 10^{-10}$	$2 \cdot 10^{-15}$	$1 \cdot 10^{-15}$	–	–	–
10	0.17	0.13	0.12	0.04	0.01	0.004	0.0009	$1.7 \cdot 10^{-4}$	$2.6 \cdot 10^{-5}$
0.1	3.8	2.6	0.6	0.08	0.006	$1.6 \cdot 10^{-5}$	$2 \cdot 10^{-7}$	$2.5 \cdot 10^{-9}$	$2 \cdot 10^{-11}$

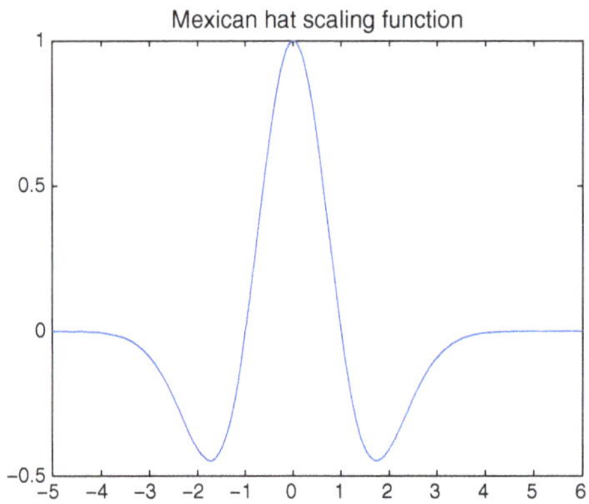

Fig. 2.5: Mexican hat $f(x) = (1 - x^2)\exp(-\alpha x^2)$, $\alpha = 1$.

Hence, we arrive at the bound on interpolation error of $C_M(f, \mathfrak{h})$,

$$\|f - C_M(f, \mathfrak{h})\|_\infty \le C \left[\frac{e^{-\pi\delta/\mathfrak{h}}}{2\pi\delta} N(f, D_\delta) + \frac{e^{-a\mathfrak{h}M}}{ab\mathfrak{h}} \exp(-be^{a\mathfrak{h}M}) \right] .$$

which proves (2.40) by substitution of $\mathfrak{h}$. □

We present some exercises demonstrating the exponentially fast convergence of the sinc approximation.

Exercise 2.47. For numerical approximation of the integral $\int_{-\infty}^{\infty} \exp(-x^2)dx = \sqrt{\pi}$, show that with the choice $h = (\pi/M)^{1/2}$,

$$\left| \int_{-\infty}^{\infty} \exp(-x^2)dx - h \sum_{-M}^{M} e^{-k^2h^2} \right| \le C \exp(-\pi M) .$$

Calculate the approximation to $\sqrt{\pi}$ for $M = 4, 8, 12$; the latter will be accurate to 15 digits. Calculate the sinc interpolant to the Gaussian $\exp(-\lambda x^2)$, $\lambda > 0$ and to the Mexican hat scaling function

$$f(x) = (1 - x^2)\exp(-\alpha x^2) , \quad \alpha > 0 .$$

The example in Table 2.2 illustrates the error of sinc interpolation on $\mathbb{R}$ applied to the Mexican hat scaling function; see Figure 2.5.

We observe fast exponential decay in the error of sinc interpolant for different values of exponential α specifying the the Mexican hat function.

2.3.5 Sinc interpolation on an interval (a, b)

To apply Theorem 2.44 in the case of interval $\Omega = (a, b)$ (say, $\Omega = \mathbb{R}_+ := (0, \infty)$) one has to substitute the variable $x \in \Omega$ by $x = \varphi(\zeta)$, such that $\varphi\colon \mathbb{R} \to (a, b)$ is a bijection. This changes $f\colon (a, b) \to \mathbb{R}$ into

$$f_1 := \varphi' \cdot (f \circ \varphi) : \mathbb{R} \to \mathbb{R} \quad \text{(quadrature case)}\,,$$
$$f_1 := f \circ \varphi \quad \text{(interpolation case)}\,.$$

Assuming $f_1 \in H^1(D_\delta)$, one can apply (2.34)–(2.35) to the transformed function f_1; see [334] for the detailed discussion.

Example 2.48. In the case of finite interval (a, b) the appropriate option is

$$\varphi^{-1}(z) = \log[(z-a)/(b-z)]\ , \qquad \operatorname{Re} z = x\,.$$

As an important special case, we consider sinc quadratures on $\mathbb{R}_+$ in the situation with polynomial or exponential decay at $x \to \infty$.

Polynomial decay. Let us set $\Omega = \mathbb{R}_+$, define $\varphi^{-1}(z) = \log(z)$, i.e., $\varphi(\zeta) = \mathrm{e}^{\zeta}$, and assume:

(i) f can be analytically extended from $\mathbb{R}_+$ into the sector (see Figure 2.6, left)

$$D_\delta^{(1)} = \{z \in \mathbb{C}\colon |\arg(z)| < \delta\} \quad \text{for some } 0 < \delta < \pi/2\,.$$

Actually, $\varphi^{-1}\colon D_\delta^{(1)} \to D_\delta$ is the conformal map.

(ii) f satisfies the inequality

$$|f(z)| \le c|z|^{\alpha-1}(1+|z|)^{-\alpha-\beta} \quad \text{for some } 0 < \alpha, \beta \le 1 \text{ and } \forall\, z \in D_\delta^{(1)}.$$

Let $\alpha = 1$ in condition (ii). Choosing any $M \in \mathbb{N}$ and taking

$$h^{(1)} = \sqrt{2\pi\delta/(\beta M)}\,,$$

we define the corresponding quadrature rule for $f_1(\zeta)$ (with substitution $x = \varphi(\zeta) = \mathrm{e}^{\zeta}$)

$$T_M^{(1)}(f_1) = h^{(1)} \sum_{k=-\beta M}^{M} c_k f(z_k)\,, \qquad z_k = \mathrm{e}^{kh^{(1)}}\,, \qquad c_k = \mathrm{e}^{kh^{(1)}}\,,$$

possessing the exponential convergence rate

$$\left|I - T_M^{(1)}(f_1)\right| \le C\mathrm{e}^{-\sqrt{2\pi\delta\beta M}}\,,$$

where a positive constant C does not dependent on M.

Remark 2.49. Note that the results for polynomial decay are in sharp contrast to the error in polynomial based approximation with algebraic singularities. For example, for the function $f(x) = x^{\alpha}(1-x)^{\alpha}$, $\alpha = 1/2$, the best approximation by polynomials of degree M converges only with the rate $C\,M^{-\alpha}$.

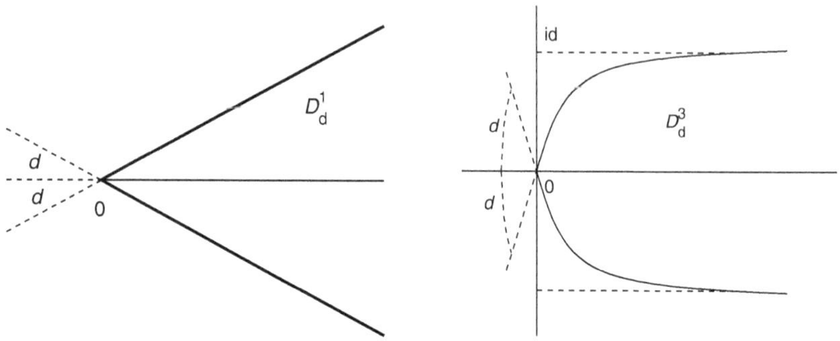

Fig. 2.6: The analyticity sector $D_\delta^{(1)}$ (left) and the 'bullet shaped' domain $D_\delta^{(3)}$ (right).

Exponential decay. Assume that the integrand f on $\mathbb{R}_+$ can be analytically extended into the 'bullet shaped' domain (Figure 2.6, right)

$$D_\delta^{(3)} = \{z \in \mathbb{C} : |\arg(\sinh z)| < \delta\}, \quad 0 < \delta < \pi/2,$$

and that f satisfies

$$|f(z)| \le C\left(\frac{|z|}{1+|z|}\right)^{\alpha-1} \mathrm{e}^{-\beta \operatorname{Re} z} \quad \text{in } D_\delta^{(3)}, \quad \alpha, \beta \in (0,1]. \tag{2.41}$$

Introduce the conformal map $\varphi^{-1}(z) = \log[\sinh(z)]$ that maps $\varphi^{-1} : D_\delta^{(3)} \to D_\delta$. Again, setting $\alpha = 1$ and choosing $h^{(2)} = h^{(1)}$ and the number $M \in \mathbb{N}$, we obtain the quadrature rule for $f(f_2)$,

$$T_M^{(2)}(f_2) = h^{(2)} \sum_{k=-\beta M}^{M} c_k^{(2)} f(z_k^{(2)}), \quad z_k^{(2)} = \log[\mathrm{e}^{kh^{(2)}} + \sqrt{1 + \mathrm{e}^{2kh^{(2)}}}], \quad c_k^{(2)} = 1 + \mathrm{e}^{-2kh^{(2)}}$$

possessing the exponential convergence rate

$$\left|I - T_M^{(2)}(f_2)\right| \le C\mathrm{e}^{-\sqrt{2\pi\delta\beta M}}.$$

Note that for practical purposes in both cases one can simplify the error estimates by inserting $\delta = \pi/2$.

2.3.6 Numerics for the sinc interpolation on (a, b) and $\mathbb{R}_+$

Figure 2.7 illustrates separable sinc approximation to the function

$$g(x, y) = \|x\|^\lambda \operatorname{sinc}(\|x\| \, \|y\|), \quad \lambda \in (-3, 1],$$

arising in the representation of kernel functions for the Boltzmann equation, $x, y \in \mathbb{R}^3$; see [207]. We observe that the sinc interpolant provides exponentially convergent

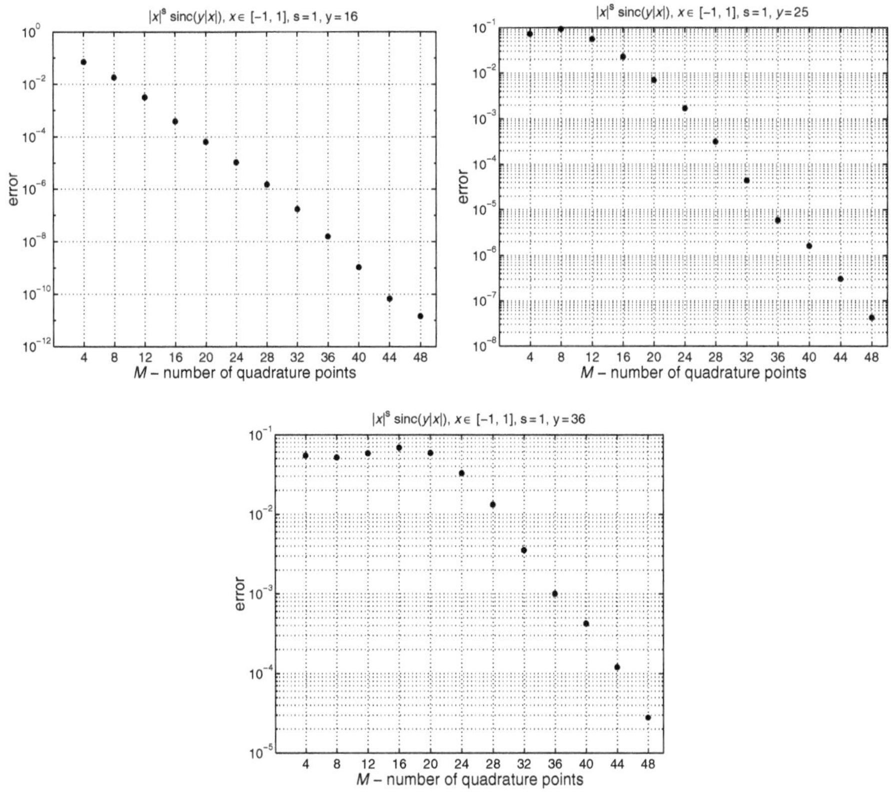

Fig. 2.7: L^∞ error of the sinc interpolation to $|x|^\lambda \operatorname{sinc}(|x|y)$, $x \in [-1, 1]$, $y = 16, 25, 36$, $\lambda = 1$.

separable approximation to $g(x, y)$ in $(x, y) \in [a, b]^3 \times [c, d]^3$ for different values of y. The main problem with such an approximation is the strong dependence of the convergence rate on the size of the computational domain in y.

To illustrate the sinc interpolation on $\mathbb{R}_+$, we consider the function

$$g(x, y) = \exp(-xy)\,, \quad x, y \geq 0\,.$$

Introduce the auxiliary function [142]

$$f(x, y) = \frac{x}{1 + x}\exp(-xy)\,, \quad x \in \mathbb{R}_+\,, \quad y \in [1, R]\,,$$

which satisfies all the conditions above with $\alpha = \beta = 1$ (exponential decay). With the choice of interpolation points $x_k := \log[e^{k\mathfrak{h}} + \sqrt{1 + e^{2k\mathfrak{h}}}] \in \mathbb{R}_+$, it can be approximated with exponential convergence, as shown in Figure 2.8.

The multiplication with the singular factor $\frac{x+1}{x}$ does not change the separation rank.

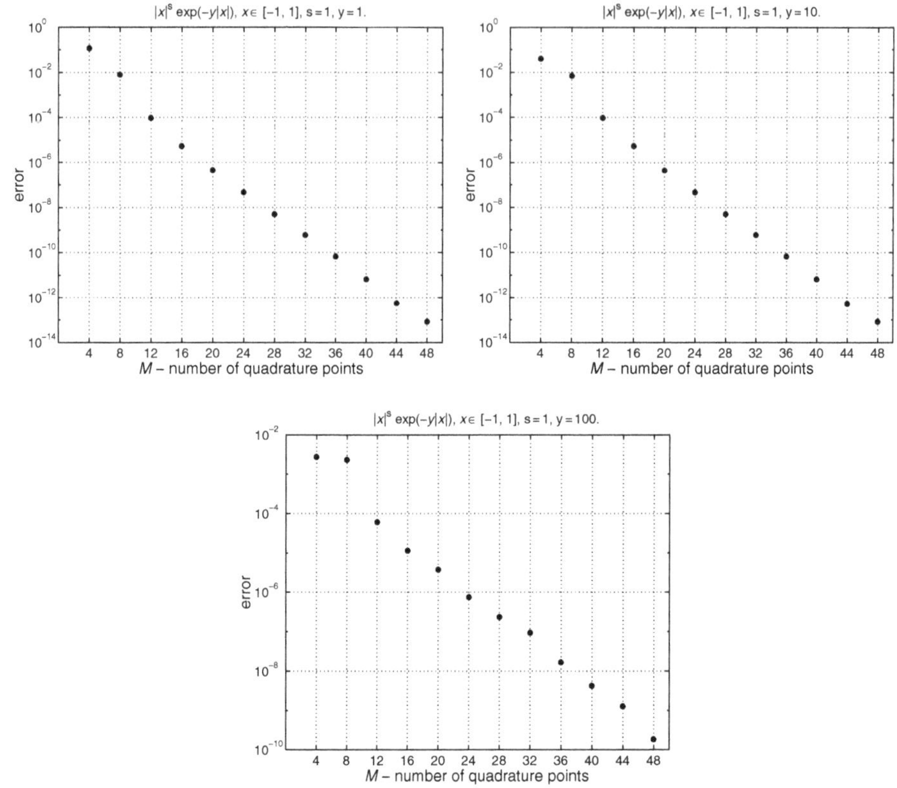

Fig. 2.8: L^∞ error of the sinc interpolation to $\exp(-|x|y)$, $x \in [-1, 1]$, $y = 1, 10, 100$.

2.4 Low rank sinc approximation to the Green kernels

In this section, we consider an approach to the Tucker type and canonical approximation of multivariate functions based on the tensor product sinc interpolation and sinc quadrature methods.

We prove the exponential convergence rate of tensor product sinc interpolant on a class of analytic functions by estimating its Lebesque constant. We focus on separable approximation to integral operators and related kernels. Special attention is paid to the case of convolution integrals in $\mathbb{R}^d$ defined by shift invariant kernels, including elliptic Green functions for the Laplace, Helmholtz, and Yukawa operators. In this way we analyze the sinc approximation error for basic examples of functions with point singularities,

$$\frac{1}{x_1 + \cdots + x_d}, \quad \frac{1}{\|x\|}, \quad \frac{e^{-\|x\|}}{\|x\|}, \qquad x \in \mathbb{R}^d,$$

and conclude the discussion by numerical illustrations.

2.4.1 Sinc interpolation of multivariate functions

First, let us recall the standard constructions of tensor product interpolation (Section 2.2). Given $N \in \mathbb{N}$, the set of interpolating functions $\{\varphi_j(x)\}$, $x \in B := [-a, a]$, and sampling points $\xi_j \in B$, such that the interpolation property is satisfied $\varphi_j(\xi_i) = \delta_{ij}$, $(i, j = 1, \ldots, N)$.

The *Lagrangian interpolant* $\mathcal{I}_N : C(B) \mapsto \operatorname{span}\{\varphi_j(x)\}_{j=1}^N$ applied to a continuous function $f : B \to \mathbb{R}$ has the form (2.20), i.e.,

$$\mathcal{I}_N f := \sum_{j=1}^{N} f(\xi_j)\varphi_j(x) , \quad f \in C(B) , \tag{2.42}$$

providing the interpolation property $(\mathcal{I}_N f)(\xi_j) = f(\xi_j)$, $(j = 1, \ldots, N)$.

The *tensor product interpolant* $\mathbf{I}_N$ in d spatial variables takes the form

$$\mathbf{I}_N f := \mathcal{I}_N^1 \times \cdots \times \mathcal{I}_N^d f = \sum_{j_1=1}^{N} \cdots \sum_{j_d=1}^{N} f(\xi_{j_1}, \ldots, \xi_{j_d})\varphi_{j_1}^{(1)}(x_1) \ldots \varphi_{j_d}^{(d)}(x_d) ,$$

where $f : B^d \to \mathbb{R}$, $B^d = B_1 \times \cdots \times B_d$, and $\mathcal{I}_N^\ell f$ is the univariate interpolation operator in $x_\ell \in B_\ell = [-a_\ell, a_\ell]$, $(1 \le \ell \le d)$ as in (2.42). This interpolant provides the explicit Tucker representation with the directional ranks bounded by N.

We consider the special case of sinc interpolation of multivariate functions on $B^d = \mathbb{R}^d$, $B = \mathbb{R}$. Extension to the case $B = \mathbb{R}_+$ or $B = (a, b)$ is straightforward. The *tensor product sinc interpolant*, $\mathbf{C}_M$, in d variables takes the form

$$\mathbf{C}_M f := C_M^1 \times \cdots \times C_M^d f , \quad f : \mathbb{R}^d \to \mathbb{R} , \tag{2.43}$$

where $C_M^\ell f = C_M^\ell(f, \mathfrak{h})$, $1 \le \ell \le d$, defines the univariate sinc interpolant at $2M + 1$ sampling points with step size $\mathfrak{h}$ in the spacial variable $x_\ell \in \mathbb{R}$,

$$C_M^\ell(f, \mathfrak{h}) = \sum_{k=-M}^{M} f(x_1, \ldots, k\mathfrak{h}, \ldots, x_d) S_{k,\mathfrak{h}}(x_\ell) , \tag{2.44}$$

constructed in Section 2.3.

2.4.2 Error bound for tensor product sinc interpolant

Stability of the Lebesgue constant of the sinc interpolant is essential for estimation of the error $f - \mathbf{C}_M f$. The Lebesgue constant $\Lambda_M \ge 1$ of the univariate interpolant is defined as the minimal number Λ_M from the inequality

$$\|C_M(f, \mathfrak{h})\|_\infty \le \Lambda_M \|f\|_\infty \quad \text{for all } f \in C(\mathbb{R}) . \tag{2.45}$$

The following inequality was proven in [334]:

$$\Lambda_M := \max_{x \in \mathbb{R}} \sum_{k=-M}^{M} |S_{k,\mathfrak{h}}(x)| \le \frac{2}{\pi}(3 + \log(M)) . \tag{2.46}$$

Note that we also have orthogonality

$$\sum_{k=-\infty}^{\infty} |S_{k,\mathfrak{h}}(x)|^2 = 1 \quad (x \in \mathbb{R}) \, ,$$

which indicates $\Lambda_M = 1$ with respect to the L^2 norm.

For each fixed $\ell \in \{1, \dots, d\}$, choose $\zeta_\ell \in B_\ell$ and define the 'single hole' parameter set by

$$Y_\ell := B_1 \times \cdots \times B_{\ell-1} \times B_{\ell+1} \times \cdots \times B_d \in \mathbb{R}^{d-1} \, .$$

To prove the sinc interpolation error let us introduce the univariate (parameter dependent) function

$$F_\ell(\cdot, y) \colon B_\ell \to \mathbb{R} \, , \quad y \in Y_\ell \, , \quad \ell = 1, \dots, d \, ,$$

that is the restriction of f onto B_ℓ. Recall the Hardy space $H^1(D_\delta)$ introduced in Section 2.3.3.

Theorem 2.50 ([141, 142]). *Define Λ_M by (2.46). For each $\ell = 1, \dots, d$, and for any fixed $y \in Y_\ell$, assume that $F_\ell(\cdot, y)$ satisfies*

(a) *$F_\ell(\cdot, y) \in H^1(D_\delta)$ with $N(F_\ell, D_\delta) \le N_0 < \infty$ uniformly in y;*

(b) *$F_\ell(\cdot, y)$ has hyperexponential decay with $a = 1$, $C, b > 0$ for all $y \in Y_\ell$.*

Then, for all $y \in Y_\ell$, the 'quasioptimal' choice $\mathfrak{h} := \frac{\log M}{M}$, yields

$$\|f - \mathbf{C}_M(f, \mathfrak{h})\|_\infty \le \frac{C}{2\pi\delta} \Lambda_M^d N_0 \, \mathrm{e}^{\frac{-\pi\delta M}{\log M}} \, . \tag{2.47}$$

Proof. Similar to the case of polynomial interpolation, the multiple use of (2.45) and the triangle inequality lead to

$$\begin{aligned}
|f - \mathbf{C}_M f| &\le |f - C_M^1 f| + |C_M^1(f - C_M^2 \dots C_M^d f)| \\
&\le |f - C_N^1 f| + |C_M^1(f - C_M^2 f)| + \\
&\quad + |C_M^1 C_M^2(f - C_M^3 f)| + \cdots + |C_M^1 \dots C_M^{d-1}(f - C_M^d f)| \\
&\le (1 + \Lambda_M)[N_1 + \Lambda_M N_2 + \cdots + \Lambda_M^{d-1} N_d] \frac{1}{2\pi\delta} \mathrm{e}^{\frac{-\pi\delta M}{\log M}} \\
&\le (1 + \Lambda_M) \frac{1 + \Lambda_M + \cdots + \Lambda_M^{d-1}}{2\pi\delta} \max_{\ell=1,\dots,d} N(F_\ell, D_\delta) \, \mathrm{e}^{\frac{-\pi\delta M}{\log M}} \, .
\end{aligned}$$

Hence, (2.47) follows since for $\lambda = \Lambda_N > 1$, we have $\frac{(1+\lambda)(\lambda^n - 1)}{\lambda - 1} \le \lambda^n$. □

Note that the width of the analyticity strip, δ, can be chosen close to $\pi/2$. The choice of δ affects only the error estimate, while $\mathbf{C}_M f$ does not depend on δ.

Theorem 2.50 applies to a class of spherically symmetric functions with point singularities. For instance, it can be shown that $\mathbf{C}_M f$ converges exponentially fast in M in the following examples:

$$f(x) = \|x\|^\alpha \, , \quad f(x) = \mathrm{e}^{-\kappa\|x\|^\gamma} \, , \quad f(x) = \frac{\operatorname{erf}(\|x\|)}{\|x\|} \, , \quad x \in \mathbb{R}^d \, , \quad \alpha, \kappa > 0 \, .$$

This provides the explicit construction of the low rank Tucker approximation to above functions (see Sections 2.4.4–2.4.8 for further details).

In the next sections, we consider in more detail the low rank separable approximation to the following multivariate functions:

$$(a)\ \frac{1}{x_1^2+\cdots+x_d^2}\,, \quad (b)\ \frac{1}{\sqrt{x_1^2+\cdots+x_d^2}}\,, \quad (c)\ \frac{e^{-\lambda\|x\|}}{\|x\|}\,, \quad (d)\ e^{-\lambda\|x\|}\,; \quad x\in\mathbb{R}^d\,,$$

arising in various applications in scientific computing.

2.4.3 Application to the function $\frac{1}{x_1^2+\cdots+x_d^2}$

The low rank canonical decomposition of the function $1/(x_1^2+\cdots+x_d^2)$ leads in a straightforward way to the low rank representation of the discrete Laplacian inverse in d dimensions.

In this case the sinc quadrature method applies to the Laplace integral transform

$$\frac{1}{\rho} = \int_{\mathbb{R}_+} e^{-\rho t}dt \quad (\rho = x_1^2+\cdots+x_d^2 \in [1,R],\ R>1) \tag{2.48}$$

with the integrand $f(t) = e^{-\rho t}$. The canonical rank estimate for $1/\rho$ can be derived as a special case of Theorem 2.46, as considered in [141, 142].

The substitution $t = \phi(u) := \log(1+e^u)$ maps $\phi\colon \mathbb{R}\mapsto\mathbb{R}_+$. This results in

$$\frac{1}{\rho} = \int_{\mathbb{R}} f_1(u)du\,, \quad f_1(u,\rho) = \frac{e^{-\rho\log(1+e^{-u})}}{1+e^{-u}}\,, \quad (\rho\ge 1)\,. \tag{2.49}$$

Lemma 2.51. *Let $\rho\in[1,R]$, and choose $\delta = \alpha^2\pi/2$ with arbitrary $0<\alpha<1$ and define $\mathfrak{h} = \frac{\alpha\pi}{\sqrt{M}}$, then the uniform in R error bound holds*

$$\left|\frac{1}{\rho} - T_M(f_1,\mathfrak{h})\right| \lesssim Ce^{-\alpha\pi\sqrt{M}}\,. \tag{2.50}$$

Proof. The function $f_1(z)$ belongs to $H^1(D_\delta)$, with $\delta<\pi/2$, and we have $N(f_1,D_\delta)<\infty$ independent of R. $f_1(x)$ exhibits exponential decay for $|x|\to\infty$ with $b=1$. Hence, application of Theorem 2.44 and Corollary 2.45 proves the result. □

A more detailed analysis of the analytic properties of function $f_1(z)$ can be found in [141].

The improved quadrature applies by using substitutions

$$t = \log(1+e^u) \quad\text{and}\quad u = \sinh(w)\,, \tag{2.51}$$

resulting in

$$\frac{1}{\rho} = \int_{\mathbb{R}} f_2(w)dw\,, \quad \text{with } f_2(w) = \frac{\cosh(w)}{1+e^{-\sinh(w)}}e^{-\rho\log(1+e^{\sinh(w)})}\,. \tag{2.52}$$

Lemma 2.52. *Let $\rho \in [1, R]$, then the choice $\delta = \delta(R) = O(1/\log(R))$, $a = 1$, $b = 1/2$, in Theorem 2.46 implies the uniform quadrature error bound by setting $\mathfrak{h} = \log(4\pi M)/M$,*

$$\left|\frac{1}{\rho} - T_M(f_2, \mathfrak{h})\right| \lesssim C e^{-\frac{\pi^2 M}{(C_1 \log(R))\log(\pi^2 M)}} . \tag{2.53}$$

Proof. The function $f_2(w)$ belongs to $H^1(D_\delta)$, with $\delta = O(1/\log(R))$, and we have $N(f_2, D_\delta) < \infty$ independent of $\rho \le R$. Double exponential decay of $f_2(w)$ on $w \in (-\infty, \infty)$ is due to asymptotic behavior

$$f_2(w) \approx \frac{1}{2} e^{w - \frac{\rho}{2} e^w} \quad \text{as } w \to \infty ; \quad f_2(w) \approx \frac{1}{2} e^{|w| - \frac{1}{2} e^{|w|}} \quad \text{as } w \to -\infty ,$$

corresponding to $C = \frac{1}{2}$, $b = \min\{1, \rho\}/2$, $a = 1$, in notations of Theorem 2.46. □

Let us present some numerical illustrations for the function $1/\rho = \frac{1}{x_1^2 + \cdots + x_d^2}$. Estimate (2.53) implies that the accuracy $\varepsilon > 0$ can be achieved with

$$M \le \mathcal{O}\left(\log(\tfrac{1}{\varepsilon}) \cdot \log R\right), \tag{2.54}$$

provided that $1 \le \rho \le R$ (can be controlled by a proper scaling). The numerical results support the even better bound $M \le \mathcal{O}(\log(\frac{1}{\varepsilon}) + \log R)$; see Figures 2.9, 2.10 representing the quadrature error in the case of long intervals $[1, R]$.
Lemma 2.52 indicates that the separation rank $r = 2M + 1$ depends only linear logarithmically on both the tolerance $\varepsilon > 0$ and the upper bound on R. We observe the stable error behavior for large variations in the interval, $1 \le \rho \le R$.

It is worth noting that the separation rank r *does not depend on the dimension d.*

Remark 2.53. The choice of δ only affects the theoretical error bound but not the quadrature itself. Hence, for practical needs one can simplify the estimate by setting $\delta = \pi/2$.

In boundary element methods, one is interested in a low separation rank representation to the kernel function $s(x, y) = \log(x + y)$, $x \in [0, 1]$, $y \in [h, 1]$ with some small mesh size parameter $h > 0$. The straightforward application of the above rank decompositions allows us to treat the 2D Laplacian Green function $\log(x+y)$. A representation like

$$\frac{1}{x + y} = \sum_{m=1}^{k} \Phi_m(x)\Psi_m(y) + \delta_k \quad \text{with } |\delta_k| \le \varepsilon \tag{2.55}$$

can be constructed by means of the quadrature applied to the integral (2.52) with $\rho = x+y$ and $k = 2M+1$. Indeed, let ψ_m be the antiderivatives of a function Ψ_m. Integration

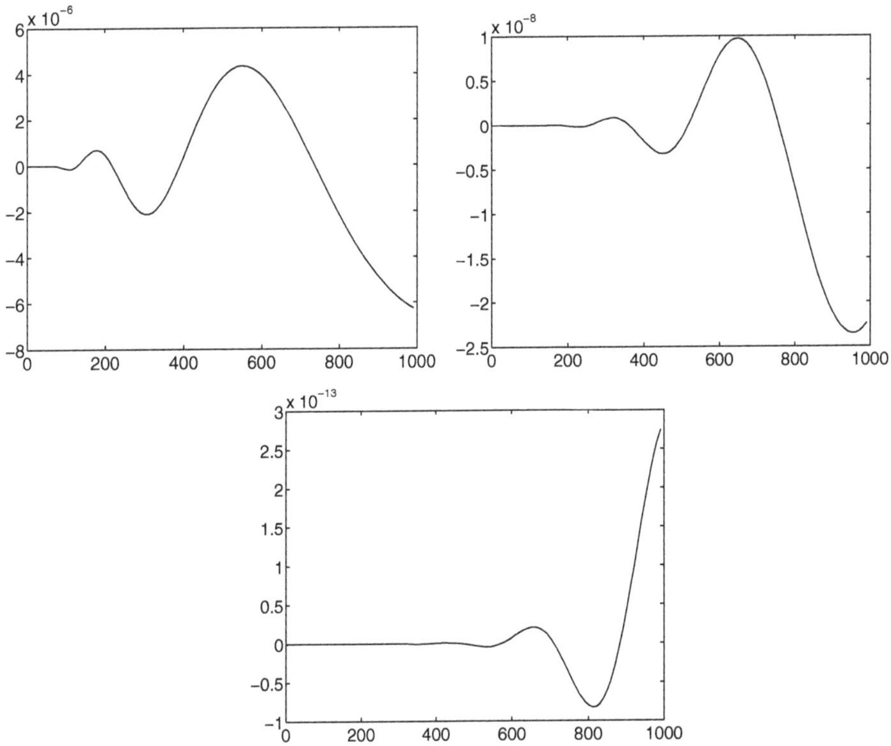

Fig. 2.9: The quadrature error versus ρ related to (2.53) with $1 \le \rho \le 10^3$, and $M = 16$ (upper left), $M = 32$ (upper right), $M = 64$ (bottom).

of (2.55) yields

$$
\begin{aligned}
\log(x+y) &= \int_{1-x}^{y} \frac{dt}{x+t} = \int_{1-x}^{y} \left(\sum_{m=1}^{k} \Phi_m(x)\Psi_m(t) + \Delta_k \right) dt \\
&= \sum_{m=1}^{k} \Phi_m(x)[\psi_m(y) - \psi_m(1-x)] + S_k \\
&= \Phi_0(x) + \sum_{m=1}^{k} \Phi_m(x)\psi_m(y) + S_k
\end{aligned}
$$

with $\Phi_0(x) = -\sum_{m=1}^{k} \Phi_m(x)\psi_m(1-x)$ and $|S_k| = |\int_{1-x}^{y} \delta_k dt| \le \varepsilon$. The resulting representation of $\log(x+y)$ has the separation rank $k+1$ and the same absolute accuracy ε as (2.55).

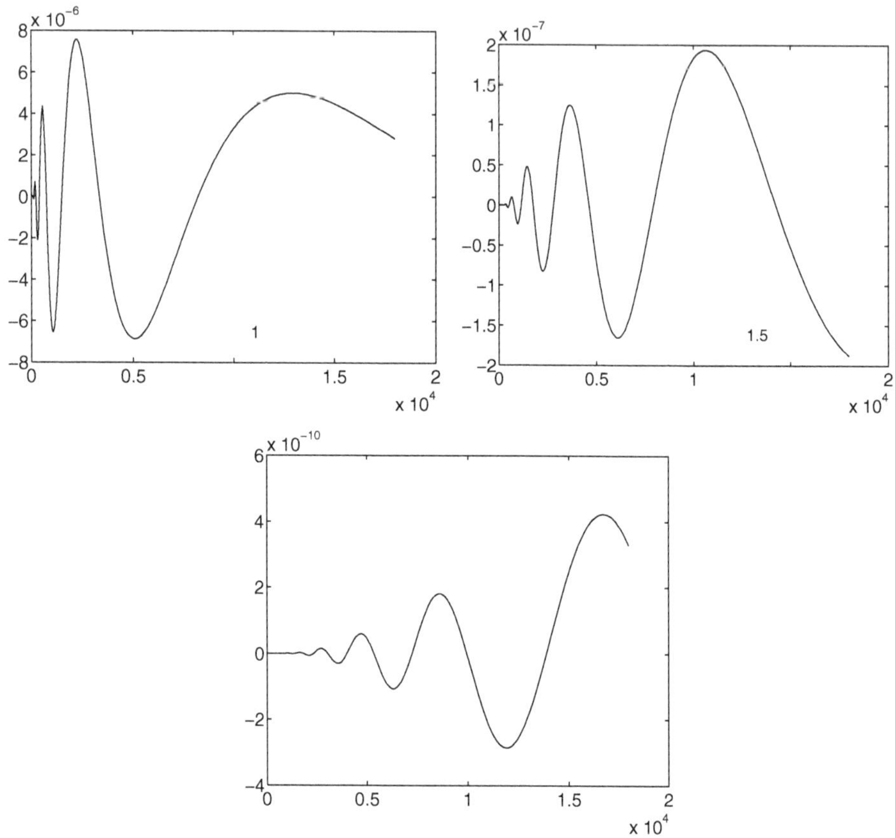

Fig. 2.10: The quadrature error related to (2.53) with $1 \le \rho \le 18\,000$, and $M = 16$ (upper left), $M = 32$ (upper right), $M = 64$ (bottom).

2.4.4 Application to the generalized Newton kernel $\frac{1}{\sqrt{x_1^2+\cdots+x_d^2}}$

In what follows, we derive the canonical decomposition via sinc quadrature for the function

$$\frac{1}{\rho} = 1\Big/\sqrt{x_1^2 + \cdots + x_d^d}\,, \quad \rho = \sqrt{x_1^2 + \cdots + x_d^d}\,,$$

which for $d = 3$ defines the classical Newton potential, associated with Green's kernel for the Laplacian in $\mathbb{R}^3$.

The sinc quadrature applies to the Gauss integral representation

$$\frac{1}{\rho} = \frac{1}{\sqrt{\pi}} \int_{\mathbb{R}} e^{-\rho^2 t^2} dt \quad (\rho \in [1, R]) \,. \tag{2.56}$$

To maintain robustness in ρ, we rewrite the Gauss integral (2.56) using substitutions (2.51) to obtain

$$\frac{1}{\rho} = \int_{\mathbb{R}} f(w)\,dw \quad \text{with} \quad f(w) := \cosh(w)F(\sinh(w))\,, \tag{2.57}$$

where

$$F(u) := \frac{2}{\sqrt{\pi}} \frac{e^{-\rho^2 \log^2(1+e^u)}}{1+e^{-u}}\,, \qquad u \in (-\infty, \infty)\,.$$

Lemma 2.54 ([141]). *Let $\delta < \pi/2$ and $\rho \geq 1$. Then for the function f in (2.57) we have $f \in H^1(D_\delta)$.*

The standard $(2M+1)$-point sinc quadrature with the choice $\mathfrak{h} = \sqrt{2\pi\delta/bM}$ exhibits the error bound

$$\left|\frac{1}{\rho} - I_M(f, \mathfrak{h})\right| \leq C e^{-\pi\sqrt{bM}}\,. \tag{2.58}$$

The improved $(2M+1)$-point quadrature with the choice $\delta(\rho) = \frac{\pi}{C+\log(\rho)}$, provides the error bound

$$\left|\frac{1}{\rho} - I_M(f, \mathfrak{h})\right| \leq C_1 \exp\left(-\frac{\pi^2 M}{(C+\log(\rho))\log M}\right). \tag{2.59}$$

Proof. It is easy to check that f is holomorphic in D_δ and $N(f, D_\delta) < \infty$ uniformly in ρ (with the choice $\delta = \delta(\rho)$). Application of Corollary 2.45 proves (2.58).

Furthermore, we check the double exponential decay of the integrand in (2.38) as $|w| \to \infty$ that satisfies with $a = 1$, and then apply Theorem 2.46 with the ρ-dependent δ,

$$\delta = \delta(\rho) = \frac{\pi}{C + \log(\rho)}\,, \quad \text{and} \quad \mathfrak{h} = \frac{\log(\pi M/b)}{M}$$

with some $C > 0$. This completes the proof. □

For given threshold $\varepsilon > 0$, application of (2.59) leads to the bound on M that is logarithmic in both ε and R (but not dependent on d),

$$M \leq \mathcal{O}\left(\log(\tfrac{1}{\varepsilon}) \cdot \log R\right). \tag{2.60}$$

Figure 2.11 presents numerical illustrations for sinc quadrature with values $\rho \in [1, R]$, $R \leq 5000$. We observe the very weak error increase in ρ. Similar results were obtained in the case $R > 5000$, manifesting a rather stable behavior of the quadrature error in R.

In Chapter 5, we shall modify the results of Lemma 2.54 for the construction of efficient tensor representation to the 3D convolution transform with the Newton kernel $1/\|x\|$, $x \in \mathbb{R}^3$ (see also Section 2.4.6).

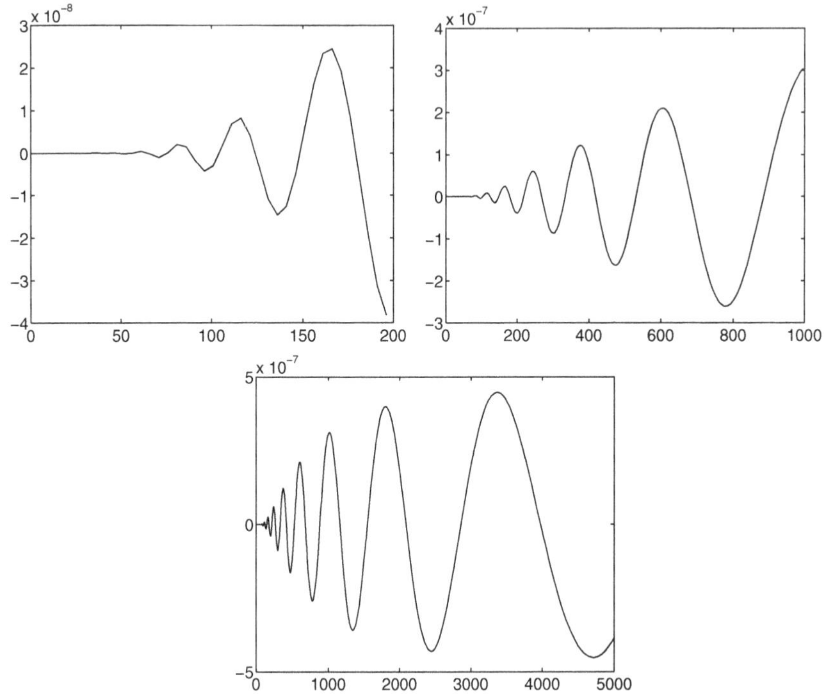

Fig. 2.11: The quadrature error for $M = 64$: $R = 200$ (upper left), $R = 1000$ (upper right), $R = 5000$ (bottom).

2.4.5 Sinc approximation of the Slater function

We consider a class of spherically symmetric convolving kernels $g\colon \mathbb{R}^d \to \mathbb{R}$,

$$g(y) = G(\rho(y)) \equiv G(\rho) \quad \text{with} \quad \rho \equiv \rho(y) = y_1^2 + \cdots + y_d^2\,,$$

where $G\colon \mathbb{R}_+ \to \mathbb{R}$ is represented via the generalized Laplace transform

$$G(\rho) = \int_{\mathbb{R}_+} \widehat{G}(\tau) e^{-\rho\tau^2}\, d\tau\,. \tag{2.61}$$

The Slater function $G(\rho) = e^{-2\sqrt{\alpha\rho}}$ has the principal significance in electronic structure calculations (say, based on the Hartree–Fock equation) since it represents the cusp behavior of electron density in the local vicinity of nuclei. This function (or its approximation) is considered as the best candidate for the localized basis function for atomic orbitals basis sets. The main limitation, however, is due to the presence of a singularity in the nonseparable form. Hence, our goal is the construction of low rank approximation to the Slater function in the form of sum of Gaussians.

The Laplace transform representation for the Slater function $G(\rho) = e^{-2\sqrt{\alpha\rho}}$ reads

$$G(\rho) = e^{-2\sqrt{\alpha\rho}} = \frac{\sqrt{\alpha}}{\sqrt{\pi}} \int_{\mathbb{R}_+} \tau^{-3/2} \exp(-\alpha/\tau - \rho\tau) d\tau \,,$$

which corresponds to the choice $\widehat{G}(\tau) = \frac{\sqrt{\alpha}}{\sqrt{\pi}} \tau^{-3/2} e^{\alpha/\tau}$ in (5.62).

Lemma 2.55 ([206]). *For given threshold $\varepsilon > 0$ let $\rho \in [1, R]$. Then the $2M + 1$-term sinc quadrature approximation provides the error of order $O(\varepsilon)$ with*

$$M = O(|\log \varepsilon|(|\log \varepsilon| + \log R)) \,.$$

Proof. Via substitution $\tau = e^t$ (i.e., $\varphi(z) = e^z$), we obtain

$$e^{-2\sqrt{\alpha\rho}} = \frac{\sqrt{\alpha}}{\sqrt{\pi}} \int_{\mathbb{R}} f(t; \alpha, \rho) dt \quad \text{with } f(t; \alpha, \rho) = \exp(-t/2 - \alpha e^{-t} - \rho e^t) \,.$$

The decay of the integrand $f(t; \alpha, \rho)$ on the real axis is

$$f(t; \alpha, \rho) \approx e^{-t/2 - \rho e^t} \quad \text{as } t \to \infty \,, \qquad f(t; \alpha, \rho) \approx e^{|t|/2 - \alpha e^{|t|}} \quad \text{as } t \to -\infty \,,$$

corresponding to $a = 1$, $b = \min\{\alpha, \rho\}$, $C = 1$ in (2.38). Moreover, it is easy to check that $f(z) \in H^1(D_\delta)$, $\delta < \pi/2$ with uniformly bounded constant $N(f, D_\delta)$ in both α and ρ. It is known that the choice $\mathfrak{h} = \log(\frac{2\pi a M}{b})/(aM)$ leads to ([109])

$$\left| \int_{\mathbb{R}} f(\xi) d\xi - T_M(f, \mathfrak{h}) \right| \le C\, N(f, D_\delta)\, e^{-2\pi\delta a M/\log(2\pi a M/b)} \,. \tag{2.62}$$

Hence (2.62) holds with $C = 1$, $a = 1$, $N(f, D_\delta) = O(1)$, $b = \alpha$ (usually, $\alpha \ll R$) implying the bound on the number of term $M = O(|\log \varepsilon|(|\log \varepsilon| + \log b))$ (Theorem 2.46). Here we set $\delta = \pi/2$. This proves the existence of a rank-$(2M+1)$ canonical ε-approximation to the Slater function. □

Figures 2.12 and 2.13 illustrate the rank-$\mathbf{r}$ Tucker approximation to the generalized Slater function $\exp(-\|x\|^\gamma)$, $d = 3$, $\|x\| \le 10$ and the corresponding orthogonal vectors. The Tucker decomposition is computed for the tensor sampled on the uniform $n \times n \times n$ grid with $n = 64$. We observe the fast exponential decay of the approximation error in the Tucker rank.

Figure 2.14 demonstrates the convergence of rank-$\mathbf{r}$ Tucker approximation to the multi-centered Slater potential $\sum_{k=1}^{64} c_k \exp(-\|x - x_k\|)$, $x, x_k \in \mathbb{R}^3$, where the 'nuclei' centers are displaced on the uniform $4 \times 4 \times 4$ spacial lattice. It is seen that the error decay is merely the same as for the single Slater function. This effect will be studied in Chapter 5.

The effect of small random perturbations can be observed in Figure 2.15, demonstrating the stagnating convergence of the rank decomposition on the level of random perturbation.

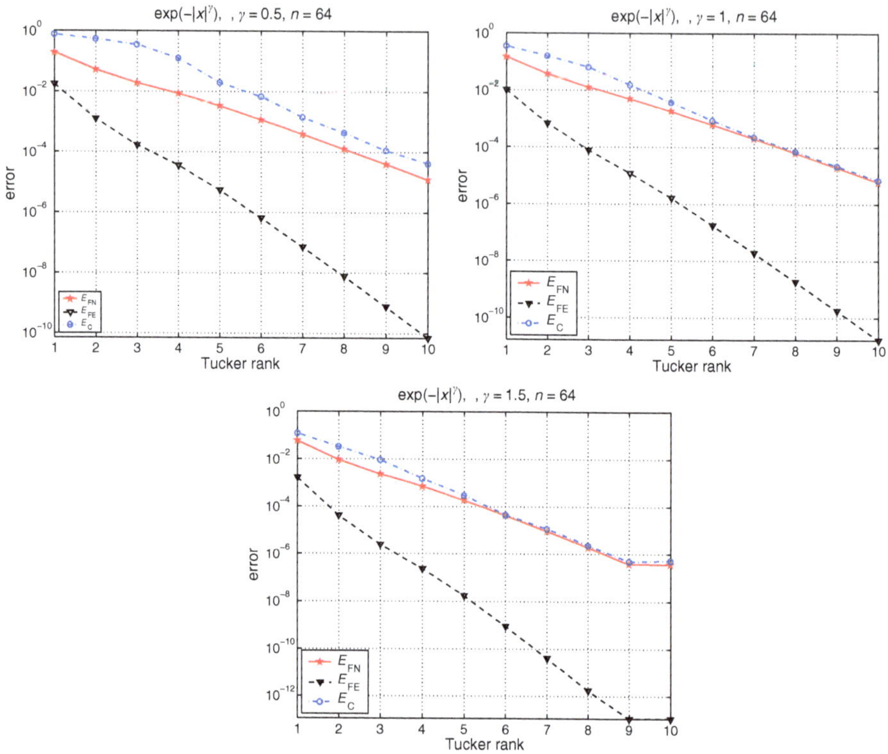

Fig. 2.12: Approximation error versus Tucker rank for the Slater potential.

2.4.6 Tucker and canonical approximation of integral operators

The tensor product sinc interpolant $\mathbf{C}_M f$ in (2.43), (2.44) can be applied to the rank approximation of multidimensional integral operators

$$(\mathcal{G}u)(x) := \int_{\Omega} g(x, y)u(y)dy\,, \quad x, y \in \mathbb{R}^d$$

in the framework of the Nyström, collocation, or Galerkin discretization schemes.

The most of practically interesting kernel functions $g(x, y)$ have a singularity at the diagonal $x = y$. Hence the Tucker type separable approximation of such singular kernels can be constructed only for coupled variables

$$g(x, y) \approx g_{\mathbf{r}}(x, y) := \sum_{\mathbf{k}=1}^{\mathbf{r}} b_{\mathbf{k}} \Phi^{(1)}_{k_1}(x_1, y_1) \ldots \Phi^{(d)}_{k_d}(x_d, y_d)\,,$$

where $\{\Phi^{\ell}_k(\cdot, \cdot)\}$ is a fixed set of bivariate functions arising, e.g., from the sinc interpolation or sinc quadrature representations.

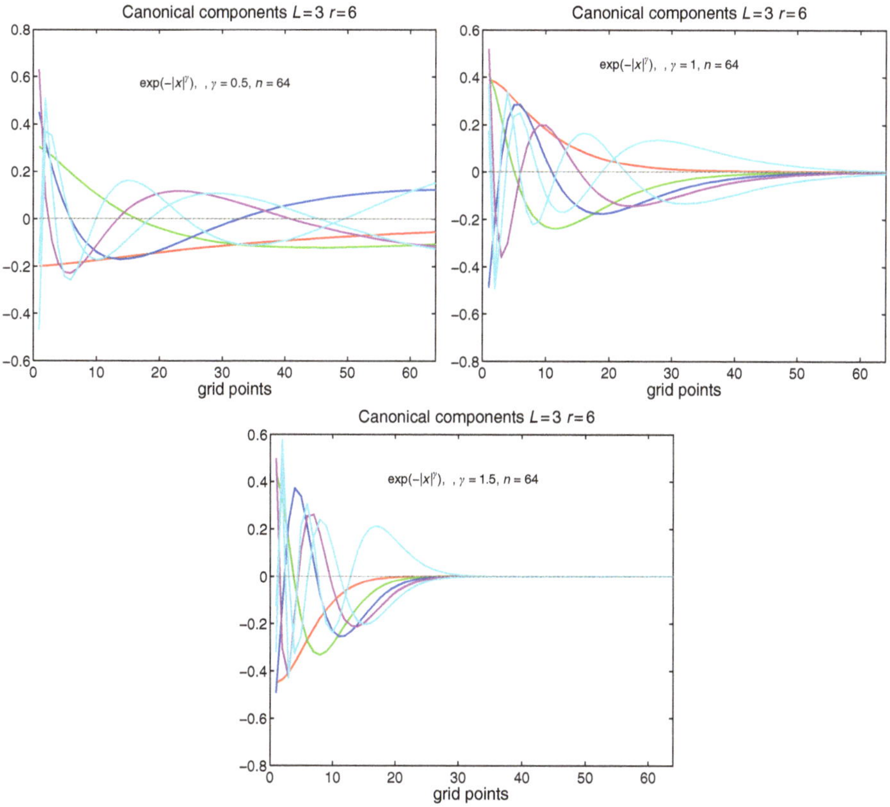

Fig. 2.13: Tucker orthogonal vectors for the Slater type potentials.

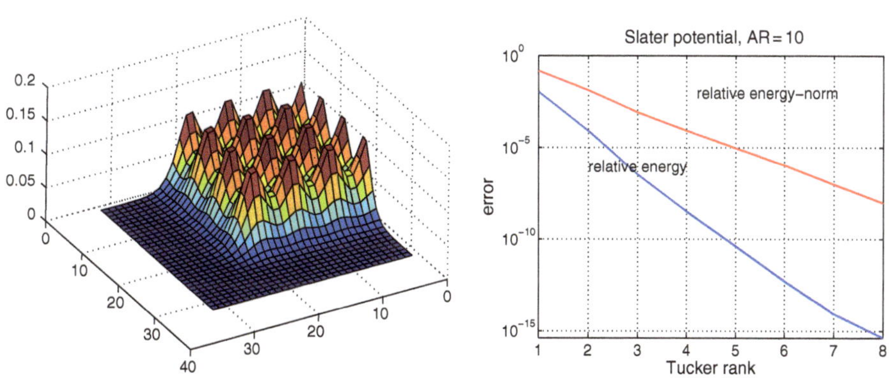

Fig. 2.14: Multicentered Slater potential.

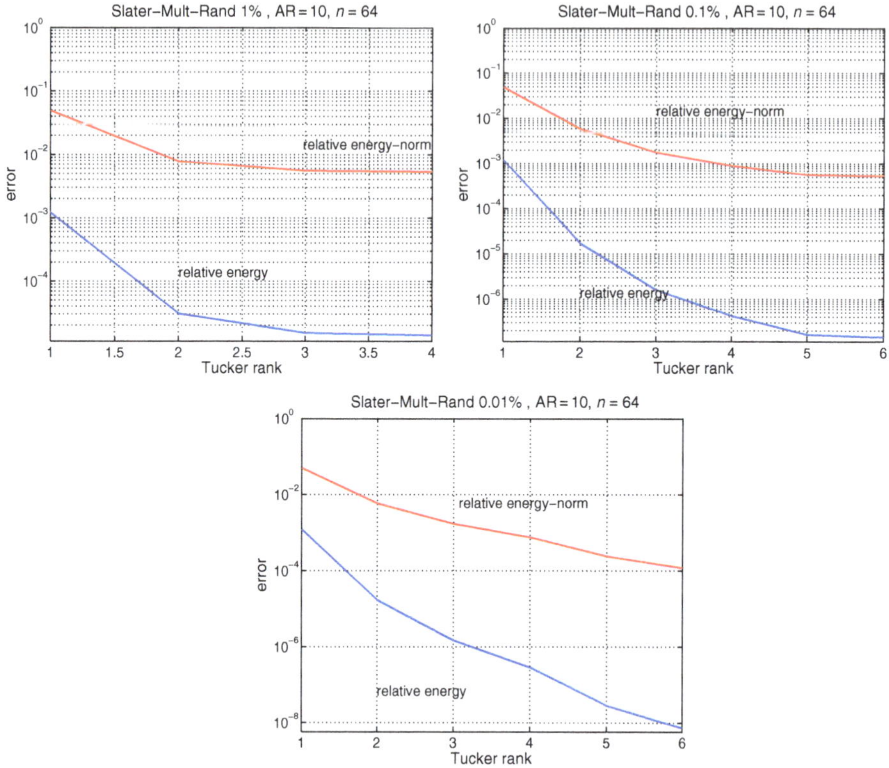

Fig. 2.15: Tucker ranks for multicentered randomly perturbed Slater potential.

In the important case of shift invariant singular kernels, $g(x, y) = g(\|x-y\|)$, we obtain

$$g(x, y) = g(\|x - y\|) \Rightarrow G(\zeta_1, \dots, \zeta_d) \equiv G\left(\sqrt{\zeta_1^2 + \cdots + \zeta_d^2}\right),$$

where $\zeta_\ell = |x_\ell - y_\ell| \geq 0$, $\ell = 1, \dots, d$. Now the sinc interpolant applies with respect to the d coupled variables $\zeta_1, \dots, \zeta_d$. In that case the function G has the only one point singularity at the origin.

Analytic methods of canonical approximation via *separation by integration* are well suited for analytic, shift invariant kernels $g(\|x - y\|) = G(\rho)$, $\rho > 0$. They apply to both collocation and Galerkin discretizations. The R-term sinc quadrature for the Laplace *integral representation* of $G(\rho)$, $\rho = \|x - y\|^2 \in [a, b]$, $a > 0$, reads as

$$G(\rho) = \int_{\mathbb{R}} f(t) e^{-t\rho} dt \approx \sum_{k=1}^{R} c_k f(t_k) \prod_{\ell=1}^{d} e^{-t_k \|x_\ell - y_\ell\|^2}, \quad \rho \in [a, b].$$

For example, in the case of a collocation projection scheme for approximation of convolution integrals one computes a rank-R tensor at $x = 0$,

$$g_{\mathbf{i}} = \langle G(\rho), \phi_{\mathbf{i}} \rangle \approx \sum_{k=1}^{r} c_k f(t_k) \prod_{\ell=1}^{d} \left\langle \mathrm{e}^{-t_k \|y_\ell\|^2}, \phi_{i_\ell}(y_\ell) \right\rangle , \quad \mathbf{i} \in \mathcal{I} ,$$

where $\phi_{\mathbf{i}} = \prod \phi_{i_\ell}$, $\mathbf{i} \in \mathcal{I}$ is the set of tensor product basis functions.

In many applications the sinc method provides asymptotically optimal bounds on both the canonical and Tucker ranks (see the discussion in Sections 2.4.7 and 2.4.8),

$$R = O(\log n |\log \varepsilon|) , \quad \mathbf{r} = (R, \ldots, R) .$$

It is worth noting that there are many successful applications of the sinc method including the efficient implementations in computational quantum chemistry and in many particle calculations. In particular, we shall consider the following problem classes:

- Trilinear approximation to third/sixth order tensors generated by the classical *Green kernels*, $x, y \in \mathbb{R}^3$,

$$\frac{1}{\|x-y\|} , \quad \frac{\mathrm{e}^{-\mu\|x-y\|}}{\|x-y\|} , \quad (\mu \in \mathbb{R}_+) , \quad \frac{\mathrm{e}^{-i\kappa^2\|x-y\|}}{\|x-y\|} , \quad (\kappa^2 \in \mathbb{R}) .$$

- Multicentered potential as a prototype of electron density for large molecules, $\sum_k c_k \mathrm{e}^{-\alpha_k\|x-x_k\|}$, $x \in \mathbb{R}^3$ (see numerics below).
- Electron density, the Hartree and exchange potentials in $\mathbb{R}^3$ for solving the Hartree–Fock equation in tensor format (Chapters 3 and 5).
- Tensor summation methods for electrostatic and other long range interaction potentials for many particle systems (Chapter 5).
- Traditional FEM/BEM applications including:
  - elliptic inverse in $\mathbb{R}^d$,
  - the elliptic boundary value, spectral and transient problems in tensor format.
- Preconditioning for solving stochastic/parametric PDEs in tensor format (Chapter 5).
- Isogeometric analyses (IGA) for 3D problems (Chapter 5).

2.4.7 Sinc method for the Yukawa potential by projection collocation

In this section we describe an example of low rank canonical approximation applied to the discrete tensor representation of the spherically symmetric convolving kernel given by the Yukawa potential.

In the particular case of Yukawa potential $\frac{\mathrm{e}^{-\kappa\sqrt{\rho}}}{\sqrt{\rho}}$, $\rho(y) = y_1^2 + \cdots + y_d^2$, for $\kappa \in [0, \infty)$, we apply the Gauss transform

$$G(\rho) = \frac{\mathrm{e}^{-\kappa\sqrt{\rho}}}{\sqrt{\rho}} = \frac{2}{\sqrt{\pi}} \int_{\mathbb{R}_+} \exp(-\rho\tau^2 - \kappa^2/\tau^2) d\tau , \tag{2.63}$$

corresponding to the choice $\widehat{G}(\tau^2) = \frac{2}{\sqrt{\pi}} e^{-\kappa^2/\tau^2}$. Under the change of variables $\tau = e^t$, the integrand in (2.63) transforms to $f(t) = e^t \exp(-\rho e^{2t} - \kappa^2 e^{-2t})$.

The projection collocation discretization method applies in the computational box $[-A, A]^3$. Introduce tensor grid and a product basis set.

Tensor grid: Let $\omega_{\mathbf{d}} := \omega_1 \times \cdots \times \omega_d$ be the equidistant tensor grid of collocation points $\{x_{\mathbf{m}}\}$ in Ω, $\mathbf{m} \in \mathcal{M} := \{1, \ldots, n+1\}^d$, $\omega_\ell := \{-A + (m-1)h\colon m = 1, \ldots, n+1\}$ $(\ell = 1, \ldots, d)$, $h = 2A/n$.

Product basis: For given piecewise constant basis functions $\{\phi_{\mathbf{i}}\}$,

$$\phi_{\mathbf{i}}(x) = \prod_{\ell=1}^{d} \phi_{i_\ell}(x_\ell), \quad \phi_{i_\ell}(\cdot) = \phi(\cdot + (i_\ell - 1)h), \quad \mathbf{i} \in \mathcal{I} := \{1, \ldots, n\}^d,$$

related to $\omega_{\mathbf{d}}$, let $f_{\mathbf{i}}$ be the representation coefficients of f in the basis set $\{\phi_{\mathbf{i}}\}$,

$$f(y) \approx \sum_{\mathbf{i} \in \mathcal{I}} f_{\mathbf{i}} \phi_{\mathbf{i}}(y), \quad f_{\mathbf{i}} = f(P_{\mathbf{i}}),$$

where $P_{\mathbf{i}}$ are the centers of $\operatorname{supp}(\phi_{\mathbf{i}})$. Now sinc quadrature method applies to the collocation coefficients tensor $\mathcal{G} = [g_{\mathbf{i}}]_{\mathbf{i} \in \mathcal{I}}$, with entries given by

$$g_{\mathbf{i}} = \int_{\mathbb{R}^d} \phi_{\mathbf{i}}(y) G(y) dy, \quad \mathbf{i} \in \mathcal{I}.$$

This results in the rank-$(2M+1)$ canonical decomposition

$$g_{\mathbf{i}} \approx \sum_{k=-M}^{M} w_k \widehat{G}(\tau_k^2) \prod_{\ell=1}^{d} \int_{\mathbb{R}} e^{-y_\ell^2 \tau_k^2} \phi_{i_\ell}(y_\ell) dy_\ell, \quad \mathbf{i} \in \mathcal{I},$$

with the suitably chosen weights $w_k \in \mathbb{R}$ and quadrature points $\tau_k \in \mathbb{R}_+$.

In what follows, we consider the error bound for sinc quadrature approximation; see [204].

Theorem 2.56. *For given $G(\rho)$ in (2.63) with fixed $\kappa > 0$, we set*

$$w_k = \mathfrak{h}_M \widehat{G}(\tau_k^2) \quad \text{and} \quad \tau_k = e^{t_k}, \quad t_k = k\mathfrak{h}_M,$$

where $\mathfrak{h}_M = C_0 \log(M)/M$ for some $C_0 > 0$. Then for the rank-$(2M+1)$ collocation coefficients tensor $\mathcal{G} = [g_{\mathbf{i}}]_{\mathbf{i} \in \mathcal{I}}$, we have

$$\left\| g_{\mathbf{i}} - \sum_{k=-M}^{M} w_k \widehat{G}(\tau_k^2) \prod_{\ell=1}^{d} \int_{\mathbb{R}} e^{-y_\ell^2 \tau_k^2} \phi_{i_\ell}(y_\ell) dy_\ell \right\| \le C e^{-\pi^2 M/(C + \log(M))}. \tag{2.64}$$

Proof. Following [334], we choose the analyticity domain for the integrand in (2.63) as a sector $S_\delta := \{w \in \mathbb{C}\colon |\arg(w)| < \delta\}$ with apex angle $0 < 2\delta < \pi/2$, and then use the conformal map

$$\varphi^{-1}\colon S_\delta \to D_\delta \quad \text{with} \quad w = \varphi(z) = e^z, \quad \varphi^{-1}(w) = \log(w).$$

Applying the change of variables $\tau = e^t$ leads to

$$G(\rho) = \int_{\mathbb{R}} f(t;\rho) dt \quad \text{with} \quad f(t;\rho) = \mathcal{Q}(t) e^{-\rho e^{2t}}, \quad \mathcal{Q}(t) = \frac{2}{\sqrt{\pi}} e^{t - \kappa^2 e^{-2t}}.$$

Here f can be analytically extended into the strip D_δ. By definition,

$$g_{\mathbf{i}} = \langle G(\rho), \phi_{\mathbf{i}}\rangle = \int_{\mathbb{R}} \langle f(t;\rho), \phi_{\mathbf{i}}\rangle dt \equiv \int_{\mathbb{R}} p_{\mathbf{i}}(t)dt\,, \quad p_{\mathbf{i}}(t) = \Omega(t)\prod_{\ell=1}^{d}\int_{\mathbb{R}} e^{-y_\ell^2 e^{2t}}\phi_{i_\ell}dy_\ell\,.$$

Now we apply the sinc quadrature directly to the integral above. The rank-1 (separable in $\mathbf{i}$) function $p_{\mathbf{i}}\colon \mathbb{R} \to \mathbb{R}$ can be analytically extended into the strip D_δ with $0 < \delta < \pi/4$, and this extension belongs to the Hardy space $H^1(D_\delta)$. In fact, using the error function $\operatorname{erf}\colon \mathbb{R} \to \mathbb{R}$,

$$\operatorname{erf}(t) := \frac{2}{\sqrt{\pi}}\int_0^t e^{-\tau^2}d\tau\,,$$

we calculate the explicit representation

$$\int_{\mathbb{R}} e^{-y^2e^{2t}}\phi_i(y)dy = \frac{1}{2t}\{\operatorname{erf}(tih) - \operatorname{erf}(t(i-1)h)\}\,, \tag{2.65}$$

with $h = 2A/n$ (uniform grid spacing) for $i = 1, \ldots, n$. Since $\operatorname{erf}(z)/z$ is an entire function one proves the required analyticity of $p_{\mathbf{i}}$.

To estimate $N(p_{\mathbf{i}}, D_\delta)$, we let $H_{\mathbf{i}} = h(i_1 - 1, \ldots, i_d - 1)^{\mathrm{T}} \in \mathbb{R}^d$ with $H_\ell = |i_\ell|h \le R$, denote $\psi_{\mathbf{i}}(y) = \phi(y + H_{\mathbf{i}})$, and let

$$f * g(u) = \int_{\mathbb{R}} f(x)g(u-x)dx$$

be the convolution product in $\mathbb{R}^d$ provided that $q(x) = |f| * |g|$ is locally integrable. Now, using the shift property of convolution, $f(\cdot + C) * g(\cdot) = f * g(\cdot + C)$ and applying the Fubini theorem in the form

$$(f * g, \mu)_{L^2} = \int_{\mathbb{R}^d\times\mathbb{R}^d} f(x)g(y)\mu(x+y)dxdy\,, \quad \mu \in \mathcal{D}(\mathbb{R}^d)\,,$$

we obtain (note that ψ is an even function)

$$\begin{aligned}\int_{\mathbb{R}^d} e^{-w^2|y|^2}\phi(y + H_{\mathbf{i}})dy &= \int_{\mathbb{R}^d\times\mathbb{R}^d} e^{-w^2|u|^2}\phi(x)\psi(u - x + H_{\mathbf{i}})dxdu\\ &= \int_{\mathbb{R}^d} e^{-w^2|u|^2}\phi(\cdot) * \psi(\cdot + H_{\mathbf{i}})(u)du\\ &= \int_{\mathbb{R}^d} e^{-w^2|u|^2}(\phi * \psi)(u + H_{\mathbf{i}})du\\ &= \int_{\mathbb{R}^d} e^{-w^2|v-H_{\mathbf{i}}|^2}\phi(v)dv\,,\end{aligned}$$

taking into account that ϕ has compact support $[-h, h]^d$.

Note that $|\Omega(\zeta\exp(i\delta))| \le C_0 < \infty$ for $\zeta \in [0, \infty)$, leading to the bound

$$N(p_{\mathbf{i}}, D_\delta) = \int_{\partial S_\delta} |p_{\mathbf{i}}(w)|\,|dw|$$

$$= \int_{\partial S_\delta} \int_{\mathbb{R}^d} \left|\Omega(w)\right| \left|e^{-w^2|y|^2}\phi(y + H_{\mathbf{i}})dy\right| \left|dw\right|$$

$$\leq 2 \int_{\mathbb{R}_+} \int_{\mathbb{R}^d} \left|\Omega(\zeta e^{i\delta})\right| \left|e^{-\zeta^2 \exp(2i\delta)|u-H_{\mathbf{i}}|^2}\phi(u)du\right| d\zeta$$

$$\leq 2C_0 \int_{\mathbb{R}^d} \int_{\mathbb{R}_+} \left|e^{-\zeta^2 \exp(2i\delta)|u+H_{\mathbf{i}}|^2}\right| d\zeta \left|\phi(u)\right| du$$

$$= 2C_0 \int_{\mathbb{R}^d} \int_{\mathbb{R}_+} e^{-\zeta^2 \cos(2\delta)|u+H_{\mathbf{i}}|^2} d\zeta \left|\phi(u)\right| du$$

$$= \frac{2C_0}{\sqrt{\cos(2\delta)}} \int_{\mathbb{R}^d} \frac{|\phi(u)|}{|u - H_{\mathbf{i}}|^2} du \,.$$

The latter integral is uniformly bounded with respect to $H_{\mathbf{i}}$, $\mathbf{i} \in \mathcal{I}$.

Asymptotic behavior of the integrand $p_{\mathbf{i}}(t) := p_{\mathbf{i}}(t;\kappa)$ on the real axis given by

$$p_{\mathbf{i}}(t;\kappa) \approx e^{t-\kappa^2 e^{2t}} \quad \text{as} \quad t \to \infty\,,$$
$$p_{\mathbf{i}}(t;\kappa) \approx e^{t-\kappa^2 e^{2|t|}} \quad \text{as} \quad t \to -\infty\,,$$

corresponds to the choice $a = 2$, $b = \kappa^2$, and $C = 1$ for the double exponential decay.

Finally, we apply Theorem 2.46, providing the exponential convergence in M of the rank-r sinc quadrature approximation as in (2.64), with $r = 2M + 1$,

$$g_{\mathbf{i}} = \int_{\mathbb{R}} p_{\mathbf{i}}(t;\kappa)dt \approx \mathfrak{h}_M \sum_{k=-M}^{M} p_{\mathbf{i}}(t_k;\kappa)\,,$$

which proves (2.64). □

Figure 2.16 represents the convergence history for the best orthogonal Tucker versus canonical approximations (via sinc quadrature) of the Newton and Yukawa potentials on $n \times n \times n$ grid for $n = 2048$, $\|x\| \leq 10$.

Note that the sinc quadrature approximation of the Galerkin tensor for the Newton kernel is analyzed in [206].

2.4.8 Helmholtz kernel revisited

One can try to approximate the Helmholtz kernel in $\mathbb{R}^d$, $d \geq 3$, by sinc interpolation in the Tucker format. We consider the singularity function corresponding to the Helmholtz operator in $\mathbb{R}^d$, $d \geq 3$; see the discussion in Section 2.2.5. Given $\kappa \in \mathbb{R}$, consider the real part of the Helmholtz kernel

$$g(x,y) := \frac{\cos(\kappa\|x-y\|)}{\|x-y\|} = \operatorname{Re}\frac{e^{i\kappa\|x-y\|}}{\|x-y\|} \quad \text{for } (x,y) \in [1,R]^d \times [1,R]^d$$

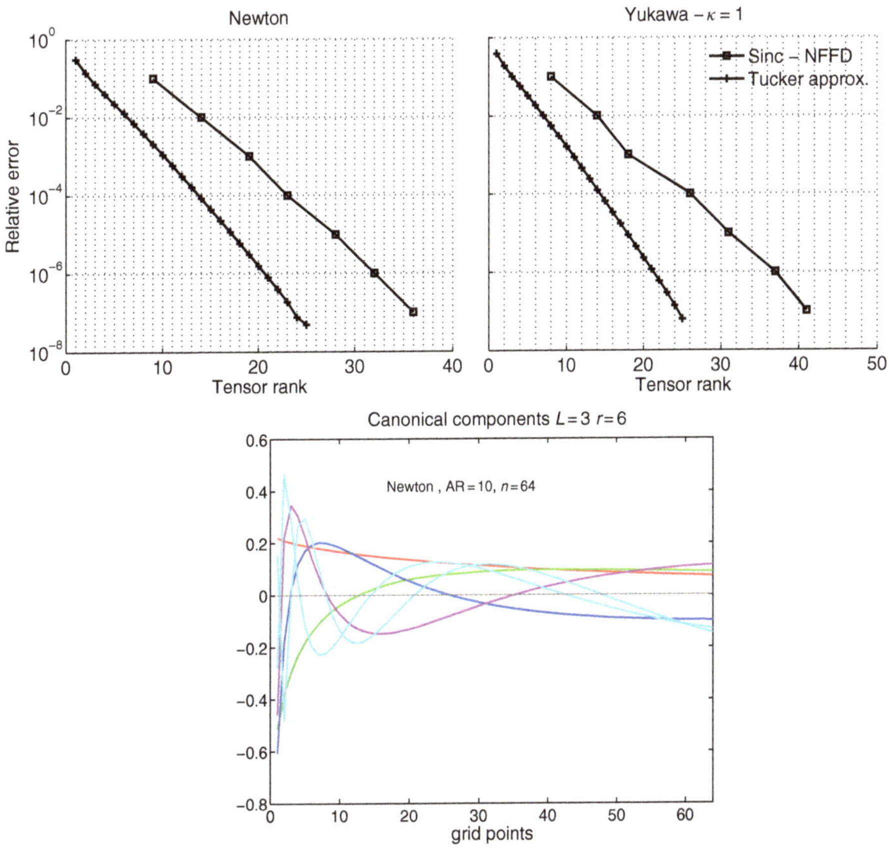

Fig. 2.16: The Tucker versus canonical approximation of the Newton/Yukawa kernels.

in Cartesian coordinates $x, y \in \mathbb{R}^d$.

The sinc interpolation applies to the modified kernel as described in Section 2.2.5. For this example, we have $N_0(F, D_\delta) = \mathcal{O}(e^\kappa)$, hence in order to compensate this factor for large frequency parameter κ the number of terms in the sinc interpolation should increase to $M \sim \kappa + |\log \varepsilon| \log R$. This implies that the separation rank $r = 2M + 1$ for Tucker approximation scales linearly in κ, while the maximal canonical rank is estimated by r^{d-1}.

This leads to unsatisfactory complexity, e.g., $O(\kappa^{d-1} n)$ for the univariate grid size n. It is still an open question whether the good sinc quadrature approximations to oscillating kernels exist.

However, we note that the result in Section 2.2.5 based on the Chebyshev interpolation leads to the following bound on the canonical rank $r = O(d(\kappa + |\log \varepsilon| \log n))$ that scales linearly in the frequency κ. This is attractive in the practically interesting case $d = 3$.

3 Multilinear algebra and nonlinear tensor approximation

3.1 Traditional numerics meets higher dimensions

In this section, we review the basic multidimensional mathematical models in scientific computing and multilinear algebra. Since numerical methods in higher dimensions include many essential ingredients of the traditional numerical methods, we first consider the classical techniques based on the separation of two variables, i.e., for $d = 2$ (matrix case). In particular, we review the main properties of rank-R matrices and approximation methods in this matrix class including the truncated SVD, reduced truncated SVD, Cholesky decomposition of matrices, and adaptive cross approximation (ACA). We briefly sketch the block low rank representation of matrices using $\mathcal{H}$ matrix format applicable in low dimensions $d \leq 3$, and outline its advantages and limitations.

Further examples include fast Fourier transform (FFT) and fast convolution transform of multidimensional arrays. We conclude by pointing out that supercomputing on parallel systems, in general, does not diminish the curse of dimensionality and thus it cannot substitute the numerical methods based on low parametric separable representations in many dimensions.

3.1.1 Multidimensional PDEs in modern applications

A wide range of applications are constituted by the stationary mathematical models governed by the elliptic equations for boundary value or eigenvalue problems in $\mathbb{R}^d$. This class of temporal phenomena is often described by the high dimensional parabolic/hyperbolic dynamical equations. In what follows, we shall discuss the recently developed tensor numerical methods applicable to the different classes of multidimensional partial differential equations (PDEs):

- Elliptic (parameter dependent) equations: Find $u \in H_0^1(\Omega)$, such that

$$\mathcal{H}u := -\operatorname{div}(A \operatorname{grad} u) + Vu = F \quad \text{in } \Omega \in \mathbb{R}^d .$$

- Eigenvalue problems: Find a pair $(\lambda, u) \in \mathbb{R} \times H_0^1(\Omega)$, satisfying

$$\mathcal{H}u = \lambda u \quad \text{in } \Omega \in \mathbb{R}^d , \quad \text{and } \langle u, u \rangle = 1 .$$

- Time dependent parabolic type equations: Find $u : \mathbb{R}^d \times (0, \infty) \to \mathbb{C}$, such that

$$u(x, 0) \in H^2(\mathbb{R}^d) : \quad \sigma \frac{\partial u}{\partial t} + \mathcal{H}u = 0 , \quad \mathcal{H} = \Delta + \mathcal{D} + V(\cdot) ,$$

https://doi.org/10.1515/9783110365917-003

where $\sigma \in \{1, i\}$ specifies the real or complex time dynamics, Δ is the d-dimensional Laplacian, $\mathcal{D}$ is the convection term, and V represents the interaction potential in $\mathbb{R}^d$.

The following *specific features* may lead to the 'curse of dimensionality' and other computational challenges:

- High spacial dimension in physical variables: $\Omega = (-b, b)^d \in \mathbb{R}^d$, $d = 2, 3, \ldots, 100, \ldots$
- Parameter dependent differential equations with high dimensional parametric space such that $A(y, x)$, $u(y, x)$, $y \in \mathbb{R}^M$, $M = 1, 2, \ldots, 100, \ldots$
- The presence of nonlinear, convolution type nonlocal (integral) operators, singular potentials $V = V(x, u)$, multiple cusps in the solution, or highly varying/oscillating coefficients.

The basic principles of tensor numerical methods were first understood theoretically and rigorously tested numerically for problems arising in approximation of multivariate functions and operators. In particular, they include examples of matrix valued functions of the discrete elliptic operator and other transforms to be discussed in detail in forthcoming sections:

- Tensor approximation to the discrete elliptic operator inverse A^{-1}, matrix exponential $\exp(-tA)$, and the Cayley transform $\frac{I+A}{I-A}$, which are applicable to efficient preconditioning in elliptic problem solvers and to the solution of parabolic and hyperbolic equations.
- Superfast tensor product convolution and FFT in $\mathbb{R}^d$ $(d \geq 2)$ of linear, $O(dn \log n)$, or even logarithmic $O(d \log^q n)$-complexity. The initial applications are related to the Poisson solver for d-Laplacian,
$$u(x) = \int_{\mathbb{R}^d} \frac{f(y)}{\|x - y\|^{d-2}} dy\,, \quad x \in \mathbb{R}^d\,,$$
and for computation of the electron density and exchange operator in electronic structure calculations (for $d = 3$); see Section 5.2
- Tensor approximation to the matrix valued integral
$$X = \int_0^\infty \mathrm{e}^{-tA} G \mathrm{e}^{-tB} dt$$
applicable to the solution of matrix Lyapunov equations (arising in control theory) and to parabolic problems.

In the following sections, we discuss the recent advances in tensor methods and demonstrate their efficiency on the examples of some challenging problems arising in computational quantum chemistry, stochastic/parametric PDEs, multiconfiguration dynamics, and homogenization theory, which include:

(A) Boundary value and spectral problems for the elliptic operators in $\mathbb{R}^d$.
(B) Multiparametric elliptic equations providing the deterministic representation of stochastic PDEs.
(C) The Hartree–Fock, Kohn–Sham, and post-Hartree–Fock approaches as the basic reduced models for the electronic Schrödinger equation for many-particle systems.
(D) Computation of excitation energies for large molecules.
(E) The Fokker–Planck and chemical master equations describing the stochastic dynamics in many-particle modeling, as well as the molecular Schrödinger equation and the Boltzmann equation for dilute gas.
(F) The Poisson–Boltzmann equation describing the electrostatic potential in biomolecules.
(G) The elliptic equations with highly oscillating coefficients in homogenization theory and the tensor approach to isogeometric analysis.

3.1.2 Numerical methods for low dimensions as the building block

In *low dimensions*, i.e., for $d = 1, 2, 3$, the goal of the *traditional* numerical methods for PDEs is the construction of fast algorithms that scale linearly in the number of discretization parameters, $O(N)$. Here N is usually understood as the size of a d dimensional grid in the computational domain, i.e., $N = O(n^d)$, where n is the univariate grid size.

The main optimization principles of traditional methods are based on the use of *hierarchical and multigrid* structures, *low rank* or other *data sparse* patterns in the system matrix, sparse grids, as well as on the *recursive* structure of algorithms. In particular, the efficient numerical methods combine the following approximation and solution techniques:

- General purpose numerical linear algebra.
- Finite element/finite difference approximation methods (FEM/FDM).
- Multiresolution representation via wavelets, sparse grid techniques, or h-p methods providing data and matrix compression to the order of $O(N \log^q N)$ operations.
- Multigrid principle leading to $O(N)$ elliptic problem solvers.
- Richardson extrapolation techniques on a sequence of grids toward higher order approximation.
- Spectrally close multilevel or domain decomposition preconditioners for enhancing the convergence of iterative solvers.
- Classical Fourier methods: FFT in $O(N \log N)$ operations, FFT based circulant convolution and the respective treatment of Toeplitz and Hankel matrices.
- Fast multipole, panel clustering, and $\mathcal{H}$ matrix methods, which are well suited for approximation of integral (nonlocal) operators and matrix resolvent in FEM/BEM applications for $d = 1, 2, 3$.

The key point of the rank structured tensor methods is based on the global separation of variables for large d (though tensor numerical methods usually require the basic ingredients of traditional computational schemes developed for $d = 1, 2, 3$). The separable representation techniques can be understood as an extension of the stable SVD based *low rank matrix approximation* algorithms to the case of many spacial dimensions. In what follows, we shall review the main matrix decompositions [78, 132, 163, 176, 345] that are used in the construction of the low rank tensor approximation in many dimensions.

3.1.3 Matrix SVD and rank-r matrices

Matrix singular value decomposition (SVD) is one of the most important algebraic tools that make the error control in tensor approximations efficient and robust [132, 163]. To simplify the discussion, we mainly consider the case of real valued matrices in $\mathbb{R}^{\tau\times\sigma}$, $\tau, \sigma \in \mathbb{N}$. The extension to the case of complex matrices in $\mathbb{C}^{\tau\times\sigma}$ is straightforward.

Lemma 3.1. *(Matrix SVD.) Every real (complex) $\tau \times \sigma$ matrix M can be factorized as*

$$M = U\,S\,V^{\mathrm{T}}\,, \quad \textit{where}$$

1. $U = [\mathbf{u}_1\mathbf{u}_2 \ldots \mathbf{u}_\tau]$ *is a unitary $\tau \times \tau$ matrix,*
2. $V = [\mathbf{v}_1\mathbf{v}_2 \ldots \mathbf{v}_\sigma]$ *is a unitary $\sigma \times \sigma$ matrix,*
3. *S is an $\tau \times \sigma$ matrix (compare to the Tucker core tensor) with the properties of*
   (i) *pseudodiagonality:* $S = \mathrm{diag}\{\sigma_1, \sigma_2, \ldots, \sigma_{\min(\tau,\sigma)}\}$,
   (ii) *ordering:* $\sigma_1 \geq \sigma_2 \geq \cdots \geq \sigma_{\min(\tau,\sigma)} \geq 0$.
4. *The σ_i are singular values of M, and the column vectors $\mathbf{u}_i \in \mathbb{R}^\tau$ and $\mathbf{v}_j \in \mathbb{R}^\sigma$ are, respectively, ith left and jth right singular vectors of M. This means that the matrix U constitutes the orthogonal eigenvectors of MM^{T}, V agglomerates the orthogonal eigenvectors of $M^{\mathrm{T}}M$, and σ_i are the square roots of the respective eigenvalues (compare to the Schmidt decomposition).*

The set of rank $\leq k$ matrices in $\mathbb{R}^{\tau\times\sigma}$ will be called $\mathcal{R}_k$ matrices, i.e., $\mathrm{rank}(M) \leq k$ for $M \in \mathcal{R}_k$. In terms of separation of variables the class of rank-k matrices $\mathcal{R}_k$ specifies the closed subset in the set of $\tau \times \sigma$ matrices.

Each $M \in \mathcal{R}_k$ can be represented in the form

$$M = A\,B^{\mathrm{T}}\,, \quad A \in \mathbb{R}^{\tau\times k}\,, \quad B \in \mathbb{R}^{\sigma\times k}\,. \tag{3.1}$$

In the following, we often use the Frobenius matrix norm, defined for an arbitrary matrix $M \in \mathbb{R}^{\tau\times\sigma}$ by

$$\|A\|_F := \left(\sum_{(i,j)\in\tau\times\sigma} a_{ij}^2 \right)^{1/2}\,.$$

This matrix norm is not associated with any vector norm. Considering a matrix as an element of the Euclidean vector space implies that $\|\cdot\|_F$ is generated by the matrix scalar product,

$$\langle A, B\rangle = \sum_{i\in\tau, j\in\sigma} a_{ij}b_{ij} = \mathrm{trace}(AB^{\mathrm{T}}) = \mathrm{trace}(B^{\mathrm{T}}A)\,.$$

The next statement collects some attractive computational features of the rank-k matrices, which are well known in numerical linear algebra.

Proposition 3.2. *$\mathcal{R}_k$ matrices exhibit the following properties:*
1. *The set $\mathcal{R}_k$ is closed (nontrivial result in linear algebra).*
2. *Only $k(\tau+\sigma)$ numbers are required to store an $\mathcal{R}_k$ matrix M.*
3. *The matrix vector multiplication $\mathbf{x}\mapsto\mathbf{y} := M\mathbf{x}$, $\mathbf{x}\in\mathbb{R}^\sigma$ can be performed in $2k(\sigma+\tau)$ operations by the two step algorithm:*
$$\mathbf{y}' := B^{\mathrm{T}}\mathbf{x}\in\mathbb{R}^k\,,\quad \textit{and}\quad \mathbf{y} := A\mathbf{y}'\in\mathbb{R}^\tau\,.$$
4. *The sum of two $\mathcal{R}_k$ matrices $R_1 = A_1B_1^{\mathrm{T}}$, $R_2 = A_2B_2^{\mathrm{T}}$ is an $\mathcal{R}_{2k}$ matrix,*
$$R_1 + R_2 = [A_1|A_2][B_1|B_2]^{\mathrm{T}}\,,\quad [A_1|A_2]\in\mathbb{R}^{\tau\times 2k}\,,\quad [B_1|B_2]\in\mathbb{R}^{\sigma\times 2k}\,,$$
where matrices $[A_1|A_2]$ and $[B_1|B_2]$ are constructed by concatenation of the respective matrix blocks.
5. *The multiplication of $R\in\mathcal{R}_k$ by an arbitrary matrix M of the proper size gives again an $\mathcal{R}_k$ matrix:*
$$RM = A(M^{\mathrm{T}}B)^{\mathrm{T}}\,,\qquad MR = (MA)B^{\mathrm{T}}\,.$$
6. *The best rank-k approximation of a matrix $M\in\mathbb{R}^{\tau\times\sigma}$ in the Frobenius norm can be calculated by the truncated SVD (described in the next section).*

Another method closely related to SVD matrix decomposition is the so called QR factorization. Let $M\in\mathbb{R}^{\tau\times\sigma}$. Then there exists an orthogonal matrix $Q\in\mathbb{R}^{\tau\times\tau}$ and an upper triangular matrix $R\in\mathbb{R}^{\tau\times\sigma}$, such that

$$M = QR\,.$$

The numerical cost is of the order of $\mathcal{O}(\tau\sigma\min\{\tau,\sigma\})$, but with a smaller constant than in the case of matrix SVD.

If $r = \mathrm{rank}(M) < \min\{\tau,\sigma\}$, the size of QR factors Q and R can be reduced,

$$M = Q_rR_r\quad (Q_r\in\mathbb{R}^{\tau\times r}, R_r\in\mathbb{R}^{r\times\sigma})\,,$$

which can be easily seen by QR factorization of a matrix $R_k = \Sigma_k V_k^{\mathrm{T}}$ involved in the truncated SVD of M.

3.1.4 Reduced truncated SVD

To compute the best rank-k approximation of an arbitrary matrix $M \in \mathbb{R}^{\tau\times\sigma}$ in the Frobenius norm one can use the so called truncated SVD that is the discrete version of the Schmidt decomposition. Denote $n = \min(\tau, \sigma)$.

Algorithm 3.1 (Truncated SVD). For given $M \in \mathbb{R}^{\tau\times\sigma}$ and $k \in \mathbb{N}$:

(i) Compute the SVD of M, $M = U\Sigma V^{\mathrm{T}}$, where $\Sigma = \operatorname{diag}\{\sigma_1, \dots \sigma_k, \dots, \sigma_n\}$ with $\sigma_1 \geq \sigma_2 \geq \cdots \geq \sigma_n \geq 0$, while $U = [\mathbf{u}_1, \dots, \mathbf{u}_k, \mathbf{u}_{k+1}, \dots, \mathbf{u}_\tau]$ and $V = [\mathbf{v}_1, \dots, \mathbf{v}_k, \mathbf{v}_{k+1}, \dots, \mathbf{v}_\sigma]$ are unitary matrices.

(ii) Set $\Sigma_{k,0} := \operatorname{diag}\{\sigma_1, \dots, \sigma_k, 0, \dots, 0\}$, $U_k := [\mathbf{u}_1, \dots, \mathbf{u}_k]$, and $V_k := [\mathbf{v}_1, \dots, \mathbf{v}_k]$.

(iii) Calculate the rank-k approximating matrix

$$M_k := U\Sigma_{k,0}V^{\mathrm{T}} = U_k\Sigma_k V_k^{\mathrm{T}} \approx M\,,$$

where $\Sigma_k := \operatorname{diag}\{\sigma_1, \dots, \sigma_k\}$. The approximation error is estimated by

$$\|M_k - M\|_F^2 \leq \sum_{j=k+1}^{n} \sigma_j^2\,.$$

The computational complexity of the truncated SVD is dominated by the cost of full matrix SVD, $\mathcal{O}(\tau\sigma\min\{\tau, \sigma\})$, which may be too expensive for large τ and σ.

Is it possible to compute nearly the best rank-k matrix approximation getting rid of full matrix SVD? To answer this important question, first we consider the special version of truncated SVD that applies to the rank-m input with $m < \min\{\tau, \sigma\}$. The corresponding algebraic algorithm is called *reduced truncated SVD*.

If $M \in \mathcal{R}_m$, then its best approximation $M_k \in \mathcal{R}_k$, $k < m$ can be computed by the following QR-SVD scheme:

Algorithm 3.2 (Reduced truncated SVD). Given $M = AB^{\mathrm{T}} \in \mathcal{R}_m$, $k < m$:

(i) Calculate the QR decompositions $A = Q_A R_A$ and $B = Q_B R_B$, with the unitary matrices $Q_A \in \mathbb{R}^{\tau\times m}$, and $Q_B \in \mathbb{R}^{\sigma\times m}$, and upper triangular matrices $R_A, R_B \in \mathbb{R}^{m\times m}$.

(ii) Calculate the SVD, $R_A R_B^{\mathrm{T}} = U\Sigma V^{\mathrm{T}}$, at the cost $O(m^3)$.

(iii) Compute truncated SVD of $R_A R_B^{\mathrm{T}} = U\Sigma V^{\mathrm{T}}$ in the form $U_k\Sigma_k V_k^{\mathrm{T}}$ with $U_k := [\mathbf{u}_1, \dots, \mathbf{u}_k]$, $V_k := [\mathbf{v}_1, \dots, \mathbf{v}_k]$ and $\Sigma_k \in \mathbb{R}^{k\times k}$.

(iv) Define $M_k = A_k B_k^{\mathrm{T}}$ with $A_k := Q_A U_k \Sigma_k \in \mathbb{R}^{\tau\times k}$ and $B_k := Q_B V_k^{\mathrm{T}} \in \mathbb{R}^{\sigma\times k}$.

Algorithm 3.2 can be implemented in $\mathcal{O}(m^2(\tau+\sigma)+m^3)$ operations, i.e., it scales linear in both τ and σ, which is much cheaper than the cost of full SVD, provided that $m \ll \min\{\tau, \sigma\}$. The approximation error is controlled by SVD in Step (iii).

Exercise 3.3. Compute the rank-r, ($r = 2M + 1$) approximation to the Hilbert matrix $A = \{a_{ij}\}$, $a_{ij} = 1/(i + j)$ $(i, j = 1, \dots, n)$ for $n = 10^2, 10^3, 10^4$, using $(2M + 1)$-term sinc quadrature with $M = 64$. Apply to the resultant rank-r matrix the best rank-$k < r$

approximation via the reduced truncated SVD by Algorithm 3.2. Compare the central processing unit (CPU) time with that for truncated SVD by Algorithm 3.2 applied to the full $\tau \times \sigma$ matrix.

3.1.5 Cholesky factorization and adaptive cross approximation

Methods to compute a suboptimal low rank approximation to a large square matrix avoiding full matrix SVD can be based on the so called incomplete Cholesky decomposition [132, 163], or in a more general setting on the heuristic method of *adaptive cross approximation (ACA)* known in the literature since [19–21, 346]. These matrix approximation methods require only partial data and can be gainfully applied in the low dimensional FEM/BEM computations.

The Cholesky decomposition is defined as follows. For any symmetric positive definite matrix $M \in \mathbb{R}^{n\times n}$, there is a unique low triangular matrix $L \in \mathbb{R}^{n\times n}$ with positive diagonal entries such that

$$M = LL^{\mathrm{T}} .$$

The computation of factor L costs $\frac{1}{3}n^3$ operations.

In the case of semidefinite matrices $M \geq 0$ the pivoted version of the Cholesky decomposition algorithm leads to the factorization

$$M = PLL^{\mathrm{T}}P^{\mathrm{T}} \quad \text{(with a permutation matrix } P) .$$

If it holds that $r = \operatorname{rank}(M) < n$ then the size of Cholesky factor reduces to $L \in \mathbb{R}^{n\times r}$. Moreover, there is a pivoted version of the Cholesky algorithm used to compute a sequence of rank-1 updates resulting in the rank-r approximation to M with guaranteed precision $\varepsilon > 0$. This requires only the diagonal of M and the r rows associated with pivot elements, see for example [156, 190] for the particular versions of the pivoted Cholesky algorithm.

This truncated Cholesky algorithm will be applied in what follows to compute the so called two-electron integrals [190, 195] in electronic structure calculations; see Section 5.2.

In the case of rectangular matrices there are heuristic algorithms in the spirit of truncated Cholesky schemes, which do not provide the guaranteed precision. Many matrix decomposition algorithms can be represented as a sequence of rank-1 Wedderburn updates [361].

Specifically, for a given $m \times n$ matrix A and vectors $\mathbf{x}$, $\mathbf{y}$ of appropriate sizes, such that $\mathbf{x}^{\mathrm{T}} A\mathbf{y} \neq 0$, the matrix

$$B = A - \frac{A\mathbf{y}\mathbf{x}^{\mathrm{T}}A}{\mathbf{x}^{\mathrm{T}}A\mathbf{y}} ,$$

has $\operatorname{rank}(B) = \operatorname{rank}(A) - 1$. Hence, after r updates of the rank-r matrix $A_0 := A$ in form

$$A_k = A_{k-1} - \frac{A_{k-1}\mathbf{y}_k\mathbf{x}_k^{\mathrm{T}}A_{k-1}}{\mathbf{x}_k^{\mathrm{T}}A_{k-1}\mathbf{y}_k} , \quad \text{provided that } \mathbf{x}_k^{\mathrm{T}}A_{k-1}\mathbf{y}_k \neq 0 , \quad k = 1, \dots, r ,$$

the matrix A_r becomes zero, leading to the rank-r decomposition of A by collecting all rank-1 summands.

The question then arises: What is the reasonable strategy for the choice of vectors $\mathbf{x}_n, \mathbf{y}_n$? The heuristic ACA algorithm is a special case of Wedderburn updates based on the 'max element' pivoting strategy [21, 119]. In general, it applies to the full rank matrix A of large size, and provides some rank-k approximation to this matrix. The principal scheme of heuristic ACA algorithm can be sketched as follows [20]:

Algorithm 3.3 (ACA). Given the matrix $A \in \mathbb{R}^{m\times n}$ and the rank parameter r:

(A) Starting from $R_0 = A \in \mathbb{R}^{m\times n}$, find a nonzero pivot in R_k for $k = 1, \dots, r$, say at (i_k, j_k), and subtract a scaled outer product of the i_kth row and the j_kth column:

$$R_k := R_{k-1} - \frac{1}{(R_{k-1})_{i_k j_k}} \mathbf{u}_k \mathbf{v}_k^{\mathrm{T}}, \quad \text{with } \mathbf{u}_k = (R_{k-1})_{1:m, j_k}, \quad \mathbf{v}_k = (R_{k-1})_{i_k, 1:n},$$

where we use the notation $(R_k)_{i_k,1:n}$ and $(R_k)_{1:m,j_k}$ for the i_kth row and j_kth column of R_k, respectively.

(B) Here j_k is chosen corresponding to the maximum element in modulus of the i_kth row in R_{k-1}, i.e.,

$$|(R_{k-1})_{i_k j_k}| = \max_{j=1,\dots,n} |(R_{k-1})_{i_k j}| .$$

The choice of i_k will be similar.

(C) The matrix $S_r := \sum_{k=1}^{r} \mathbf{u}_k \mathbf{v}_k^{\mathrm{T}}$ is considered as the rank-r approximation to

$$A = S_r + R_r$$

(since $\operatorname{rank}(S_r) \le r$), which ensures the accuracy $\|R_r\|$ that could be controlled only approximately.

(D) If estimate on $\|R_r\|$ is satisfactory, apply the reduced truncated SVD to S_r for the rank optimization, otherwise proceed with $r \mapsto r + 1$.

An example of the special fast method for low rank approximation of matrices was reported in [102].

3.1.6 $\mathcal{H}$ matrix format in low dimensions $d \le 3$: a short excursus

The $\mathcal{H}$- and $\mathcal{H}^2$ matrix techniques can be viewed as the direct descendants of the fast multipole [126] and panel clustering [316] representations based on a block low rank approximation of functions and matrices. Along with almost linear storage cost, $\mathcal{H}$ matrix type formats allow data sparse matrix matrix operations.

Conventionally, $\mathcal{M}_{\mathcal{H},k}$ denotes *the class of data sparse hierarchical $\mathcal{H}$ matrices* invented in [133] and first investigated in [120, 139, 140, 149]. The $\mathcal{H}^2$ matrix format was introduced and analyzed in [149].

We refer to monographs [20, 49, 137] for the detailed discussion on $\mathcal{H}$ matrix techniques.

The construction of $\mathcal{H}$ matrices defined on the product index set $I \times I$ is based on the following ingredients:

- An $\mathcal{H}$-tree $T(I)$ of the index set I (hierarchical cluster tree).
- The admissible partitioning $\mathcal{P}$ of $I \times I$ based on a block cluster tree $T(I \times I)$.
- Low rank approximation of all large enough matrix blocks in $\mathcal{P}$.

Definition 3.4. For an admissible partitioning $\mathcal{P}$ and $k \in \mathbb{N}$, define the set $\mathcal{M}_{\mathcal{H},k}(I \times I, \mathcal{P}) \subset \mathbb{R}^{I \times I}$ of (real) $\mathcal{H}$ matrices by

$$\mathcal{M}_{\mathcal{H},k} = \mathcal{M}_{\mathcal{H},k}(I \times I, \mathcal{P}) := \{M \in \mathbb{R}^{I \times I} : \operatorname{rank}(M|_b) \le k \text{ for all } b \in \mathcal{P}\}. \tag{3.2}$$

$M|_b = (m_{ij})_{(i,j)\in b}$ denotes the matrix block of M, corresponding to $b \in \mathcal{P}$.

Figure 3.1 presents simple examples of *hierarchical partitions* $\mathcal{P}_{1/2}(I \times I)$ (a standard partitioning with admissibility constant equal to $1/2$) and $\mathcal{P}_W(I \times I)$ (weak admissible, see [147]) corresponding to different criteria defining the admissible clusters on a one dimensional geometric grid indexed by I.

The matrices in $\mathcal{M}_{\mathcal{H},k}$ are implemented by means of the list $\{M|_b : b \in \mathcal{P}\}$ of matrix blocks, where each $M|_b$ ($b = \tau \times \sigma$ with $\tau, \sigma \in T(I)$) is represented by the rank-k matrix

$$M|_b = \sum_{\nu=1}^{k} \mathbf{a}_\nu \mathbf{b}_\nu^\top \quad \text{with vectors } \mathbf{a}_\nu \in \mathbb{R}^\tau, \; \mathbf{b}_\nu \in \mathbb{R}^\sigma.$$

The number k is called the local rank.

A summary of the cost estimates of $\mathcal{H}$ matrix arithmetics is collected in the following statement. Here we denote by $N_{\mathcal{H},\text{store}}$, $N_{\mathcal{H}\oplus\mathcal{H}}$, $N_{\mathcal{H}\odot\mathcal{H}}$, and $N_{\widetilde{Inv}(\mathcal{H})}$ the storage cost, the complexity of a matrix sum, matrix product, and matrix inversion, respectively for the rank-k $\mathcal{H}$ matrices. Below $N := \#I$ means the cardinality of the index set I.

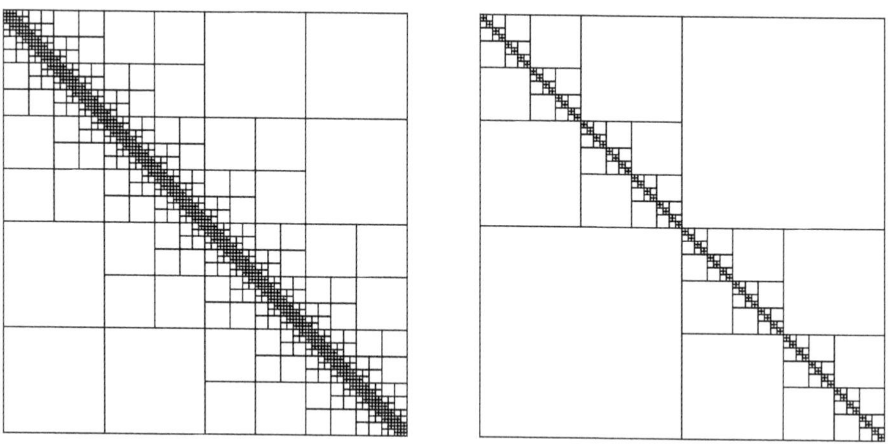

Fig. 3.1: Standard (left) and weak admissible (right) $\mathcal{H}$-partitions, $d = 1$.

Theorem 3.5 ($\mathcal{H}$ matrix arithmetic). *For $k \in \mathbb{N}$, and $\mathcal{H}$-tree $T_{I\times I}$ of depth $L > 1$, the arithmetic of $N \times N$ matrices in $\mathcal{M}_{\mathcal{H},k}(T_{I\times I}, \mathcal{P})$ has the complexity*

$$N_{\mathcal{H},store} \le 2C_{\rm sp}kLN\,, \qquad N_{\mathcal{H}\cdot v} \le 4C_{\rm sp}kLN\,,$$

$$N_{\mathcal{H}\oplus\mathcal{H}} \le C_{\rm sp}k^2N(C_1L + C_2k)\,,$$

$$N_{\mathcal{H}\odot\mathcal{H}} \le C_0C_{\rm sp}^2k^2LN\max\{k, L\}\,, \qquad N_{\widetilde{Inv}(\mathcal{H})} \le CN_{\mathcal{H}\odot\mathcal{H}}\,,$$

where $C_{\rm sp} = C_{\rm sp}(d)$ is the so called sparsity constant [140] for d dimensional FEM approximation.

The notion of sparsity constant $C_{\rm sp}$, introduced in [140], plays an important role in the complexity characterizations of $\mathcal{H}$ matrix arithmetics. Typical estimates on the sparsity constant depending on the dimension are described by $C_{\rm sp}(1) \approx 3$, $C_{\rm sp}(2) \approx 25$ and $C_{\rm sp}(3) \approx 150$, confirming the exponential increase of the computational cost in d ([139, 140]). Hence, the hierarchical approximation remains efficient only for low dimensions up to $d \le 3$.

We observe that the $\mathcal{H}$ matrix format is well suited for representation of the integral (nonlocal) operators in BEM applications ($d = 2, 3$). A closely related approach via the hierarchical LU decomposition applies in FEM techniques [20].

We conclude by collecting the favorable features of the hierarchical $\mathcal{H}$ matrix and LU representations making them attractive in low dimensions $d \le 3$:
- Matrix arithmetic in $O(N \log^q N)$ complexity.
- Accurate and efficient approximation on a class of nonlocal (integral) operators with asymptotically smooth kernels, and operator valued functions $\mathcal{F}(\mathcal{L})$ such as the elliptic operator inverse $\mathcal{L}^{-1}$ [20, 22, 110] and operator exponential $e^{-t\mathcal{L}}$ [111].
- Rigorous theoretical analysis.

At the same time the $\mathcal{H}$ matrix techniques have the inherent limitations:
- Not applicable in high dimensions, i.e., to $d > 3$.
- Complicated data structure for $d = 3$ due to large sparsity constant.
- Difficulties with technical implementation in the case of varying cluster tree (local greed refinement).

Note that further optimization of the $\mathcal{H}$ matrix representations is based on the so called $\mathcal{H}^2$ matrix formats introduced and analyzed in [149]; see also [49].

Some applications of the $\mathcal{H}$ matrix techniques and, in particular, low rank decompositions can be found in [26, 110, 111, 121, 199] and in [347, 349, 352].

The new approach to data sparse approximation of the elliptic equations via the boundary concentrated FEM/BEM was presented in [148, 202, 217, 218].

3.1.7 Fast Fourier transform

This section discusses the most commonly used exact algorithms based on the fast Fourier transform (FFT). Furthermore, in the forthcoming sections, we describe the superfast FFT of logarithmic complexity using the rank structured tensor methods (the so called QTT tensor approximation, see Chapter 4).

Let S_N be the space of complex sequences $f = \{f[n]\}_{0\le n<N}$ of period N, that is the Euclidean space with the scalar product

$$\langle f, g\rangle = \sum_{n=0}^{N-1} f[n] g^*[n] \,.$$

Lemma 3.6. *The family* $\{e_k[n] = \exp(\frac{2i\pi kn}{N})\}_{0\le k<N}$ *is an orthogonal basis of* S_N *with* $\|e_k\|^2 = N$. *Any* $f \in S_N$ *can be represented by the orthogonal projection*

$$f = \sum_{k=0}^{N-1} \frac{\langle f, e_k\rangle}{\|e_k\|^2} e_k \,. \tag{3.3}$$

Definition 3.7. The discrete Fourier transform (DFT) of a sequence f is

$$\widehat{f}[k] := \langle f, e_k\rangle = \sum_{n=0}^{N-1} f[n] \exp\left(\frac{-2i\pi kn}{N}\right), \quad k = 0, \dots, N-1 \;(N^2 \text{ complex multiplications}).$$

Due to (3.3) an inverse DFT is given by

$$f[n] := \frac{1}{N} \sum_{k=0}^{N-1} \widehat{f}[k] \exp\left(\frac{2i\pi kn}{N}\right).$$

Now let us consider the fast Fourier transform (FFT) matrix representation and the respective fast evaluation scheme. The idea of FFT traces back to Gauss (1805), while first computer implementation of FFT was designed by Coolly and Tukey [73].

The FT is represented by a matrix $F_N = \{f_{k,n}\}_{k,n=1}^N$ given by

$$f_{k,n} := \exp\left(\frac{-2i\pi kn}{N}\right) = W^{-nk}, \quad W = e^{2i\pi/N} \,.$$

Suppose that $N = 2^L$. The FFT recursion connects the M-point transform to two copies of the $M/2$-point transforms starting with $M = N$ as follows:

$$F_N = \begin{pmatrix} I_{N/2} & D_{N/2} \\ I_{N/2} & -D_{N/2} \end{pmatrix} \begin{pmatrix} F_{N/2} & 0 \\ 0 & F_{N/2} \end{pmatrix} \begin{pmatrix} \text{even} \\ \text{odd} \end{pmatrix}.$$

Here $I_{N/2}$ is the identity matrix, $D_{N/2}$ is the diagonal matrix with diagonal entries $\{1, W^{-1}, \dots, W^{-N/2}\}$. The permutation matrix at the third place transforms the input vector into its ‘even’ and ‘odd’ parts. Finally, the FFT algorithm keeps going *recursively*:

$$F_N \to F_{N/2} \to \cdots \to F_1 \,.$$

The FFT(N) can be calculated by reduction to two FFT($N/2$), plus $C_F N$ operations to compute $f_{\rm ev}[n]$ and $f_{\rm od}[n]$, $n = 0, \dots, N/2 - 1$. One obtains the recursion for the complexity count

$$\mathcal{N}_{\rm FFT}(N) = 2\mathcal{N}_{\rm FFT}(N/2) + C_F N \quad \text{with} \quad \mathcal{N}_{\rm FFT}(1) = 0 \,.$$

Setting $N = 2^L$, $L \in \mathbb{N}$, and introducing $Q(L) = \mathcal{N}_{\rm FFT}(N)/N$, we get

$$Q(L) = Q(L-1) + C_F \quad \text{with} \quad Q(0) = 0 \,,$$

which implies $Q(L) = C_F L$. Hence computation of FFT costs

$$W_{\rm FFT}(N) = C_F N \log_2 N$$

operations. The existing fast implementations exhibit a small constant $C_F \approx 4$.

The inverse FFT of $\hat{f}$ can be derived from the forward FFT of its complex conjugate $\hat{f}^*$ due to

$$f^*[n] := \frac{1}{N} \sum_{k=0}^{N-1} \hat{f}^*[k] \exp\left(\frac{-2i\pi kn}{N} \right) ,$$

implying $F_N^{-1} = F_N^*$.

Note that the multidimensional FFT matrix $F_N^{(d)}$ can be implemented by the tensorization process with the linear logarithmic cost $O(N \log_2 N)$, with $N = n_1 \dots n_d$, by using the rank-1 separable factorization

$$F_N^{(d)} = F_{n_1}^{(1)} \otimes \dots \otimes F_{n_d}^{(1)} , \quad F_{n_k}^{(1)} \in \mathbb{R}^{n_k \times n_k} \,.$$

3.1.8 Discrete convolution via FFT

Let g be the discrete convolution of two signals f, h supported only by the indices $0 \le n \le M-1$,

$$g[n] = (f * h)[n] = \sum_{k=-\infty}^{\infty} f[k] h[n-k] \,.$$

The naive implementation requires $M(M+1)$ operations.

The fast algorithm is based on a simple observation that the discrete convolution can be represented as a matrix-by-vector product with the Toeplitz matrix

$$T = \{h[n-k]\}_{0 \le n,k < M} \in \mathbb{R}^{M \times M} , \quad g = Tf \,.$$

Extending f and h over M samples by

$$\tilde{h}[M] = 0 \,, \quad \tilde{h}[2M - i] = h[i] \,, \quad i = 1, \dots, M-1 \,,$$

$$\tilde{f}[n] = 0 \,, \quad n = M, \dots, 2M-1 \,,$$

we reduce the problem to the matrix-by-vector product with a circulant matrix $\mathcal{C} \in \mathbb{R}^{2M\times 2M}$ specified by the first row $\tilde{h} \in \mathbb{R}^{2M}$. Hence, the problem is reduced to a circulant convolution to be performed by the FFT.

An $n \times n$ Toeplitz matrix $\mathcal{C}$ is called *circulant* if it has the form

$$\mathcal{C} = \operatorname{circ}\{c_1, \ldots, c_n\} := \begin{pmatrix} c_1 & c_2 & \ldots & c_n \\ c_n & c_1 & \ldots & c_{n-1} \\ \vdots & \vdots & \ddots & \vdots \\ c_2 & \ldots & c_n & c_1 \end{pmatrix}, \quad c_i \in \mathbb{C}\,.$$

The set of all $n \times n$ circulant matrices is closed with respect to addition and multiplication by a constant. Any circulant matrix $\mathcal{C}$ is *associated with the polynomial*

$$p_c(z) := c_1 + c_2 z + \cdots + c_n z^{n-1}\,, \quad z \in \mathbb{C}\,.$$

Matrix $\mathcal{C}$ has a diagonal representation in the Fourier basis,

$$\mathcal{C} = F_n^{\mathrm{T}} \Lambda_c F_n$$

with

$$\Lambda_c = \operatorname{diag}\{p_c(1), \ldots, p_c(\omega^{n-1})\}\,, \quad \omega = \mathrm{e}^{i\pi/n}\,.$$

The eigenvector $\vec{\omega}_j$ corresponding to the eigenvalue $p_c(\omega^{j-1})$ is given by the jth column of F_n,

$$\vec{\omega}_j = \frac{1}{\sqrt{n}} \left\{\omega^{(k-1)(j-1)}\right\}_{k=1}^{n}\,.$$

The matrix vector product with the circulant $\mathcal{C}$ costs $2C_F n \log_2 n + O(n)$ operations and so is used for the discrete convolution.

3.1.9 A new paradigm: tensor methods beat supercomputers

In the following discussion, we argue that the huge problems in high dimensions could not be solved efficiently by the traditional methods in 'exact' arithmetics on supercomputers. In fact, suppose that the target problem is discretized on a $n^{\otimes d}$ grid where n may vary from several hundred to several thousand. Then severe computational problems may arise due to the following bottlenecks:

- Asymptotically optimal linear cost of standard algebraic operations in exact arithmetics, $O(N)$, with $N = n^d$ satisfactory only for small d (speed up of the multiprocessor system n times is equivalent to the reduction of dimension by one, $d \mapsto d-1$).
- Complexity of QR, SVD, and matrix inversion in full arithmetics is of the order of $O(N^3)$, which might be too large already for $d = 3$, i.e., for $N = n^3$, implying $N^3 = n^9$ (it is large even for the moderate grid size $n = 10^2 - 10^3$).

– Traditional 'asymptotically optimal' numerical methods (say, multigrid, multiresolution, domain decomposition) of the complexity $O(N)$ suffer from the 'curse of dimensionality' unless the number of processors increases exponentially in d.

Hence, we formulate a paradigm of the up-to-date large scale numerical simulations:

The higher computer capacities do not resolve the curse of dimensionality.

The attractive remedy is the identification and efficient implementation of the low rank tensor structured representations (parametrizations), which allow almost linear complexity scaling in the dimension d.

In what follows, we discuss the basic approaches to the rank structured tensor decompositions of multidimensional data arrays.

3.2 Introduction to canonical and Tucker tensor formats

3.2.1 Preliminary discussion

In this section, we discuss the rank structured tensor formats based on the traditional canonical and Tucker representations.

First, we recall the tensor product of finite dimensional Hilbert spaces constituted by multidimensional vectors, and endorsed by the Euclidean scalar product, as the particular case of a functional construction in Section 2.1. We describe the important bilinear operations on tensors such as the scalar product and matrix unfolding, as well as contracted and convolution products.

The simplest and probably most popular low parametric tensor format is based on the canonical sum of rank-1 elements. It has been extensively used in multiway analysis since [164, 165, 232, 332]. As a nonstandard example, we show that rank decomposition can be useful also in numerical linear algebra leading to $O(n^{\log_2 7})$ – Strassen algorithm of fast matrix multiplication [336].

The orthogonal Tucker decomposition is a powerful tool for the low parametric representation of rather small data arrays in moderate dimensions [248, 350]. In [206] it was proven that this orthogonal decomposition can be very efficient for the low rank representation of tensors obtained by sampling of functions on large tensor grids in $\mathbb{R}^d$, demonstrating very low approximation ranks. In the case of moderate dimensions d, for example in 3D, the Tucker tensors can be coupled with the canonical decomposition of its small core tensor in the form of a mixed Tucker-canonical format. This allows us to relax some limitations of both tensor models.

The principal ingredient of tensor calculus is the efficient implementation of linear and multilinear operations on 'formatted tensors' by their reduction to univariate operations. We begin this discussion with the simple example of canonical and Tucker tensors. Finally, we address a general problem of the best nonlinear approximation in rank structured tensor formats.

3.2.2 Tensor product of finite dimensional Hilbert spaces

Multidimensional data arrays are considered as elements of a finite dimensional tensor product Hilbert space, that is a special case of functional tensor product spaces, discussed in Section 2.1. Let $\mathbb{H} = H_1 \otimes \cdots \otimes H_d$ be a tensor product Hilbert space, where H_ℓ is a real Euclidean space of vectors

$$H_\ell = \mathbb{R}^{n_\ell}, \quad n_\ell \in \mathbb{N}, \quad n_\ell := \dim H_\ell, \quad \ell = 1, \ldots, d.$$

dth order tensor $\mathbf{A} \in \mathbb{H}$ of size $\mathbf{n} = (n_1, \ldots, n_d)$ is a function of d discrete arguments, that is a multidimensional array (vector) over $\mathcal{I} := I_1 \times \cdots \times I_d$, $I_\ell = \{1, \ldots, n_\ell\}$, i.e.,

$$\mathbf{A}: I_1 \times \cdots \times I_d \to \mathbb{R}, \quad \text{with} \quad \dim \mathbb{H} = |\mathbf{n}| := n_1 \ldots n_d.$$

In the literature there are several notations for the coordinate representation of **A**,

$$\mathbf{A} := [a_{\mathbf{i}}] = [a_{i_1 \ldots i_d}] = [a(i_1, \ldots, i_d)] \in \mathbb{R}^{\mathcal{I}}.$$

The dimension directions ℓ are called *modes*, hence tensor is a union of *ℓ-mode fibers*, $a(i_1, \ldots, i_{\ell-1}, :, i_{\ell+1}, \ldots, i_d)$, for each $\ell = 1, \ldots, d$.

Choose a basis $\{\phi_k^{(\ell)} : 1 \le k \le n_\ell\}$ of H_ℓ, then the set $\{\phi_{k_1}^{(1)} \otimes \phi_{k_2}^{(2)} \otimes \cdots \otimes \phi_{k_d}^{(d)}\}$ $(1 \le k_\ell \le n_\ell, 1 \le \ell \le d)$ is the basis in $\mathbb{H}$ implying $\dim \mathbb{H} = \prod_{\ell=1}^d n_\ell$.

The *Euclidean scalar product* of tensors $\mathbf{A}, \mathbf{B} \in \mathbb{H}$ takes the form

$$\langle \mathbf{A}, \mathbf{B} \rangle := \sum_{(i_1,\ldots,i_d)\in\mathcal{I}} a_{i_1 \ldots i_d} b_{i_1 \ldots i_d},$$

inducing the Euclidean (Frobenious) norm $\|\mathbf{A}\|_F := \sqrt{\langle \mathbf{A}, \mathbf{A} \rangle}$.

The scalar product of rank-1 tensors $\mathbf{W}, \mathbf{V} \in \mathbb{H}$ is reduced to 1D calculation, with linear complexity scaling in dimension, $O(dn)$,

$$\langle \mathbf{W}, \mathbf{V} \rangle = \langle \mathbf{w}^{(1)} \otimes \cdots \otimes \mathbf{w}^{(d)}, \mathbf{v}^{(1)} \otimes \cdots \otimes \mathbf{v}^{(d)} \rangle = \prod_{\ell=1}^{d} \langle \mathbf{w}^{(\ell)}, \mathbf{v}^{(\ell)} \rangle_{H_\ell}, \tag{3.4}$$

corresponding to the entrywise representation of a rank-1 tensor

$$\mathbf{W} = [w_{\mathbf{i}}] = \mathbf{w}^{(1)} \otimes \cdots \otimes \mathbf{w}^{(d)},$$

$$w(i_1, \ldots, i_d) = \prod_{\ell=1}^{d} w^{(\ell)}(i_\ell).$$

Clearly, the latter leads to the linear in d storage demand,

$$\mathrm{Stor}(\mathbf{W}) = n_1 + \cdots + n_d \ll \prod_{\ell=1}^{d} n_\ell.$$

The linear in d storage and scalar product costs lead to the tremendous simplification of the algebraic calculus on rank-1 elements, thus making them the building block in the efficient representation of higher order tensors.

Denote the d-fold tensor product $\mathbb{H} = H \otimes \cdots \otimes H$ by $H^{\otimes d}$ $(= \mathbb{R}^{I^d})$. The rank-1 element in $H^{\otimes d}$ is called symmetric if $\mathbf{U} = \mathbf{u} \otimes \cdots \otimes \mathbf{u}$, hence we can use the shortage in notation $\mathbf{U} = \mathbf{u}^{\otimes d}$.

Example 3.8 (Matrix case, $d = 2$). Let $A = \mathbf{a}_1 \otimes \mathbf{a}_2$, $B = \mathbf{b}_1 \otimes \mathbf{b}_2$, $\mathbf{a}_i, \mathbf{b}_i \in \mathbb{R}^n$.

$$\langle A, B \rangle = \langle \mathbf{a}_1, \mathbf{b}_1 \rangle \langle \mathbf{a}_2, \mathbf{b}_2 \rangle , \quad \|A\|_F = \sqrt{\langle \mathbf{a}_1, \mathbf{a}_1 \rangle \langle \mathbf{a}_2, \mathbf{a}_2 \rangle} .$$

3.2.3 Matrix unfolding and contracted product of tensors

For a matrix $A \in \mathbb{R}^{m \times n}$ we use the *vector representation* (vectorization or concatenation) $A \to \mathrm{vec}(A) \in \mathbb{R}^{mn}$, where $\mathrm{vec}(A)$ is an $nm \times 1$ vector obtained by 'stacking' A's columns (the FORTRAN style ordering)

$$\mathrm{vec}(A) := [a_{11}, \ldots, a_{n1}, a_{12}, \ldots, a_{n2}, \ldots, a_{1m}, \ldots, a_{nm}]^{\mathrm{T}} .$$

In this way, $\mathrm{vec}(A)$ is a rearranged version of A. *Vectorization of a tensor* is a map of a multidimensional array to a long vector by stacking its fibers.

Definition 3.9. Let $\mathbf{A} \in \mathbb{R}^{I_1 \times \cdots \times I_d}$ be a tensor, then the vectorization of $\mathbf{A}$ is recursively defined by

$$\mathrm{vec}(\mathbf{A}) = \begin{bmatrix} \mathrm{vec}([a(i_1, \ldots, i_{d-1}, 1)]) \\ \mathrm{vec}([a(i_1, \ldots, i_{d-1}, 2)]) \\ \vdots \\ \mathrm{vec}([a(i_1, \ldots, i_{d-1}, n_d)]) \end{bmatrix} \in \mathbb{R}^{|\mathbf{n}| \times 1} .$$

The element $a(i_1, \ldots, i_d)$ maps to the vector entry $(j, 1)$, where

$$j = 1 + \sum_{k=1}^{d} (i_k - 1) \prod_{\ell=1}^{k-1} n_\ell .$$

Unfolding of a tensor into a matrix (matricization) is a way to map a high order tensor into twofold arrays $\mathbb{R}^{\mathcal{I}} \mapsto \mathbb{R}^{I_\ell \times I_{(-\ell)}}$, first by rearranging (reshaping) it for some $\ell \in \{1, \ldots, d\}$ and then vectorizing the $(d-1)$-order tensors in $\mathbb{R}^{i_\ell \times I_{(-\ell)}}$ for each $i_\ell \in I_\ell$. Here the single hole index set is defined by $I_{(-\ell)} := I_1 \times \cdots \times I_{\ell-1} \times I_{\ell+1} \times \cdots \times I_d$ with the cardinality given by

$$\# I_{(-\ell)} = \bar{n}_\ell := n_1 \ldots n_{\ell-1} n_{\ell+1} \ldots n_d .$$

Definition 3.10. The unfolding of a tensor $\mathbf{A} \in \mathbb{R}^{I_1 \times \cdots \times I_d}$ along mode ℓ is defined by a matrix $A_{(\ell)} = [a_{i_\ell j}]$ of dimension $n_\ell \times \bar{n}_\ell$, so that the tensor element $a(i_1, \ldots, i_d)$ maps

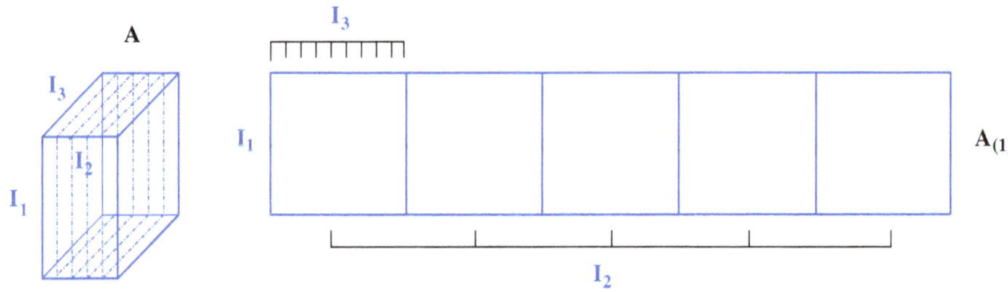

Fig. 3.2: Visualization of the matrix unfolding $A_{(1)}$ for $d = 3$.

to matrix element $a(i_\ell, j)$, $i_\ell \in I_\ell, j \in \{1, \dots, \bar{n}_\ell\}$ where

$$j = 1 + \sum_{k=1, k \neq \ell}^{d} (i_k - 1) J_k\,, \quad J_k = \prod_{m=1, m \neq \ell}^{k-1} n_m\,.$$

Figure 3.2 visualizes the procedure of matrix unfolding for $d = 3$, $\ell = 1$. By definition, the ℓ-mode fibers of $\mathbf{A}$ are the column vectors of the matrix unfolding $A_{(\ell)}$, where the latter may be also called $\text{mat}_\ell(\mathbf{A})$. Hence, $A_{(\ell)}$ can be defined by recursion over $\text{vec}(\mathbf{A})$,

$$\begin{aligned} A_{(\ell)} = [&\text{vec}[a(i_1, \dots, i_{\ell-1}, 1, i_{\ell+1}, \dots, i_d)], \dots, \\ &\text{vec}[a(i_1, \dots, i_{\ell-1}, n_\ell, i_{\ell+1}, \dots, i_d)]]^{\mathrm{T}}\,. \end{aligned}$$

Remark 3.11. Kolmogorov's decomposition (Section 2.1) is a particular way to unfold the multivariate function into 'one dimensional' representation by univariate functions.

The following definition introduces the ℓ rank of a tensor:

Definition 3.12. The ℓ rank of a tensor $\mathbf{A}$, denoted by $\text{rank}_\ell(\mathbf{A})$ ($\ell = 1, \dots, d$), is the dimension of the vector space spanned by the ℓ-mode vectors (*fibers*), i.e,

$$\text{rank}_\ell(\mathbf{A}) = \text{rank}(A_{(\ell)})\,.$$

The major difference from the matrix case, however, is the fact that ℓ ranks of a higher order tensor are not necessarily the same for different ℓ.

Here we present a simple explicit example of a matrix unfolding of a tensor.

Example 3.13. Define a tensor $\mathbf{A} \in \mathbb{R}^{3\times2\times3}$ by $a_{111} = a_{112} = a_{211} = -a_{212} = 1$, $a_{213} = a_{311} = a_{313} = a_{121} = a_{122} = a_{221} = -a_{222} = 2$, $a_{223} = a_{321} = a_{323} = 4$, $a_{113} = a_{312} = a_{123} = a_{322} = 0$. The matrix unfolding $A_{(1)}$ is given by

$$A_{(1)} = \begin{bmatrix} 1 & 1 & 0 & 2 & 2 & 0 \\ 1 & -1 & 2 & 2 & -2 & 4 \\ 2 & 0 & 2 & 4 & 0 & 4 \end{bmatrix}.$$

It can be seen that $\text{rank}_1(\mathbf{A}) = 3$.

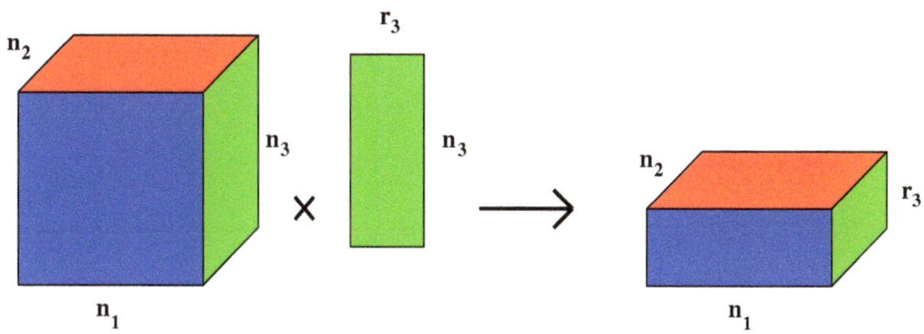

Fig. 3.3: Visualizing a contracted product of a third order tensor with a matrix.

An important tensor-tensor operation is the *contracted product* of two tensors. In the following we often use a *tensor matrix contracted product* along mode ℓ; see Figure 3.3.

Definition 3.14. Given $\mathbf{V} \in \mathbb{R}^{I_1\times\cdots\times I_d}$, and a matrix $M \in \mathbb{R}^{I_\ell \times J_\ell}$, define the mode-$\ell$ tensor matrix contracted product by a tensor

$$\mathbf{U} = \mathbf{V} \times_\ell M \in \mathbb{R}^{I_1\times\cdots\times I_{\ell-1}\times J_\ell \times I_{\ell+1}\cdots\times I_d},$$

where

$$u_{i_1,\ldots,i_{\ell-1},j_\ell,i_{\ell-1},\ldots,i_d} = \sum_{i_\ell=1}^{n_\ell} v_{i_1,\ldots,i_{\ell-1},i_\ell,i_{\ell-1},\ldots,i_d} m_{j_\ell,i_\ell}, \quad j_\ell \in J_\ell .$$

Contracted product by Definition 3.14 generalizes the procedure of matrix matrix multiplication:

$$M_{(n,m)} \times_2 M_{(p,m)} = M_{(n,m)} M^{\mathrm{T}}_{(p,m)} \mapsto M_{(n,p)} .$$

Adapting the definition of contracted product to rank-1 tensors, we conclude that a dth-order tensor $\mathbf{A}$ is the rank-1 element, $\operatorname{rank}(\mathbf{A}) = 1$, if it is a contracted product of d vectors $\mathbf{t}^{(1)}, \ldots, \mathbf{t}^{(d)}$, $\mathbf{t}^{(\ell)} \in \mathbb{R}^{I_\ell}$, $(\ell = 1, \ldots, d)$,

$$\mathbf{A} = \mathbf{t}^{(1)} \times_2 \mathbf{t}^{(2)} \cdots \times_d \mathbf{t}^{(d)}, \quad a_{i_1\ldots i_d} = t^{(1)}_{i_1} \ldots t^{(d)}_{i_d}, \quad i_\ell \in I_\ell ,$$

that is equivalent to the initial definition of a rank-1 tensor.

3.2.4 Canonical representation as a sum of rank-1 tensors

The simplest and probably the most popular low parametric tensor representation is based on finite sums of rank-1 elements, first introduced in [164, 165] and [62]. Such representation is called the canonical format of dth order tensors. The alternative notations are CANDECOMP, PARAFAC, and CP representations; see the survey paper [232].

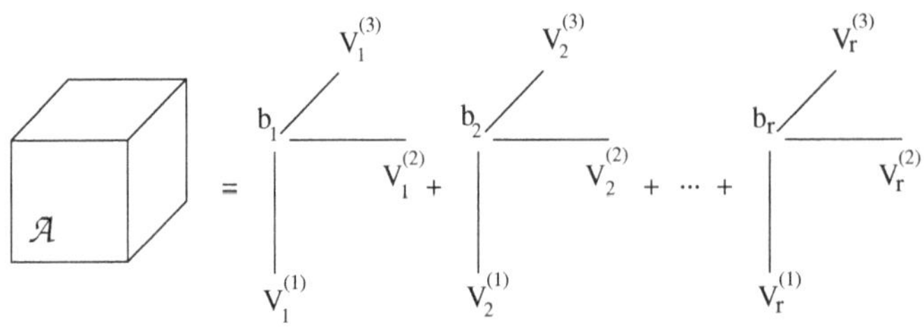

Fig. 3.4: Visualization of the canonical tensor for $d = 3$.

Definition 3.15 (Canonical format). Introduce a subset of elements in $\mathbb{H}$ that allows an R-term representation by a sum of rank-1 tensors. They form a set

$$\mathcal{C}_R = \left\{ \mathbf{A} \in \mathbb{H} \colon \mathbf{A} = \sum_{k=1}^{R} \mathbf{w}_k^{(1)} \otimes \mathbf{w}_k^{(2)} \otimes \cdots \otimes \mathbf{w}_k^{(d)}, \ \mathbf{w}_k^{(\ell)} \in H_\ell \right\} .$$

Elements $\mathbf{A} \in \mathcal{C}_R$ with $\mathbf{A} \notin \mathcal{C}_{R-1}$ are referred to as having the *canonical tensor rank* $R = \operatorname{rank}(\mathbf{A})$.

Figure 3.4 visualizes the canonical tensor for $d = 3$. Tensors $\mathbf{A} \in \mathcal{C}_R$ can be represented by the description of Rd elements $\mathbf{w}_k^{(\ell)} \in H_\ell$, which implies the linear storage cost in d, dRn.

It is remarkable that the storage complexity for an R-term canonical representation scales linear in all characteristic parameters, i.e., in the dimension d, rank R, and the mode size n. This means that the canonical tensor format is asymptotically optimal in the limit of large parameters d, R, and n, which is not the case for other types of nonlinear parametric representations commonly used in computational practice. At the end of this section we collect the advantageous features and limitations of the CP tensor format.

Recalling the notation $\overline{n}_\ell$ for the ℓ-mode single hole product of dimensions,

$$\overline{n}_\ell = n_1 \dots n_{\ell-1} n_{\ell+1} \dots n_d ,$$

we arrive at the rough upper bound for the canonical rank of a tensor.

Remark 3.16. The canonical rank R of a tensor $\mathbf{A} \in \mathbb{V}_\mathbf{n}$ is bounded by

$$R \le \min_{1 \le \ell \le d} \overline{n}_\ell = \left(\prod_{\ell=1}^{d} n_\ell \right) / \max_\ell n_\ell . \tag{3.5}$$

Proof. Consider the case $d = 3$. Let $n_1 = \max_{1 \le \ell \le d} n_\ell$ for definiteness. One can represent a tensor $\mathbf{A}$ as

$$\mathbf{A} = \sum_{k=1}^{n_3} B_k \otimes \mathbf{z}_k , \quad B_k \in \mathbb{R}^{n_1 \times n_2} , \ \mathbf{z}_k \in \mathbb{R}^{n_3} ,$$

where $B_k = a(:, :, k)$ $(k = 1, \dots, n_3)$ is the $n_1 \times n_2$ matrix slice of $\mathbf{A}$ and $\mathbf{z}_k(i) = 0$, for $i \ne k$, $\mathbf{z}_k(k) = 1$. Let $\operatorname{rank}(B_k) = r_k \le n_2$, $k = 1, \dots, n_3$, then

$$\operatorname{rank}(B_k \otimes \mathbf{z}_k) = \operatorname{rank}(B_k) \le n_2 ,$$

implying

$$\operatorname{rank}(\mathbf{A}) \le \sum_{n=1}^{n_3} \operatorname{rank}(B_k) \le n_2 n_3 = \min_{1\le \ell \le 3} \overline{n}_\ell \, .$$

The general case of $d > 3$ is treated similarly by induction. □

Note that any rank decomposition of B_k in the above argument leads to the straightforward (usually redundant) canonical representation of $\mathbf{A}$. Such a nested (direct) canonical representation can be constructed similarly for any dth order tensor.

Now assume that the target tensor is already given in the R-term canonical form

$$\mathbf{A} = \sum_{k=1}^{R} \mathbf{A}_k \,, \quad \operatorname{rank}(\mathbf{A}_k) = 1 \,, \quad \text{with} \quad \mathbf{A}_k = \mathbf{a}_k^{(1)} \otimes \cdots \otimes \mathbf{a}_k^{(d)} \, .$$

Then the refined rank estimate is possible using the computable matrix rank. Define the $(n_1 + \cdots + n_d) \times R$ matrix $M = [\widehat{\mathbf{a}_1} \dots \widehat{\mathbf{a}_R}]$, where each kth column vector is obtained by a vertical 'stacking' of skeleton vectors of $\mathbf{A}_k$, $\widehat{\mathbf{a}_k} = [(\mathbf{a}_k^{(1)})^{\mathrm{T}} (\mathbf{a}_k^{(2)})^{\mathrm{T}} \dots (\mathbf{a}_k^{(d)})^{\mathrm{T}}]^{\mathrm{T}}$.

Proposition 3.17. *Let $\mathbf{A} \in \mathcal{C}_R$, then the canonical rank estimate holds*

$$\operatorname{rank}(\mathbf{A}) \le R_0^{d-1} \,, \quad \textit{where} \quad R_0 := \operatorname{rank}(M) \, .$$

The estimate remains valid in the sense of ε ranks.

Proof. The rank-R_0 decomposition of the matrix M defined above generates the R_0^{d-1}-term canonical representation of $\mathbf{A}$. Indeed, all vectors $\widehat{\mathbf{a}_k}$, $k = 1, \dots, R$, can be represented as the linear combination of R_0 left orthogonal vectors in SVD decomposition of M. Then each rank-1 tensor $\mathbf{A}_k$ can be represented by a sum of R_0^{d-1} rank-1 terms at most, all from the same set of generating elementary tensors. □

We briefly summarize the main (beneficial) computational features of the canonical format.

Advantages:

(A) $\mathcal{C}_R$ format leads to the tremendous reduction of storage cost, removing d from the exponential, $n^d \to d\,R\,n$. The storage complexity scales linear, i.e., is asymptotically optimal, in all representation parameters.

(B) There are analytic sinc based methods of low rank approximation applied to some classes of discretized multivariate functions of a form $f = f(g_1(x_1) + \cdots + g_d(x_d))$, in particular, to Green's kernels, radial basis functions, and elliptic resolvent (Section 2.2);

(C) Multilinear algebraic operations on canonical tensors such as the scalar, Hadamard and contracted product, convolution of tensors, as well as addition, translation to other formats, etc. can be computed easily.

(D) Canonical tensors are naturally embedded into the Tucker, matrix product states (tensor train), and other more general classes of tensors. Hence, they can be effectively used in combination with other tensor formats.

Limitations: $\mathcal{C}_R$ is a nonclosed set, hence approximation algorithms in $\mathcal{C}_R$ are not robust (see the discussion in Section 3.2.5). Moreover, we do not know stable algorithms to compute the rank-R canonical approximation of an arbitrary tensor. This is a serious drawback of the canonical format from the point of view its systematic use in tensor numerical methods for solving the multidimensional PDEs, since the rank truncation procedure is the necessary ingredient of rank structured numerical algorithms in many dimensions.

3.2.5 Little analogy between the cases $d = 2$ and $d \geq 3$

In this paragraph, we focus on the main distinctions between matrices, i.e., the case $d = 2$, and multidimensional tensors. Many basic features of matrix algebra are no longer true in the case $d \geq 3$, which brings up a lot of completely new and challenging theoretical and computational problems in numerical multilinear algebra and in tensor numerical methods for PDEs.

Recall that the minimal number R in the canonical representation

$$\mathbb{R}^{\mathcal{I}} \ni \mathbf{A} = \sum_{k=1}^{R} \mathbf{v}_k^{(1)} \otimes \cdots \otimes \mathbf{v}_k^{(d)}, \quad \mathbf{v}_k^{(\ell)} \in \mathbb{R}^n, \tag{3.6}$$

is called a *tensor rank* or a canonical rank of $\mathbf{A}$, $R = \operatorname{rank}(\mathbf{A})$.

Finding of a tensor rank R and the corresponding decomposition(s) in many dimensions ($d \geq 3$) is the main issue of the multifactor analysis traced back to [62, 63, 165, 242]. The uniqueness conditions for the canonical representations were discussed, in particular, in [333]. The multifactor analysis is essentially a nonlinear approximation process. In general, computing the rank of a high order tensor is nondeterministic polynomial time (NP) hard (Hästad 1990); see also estimates in Proposition 3.17.

Remark 3.18. For $d = 2$, the definition of a tensor rank coincides with the standard definition of the matrix rank(A), which can be calculated along with the respective rank decomposition by the finite SVD algorithm in $O(n^3)$ operations. The *orthogonality requirement in SVD* ensures in general the uniqueness of a rank decomposition.

If $d \geq 3$, the situation changes dramatically as illustrated in the following:

(I) rank($\mathbf{A}$) depends on the number field (say, $\mathbb{R}$ or $\mathbb{C}$), $\operatorname{rank}_{\mathbb{C}}(\mathbf{A}) \leq \operatorname{rank}_{\mathbb{R}}(\mathbf{A})$.

(II) A set of tensors of the rank not larger than r,

$$\mathcal{C}_r(d) := \{\mathbf{A} \in H_1 \otimes \cdots \otimes H_d : \ \operatorname{rank}(\mathbf{A}) \leq r\},$$

is closed only when $d = 2$ (matrices), or if $r = 1$ (rank-1 tensors) ([373]).

(III) For $d \geq 3$, $r \neq 1$, the set $\mathcal{C}_r(d)$ is nonclosed (see Examples 3.20 and 3.21 below).

(IV) For $d \geq 3$, we do not know any finite algorithm to compute $r = \operatorname{rank}(\mathbf{A})$, except simple bounds (compare with the case $d = 2$):

$$0 \leq \operatorname{rank}(\mathbf{A}) \leq n^{d-1}.$$

(V) In the case of a canonical target, the refined rank estimate is given by Proposition 3.17.

(VI) For fixed $d \geq 3$ and n, we do not know the exact value of $\max\{\mathrm{rank}(\mathbf{A})\}$. There are few results for special cases. For example, from J. Kruskal (1977) [242]:
- For any $2 \times 2 \times 2$ tensor we have $\max\{\mathrm{rank}(\mathbf{A})\} = 3 < 4$
- For $3 \times 3 \times 3$ tensors there holds $\max\{\mathrm{rank}(\mathbf{A})\} = 5 < 9$.
- 'Probabilistic' properties of $\mathrm{rank}(\mathbf{A})$: In the set of $2 \times 2 \times 2$ tensors there are about 79% of rank-2 tensors and 21% of rank-3 tensors, while rank-1 tensors appear with probability 0. Clearly, for $n \times n$ matrices we have

$$\mathcal{P}\{\mathrm{rank}(A) = n\} = 1 .$$

It is possible to prove the uniqueness property of the canonical decomposition within the equivalence classes. Two representations like (3.6) are considered as equivalent (essential equivalence) if either

(a) they differ in the order of terms or

(b) for some set of parameters $a_k^\ell \in \mathbb{R}$ such that $\prod_{\ell=1}^d a_k^\ell = 1$ $(k = 1, \ldots, R)$, there is a transform $\mathbf{v}_k^{(\ell)} \to a_k^\ell \mathbf{v}_k^{(\ell)}$.

A simplified version of the general uniqueness result is given by the following statement:

Proposition 3.19 ([242]). *Let for each $\ell = 1, \ldots, d$, the vectors $\mathbf{v}_k^{(\ell)}$, $(k = 1, \ldots, R)$ where $R = \mathrm{rank}(\mathbf{A})$, are linear independent. If*

$$(d-2)R \geq d-1 ,$$

then the decomposition (3.6) is uniquely determined up to the equivalences (a)–(b) above.

To illustrate the noncloseness of $\mathcal{C}_r(d)$ (item III), we present two examples.

Example 3.20. Let $\mathbf{x}, \mathbf{y}$ be two linearly independent vectors in H. Consider the tensor $\mathbf{T} \in H \otimes H \otimes H = H^{\otimes 3}$,

$$\mathbf{T} := \mathbf{x} \otimes \mathbf{x} \otimes \mathbf{x} + \mathbf{x} \otimes \mathbf{y} \otimes \mathbf{y} + \mathbf{y} \otimes \mathbf{x} \otimes \mathbf{y} .$$

Then the following properties hold:
(a) $\mathrm{rank}(\mathbf{T}) = 3$; and (b) there is a sequence of rank-2 tensors approximating $\mathbf{T}$ with arbitrarily small error, which do not converge to some rank-2 tensor.

Proof. (a) See [328] for the details.
(b) Consider a sequence $\{\mathbf{S}_k\}_{k=1}^{\infty}$ in $H^{\otimes 3}$,

$$\mathbf{S}_k := \mathbf{x} \otimes \mathbf{x} \otimes (\mathbf{x} - k\mathbf{y}) + (\mathbf{x} + 1/k\mathbf{y}) \otimes (\mathbf{x} + 1/k\mathbf{y}) \otimes k\mathbf{y} .$$

Clearly, we have that $\operatorname{rank}(\mathbf{S}_k) \leq 2$ for all k. By multilinearity of $\otimes$,

$$\mathbf{S}_k = \mathbf{T} + \frac{1}{k}\mathbf{y} \otimes \mathbf{y} \otimes \mathbf{y}\,.$$

Hence, for any choice of norm on $H \otimes H \otimes H$,

$$\|\mathbf{S}_k - \mathbf{T}\| = \frac{1}{k}\|\mathbf{y} \otimes \mathbf{y} \otimes \mathbf{y}\| \to 0 \quad \text{as } k \to \infty\,,$$

which proves the statement. □

The next example demonstrates the nonconverging rank-2 approximation of the multivariate sum in an arbitrary dimension d.

Example 3.21. Given $\mathbf{x}_\ell \in H_\ell$ and all-ones vector $\mathbf{1}_\ell \in H_\ell$ defining the d-term representation of $\mathbf{x}$ below, the following rank-2 approximation holds:

$$\mathbf{x} := \mathbf{x}_1 \otimes \mathbf{1}_2 \otimes \cdots \otimes \mathbf{1}_d + \cdots + \mathbf{1}_1 \otimes \cdots \otimes \mathbf{1}_{d-1} \otimes \mathbf{x}_d = \lim_{\varepsilon \to 0} \frac{\otimes_{\ell=1}^{d}(\mathbf{1}_\ell + \varepsilon \mathbf{x}_\ell) - \otimes_{\ell=1}^{d}\mathbf{1}_\ell}{\varepsilon}\,.$$

Proof. The result follows by the straightforward calculation of the numerator on the right hand side above. Tensor $\mathbf{x}$ can be proven to have the canonical rank equal to d; see [246]. □

Both examples clearly indicate that the class of rank-r canonical tensors is a nonclosed set in the tensor product space $\mathbb{H}$.

Example 3.21 also applies to the matrix case just by substitution of tensor products of vectors $\mathbf{x}_\ell$, $\mathbf{1}_\ell$ by the Kronecker product of matrices (see examples of d-Laplacian and its inverse in Section 3.1.8 as well as the discussion in Chapter 4).

3.2.6 Strassen algorithm via rank decomposition

Constructive evaluation of the tensor rank can be a useful concept even in the classical linear algebra. As a familiar example, we consider the well known Strassen algorithm ([336]) of fast matrix matrix multiplication in $O(n^{\log_2 7})$ operations based on the canonical rank decomposition of certain tensors.

Note that the $O(n^{2+\varepsilon})$ algorithm to multiply two $n \times n$ matrices gives rise to $O(n^{2+\varepsilon})$ method for solving a system of n linear equations $Ax = b$ [336]. In this way, the principal question is the existence of the algorithm for any $\varepsilon > 0$ (numerical stability is not an issue). Until now, the best known result provides the asymptotic $O(n^{2.376})$; see [74].

Let us look at this problem from the viewpoint of principle of separation of variables. Matrix matrix multiplication can be represented in the block form

$$\begin{bmatrix} C_1 & C_2 \\ C_3 & C_4 \end{bmatrix} = \begin{bmatrix} A_1 & A_2 \\ A_3 & A_4 \end{bmatrix} \cdot \begin{bmatrix} B_1 & B_2 \\ B_3 & B_4 \end{bmatrix}$$

with

$$C_k = \sum_{i=1}^{4}\sum_{j=1}^{4} \gamma_{ijk} A_i B_j\,, \quad k = 1, \dots, 4\,,$$

where for the third order coefficients tensor of size $4 \times 4 \times 4$ we have (slicewise)

$$\Gamma = [\gamma_{ijk}] := \left\{ \begin{bmatrix} 1000 \\ 0010 \\ 0000 \\ 0000 \end{bmatrix}, \begin{bmatrix} 0100 \\ 0001 \\ 0000 \\ 0000 \end{bmatrix}, \begin{bmatrix} 0000 \\ 0000 \\ 1000 \\ 0010 \end{bmatrix}, \begin{bmatrix} 0000 \\ 0000 \\ 0100 \\ 0001 \end{bmatrix} \right\}.$$

Suppose there exists a rank-R expansion (with $R \leq 8$)

$$\gamma_{ijk} = \sum_{t=1}^{R} u_{it} v_{jt} w_{kt} \quad \text{for} \quad i, j, k = 1, \dots, 4\,.$$

Then the following representation holds:

$$C_k = \sum_{t=1}^{R} w_{kt} \sum_{i=1}^{4}\sum_{j=1}^{4} u_{it} A_i v_{jt} B_j = \sum_{t=1}^{R} w_{kt} \left(\sum_{i=1}^{4} u_{it} A_i \right) \left(\sum_{j=1}^{4} v_{jt} B_j \right).$$

If we precompute matrices $\Sigma_t = \sum_{i=1}^{4} u_{it} A_i$ and $\Delta_t = \sum_{j=1}^{4} v_{jt} B_j$, the the initial task will be reduced to R matrix matrix products of size $n/2 \times n/2$. By the construction we have the maximal rank $R \leq 8$ (because there are only eight nonzero entries in tensor Γ), however Strassen's result says that there are representations (infinitely many) of rank 7 that lead to a matrix multiplication algorithm of complexity $O(n^{\log_2 7})$. Indeed, let $n = 2^p$, then

$$n^3 \mapsto R8^{-1} n^3 \mapsto \cdots \mapsto R^p 8^{-p} n^3 = 2^{p \log_2 R} = n^{\log_2 R}\,.$$

There remains still an open problem: Is it possible to construct rank decompositions of Γ with $R < 7$? If yes, then the above result can be improved.

Exercise 3.22. Compute the canonical rank-7 decomposition of $\Gamma = [\gamma_{ijk}]$ by the MATLAB Tensor Toolbox [13]. What can be said about the stability of ALS iteration?

3.2.7 Tucker format: orthogonal subspace representation

The replacement of H_ℓ by subspaces $V_\ell \subset H_\ell$ ($1 \leq \ell \leq d$) leads to the smaller tensor product space (compare to the projection onto the Galerkin subspace)

$$\mathbb{V} = V_1 \otimes V_2 \otimes \cdots \otimes V_d \subset \mathbb{H},$$

which can be used for approximation of elements in $\mathbb{H}$ with a lower cost. Denoting

$$r_\ell := \dim V_\ell < n_\ell = \dim H_\ell$$

and choosing an orthonormal basis $\{\phi^{(\ell)}_{k_\ell} : 1 \le k_\ell \le r_\ell\}$ of V_ℓ, one can represent each $\mathbf{V} \in \mathbb{V}$ by

$$\mathbf{V} = \sum_{\mathbf{k} \in J_1 \times \cdots \times J_d} b_{\mathbf{k}} \phi^{(1)}_{k_1} \otimes \phi^{(2)}_{k_2} \otimes \cdots \otimes \phi^{(d)}_{k_d}, \quad \text{where} \quad b_{\mathbf{k}} \in \mathbb{R}^{J_1 \times \cdots \times J_d}, \tag{3.7}$$

with the multi-index $\mathbf{k} = (k_1, \ldots, k_d)$, such that $J_\ell := \{1, \ldots, r_\ell\}$, $(1 \le \ell \le d)$. On the other hand, each element in H_ℓ can be approximated by elements in V_ℓ by projection onto the orthogonal basis. This is an example of the linear approximation of a tensor in H_ℓ by 'simpler' tensors living in a smaller space.

Let $\mathbf{r} = (r_1, \ldots, r_d) \in \mathbb{N}^d$ be a d-tuple of dimensions. For a fixed rank parameter $\mathbf{r}$, let us consider all possible orthogonal rank-$\mathbf{r}$ representations as in (3.7), where the subspaces V_ℓ are not fixed but only restricted by their dimension r_ℓ. This leads to the so called Tucker tensor format introduced and rigorously analyzed in [248, 350].

Definition 3.23 (Rank-$\mathbf{r}$ Tucker format). Given $\mathbf{r}$, define a set of Tucker tensors in $\mathbb{H}$

$$\mathcal{T}_{\mathbf{r}} := \{\mathbf{V} \in V_1 \otimes V_2 \otimes \cdots \otimes V_d \quad \forall\, V_\ell \subset H_\ell \text{ with } \dim V_\ell = r_\ell,\ \ \ell = 1, \ldots, d\}.$$

Each $\mathbf{V} \in \mathcal{T}_{\mathbf{r}}$ allows the representation (3.7), where $\boldsymbol{\beta} = [b_{\mathbf{k}}] \in \mathbb{R}^{J_1 \times \cdots \times J_d}$ is the core (coefficients) tensor and $\{\phi^{(\ell)}_{k_\ell} : 1 \le k_\ell \le r_\ell\}$ represents the set of orthogonal vectors in H_ℓ.

Denote by $U^{(\ell)} = [\phi^{(\ell)}_1, \ldots, \phi^{(\ell)}_{r_\ell}] \in \mathbb{R}^{n_\ell \times r_\ell}$ the ℓ-mode orthogonal side matrix, i.e., $(U^{(\ell)})^{\mathrm{T}} U^{(\ell)} = \mathbb{1}_{r_\ell \times r_\ell}$. Then we may write $U^{(\ell)} \in \mathbb{S}_{r_\ell} = \mathbb{S}_{n_\ell, r_\ell}$, where $\mathbb{S}_{r_\ell}$ is the so called *Stiefel manifold* of orthogonal $n_\ell \times r_\ell$ matrices. Using the (orthogonal) side matrices $U^{(\ell)} \in \mathbb{S}_{r_\ell}$, one can represent the Tucker decomposition of $\mathbf{V} \in \mathcal{T}_{\mathbf{r}}$ by using tensor-by matrix contracted products,

$$\mathbf{V} = \boldsymbol{\beta} \times_1 U^{(1)} \times_2 U^{(2)} \cdots \times_d U^{(d)}, \quad \boldsymbol{\beta} \in \mathbb{R}^{J_1 \times \cdots \times J_d}.$$

Clearly, the Tucker representation of a given tensor is not unique since it is defined up to rotation of

$$U^{(\ell)} \mapsto U^{(\ell)} S^{(\ell)}, \quad \boldsymbol{\beta} \mapsto \boldsymbol{\beta} \times_1 (S^{(1)})^{\mathrm{T}} \times_2 (S^{(2)})^{\mathrm{T}} \cdots \times_d (S^{(d)})^{\mathrm{T}},$$

where $S^{(\ell)} \in \mathbb{S}_{r_\ell, r_\ell}$.

Figure 3.5 visualizes the Tucker tensor for $d = 3$.

Remark 3.24. In the case $d = 2$, the Tucker representation is a multilinear equivalent of a rank-r matrix factorization, i.e.,

$$\mathbb{R}^{n_1 \times n_2} \ni A = \boldsymbol{\beta} \times_1 U^{(1)} \times_2 U^{(2)} = U^{(1)} \cdot \boldsymbol{\beta} \cdot U^{(2)^{\mathrm{T}}},$$

where $U^{(1)} \in \mathbb{R}^{n_1 \times r}$, $U^{(2)} \in \mathbb{R}^{n_2 \times r}$, and $\boldsymbol{\beta} \in \mathbb{R}^{r \times r}$. The latter can be transformed to the matrix SVD via the diagonalization of the matrix $\boldsymbol{\beta}$.

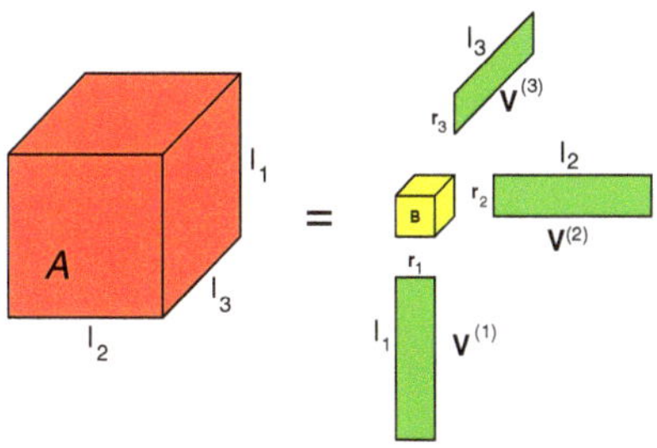

Fig. 3.5: Visualization of the Tucker tensor for $d = 3$.

For ease of exposition, let us set $n = n_\ell$, $(\ell = 1, \ldots, d)$. The storage of the Tucker tensor $\mathbf{V} \in \mathcal{T}_\mathbf{r}$ includes $\prod_{\ell=1}^d r_\ell$ reals to represent the core tensor and the storage of $\sum_{\ell=1}^d r_\ell$ vectors $\phi_k^{(\ell)} \in \mathbb{R}^n$, implying the upper bound

$$\mathrm{Stor}(\mathbf{V}) = r^d + drn\,,$$

where $r = \max r_\ell$ is the maximal Tucker rank. This representation still suffers from the curse of dimensionality, but now in a reduced form especially if r is rather small. The Tucker format appears to be numerically efficient for approximation of function related tensors in moderate dimensions if the rank parameter r_ℓ is noticeably smaller than n_ℓ. Indeed, for the class of tensors generated by the regular enough functions we usually have

$$r = O(\log n)\,,$$

which ensures the approximation error of the order of $O(\mathrm{e}^{-cr})$. The results were proven in [141, 142, 204, 206, 208] for some classes of analytic radial basis functions and Green's kernels; see also the sinc approximation method in Chapter 2, and in Sections 5.1 and 5.2. For example, this is the case for functions arising in 3D Hartree–Fock electronic structure calculations (Section 5.2).

3.2.8 Tucker orthogonality meets the canonical sparsity

In the case of function related tensors obtained by sampling of multivariate functions on a tensor grid of size $n^{\otimes d}$, the univariate grid parameter n may be rather large, say $n \approx 10^4$ (by approximation reasons). This is typically the case for high precision representation of functions with cusps and point singularities arising in computational quantum chemistry. Now the question arises of how to combine both $\mathcal{T}_{\mathbf{r},\mathbf{n}}$ and $\mathcal{C}_R$ to achieve linear complexity scaling in d, n, R.

The main idea is the two-level tensor format introduced in [206, 212], which inherits the Tucker orthogonality in *primal space* (robust decomposition) and the $\mathcal{C}_R$ structure of the core tensor in the *dual coefficients space* (small mode size in the Tucker core, $r \ll n$, say $r = O(\log n)$) implying linear scaling in all representation parameters d, n, R, r.

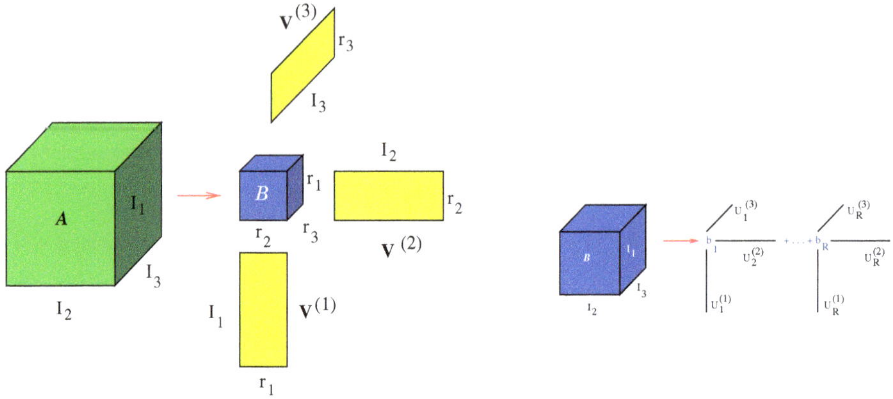

Fig. 3.6: Level I: the Tucker representation (left). Level II: canonical decomposition of $\boldsymbol{\beta}$ (right).

Definition 3.25 (Mixed Tucker-canonical model). Given the rank parameters $\mathbf{r}$, R, define a subclass $\mathcal{T}_{\mathcal{C}_{R,\mathbf{r}}} \subset \mathcal{T}_{\mathbf{r},\mathbf{n}}$ of Tucker tensors with the core tensor in the canonical format $\boldsymbol{\beta} \in \mathcal{C}_{R,\mathbf{r}} \subset \mathbb{R}^{J_1 \times \cdots \times J_d}$,

$$\mathbf{V} = \left(\sum_{\nu=1}^{R} b_\nu \mathbf{u}_\nu^{(1)} \otimes \cdots \otimes \mathbf{u}_\nu^{(d)} \right) \times_1 V^{(1)} \times_2 V^{(2)} \cdots \times_d V^{(d)} .$$

Now the storage cost is estimated by

$$\operatorname{Stor}(\mathbf{V}) = dRr + R + drn ,$$

indicating linear scaling in d, n, R and r. Note that usually $r < R$.

Figure 3.6 illustrates that mixed Tucker-canonical decomposition can be computed in two steps: first by the Tucker approximation of the target tensor and then by canonical representation of the core tensor. The multilinear algebra in the mixed Tucker-canonical format will be addressed in what follows.

Exercise 3.26. Compute the mixed decomposition of a functional tensor for the 3D oscillating function $f_{1,\kappa}$ (2.2.5) on large spacial grids. Mixed decomposition proves to be much faster and more robust than direct computing the canonical approximation of a large initial tensor.

3.2.9 Bilinear operations on formatted tensors

The main advantage of rank structured tensor representations is the opportunity to perform multilinear algebra using operations only on *one dimensional vectors* thus getting rid of the curse of dimensionality. Here we illustrate this point on the example of canonical tensors (for the sake of efficiency the canonical rank is supposed to be of moderate size). Tensor algebra with other formats will be considered in the following sections. Let

$$\mathbf{U} = \sum_{k=1}^{R_1} c_k \mathbf{u}_k^{(1)} \otimes \cdots \otimes \mathbf{u}_k^{(d)} \in \mathbb{H}\,, \quad \mathbf{V} = \sum_{m=1}^{R_2} b_m \mathbf{v}_m^{(1)} \otimes \cdots \otimes \mathbf{v}_m^{(d)} \in \mathbb{H}\,.$$

The Euclidean scalar product can be calculated by

$$\langle \mathbf{U}, \mathbf{V} \rangle := \sum_{k=1}^{R_1} \sum_{m=1}^{R_2} c_k b_m \prod_{\ell=1}^{d} \left\langle \mathbf{u}_k^{(\ell)}, \mathbf{v}_m^{(\ell)} \right\rangle ,$$

at the cost $O(dR_1R_2n) \ll n^d$.

The Hadamard product $\mathbf{C} = [c_{\mathbf{i}}] = \mathbf{U} \odot \mathbf{V}$ of tensors $\mathbf{U}$ and $\mathbf{V}$, defined by $c_{\mathbf{i}} = u_{\mathbf{i}} v_{\mathbf{i}}$, $\mathbf{i} \in \mathcal{I}$, is represented by the canonical tensor

$$\mathbf{U} \odot \mathbf{V} := \sum_{k=1}^{R_1} \sum_{m=1}^{R_2} c_k b_m \left(\mathbf{u}_k^{(1)} \odot \mathbf{v}_m^{(1)} \right) \otimes \cdots \otimes \left(\mathbf{u}_k^{(d)} \odot \mathbf{v}_m^{(d)} \right) \in \mathbb{H}\,,$$

requiring dR_1R_2n reals for storage.

Convolution transform $\mathbf{C} = [c_{\mathbf{j}}] = \mathbf{U} * \mathbf{V}$ of equal size canonical tensors is defined over the large index set as follows:

$$c_{\mathbf{j}} := \sum_{\mathbf{i} \in \mathcal{I}, \mathbf{j}-\mathbf{i}+1 \in \mathcal{I}} u_{\mathbf{i}} v_{\mathbf{j}-\mathbf{i}+1}\,, \quad \text{with} \quad \mathbf{j} \in \{1, \ldots, 2n-1\}^d\,.$$

The unconditional summation can be utilized if one assumes that the convolving tensor $\mathbf{V}$ is extended by zeros beyond the index set $\mathcal{I}$. The resultant tensor has the rank-R_1R_2 representation that includes only univariate convolution transforms

$$\mathbf{U} * \mathbf{V} = \sum_{k=1}^{R_1} \sum_{m=1}^{R_2} c_k b_m \left(\mathbf{u}_m^{(1)} * \mathbf{v}_k^{(1)} \right) \otimes \cdots \otimes \left(\mathbf{u}_m^{(d)} * \mathbf{v}_k^{(d)} \right) .$$

The numerical cost is estimated by $O(dR_1R_2 n \log n) \ll n^d \log n$ since the univariate convolutions can be calculated by 1D FFT. This outperforms the FFT(d) based convolution transform even in low dimensions $d = 2, 3$, if n is large enough. A more detailed discussion of multilinear tensor convolution for classes of function related tensors will be presented in Section 5.1.

Exercise 3.27. Prove that the tensor obtained by sampling the function

$$f(x) = \sin(x_1 + \cdots + x_d)$$

on the tensor grid in $\mathbb{R}^d$ has the maximal Tucker rank equal to 2 (Example 2.16 in Section 2.1.5). Check it by numerical tests.

Exercise 3.28. Estimate by numerical tests the canonical, Tucker, and ℓ-mode ε rank of the Hilbert tensor

$$A = [a_{ijk}]\,, \quad a_{ijk} = 1/(i+j+k)\,, \quad i,j,k = 1,\ldots,n$$

with the mode size $n = 10^2, 10^3$, corresponding to approximation error $\varepsilon = 10^{-3}$, $10^{-4}, 10^{-5}$. Do you observe the exponential convergence in the rank parameter r_ε, i.e., $r_\varepsilon = O(|\log \varepsilon|)$?

In what follows, we discuss the numerical schemes for solving the approximation problems in tensor formats. The main ingredients include the sinc approximation methods, MLA on multidimensional tensors, high order extension(s) of the truncated SVD, and nonlinear ALS type iteration, possibly combined with the multigrid acceleration. The latter gainfully applies to the case of function related tensors when discretizing functions on fine spacial grids [212].

3.3 Direct methods of low rank approximation

In this section, we focus on the direct tensor approximation methods in the CP and Tucker formats in the framework of the multilinear algebra. As was already discussed in Chapter 2, the approximation theory suggests that the tensors related to multivariate functions provide exponential decay of their Tucker decomposition error with respect to the Tucker rank. Here we consider the Tucker tensor decomposition algorithms in view of these properties. In particular, we note that the canonical-to-Tucker decomposition and the corresponding reduced higher order SVD were specifically designed for the rank reduction of function related tensors.

These two basic rank structured tensor formats are of major significance for solving moderate dimensional problems, in particular arising from FEM/FDM approximation to 3D PDEs discretized on large spatial domains. For example, the classical Hartree–Fock integrodifferential equation, arising in electronic structure calculations, is defined in the whole 3D space $\mathbb{R}^3$. The governing Fock operator includes the core potential with many local singularities as well as the integral convolution transform in $\mathbb{R}^3$ with the Newton kernel.

Our experience in the grid based tensor approximation in quantum chemistry shows that the orthogonal Tucker format is probably the most convenient rank structured data format for representation of function related tensors obtained by sampling on large tensor grids. This allows fast and robust numerical schemes for iterative nonlinear approximation of such tensors with controllable accuracy. In the typical case of singular target functions, such approximations cannot be calculated by a kind of polynomial or some other classical interpolation scheme in the problem independent basis sets. On the other hand, the canonical format is very efficient for the direct multidimensional approximation of the radial basis functions and related operators (say,

Green's kernels or elliptic resolvent) in the closed (analytic) form by means of sinc quadratures or sinc interpolation.

3.3.1 On nonlinear approximation by rank structured tensors

The important task in tensor calculations is a construction of efficient and accurate MLA in fixed tensor classes $\mathcal{S}$ getting rid of the curse of dimensionality. This can be formulated as the following approximation problems:

Problem 1. Find the best rank structured approximation of a high order tensor $\mathbf{A} \in \mathbb{V}_\mathbf{n}$ in the fixed nonlinear set $\mathcal{S} \subset \{\mathcal{T}_\mathbf{r}, \mathcal{C}_R, \mathcal{T}_{\mathcal{C}_{R,\mathbf{r}}}\}$.

Problem 2. For a fixed accuracy $\varepsilon > 0$, find efficient approximation of a high order tensor $\mathbf{A} \in \mathbb{V}_\mathbf{n}$ in $\mathcal{S}$ with adaptive rank parameters.

Since both $\mathcal{T}_\mathbf{r}$ and $\mathcal{C}_R$ are not linear spaces, we arrive at a nontrivial nonlinear approximation problem on estimation: Given $\mathbf{A} \in \mathbb{V}_\mathbf{n}$ (or more specifically, $\mathbf{A} \in \mathcal{S}_0 \subset \mathbb{V}_\mathbf{n}$ such that $\mathcal{S} \subset \mathcal{S}_0$), find

$$T_\mathcal{S}(\mathbf{A}) := \underset{\mathbf{X} \in \mathcal{S}}{\operatorname{argmin}} \|\mathbf{X} - \mathbf{A}\|, \quad \text{where} \quad \mathcal{S} \subset \{\mathcal{T}_\mathbf{r}, \mathcal{C}_R, \mathcal{T}_{\mathcal{C}_{R,\mathbf{r}}}\}. \tag{3.8}$$

The theoretical estimate of the approximation rate in (3.8) in terms of rank parameters (which specify the numerical complexity) and constructive approximation to the exact solutions are possible for the class of tensors obtained by sampling of some continuous functions on a tensor grid.

We recall the definition of function related tensors:

Definition 3.29. Given the continuous multivariate function

$$g : \Omega \in \mathbb{R}^d \to \mathbb{R}, \quad \Omega := [-L, L]^d$$

and a set of collocation points $\zeta_\mathbf{i} = (\zeta^1_{i_1}, \ldots, \zeta^d_{i_d})$, $\mathbf{i} \in \mathcal{I}^d$, specified by a tensor grid in Ω. The *function related dth order tensor* is defined by

$$\mathbf{G} \equiv \mathbf{G}(g) := [g_{i_1 \ldots i_d}] \in \mathbb{R}^{\mathcal{I}^d} \quad \text{with} \quad g_{i_1 \ldots i_d} := g(\zeta^1_{i_1}, \ldots, \zeta^d_{i_d}).$$

In the case $\mathcal{S} = \mathcal{C}_R$, the estimation of a minimum in (3.8) might be a difficult problem. For the particular illustration, we recall that the canonical d-term decomposition of a function related tensor $\mathbf{F}$ for d-variate function

$$f(x) := \sin\left(\sum_{j=1}^{d} x_j\right),$$

holds for the continuum of parameters $\alpha_k \in \mathbb{R}$, such that $\sin(\alpha_k - \alpha_j) \neq 0$ for all $j \neq k$; see (2.11) in Proposition 2.19. Representation (2.11) shows the lack of uniqueness (ambiguity) of the best rank-d tensor representation. For this example, the convergence of minimization schemes in $\mathcal{C}_R$ might be nonrobust (multiple local minima), especially for approximation with the canonical rank smaller than d.

Methods of low rank formatted approximation to higher order tensors can be classified to the direct representation, iterative nonlinear optimization, and combined approaches. The combined algorithms can be considered as the iterative nonlinear optimization with the initial guess obtained by the quasioptimal direct approximation. In turn, the direct methods can be based either on algebraic or analytic treatment of a tensor.

The direct algebraic low rank approximation in the Tucker format is based on an extension of matrix SVD to the so called higher order SVD (HOSVD), introduced in [247]. In this way, the truncated HOSVD allows the quasioptimal Tucker approximation of the full format tensors of moderate size representing multidimensional data arrays of general content.

Alternatively, the direct Tucker approximation to function related tensors can be calculated by the tensor product polynomial interpolation (Section 2.2).

In the case of a canonical target tensor with possibly large initial rank and large tensor size, we describe the algebraic canonical-to-Tucker (C2T) transform by the reduced higher order SVD (RHOSVD) ([212]) and prove the asymptotically optimal error bound. We note that such tensors may be generated, for example, by the analytic methods of canonical approximation based on sinc quadrature techniques. They may apply to different classes of tensors generated by functions like spherically symmetric or radial basis functions, Green's kernels, etc. (see Section 2.4 for more details).

Note that HOSVD requires the full size tensors, which is not feasible for numerical modeling in physics, quantum chemistry, and in multidimensional scientific computing. Hence, the HOSVD approach does not resolve the curse of dimensionality and actually it has very limited significance in computational practice. In contrast, RHOSVD approximation does not require the full format tensor and it applies to the canonical skeleton vectors only, which leads to the linearly complexity scaling in the dimension. This approach can be used for the computation of the Tucker, TT, and other MPS type tensor approximations.

When using tensor numerical methods, the calculation of multidimensional convolution operators is reduced to a combination of one dimensional Hadamard products and convolutions. Canonical tensors of large ranks and on large grids may occur during sequences of such tensor operations, then the ranks of tensors are multiplied. Hence, C2T and T2C transforms can be used for reducing the ranks of tensors in the course of the solution process.

A low rank approximation can be applied to both full format and already formatted input, that means $\mathbf{A} \in \mathcal{S}_0 \subset \mathbb{V}_\mathbf{n}$, where $\mathcal{S}_0$ is a set of rank structured tensors with larger rank parameters. We overview basic tensor transforms (approximations) acting between different formats as follows:
- Full-to-Tucker (by HOSVD),
- Canonical-to-Tucker (by RHOSVD),

- Tucker-to-Tucker (by SVD of directional matrices),
- Tucker-to-canonical (by CP approximation of the core tensor).

These operations can be understood as the conversion of tensors represented in different formats to each other.

In what follows, we also consider the relations between the canonical, Tucker, and the MPS/TT representations.

3.3.2 Higher order SVD (HOSVD)

The natural extension of the matrix SVD decomposition to dth order tensors is called the higher order SVD (HOSVD) or dth order SVD.

Theorem 3.30 (dth order SVD, [247]). *Every complex $n_1 \times n_2 \times \cdots \times n_d$-tensor $\mathbf{A}$ can be written as the contracted product*

$$\mathbf{A} = \mathbf{S} \times_1 U^{(1)} \times_2 U^{(2)} \cdots \times_d U^{(d)}\,, \quad \text{where}: \tag{3.9}$$

(1) $U^{(\ell)} = [\mathbf{u}_1^{(\ell)} \mathbf{u}_2^{(\ell)} \ldots \mathbf{u}_{n_\ell}^{(\ell)}]$ *is a unitary $n_\ell \times n_\ell$ matrix, i.e., $(U^{(\ell)})^{\mathrm{T}} U^{(\ell)} = I_{n_\ell}$ for $\ell = 1, \ldots, d$.*

(2) *$\mathbf{S}$ is a complex $n_1 \times n_2 \times \cdots \times n_d$-tensor of which the subtensors $\mathbf{S}_{i_\ell=\alpha}$, obtained by fixing the ℓth index to α, have the properties of*
   (i) *All-orthogonality: two subtensors $\mathbf{S}_{i_\ell=\alpha}$ and $\mathbf{S}_{i_\ell=\beta}$ are orthogonal for all possible values of ℓ, α, and β such that for $\alpha \neq \beta$,*

$$\langle \mathbf{S}_{i_\ell=\alpha}\,, \mathbf{S}_{i_\ell=\beta} \rangle = 0\,;$$

   (ii) *Ordering:*

$$\|\mathbf{S}_{i_\ell=1}\| \geq \|\mathbf{S}_{i_\ell=2}\| \geq \cdots \geq \|\mathbf{S}_{i_\ell=n_\ell}\| \geq 0\,, \quad \forall\, \ell = 1, \ldots, d\,.$$

   (iii) *The Frobenius norms $\sigma_i^{(\ell)} = \|\mathbf{S}_{i_\ell=i}\|$, are ℓ-mode singular values of the matrix unfolding $A_{(\ell)}$ and the vector $\mathbf{u}_i^{(\ell)} \in \mathbb{R}^{n_\ell}$, $i = 1, \ldots, n_\ell$, is an ith ℓ-mode left singular vector of $A_{(\ell)}$.*

Proof. Here, we only give a sketch of the proof for the main issues (i) and (ii) (see [247] for the detailed analysis). Rewrite the matrix representations

$$A_{(\ell)} = U^{(\ell)}\, \Sigma^{(\ell)}\, V^{(\ell)^{\mathrm{T}}}\,, \quad \Sigma^{(\ell)} = \operatorname{diag}\{\sigma_1^{(\ell)}, \ldots, \sigma_{n_\ell}^{(\ell)}\}$$

in the form

$$A_{(\ell)} = U^{(\ell)}\, S_{(\ell)}\, [U^{(1)} \otimes \cdots \otimes U^{(\ell-1)} \otimes U^{(\ell+1)} \otimes \cdots \otimes U^{(d)}]^{\mathrm{T}}\,,$$

where

$$S_{(\ell)} = \Sigma^{(\ell)}\, V^{(\ell)^{\mathrm{T}}}\, [U^{(1)} \otimes \cdots \otimes U^{(\ell-1)} \otimes U^{(\ell+1)} \otimes \cdots \otimes U^{(d)}]\,,$$

and where ⊗ is the Kronecker product of matrices (see Section 3.4 for the definition). Now the representation for the unfolding matrix $S_{(\ell)}$ implies (i) and (ii) thanks to orthogonality of matrices $U^{(\ell)}$. □

Computation of HOSVD that transforms the full format tensor to the Tucker one can be implemented in two steps:

(a) Compute $U^{(\ell)}$ ($\ell = 1, \ldots, d$) in the form of left singular matrix of $A_{(\ell)}$.

(b) Compute the core tensor $\mathbf{S}$ by bringing the matrices of singular vectors to the left hand side in (3.9):

$$\mathbf{S} = \mathbf{A} \times_1 U^{(1)^{\mathrm{T}}} \times_2 U^{(2)^{\mathrm{T}}} \cdots \times_d U^{(d)^{\mathrm{T}}} .$$

The method of computing the fixed rank Tucker approximation to the full format tensor in many dimensions, called the truncated HOSVD, is an extension of the truncated SVD of matrices to dth order tensors. The next theorem provides the error bound for the truncated HOSVD.

Theorem 3.31 (Approximation by HOSVD, [247]). *Let the HOSVD of* $\mathbf{A}$ *be given as in Theorem 3.30 and the* ℓ*-mode rank of* $\mathbf{A}$ *be equal to* R_ℓ ($\ell = 1, \ldots, d$). *For a given rank parameter* $\mathbf{r} = (r_1, \ldots, r_d)$, *define a tensor* $\tilde{\mathbf{A}}_{\mathbf{r}}$ *by discarding the smallest* ℓ*-mode singular values* $\sigma^{(\ell)}_{r_\ell+1}, \sigma^{(\ell)}_{r_\ell+2}, \ldots, \sigma^{(\ell)}_{R_\ell}$ ($\ell = 1, \ldots, d$), *i.e., by setting the corresponding parts of* $\mathbf{S}$ *equal to zero. Then we have*

$$\|\mathbf{A} - \tilde{\mathbf{A}}_{\mathbf{r}}\|^2 \leq \sum_{\ell=1}^{d} \sum_{i_\ell=r_\ell+1}^{R_\ell} [\sigma^{(\ell)}_{i_\ell}]^2 .$$

Proof. Taking into account the orthogonality of $U^{(\ell)}$, $\ell = 1, \ldots, d$, Theorem 3.30 implies

$$\begin{aligned}
\|\mathbf{A} - \tilde{\mathbf{A}}_{\mathbf{r}}\|^2 &= \sum_{i_1=1}^{R_1} \sum_{i_2=1}^{R_2} \cdots \sum_{i_d=1}^{R_d} s^2_{i_1 i_2 \ldots i_d} - \sum_{i_1=1}^{r_1} \sum_{i_2=1}^{r_2} \cdots \sum_{i_d=1}^{r_d} s^2_{i_1 i_2 \ldots i_d} \\
&\leq \sum_{i_1=r_1+1}^{R_1} \sum_{i_2=1}^{R_2} \cdots \sum_{i_d=1}^{R_d} s^2_{i_1 i_2 \ldots i_d} + \sum_{i_1=1}^{R_1} \sum_{i_2=r_2+1}^{R_2} \cdots \sum_{i_d=1}^{R_d} s^2_{i_1 i_2 \ldots i_d} \\
&\quad + \cdots + \sum_{i_1=1}^{R_1} \sum_{i_2=1}^{R_2} \cdots \sum_{i_d=r_d+1}^{R_d} s^2_{i_1 i_2 \ldots i_d} \\
&= \sum_{i_1=r_1+1}^{R_1} [\sigma^{(1)}_{i_1}]^2 + \sum_{i_2=r_2+1}^{R_2} [\sigma^{(2)}_{i_2}]^2 + \cdots + \sum_{i_d=r_d+1}^{R_d} [\sigma^{(d)}_{i_d}]^2 ,
\end{aligned}$$

which completes the proof. □

Note that the approximant obtained by truncated HOSVD gets lost only in the factor $\sqrt{d}$ compared with 'best' rank-$\mathbf{r}$ approximation. Though in general the tensor $\tilde{\mathbf{A}}_{\mathbf{r}}$ is not the best approximation of $\mathbf{A}$ under the given ℓ-mode rank constraints, it normally provides a good Tucker approximation to $\mathbf{A}$ that can be used as the initial guess for

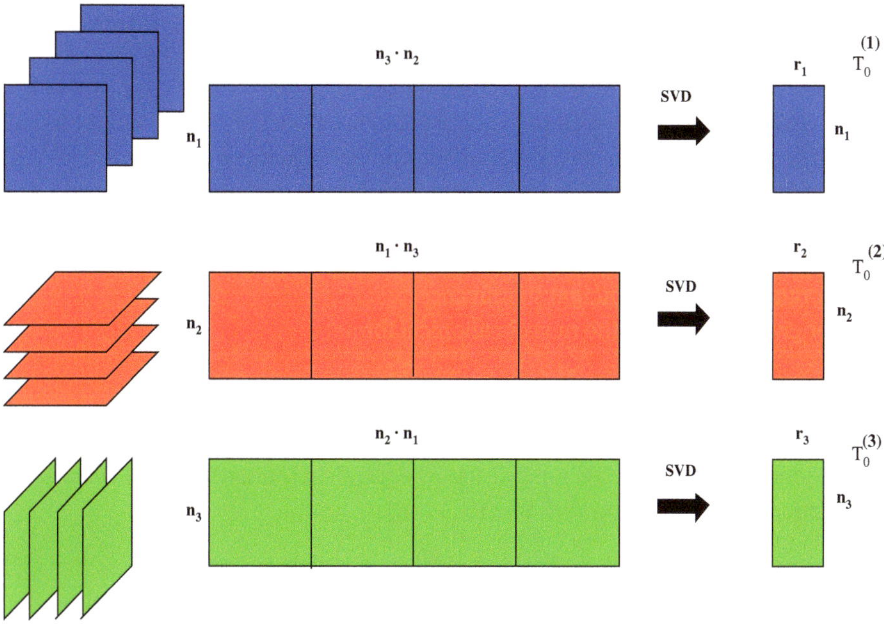

Fig. 3.7: $d = 3$: Approximation by HOSVD via truncated SVD of $A_{(\ell)}$ for $r_\ell < n_\ell$.

further iterative corrections; see Section 3.5. Figure 3.7 illustrates the scheme of the truncated HOSVD approximation to a third order tensor.

Remark 3.32. Practical application of the truncated HOSVD is limited to small dimensions d and moderate mode size n due to hard complexity scaling $O(n^{d+1})$. Therefore, the truncated HOSVD does not apply to functions arising in the solution of multidimensional PDEs.

Exercise 3.33. Compare the rank-$\mathbf{r}$ truncated HOSVD of $n \times n \times n$ Hilbert tensor (Example 3.28) for $d = 3, n = 100$ and $d = 4, n = 50$, with the 'best' rank-$\mathbf{r}$ Tucker approximation.

3.3.3 Reduced HOSVD and the canonical-to-Tucker transform

In computational schemes including bilinear tensor-tensor or matrix-tensor operations the increase of tensor ranks leads to the critical loss of efficiency. Moreover, in many applications, for example in electronic structure calculations, the canonical tensors with large rank parameters arise as the result of polynomial type or convolution transforms of some function related tensors (say, electron density, the Hartree potential, etc.). In what follows, we present the direct method of rank reduction for the canonical tensors with large initial rank, the reduced HOSVD, introduced and analyzed in [212].

The basic idea of the reduced HOSVD is that for large (function related) tensors given in the canonical format their HOSVD does not require the construction of a tensor in the full format and SVD based computation of its matrix unfolding. Instead, it is sufficient to compute the SVD of the directional matrices $U^{(\ell)}$ in (3.11) composed by only skeleton vectors of the canonical tensor in every dimension separately. This will provide the initial guess for the Tucker orthogonal basis in the given dimension. For the practical applicability, the results of the approximation theory on the low rank approximation to the multivariate functions, exhibiting exponential error decay in the Tucker rank, are of the principal significance.

For given $\mathbf{A} \in \mathcal{C}_{R,\mathbf{n}}$ in the rank-R canonical format

$$\mathbf{A} = \sum_{\nu=1}^{R} \xi_\nu \mathbf{u}_\nu^{(1)} \otimes \cdots \otimes \mathbf{u}_\nu^{(d)}, \quad \xi_\nu \in \mathbb{R}, \tag{3.10}$$

with normalized canonical vectors, we use its equivalent (nonorthogonal) Tucker representation via construction with $\mathbf{r} = (R, \dots, R)$,

$$\mathbf{A} = \boldsymbol{\xi} \times_1 U^{(1)} \times_2 U^{(2)} \cdots \times_d U^{(d)}, \quad \boldsymbol{\xi} = \operatorname{diag}\{\xi_1, \dots, \xi_R\}, \tag{3.11}$$

via ℓ-mode side matrices $U^{(\ell)} = [\mathbf{u}_1^{(\ell)} \dots \mathbf{u}_R^{(\ell)}] \in \mathbb{R}^{n\times R}$, where conventionally $\|\mathbf{u}_\nu^{(\ell)}\| = 1$ for $\ell = 1, \dots, d$, $\nu = 1, \dots, R$.

How can we simplify the HOSVD Tucker approximation in the case of canonical input? The problem can be solved by the method of reduced HOSVD (RHOSVD) proposed in [212].

To fix the idea, suppose that $n \le R$ and let $U^{(\ell)} \approx W^{(\ell)} := Z_0^{(\ell)} D_{\ell,0} V_0^{(\ell)^{\mathrm{T}}}$ be the truncated SVD of the side matrix $U^{(\ell)}$ ($\ell = 1, \dots, d$), where for fixed Tucker rank parameters $\mathbf{r} = (r_1, \dots, r_d)$,

$$D_{\ell,0} = \operatorname{diag}\{\sigma_{\ell,1}, \sigma_{\ell,2}, \dots, \sigma_{\ell,r_\ell}\} \text{ and } Z_0^{(\ell)} = [\mathbf{z}_1^{(\ell)}, \dots, \mathbf{z}_{r_\ell}^{(\ell)}] \in \mathbb{R}^{n\times r_\ell},\ V_0{}^{(\ell)} \in \mathbb{R}^{R\times r_\ell}$$

represent the respective orthogonal factors so that the complete SVD of the side matrix $U^{(\ell)}$ is given by

$$U^{(\ell)} = Z^{(\ell)} D_\ell W^{(\ell)^{\mathrm{T}}} = \sum_{k=1}^{n} \sigma_{\ell,k} \mathbf{z}_k^{(\ell)} \mathbf{w}_k^{(\ell)^{\mathrm{T}}}, \quad \mathbf{z}_k^{(\ell)} \in \mathbb{R}^n, \quad \mathbf{w}_k^{(\ell)} \in \mathbb{R}^R,$$

with the orthogonal matrices $Z^{(\ell)} = [\mathbf{z}_1^{(\ell)}, \dots, \mathbf{z}_n^{(\ell)}]$, and $W^{(\ell)} = [\mathbf{w}_1^{(\ell)}, \dots, \mathbf{w}_n^{(\ell)}]$, $\ell = 1, \dots, d$. We use the following notations for the vector entries: $w_k^{(\ell)}(\nu) = w_{k,\nu}^{(\ell)}$ ($\nu = 1, \dots, R$).

Definition 3.34 (Reduced HOSVD, [212]). For given $\mathbf{A} \in \mathcal{C}_{R,\mathbf{n}}$ and the truncation rank $\mathbf{r}$, ($r_\ell \le R$), the RHOSVD approximation of $\mathbf{A}$ is defined by the rank-$\mathbf{r}$ Tucker tensor

$$\mathbf{A}^0_{(\mathbf{r})} := \boldsymbol{\xi} \times_1 \left[Z_0^{(1)} D_{1,0} V_0^{(1)^{\mathrm{T}}}\right] \times_2 \cdots \times_d \left[Z_0^{(d)} D_{d,0} V_0^{(d)^{\mathrm{T}}}\right] \in \mathcal{T}_{\mathbf{r}}, \tag{3.12}$$

obtained by the projection of canonical side matrices $U^{(\ell)}$ onto the left singular matrices $Z_0^{(\ell)}$.

The following theorem proves the error bound for the RHOSVD approximation:

Theorem 3.35 (RHOSVD error, [212]). *For given $\mathbf{A} \in \mathcal{C}_{R,\mathbf{n}}$ in (3.10), let $\sigma_{\ell,1} \geq \sigma_{\ell,2} \cdots \geq \sigma_{\ell,\min(n,R)}$ be the singular values of ℓ-mode side matrices $U^{(\ell)} \in \mathbb{R}^{n\times R}$ $(\ell = 1,\ldots,d)$ with normalized canonical vectors. Then the error of RHOSVD approximation, $\mathbf{A}^0_{(\mathbf{r})}$, is bounded by*

$$\|\mathbf{A} - \mathbf{A}^0_{(\mathbf{r})}\| \leq \|\boldsymbol{\xi}\| \sum_{\ell=1}^{d} \left(\sum_{k=r_\ell+1}^{\min(n,R)} \sigma^2_{\ell,k} \right)^{1/2}, \quad \|\boldsymbol{\xi}\| = \sqrt{\sum_{\nu=1}^{R} \xi_\nu^2}\,. \tag{3.13}$$

Proof. Using the contracted product representations of $\mathbf{A} \in \mathcal{C}_{R,\mathbf{n}}$ and $\mathbf{A}^0_{(\mathbf{r})} \in \mathcal{T}_{\mathbf{r}}$, leads to the following expansion for the approximation error:

$$\begin{aligned}
\mathbf{A} - \mathbf{A}^0_{(\mathbf{r})} &= \boldsymbol{\xi} \times_1 U^{(1)} \times_2 U^{(2)} \cdots \times_d U^{(d)} \\
&\quad - \boldsymbol{\xi} \times_1 \left[Z_0^{(1)} D_{1,0} V_0^{(1)^{\mathrm{T}}} \right] \times_2 \left[Z_0^{(2)} D_{2,0} V_0^{(2)^{\mathrm{T}}} \right] \cdots \times_d \left[Z_0^{(d)} D_{d,0} V_0^{(d)^{\mathrm{T}}} \right] \\
&= \boldsymbol{\xi} \times_1 \left[U^{(1)} - Z_0^{(1)} D_{1,0} V_0^{(1)^{\mathrm{T}}} \right] \times_2 \left[Z_0^{(2)} D_{2,0} V_0^{(2)^{\mathrm{T}}} \right] \cdots \times_d \left[Z_0^{(d)} D_{d,0} V_0^{(d)^{\mathrm{T}}} \right] \\
&\quad + \boldsymbol{\xi} \times_1 U^{(1)} \times_2 \left[U^{(2)} - Z_0^{(2)} D_{2,0} V_0^{(2)^{\mathrm{T}}} \right] \cdots \times_d \left[Z_0^{(d)} D_{d,0} V_0^{(d)^{\mathrm{T}}} \right] + \ldots \\
&\quad + \boldsymbol{\xi} \times_1 U^{(1)} \times_2 U^{(2)} \cdots \times_d \left[U^{(d)} - Z_0^{(d)} D_{d,0} V_0^{(d)^{\mathrm{T}}} \right].
\end{aligned}$$

Introducing the ℓ-mode residual

$$\Delta^{(\ell)} = U^{(\ell)} - Z_0^{(\ell)} D_{\ell,0} V_0^{(\ell)^{\mathrm{T}}}, \quad \{\Delta^{(\ell)}\}_\nu = \sum_{k=r_\ell+1}^{n} \sigma_{\ell,k} \mathbf{z}_k^{(\ell)} v_{k,\nu}^{\ell}, \quad \nu = 1,\ldots,R,$$

with notations

$$V_0^{(\ell)} = [\mathbf{v}_1^{(\ell)},\ldots,\mathbf{v}_{r_\ell}^{(\ell)}]^{\mathrm{T}}, \quad \mathbf{v}_k^{(\ell)} = \{v_{k,\nu}^{\ell}\}_{\nu=1}^{R} \in \mathbb{R}^R,$$

we then represent the ℓth summand in the right hand side of $\|\mathbf{A} - \mathbf{A}^0_{(\mathbf{r})}\|$ in the form

$$\mathbf{B}_\ell = \boldsymbol{\xi} \times_1 U^{(1)} \cdots \times_{\ell-1} U^{(\ell-1)} \times_\ell \Delta^{(\ell)} \times_{\ell+1} W^{(\ell+1)} \cdots \times_d W^{(d)}.$$

This leads to the error bound (by the triangle inequality)

$$\begin{aligned}
\|\mathbf{A} - \mathbf{A}^0_{(\mathbf{r})}\| \leq \sum_{\ell=1}^{d} \|\mathbf{B}_\ell\| &= \|\boldsymbol{\xi} \times_1 \Delta^{(1)} \times_2 W^{(2)} \cdots \times_d W^{(d)}\| \\
&\quad + \|\boldsymbol{\xi} \times_1 U^{(1)} \times_2 \Delta^{(2)} \cdots \times_d W^{(d)}\| + \ldots \\
&\quad + \|\boldsymbol{\xi} \times_1 U^{(1)} \times_2 U^{(2)} \cdots \times_d \Delta^{(d)}\|,
\end{aligned}$$

where the ℓth term $\mathbf{B}_\ell$ is represented by

$$\sum_{\nu=1}^{R} \xi_\nu \left[\mathbf{u}_\nu^{(\ell)} \cdots \times_{\ell-1} \mathbf{u}_\nu^{(\ell-1)} \times_\ell \{\Delta^{(\ell)}\}_\nu \times_{\ell+1} \sum_{k=1}^{r_{\ell+1}} \sigma_{\ell+1,k} \mathbf{z}_k^{(\ell+1)} v_{k,\nu}^{\ell+1} \times_{\ell+2} \cdots \times_d \sum_{k=1}^{r_d} \sigma_{d,k} \mathbf{z}_k^{(d)} v_{k,\nu}^{d} \right],$$

providing the estimate (in view of $\|\mathbf{u}_\nu^{(\ell)}\| = 1$, $\ell = 1, \dots, d$, $\nu = 1, \dots, R$)

$$\|\mathbf{B}_\ell\| \le \sum_{\nu=1}^{R} |\xi_\nu| \left(\sum_{k=r_\ell+1}^{n} \sigma_{\ell,k}^2 (v_{k,\nu}^\ell)^2 \right)^{1/2} \cdot \left(\sum_{k=1}^{r_{\ell+1}} \sigma_{\ell+1,k}^2 (v_{k,\nu}^{\ell+1})^2 \right)^{1/2} \cdots \left(\sum_{k=1}^{r_d} \sigma_{d,k}^2 (v_{k,\nu}^d)^2 \right)^{1/2} .$$

Furthermore, since $U^{(\ell)}$ has normalized columns, i.e.,

$$1 = \|\mathbf{u}_\nu^{(\ell)}\| = \left\| \sum_{k=1}^{n} \sigma_{\ell,k} \mathbf{z}_k^{(\ell)} v_{k,\nu}^\ell \right\| , \quad \ell = 1, \dots, d ,$$

we obtain $\sum_{k=1}^n \sigma_{\ell,k}^2 (v_{k,\nu}^\ell)^2 = 1$ for $\ell = 1, \dots, d$ $\nu = 1, \dots, R$. Now the error estimate can be finalized:

$$\begin{aligned}\|\mathbf{A} - \mathbf{A}_{(\mathbf{r})}^0\| &\le \sum_{\ell=1}^{d} \sum_{\nu=1}^{R} |\xi_\nu| \left(\sum_{k=r_\ell+1}^{n} \sigma_{\ell,k}^2 (v_{k,\nu}^\ell)^2 \right)^{1/2} \\ &\le \sum_{\ell=1}^{d} \left(\sum_{\nu=1}^{R} \xi_\nu^2 \right)^{1/2} \left(\sum_{\nu=1}^{R} \sum_{k=r_\ell+1}^{n} \sigma_{\ell,k}^2 (v_{k,\nu}^\ell)^2 \right)^{1/2} \\ &= \sum_{\ell=1}^{d} \|\boldsymbol{\xi}\| \left(\sum_{k=r_\ell+1}^{n} \sigma_{\ell,k}^2 \sum_{\nu=1}^{R} (v_{k,\nu}^\ell)^2 \right)^{1/2} = \|\boldsymbol{\xi}\| \sum_{\ell=1}^{d} \left(\sum_{k=r_\ell+1}^{n} \sigma_{\ell,k}^2 \right)^{1/2} .\end{aligned}$$

The case $R < n$ can be analyzed along the same line. □

The error estimate in Theorem 3.35 differs from the case of complete HOSVD by the extra factor $\|\boldsymbol{\xi}\|$ (compared to HOSVD), which is the payoff for the lack of orthogonality.

The result of Theorem 3.35 does not provide, in general, the stable control of relative error since for the general canonical tensors there is no upper bound on the constant C in the estimate

$$\|\boldsymbol{\xi}\| \le C\|\mathbf{A}\| .$$

The problem is that it applies to the nonorthogonal canonical decomposition. The stable decomposition can be proven in the case of partially orthogonal or monotone decompositions; see [188].

Corollary 3.36. *Assume the conditions of Theorem 3.35 are satisfied. (a) Suppose that one of the matrices $U^{(\ell)}$, say $U^{(1)}$, is orthogonal. Then the RHOSVD error can be bounded by*

$$\|\mathbf{A} - \mathbf{A}_{(\mathbf{r})}^0\| \le C\|\mathbf{A}\| \sum_{\ell=1}^{d} \left(\sum_{k=r_\ell+1}^{\min(n,R)} \sigma_{\ell,k}^2 \right)^{1/2} , \tag{3.14}$$

with the constant $C = 1$. (b) Let decomposition (3.10) be monotone, i.e., all coefficients and skeleton vectors have nonnegative values. Then (3.14) holds with the constant C that does not depend on $\mathbf{A}$.

Proof. The partial orthogonality assumption combined with normalization constraints imply

$$\|\boldsymbol{\xi}\|^2 = \sum_{\nu=1}^{R} \xi_\nu^2 = \|\mathbf{A}\|^2 ,$$

then the result follows by (3.13). We refer to [188] where the case of monotone canonical sums was discussed in detail. □

Clearly, the orthogonality assumption may lead to slightly higher separation rank, however this constructive decomposition can stabilize the approximation methods in the canonical format. Some very special cases of the orthogonal canonical tensor decomposition were considered in [232]. This issue will be discussed in Section 3.5 in more detail.

The case of monotone canonical sums typically arises in the sinc based canonical approximation to Green's kernels by a sum of Gaussians; see Section 5.2.

3.3.4 Other direct methods of approximation and general overview

In this section, we consider the situation opposite to the canonical-to-Tucker approximation. The next lemma describes the approximation of the Tucker tensor by using canonical representation [211].

Lemma 3.37 (Mixed Tucker-to-canonical approximation).

(A) *Let the target tensor* $\mathbf{A}$ *have the form*

$$\mathbf{A} = \boldsymbol{\beta} \times_1 V^{(1)} \times_2 \cdots \times_d V^{(d)} \in \mathcal{T}_{\mathbf{r},\mathbf{n}} ,$$

with the orthogonal side matrices $V^{(\ell)} = [\mathbf{v}_1^{(\ell)} \dots \mathbf{v}_{r_\ell}^{(\ell)}] \in \mathbb{R}^{n \times r_\ell}$ *and* $\boldsymbol{\beta} \in \mathbb{R}^{r_1 \times \cdots \times r_d}$. *Then, for a given* $R \leq \min_{1\leq\ell\leq d} \overline{r}_\ell$,

$$\min_{\mathbf{Z}\in\mathcal{C}_{R,\mathbf{n}}} \|\mathbf{A} - \mathbf{Z}\| = \min_{\boldsymbol{\mu}\in\mathcal{C}_{R,\mathbf{r}}} \|\boldsymbol{\beta} - \boldsymbol{\mu}\| . \tag{3.15}$$

(B) *Assume that there exists the best rank-R approximation* $\mathbf{A}_{(R)} \in \mathcal{C}_{R,\mathbf{n}}$ *of* $\mathbf{A}$, *then there is the best rank-R approximation* $\boldsymbol{\beta}_{(R)} \in \mathcal{C}_{R,\mathbf{r}}$ *of* $\boldsymbol{\beta}$, *such that*

$$\mathbf{A}_{(R)} = \boldsymbol{\beta}_{(R)} \times_1 V^{(1)} \times_2 \cdots \times_d V^{(d)} . \tag{3.16}$$

Proof.

(A) Note that the canonical vectors $\mathbf{y}_k^{(\ell)}$ of any test element in the left hand side of (3.15),

$$\mathbf{Z} = \sum_{k=1}^{R} \lambda_k \, \mathbf{y}_k^{(1)} \otimes \cdots \otimes \mathbf{y}_k^{(d)} \in \mathcal{C}_{R,\mathbf{n}} , \tag{3.17}$$

can be chosen in $\operatorname{span}\{\mathbf{v}_1^{(\ell)}, \dots, \mathbf{v}_{r_\ell}^{(\ell)}\}$, that means

$$\mathbf{y}_k^{(\ell)} = \sum_{m=1}^{r_\ell} \mu_{k,m}^{(\ell)} \mathbf{v}_m^{(\ell)} , \quad k = 1, \dots, R, \ \ell = 1, \dots, d . \tag{3.18}$$

Indeed, assuming

$$\mathbf{y}_k^{(\ell)} = \sum_{m=1}^{r_\ell} \mu_{k,m}^{(\ell)} \mathbf{v}_m^{(\ell)} + \mathbf{e}_k^{(\ell)} \quad \text{with} \quad \mathbf{e}_k^{(\ell)} \perp \operatorname{span}\{\mathbf{v}_1^{(\ell)}, \ldots, \mathbf{v}_{r_\ell}^{(\ell)}\},$$

we conclude that $\mathbf{e}_k^{(\ell)}$ does not effect the cost function in (3.15) because of the orthogonality of $V^{(\ell)}$. Hence, setting $\mathbf{e}_k^{(\ell)} = 0$, and plugging (3.18) into (3.17), we arrive at the desired Tucker decomposition of $\mathbf{Z}$,

$$\mathbf{Z} = \boldsymbol{\beta}_z \times_1 V^{(1)} \times_2 \cdots \times_d V^{(d)}, \quad \boldsymbol{\beta}_z \in \mathcal{C}_{R,\mathbf{r}}.$$

This implies

$$\|\mathbf{A} - \mathbf{Z}\|^2 = \|(\boldsymbol{\beta}_z - \boldsymbol{\beta}) \times_1 V^{(1)} \times_2 \cdots \times_d V^{(d)}\|^2 = \|\boldsymbol{\beta} - \boldsymbol{\beta}_z\|^2 \geq \min_{\boldsymbol{\mu} \in \mathcal{C}_{R,\mathbf{r}}} \|\boldsymbol{\beta} - \boldsymbol{\mu}\|^2.$$

On the other hand, we have

$$\min_{\mathbf{Z} \in \mathcal{C}_{R,\mathbf{n}}} \|\mathbf{A} - \mathbf{Z}\|^2 \leq \min_{\boldsymbol{\beta}_z \in \mathcal{C}_{R,\mathbf{r}}} \|(\boldsymbol{\beta} - \boldsymbol{\beta}_z) \times_1 V^{(1)} \times_2 \cdots \times_d V^{(d)}\|^2 = \min_{\boldsymbol{\mu} \in \mathcal{C}_{R,\mathbf{r}}} \|\boldsymbol{\beta} - \boldsymbol{\mu}\|^2.$$

This proves (3.15).

(B) Likewise, for any minimizer $\mathbf{A}_{(R)} \in \mathcal{C}_{R,\mathbf{n}}$ in the right hand side of (3.15), one obtains

$$\mathbf{A}_{(R)} = \boldsymbol{\beta}_{(R)} \times_1 V^{(1)} \times_2 V^{(2)} \cdots \times_d V^{(d)}$$

with the respective rank-R core tensor

$$\boldsymbol{\beta}_{(R)} = \sum_{k=1}^{R} \lambda_k \mathbf{u}_k^{(1)} \otimes \cdots \otimes \mathbf{u}_k^{(d)} \in \mathcal{C}_{R,\mathbf{r}}.$$

Here $\mathbf{u}_k^{(\ell)} = \{\mu_{k,m_\ell}^{(\ell)}\}_{m_\ell=1}^{r_\ell} \in \mathbb{R}^{r_\ell}$ are calculated by plugging representation (3.18) into (3.17), and then by changing the order of summation,

$$\begin{aligned}
\mathbf{A}_{(R)} &= \sum_{k=1}^{R} \lambda_k \mathbf{y}_k^{(1)} \otimes \cdots \otimes \mathbf{y}_k^{(d)} \\
&= \sum_{k=1}^{R} \lambda_k \left(\sum_{m_1=1}^{r_1} \mu_{k,m_1}^{(1)} \mathbf{v}_{m_1}^{(1)} \right) \otimes \cdots \otimes \left(\sum_{m_d=1}^{r_d} \mu_{k,m_d}^{(d)} \mathbf{v}_{m_d}^{(d)} \right) \\
&= \sum_{m_1=1}^{r_1} \cdots \sum_{m_d=1}^{r_d} \left\{ \sum_{k=1}^{R} \lambda_k \prod_{\ell=1}^{d} \mu_{k,m_\ell}^{(\ell)} \right\} \mathbf{v}_{m_1}^{(1)} \otimes \cdots \otimes \mathbf{v}_{m_d}^{(d)}.
\end{aligned}$$

Now (3.16) implies that

$$\|\mathbf{A} - \mathbf{A}_R\| = \|\boldsymbol{\beta} - \boldsymbol{\beta}_{(R)}\|,$$

since the ℓ-mode multiplication with the orthogonal side matrices $V^{(\ell)}$ does not change the cost functional. Inspection of the left hand side in (3.15) indicates that the latter equation ensures that $\boldsymbol{\beta}_{(R)}$ is, in fact, the minimizer of the right hand side in (3.15). □

Combination of Theorem 3.35 and Lemma 3.37 opens the way to the rank optimization of canonical tensors with the large mode size arising, for example, in grid based numerical methods for multidimensional PDEs with nonregular (singular) solutions. In such applications the univariate grid size (i.e., the mode size) may be about $n = 10^4$ or even larger.

In some cases the rank reduction in Tucker format is required. This calculation can be done by the truncated HOSVD applied to the respective Tucker core as in the following scheme: $\boldsymbol{\beta} \mapsto \boldsymbol{\beta}_0 \in \mathcal{T}_{\mathbf{q},\mathbf{r}} \subset \mathbb{R}^{q_1 \times \cdots \times q_d}, \quad q_\ell < r_\ell$,

$$
\begin{aligned}
\mathcal{T}_{\mathbf{n},\mathbf{r}} \ni \mathbf{V} &= \boldsymbol{\beta} \times_1 V^{(1)} \times_2 V^{(2)} \cdots \times_d V^{(d)} \\
&\mapsto \left(\boldsymbol{\beta}_0 \times_1 B^{(1)} \times_2 B^{(2)} \cdots \times_d B^{(d)}\right) \times_1 V^{(1)} \times_2 V^{(2)} \cdots \times_d V^{(d)} \\
&= \boldsymbol{\beta}_0 \times_1 (V^{(1)} B^{(1)}) \times_2 (V^{(2)} B^{(2)}) \cdots \times_d (V^{(d)} B^{(d)}) \in \mathcal{T}_{\mathbf{n},\mathbf{q}} .
\end{aligned}
$$

Note that the Tucker (for moderate d) and canonical formats allow us to perform basic multilinear algebra using *one dimensional operations*, thus reducing the exponential scaling in d. Rank truncated transforms between different formats can be applied in multilinear algebra on mixed tensor representations as well; see Section 3.2.8 concerning the mixed Tucker-canonical format. The particular application to tensor convolution in many dimensions was discussed in [201]; see also Section 5.1.

In summary, the direct methods of tensor approximation can be classified by:

(1) Analytic Tucker approximation to some classes of function related dth order tensors ($d \geq 2$), see Section 2.2.4 on multivariate polynomial interpolation.
(2) Sinc quadrature based approximation methods in the canonical format applied to a class of function related tensors.
(3) Combination of ACA type and truncated SVD in the matrix case ($d = 2$).
(4) Truncated HOSVD and truncated RHOSVD, for quasioptimal Tucker approximation of the full-format and canonical tensors respectively.
(5) In all cases the algebraic methods for rank reduction by the ALS type iterative Tucker and/or canonical approximation can be applied (convergence theory is far from complete).

Direct analytic approximation methods by sinc quadratures/interpolation are of principal importance. Basic examples are given by the tensor representation of Green's kernels, and the elliptic operator inverse and analytic matrix valued functions; see also Sections 2.4, 3.5 and 5.2.

The SVD based approximation is a powerful tool that applies well to low dimensions. However, the bottlenecks and limitations are well recognized:

(a) lack of robust and efficient algebraic methods for the robust canonical tensor decomposition of multidimensional tensors (for $d \geq 3$);
(b) high numerical cost of the HOSVD type algorithms applied to the full format target.

Further improvement and enhancement of algebraic tensor approximation methods may be based on the advanced nonlinear iteration, multigrid tensor methods, greedy algorithms, combined tensor representations, and the use of new problem adapted tensor formats. In the next section, we consider the ALS type Tucker and canonical iterative approximation schemes.

3.4 Tensor approximation by nonlinear ALS iteration

In this section, we discuss in more detail the nonlinear approximation and tensor truncation algorithms in the canonical and Tucker formats. We consider the problem of the best approximation of a high order tensor $\mathbf{A} \in \mathbb{V}_{\mathbf{n}}$ in the following formats: $\mathcal{T}_{\mathbf{r}}$, $\mathcal{C}_R$, and $\mathcal{T}_{\mathcal{C}_{R,\mathbf{r}}}$, with the fixed rank parameters.

For given target $\mathbf{A} \in \mathcal{S}_0 \subset \mathbb{V}_{\mathbf{n}}$, this leads to the nonlinear approximation problem (3.8) on estimation $T_{\mathcal{S}}(\mathbf{A}) := \operatorname{argmin}_{\mathbf{X} \in \mathcal{S}} \|\mathbf{A}-\mathbf{X}\|$ where $\mathcal{S} \subset \{\mathcal{T}_{\mathbf{r}}, \mathcal{C}_R, \mathcal{T}_{\mathcal{C}_{R,\mathbf{r}}}\}$. Solution of this problem defines the *tensor truncation* operator

$$T_{\mathcal{S}} : \mathcal{S}_0 \subset \mathbb{V}_{\mathbf{n}} \mapsto \mathcal{S}\,, \quad \mathcal{S} \subset \mathcal{S}_0\,.$$

Even if the extremal point (local or global minimum) exists, in computational practice the particular minimizer can be calculated only approximately up to a certain accuracy $\varepsilon > 0$ (ε-approximation).

The important step is the reduction of the rank-$\mathbf{r}$ Tucker approximation to the dual maximization problem of finding the optimal dominating subspaces. First, we discuss in more detail the alternating least square (ALS) iteration for computation of the best rank-1 approximation and then proceed with the ALS iteration for the general orthogonal Tucker decomposition. The termination criteria in the ALS Tucker algorithm can be adapted to either the fixed rank-$\mathbf{r}$ or to the ε approximation strategy.

The general canonical rank-R approximation by the ALS iteration is considered including the special case of partially orthogonal decompositions. In the case of a canonical input the two-level canonical-to-Tucker approximation provides the efficient numerical schemes with the initial guess computed by the reduced HOSVD; see Theorem 3.35.

In application to function related tensors obtained by 'sampling' of the continuous function on a large spacial grid the enhanced version of the ALS iteration is based on the multigrid accelerated Tucker approximation. This provides fast convergence due to the good choice of the initial guess extrapolated from the coarser grid, combined with the construction of the so called most important fibers on the coarse level.

Finally, we present numerical illustrations related to the Hartree–Fock calculations in electronic structure modeling.

3.4.1 Approximation on Tucker manifold by dual maximization

Let us consider the approximation of a tensor in the fixed rank Tucker format formulated as the minimization problem

$$\mathbf{A} \in \mathcal{S}_0 \subset \mathbb{V}_\mathbf{n}: \quad \mathbf{A}_{(\mathbf{r})} = \operatorname{argmin}_{\mathbf{X} \in \mathcal{T}_\mathbf{r}} \|\mathbf{A} - \mathbf{X}\|_{\mathbb{V}_\mathbf{n}} . \tag{3.19}$$

This problem was first analyzed in [248] and further addressed in [206, 211, 212].

The next lemma proves the well known result on quadratic convergence for the difference of norms.

Lemma 3.38 (Quadratic convergence of norm). *Let* $\mathbf{A}_{(\mathbf{r})} = \boldsymbol{\beta} \times_1 V^{(1)} \times_2 V^{(2)} \cdots \times_d V^{(d)} \in \mathbb{R}^{I_1 \times \cdots \times I_d}$ *solve the minimization problem* (3.19), *then*

$$\|\boldsymbol{\beta}\| = \|\mathbf{A}_{(\mathbf{r})}\| \leq \|\mathbf{A}\| .$$

Moreover, we have the 'quadratic' relative error bound on the norm of a tensor,

$$\frac{\|\mathbf{A}\| - \|\mathbf{A}_{(\mathbf{r})}\|}{\|\mathbf{A}\|} \leq \frac{\|\mathbf{A}_{(\mathbf{r})} - \mathbf{A}\|^2}{\|\mathbf{A}\|^2} . \tag{3.20}$$

Proof. The Tucker orthogonality implies $\|\boldsymbol{\beta}\| = \|\mathbf{A}_{(\mathbf{r})}\|$. Relation (3.19) is merely a linear least square problem with respect to $\boldsymbol{\beta} \in \mathbb{R}^{r_1 \times \cdots \times r_d}$,

$$g(\boldsymbol{\beta}) := \langle \mathbf{A}, \mathbf{A} \rangle - 2\langle \mathbf{A}, \boldsymbol{\beta} \times_1 V^{(1)} \times_2 \cdots \times_d V^{(d)} \rangle + \langle \boldsymbol{\beta}, \boldsymbol{\beta} \rangle \to \min . \tag{3.21}$$

The corresponding minimization condition

$$g(\boldsymbol{\beta} + \Delta\boldsymbol{\beta}) - g(\boldsymbol{\beta}) \geq 0 \quad \forall\, \Delta\boldsymbol{\beta} \in \mathbb{R}^{r_1 \times \cdots \times r_d}$$

leads to the following variational equations for the minimizer:

$$\begin{aligned} -\langle \mathbf{A}, \Delta\boldsymbol{\beta} \times_1 V^{(1)} \times_2 \cdots \times_d V^{(d)} \rangle + \langle \boldsymbol{\beta}, \Delta\boldsymbol{\beta} \rangle &= 0 \quad \forall\, \Delta\boldsymbol{\beta} \in \mathbb{R}^{r_1 \times \cdots \times r_d} , \\ \langle -\mathbf{A} \times_1 V^{(1)^{\mathrm{T}}} \times_2 \cdots \times_d V^{(d)^{\mathrm{T}}} + \boldsymbol{\beta}, \Delta\boldsymbol{\beta} \rangle &= 0 \quad \forall\, \Delta\boldsymbol{\beta} \in \mathbb{R}^{r_1 \times \cdots \times r_d} , \\ \boldsymbol{\beta} - \mathbf{A} \times_1 V^{(1)^{\mathrm{T}}} \times_2 \cdots \times_d V^{(d)^{\mathrm{T}}} &= 0 . \end{aligned} \tag{3.22}$$

Next, we derive

$$\begin{aligned} \|\mathbf{A}_{(\mathbf{r})} - \mathbf{A}\|^2 &= \|\mathbf{A}_{(\mathbf{r})}\|^2 - 2\langle \boldsymbol{\beta} \times_1 V^{(1)} \times_2 \cdots \times_d V^{(d)}, \mathbf{A} \rangle + \|\mathbf{A}\|^2 \\ &= \|\mathbf{A}_{(\mathbf{r})}\|^2 + \|\mathbf{A}\|^2 - 2\left\langle \boldsymbol{\beta}, \mathbf{A} \times_1 V^{(1)^{\mathrm{T}}} \times_2 \cdots \times_d V^{(d)^{\mathrm{T}}} \right\rangle , \\ &= \|\mathbf{A}\|^2 - \|\boldsymbol{\beta}\|^2 . \end{aligned}$$

Hence it follows that $\|\mathbf{A}\|^2 - \|\mathbf{A}_{(\mathbf{r})}\|^2 = \|\mathbf{A}_{(\mathbf{r})} - \mathbf{A}\|^2$, and then

$$\frac{\|\mathbf{A}\| - \|\mathbf{A}_{(\mathbf{r})}\|}{\|\mathbf{A}\|} = \frac{\|\mathbf{A}_{(\mathbf{r})} - \mathbf{A}\|^2}{(\|\mathbf{A}_{(\mathbf{r})}\| + \|\mathbf{A}\|)\|\mathbf{A}\|} \leq \frac{\|\mathbf{A}_{(\mathbf{r})} - \mathbf{A}\|^2}{\|A\|^2} ,$$

justifying the inequality (3.20). □

Now consider the algorithm for computation of the best orthogonal rank-$(r_1, \dots, r_d)$ Tucker approximation by reduction to the optimization problem over the Stiefel manifold of orthogonal $n \times r$ matrices,

$$\mathbb{S}_{n,r} := \{Y \in \mathbb{R}^{n \times r} : Y^T Y = I_{r \times r}\} .$$

Theorem 3.39. *The minimization problem* (3.19) *is equivalent to the dual maximization problem*

$$[U^{(1)}, \dots, U^{(d)}] = \operatorname{argmax} \left\| \mathbf{A} \times_1 V^{(1)^T} \times_2 \cdots \times_d V^{(d)^T} \right\|_F^2 \tag{3.23}$$

over the product of (compact) Stiefel manifolds, $V^{(\ell)} = [\mathbf{v}_\ell^1 \dots \mathbf{v}_\ell^{r_\ell}] \in \mathbb{S}_{n_\ell, r_\ell}$, $\ell = 1, \dots, d$*. For given maximizing matrices* $U^{(m)}$ $(m = 1, \dots, d)$*, the core tensor* $\boldsymbol{\beta}$ *minimizing* (3.19) *is represented by*

$$\boldsymbol{\beta} = \mathbf{A} \times_1 U^{(1)^T} \times_2 \cdots \times_d U^{(d)^T} \in \mathbb{R}^{r_1 \times \dots \times r_d} . \tag{3.24}$$

Under the compatibility condition

$$r_m \le \bar{r}_m := r_1 \dots r_{m-1} r_{m+1} \dots r_d , \quad m = 1, \dots, d , \tag{3.25}$$

there is at least one solution of (3.23).

Proof. Plugging $\boldsymbol{\beta}$ from (3.22) into (3.21) leads to the equivalent minimizing equation

$$\langle \mathbf{A}, \mathbf{A} \rangle - \langle \mathbf{A} \times_1 V^{(1)^T} \times_2 \cdots \times_d V^{(d)^T}, \mathbf{A} \times_1 V^{(1)^T} \times_2 \cdots \times_d V^{(d)^T} \rangle \to \min ,$$

which proves (3.23). Finally, (3.22) yields (3.24).

Size consistency of arising tensors requires the compatibility conditions (3.25). Then the dual maximization problem (3.23) posed on the compact manifold $\mathbb{S}_{n_\ell, r_\ell}$ possesses at least one global maximum. □

The rotational nonuniqueness of the maximizer in (3.23) can be avoided if one solves this maximization problem in a product of the so called Grassmann manifolds $\mathcal{G}_\ell$, $\ell = 1, \dots, d$. The latter is the factor space of $\mathbb{S}_{n_\ell, r_\ell}$ with respect to the rotational transforms.

3.4.2 Best rank-1 approximation

Let us consider the alternating least square (ALS) iteration to compute the best orthogonal Tucker approximation. First, we look in more detail at the simplest special case of the Tucker/canonical model, that is the best rank-1 approximation. It is an important ingredient in typical multilinear algebra algorithms, in particular in the greedy approximation method.

To derive the corresponding Lagrange equations, we note that due to the normalization

$$\|\beta V^{(1)^{\mathrm{T}}} \times_2 \cdots \times_d V^{(d)^{\mathrm{T}}}\|^2 = \beta^2 , \qquad V^{(\ell)} \in \mathbb{R}^{n_\ell \times 1} ,$$

the dual problem of maximizing the *generalized Rayleigh quotient* over the unit norm vectors (eliminates the scalar β), reads as

$$\left| \mathbf{A} \times_1 V^{(1)^{\mathrm{T}}} \times_2 \cdots \times_d V^{(d)^{\mathrm{T}}} \right|^2 - \sum_{\ell=1}^{d} \lambda^{(\ell)} \left(\|V^{(\ell)}\|^2 - 1 \right) \to \max . \tag{3.26}$$

For any solution of this problem, the corresponding scalar β can be chosen as $\beta = \mathbf{A} \times_1 V^{(1)^{\mathrm{T}}} \times \cdots \times_d V^{(d)^{\mathrm{T}}}$.

Differentiating (3.26) with respect to $V^{(m)}$ for some fixed m, $(1 \leq m \leq d)$ leads to the equation

$$\beta \mathbf{A} \times_1 V^{(1)^{\mathrm{T}}} \cdots \times_{m-1} V^{(m-1)^{\mathrm{T}}} \times_{m+1} V^{(m+1)^{\mathrm{T}}} \cdots \times_d V^{(d)^{\mathrm{T}}} = \lambda^{(m)} V^{(m)} ,$$

which implies $\lambda^{(m)} = \beta^2$. Finally, the Lagrange equations read as

$$\begin{aligned} \mathbf{A} \times_1 V^{(1)^{\mathrm{T}}} \cdots \times_{m-1} V^{(m-1)^{\mathrm{T}}} \times_{m+1} V^{(m+1)^{\mathrm{T}}} \cdots \times_d V^{(d)^{\mathrm{T}}} &= \beta V^{(m)} , \\ \mathbf{A} \times_1 V^{(1)^{\mathrm{T}}} \times_2 \cdots \times_d V^{(d)^{\mathrm{T}}} &= \beta , \\ \|V^{(m)}\| &= 1 \end{aligned}$$

for each $1 \leq m \leq d$. The above system of Lagrange equations can be solved by an alternating least squares algorithm: At each iterative step m an approximation to the scalar β and the estimate of the vector $V^{(m)}$ are optimized, while the rest of the vector components with $\ell \neq m$ are kept constant $(m = 1, \ldots, d)$.

Under certain conditions, the ALS method of the best rank-1 approximation is expected to provide a locally linear convergence rate [358, 373]. The Newton's type methods, theoretically providing locally quadratic convergence, have been analyzed in [2, 94, 95, 172, 173, 358, 373].

3.4.3 Best rank-r Tucker approximation of full format target

The ALS iteration to compute the best rank-$\mathbf{r}$ Tucker approximation is also known as *higher order orthogonal iteration* (HOOI) [248]. The corresponding algorithm includes the following steps:

Algorithm (Algorithm Tucker-ALS ($\mathbb{V}_{\mathbf{n}} \to \mathcal{T}_{\mathbf{r},\mathbf{n}}$)). Given the input tensor $\mathbf{A} \in \mathbb{V}_{\mathbf{n}}$:

1. Compute an initial guess $V_0^{(\ell)}$ $(\ell = 1, \ldots, d)$ for the ℓ-mode side matrix by the 'truncated' SVD applied to a $n \times n^{d-1}$ unfolding matrix $A_{(\ell)}$, i.e., apply HOSVD, (cost $O(n^{d+1})$).

2. For each $q = 1, \ldots, d$, fix the side matrices $V^{(\ell)} = [\mathbf{v}_1^{(\ell)}, \ldots, \mathbf{v}_{r_\ell}^{(\ell)}] \in \mathbb{R}^{n \times r_\ell}$, $\ell \neq q$, and perform the ALS iteration to optimize the q-mode matrix $V^{(q)}$ via computing the dominating r_q-dimensional subspace (the truncated SVD) for the unfolding matrix $B_{(q)} \in \mathbb{R}^{n \times \bar{r}_q}$, corresponding to the q-mode contracted product

$$\mathbf{B}_{(q)} = \mathbf{A} \times_1 V^{(1)^{\mathrm{T}}} \times_2 \cdots \times_{q-1} V^{(q-1)^{\mathrm{T}}} \times_{q+1} V^{(q+1)^{\mathrm{T}}} \cdots \times_d V^{(d)^{\mathrm{T}}} \in \mathbb{R}^{r_1 \times \ldots r_{q-1} \times n_q \times r_{q+1} \cdots \times r_d},$$

where the 'single hole' index is defined by

$$\bar{r}_q = r_1 \ldots r_{q-1} r_{q+1} \ldots r_d = O(r^{d-1}) . \tag{3.27}$$

Each iteration has the numerical cost $O(dr^{d-1} n \max\{r^{d-1}, n\})$.
3. Check stopping criteria on updated directional matrices $\{V^{(\ell)}\}$.
4. Compute the core tensor $\boldsymbol{\beta}$ as the representation coefficients of the orthoprojection of $\mathbf{A}$ onto the subspace $\otimes_{\ell=1}^{d} \operatorname{span}\{\mathbf{v}_1^{(\ell)}, \ldots, \mathbf{v}_{r_\ell}^{(\ell)}\}$, (at the expense $O(rn^d)$),

$$\boldsymbol{\beta} = \mathbf{A} \times_1 V^{(1)^{\mathrm{T}}} \times_2 \cdots \times_d V^{(d)^{\mathrm{T}}} \in \mathbb{R}^{r_1 \times \cdots \times r_d} .$$

In general, the convergence theory for the Tucker-ALS algorithm is an open question.

We conclude that the Tucker-ALS algorithm ($\mathbb{V}_{\mathbf{n}} \to \mathcal{T}_{\mathbf{r},\mathbf{n}}$) can be applied to the case of moderate d and very moderate n because of the exponential complexity scaling in d, $O(n^{d+1})$. Hence, the Tucker-ALS scheme for full format tensors is practically not applicable in tensor numerical methods for PDEs.

3.4.4 Remarks on rank-R canonical approximation by ALS iteration

The R-term canonical approximation of a full format dth order tensor is a challenging numerical task. Several enhanced ALS type schemes have been considered in the literature [94, 231–233, 358]. The main problem is the nonstable convergence of the ALS type iterations, degeneracy in the large governing matrices, and the absence of rigorous theoretical analysis.

The approximation problem is formulated as follows: Given $\mathbf{A} \in \mathbb{R}^{n_1 \times \cdots \times n_d}$, find

$$T_{\mathcal{S}}(\mathbf{A}) := \operatorname*{argmin}_{\mathbf{X} \in \mathcal{S}} \|\mathbf{X} - \mathbf{A}\| , \quad \text{where} \quad \mathcal{S} = \mathcal{C}_R . \tag{3.28}$$

We denote by $U^{(m)} \in \mathbb{R}^{n \times R}$ ($m = 1, \ldots, d$), and $\boldsymbol{\xi} = \{\xi_k\}_{k=1}^{R}$, the side (factor) matrices with normalized column vectors, and respective weights of the R-term approximation $\mathbf{X} \in \mathcal{C}_R$ of $\mathbf{A}$; see (3.10). We sketch the general scheme of ALS iteration to compute the rank-R approximation to solve (3.28); see [62, 63].

Given initial side matrices $U^{(m)}$ ($m = 1, 2, \ldots, d$), the ALS scheme fixes all $U^{(m)}$ ($m = 2, \ldots, d$) and solves (3.28) for $U^{(\ell)}$, $\ell = 1$, then fixes $U^{(m)}$ ($m = 1, 3, \ldots, d$) to solve (3.28) for $U^{(\ell)}$, $\ell = 2$, and so on for $\ell = 1, 2, \ldots, d$, and continues to repeat such a global iteration loop until some convergence/stopping criteria is satisfied. Having fixed all but one side matrix, the numerical task reduces to a linear least squares minimization problem.

A convenient description of the ALS scheme in canonical format requires the so called Khatri–Rao product of matrices, which is the 'matching columnwise' Kronecker product.

Definition 3.40. Given matrices $A = [a_{ij}] \in \mathbb{R}^{m\times n}$, $B \in \mathbb{R}^{h\times n}$, their Khatri–Rao product is denoted by $A \boxdot B$. It has the size $mh \times n$ and is defined by

$$A \boxdot B = [a_1 \otimes b_1 \; a_2 \otimes b_2 \; \dots \; a_n \otimes b_n] .$$

If A and B are vectors, then the Khatri–Rao and Kronecker products are identical, i.e., $A \otimes B = A \boxdot B$.

Note that the full unfolding matrices $U_{(\ell)}$ for a canonical tensor can be represented in terms of the Khatri–Rao product of side matrices $U^{(\ell)}$; see [232].

Proposition 3.41. *The mode-ℓ unfolding matrix of a canonical tensor in* (3.10) *is given by*

$$U_{(\ell)} = U^{(\ell)} \operatorname{diag}\{\boldsymbol{\xi}\}(U^{(d)} \boxdot \cdots \boxdot U^{(\ell+1)} \boxdot U^{(\ell-1)} \boxdot \cdots \boxdot U^{(1)})^{\mathrm{T}} , \quad \ell = 1, \dots, d .$$

For ease of presentation, here we illustrate the ALS algorithm for the instructive case $d = 3$, which is of major importance in many applications, in particular 3D electronic structure calculations. The discussion of an ALS iteration scheme for approximation in the canonical format for arbitrary d is presented, for example, in [71, 232].

Let $d = 3$ and denote by $A_{(1)}, A_{(2)}, A_{(3)}$ the three unfolding matrices of $\mathbf{A}$. We use the representation (3.10) for $\mathbf{X}$,

$$\mathbf{X} = \sum_{\nu=1}^{R} \xi_\nu \mathbf{u}_1^\nu \otimes \cdots \otimes \mathbf{u}_d^\nu , \quad \xi_\nu \in \mathbb{R} ,$$

with the side matrices $U^{(\ell)}$ and normalized canonical vectors $\mathbf{u}_\ell^\nu$.

For fixed side matrices $U^{(2)}, U^{(3)}$, the ALS iteration step for $\ell = 1$ reads as the minimization problem (3.28) (in Frobenius norm) for $U^{(1)}$ and $\boldsymbol{\xi}$:

$$\min_{\tilde{U}} \|A_{(1)} - \tilde{U}(U^{(3)} \boxdot U^{(2)})^{\mathrm{T}}\|_F^2 , \quad \text{where} \quad \tilde{U} = U^{(1)} \operatorname{diag}\{\boldsymbol{\xi}\} .$$

The optimal solution (minimizer) is then given by

$$\tilde{U} = A_{(1)} \left[(U^{(3)} \boxdot U^{(2)})^{\mathrm{T}}\right]^{\dagger} , \tag{3.29}$$

where $B^\dagger$ denotes the Moore–Penrose pseudoinverse of a matrix B. Updates for the vector $\boldsymbol{\xi}$ and for the matrix U are calculated by normalizing columns of $\tilde{U}$. Note that the matrix $B = (U^{(3)} \boxdot U^{(2)})^{\mathrm{T}}$ in (3.29) has the size $n_2 n_3 \times R$ indicating the $O(n^2 R^2)$

complexity for computation of the Moore–Penrose pseudoinverse. In higher dimensions, this cost scales exponentially in d. In the case of large mode sizes and large R, this is the severe limitation even for $d = 3$.

The more convenient form of the solution is given by [232]

$$\tilde{U} = A_{(1)}(U^{(3)} \boxdot U^{(2)}) \left((U^{(3)})^{\mathrm{T}} U^{(3)} \odot (U^{(2)})^{\mathrm{T}} U^{(2)}\right)^{\dagger} .$$

The pseudoinverse $B^{\dagger}$ of the matrix in this case is only of size $R \times R$.

Clearly, the ALS algorithm is simple to implement, but convergence may be very slow. The convergence theory for the CP-ALS iteration is far from complete [358, 373]. Different enhancement strategies have been advocated in the literature [95, 232, 233].

In the following, a practically interesting situation of the canonical input tensor will be addressed. The ALS algorithm can be interpreted as the rank reduction in the canonical format. In this case, in view of Proposition 3.41, the minimization problem and its solution read

$$\min_{\tilde{U}} \|A_{(1)} - \tilde{U}(U^{(3)} \boxdot U^{(2)})^{\mathrm{T}}\|_F^2 , \quad \text{and} \quad \tilde{U} = A^{(1)} \ \operatorname{diag}\{\boldsymbol{\xi}\} \left[(A^{(3)} \boxdot A^{(2)})^{\mathrm{T}}\right]^{\dagger} \tag{3.30}$$

respectively, where again $\tilde{U} = U^{(1)} \operatorname{diag}\{\boldsymbol{\xi}\}$. Here the computational cost is much smaller compared with the general case (3.29). This CP-ALS rank reduction scheme was applied in [227] to the rank structured solution of stochastic/parametric PDEs (in a high dimensional parametric space) by solving the approximation problems for the increasing sequence of canonical rank parameters $r = 1, 2, \ldots, R$. This stabilizes the convergence of the ALS iteration by lifting the solution for a lower rank $r = k$ as the initial guess for the larger rank, $r = k + 1$.

In the next sections, we focus on a special case of partially orthogonal decompositions. In this case the idea of stabilizing mechanism is based on the use of the Tucker approximation at the intermediate stage. Approximation algorithms for canonical tensors with large initial ranks of the order of several thousand based on the RHOSVD approach have been applied in the Hartree–Fock calculations; see [191, 212].

3.4.5 Two-level Tucker-to-canonical approximation to the CP input

In this section, we consider the numerical approximation by using the two-level Tucker-canonical format applied to the canonical input tensor.

In the iterative solution of multidimensional PDEs the typical situation may arise when the target tensor is already presented in the rank-R canonical format, $\mathbf{A} \in \mathcal{C}_{R,\mathbf{n}}$, but with large R and large n. Here n plays the role of the univariate grid size that might be of the order of 10^3–10^4 or even larger. This is the case, for example, in electronic structure calculations based on the 3D Hartree–Fock equation [191, 212], where high precision approximation of molecular orbitals requires large grids due to the presence of multiple singularities.

We consider the minimization problem

$$\mathbf{A} \in \mathcal{C}_{R,\mathbf{n}} \subset \mathbb{V}_{\mathbf{n}}: \quad \mathbf{A}_{(\mathbf{r})} = \underset{\mathbf{X} \in \mathcal{T}_{\mathbf{r},\mathbf{n}}}{\operatorname{argmin}} \|\mathbf{A} - \mathbf{X}\|^2_{\mathbb{V}_{\mathbf{n}}}, \tag{3.31}$$

where the minimizer $\mathbf{X} = \sum_{\mathbf{k}=1}^{\mathbf{r}} \mu_{\mathbf{k}} \mathbf{x}^{(1)}_{k_1} \otimes \cdots \otimes \mathbf{x}^{(d)}_{k_d}$ belongs to the class of rank-$\mathbf{r}$ Tucker tensors in $\mathcal{T}_{\mathbf{r},\mathbf{n}}$. In this case, a two-level multigrid approximation can be applied [211, 212] such that the computational cost can be reduced dramatically. In particular, one step of the corresponding two-level canonical-to-Tucker ALS iteration with the Tucker rank $r \ll R$ can be implemented in $O(dn^2Rr^2)$ complexity up to low order terms (Section 2.4.2 in [212]). The corresponding approximation scheme can be described as the following two-level chain of rank structured approximations:

$$\mathcal{C}_{R,\mathbf{n}} \mapsto \mathcal{T}_{\mathcal{C}_{R},\mathbf{r}} \mapsto \mathcal{T}_{\mathcal{C}_{R'},\mathbf{r}} \mapsto \mathcal{C}_{R',\mathbf{n}}\,. \tag{3.32}$$

At Level I, the best orthogonal Tucker approximation applies to an input in $\mathcal{C}_{R,\mathbf{n}}$, so that the core tensor of the resultant Tucker approximation is represented in the $\mathcal{C}_{R,\mathbf{r}}$ canonical format. In that case the two-level Tucker representation remains free of the curse of dimensionality.

At Level II, the 'small size' Tucker core in $\mathcal{C}_{R,\mathbf{r}}$ is approximated by an element in $\mathcal{C}_{R',\mathbf{r}}$ with $R' \ll R$ by using the ALS iteration described above. Since directional Tucker ranks are supposed to be small, the computational cost of such an ALS scheme appears to be much lower compared with that for the ALS applied to the initial tensor.

The next statement describes the results on *solvability and tensor structure* of the Level I scheme in (3.31), and provides the key to construct its efficient numerical implementation.

Theorem 3.42 (Canonical-to-Tucker approximation [212]).

(a) *For* $\mathbf{A} = \sum_{\nu=1}^{R} \xi_\nu \mathbf{u}_\nu^{(1)} \otimes \cdots \otimes \mathbf{u}_\nu^{(d)} \in \mathcal{C}_{R,\mathbf{n}}$, *the minimization problem* (3.31) *is equivalent to the dual maximization problem*

$$[V^{(1)}, \ldots, V^{(d)}] = \operatorname{argmax}_{X^{(\ell)} \in \mathcal{G}_\ell[\mathcal{S}_{r_\ell}]} \left\| \sum_{\nu=1}^{R} \xi_\nu \left(X^{(1)^{\mathrm{T}}} \mathbf{u}_\nu^{(1)} \right) \otimes \cdots \otimes \left(X^{(d)^{\mathrm{T}}} \mathbf{u}_\nu^{(d)} \right) \right\|^2_{\mathbb{V}_{\mathbf{r}}}. \tag{3.33}$$

(b) *The compatibility condition* (3.25) *in Theorem 3.39 is simplified to*

$$r_\ell \le \operatorname{rank}(U^{(\ell)}) \quad \textit{with} \quad U^{(\ell)} = [\mathbf{u}_1^{(\ell)} \ldots \mathbf{u}_R^{(\ell)}] \in \mathbb{R}^{n \times R},$$

ensuring the solvability of (3.33). *The maximizer is given by orthogonal matrices* $V^{(\ell)} = [\mathbf{v}_1^{(\ell)} \ldots \mathbf{v}_{r_\ell}^{(\ell)}] \in \mathbb{R}^{n \times r_\ell}$, *computed as in the ALS-Tucker algorithm, where the initial guess is calculated by RHOSVD approximation instead of HOSVD.*

(c) *Given* $\mathcal{V}_\ell = \operatorname{span}\{V^{(\ell)}\}$, *the minimizer in* (3.31) *is computed by the orthoprojection onto* $\mathbb{V}_{\mathbf{n},\mathbf{r}} = \mathcal{V}_1 \otimes \cdots \otimes \mathcal{V}_d$,

$$\mathbf{A}_{(\mathbf{r})} = \sum_{\mathbf{k}=1}^{\mathbf{r}} \sum_{k_1=1}^{r_1} \cdots \sum_{k_d=1}^{r_d} \mu_{\mathbf{k}} \mathbf{v}^{(1)}_{k_1} \otimes \cdots \otimes \mathbf{v}^{(d)}_{k_d},$$

where the core tensor $\boldsymbol{\mu} = [\mu_{\mathbf{k}}]$ *is represented in the rank-R canonical format*

$$\boldsymbol{\mu} = \sum_{\nu=1}^{R} \xi_\nu \left(V^{(1)^{\mathrm{T}}} \mathbf{u}_\nu^{(1)} \right) \otimes \cdots \otimes \left(V^{(d)^{\mathrm{T}}} \mathbf{u}_\nu^{(d)} \right) \in \mathcal{C}_{R,\mathbf{r}} . \tag{3.34}$$

Proof. The statement is a direct consequence of the general Theorem 3.39:

(a) The generic dual maximization problem (3.23) with $\mathbf{A} \in \mathcal{C}_{R,\mathbf{n}}$ takes the form (3.33) due to the relation

$$\left\langle \mathbf{v}_{k_1}^{(1)} \otimes \cdots \otimes \mathbf{v}_{k_d}^{(d)}, \mathbf{A} \right\rangle = \sum_{\nu=1}^{R} \xi_\nu \left\langle \mathbf{v}_{k_1}^{(1)}, \mathbf{u}_\nu^{(1)} \right\rangle \cdots \left\langle \mathbf{v}_{k_d}^{(d)}, \mathbf{u}_\nu^{(d)} \right\rangle .$$

(b) The compatibility condition (3.25) ensures size consistency of all unfolding matrices.

(c) Formula for $\mathbf{A}_{(\mathbf{r})}$ is a special case of the general orthoprojection representation (3.24) for the Tucker core. □

Note that the approximation error $\|\mathbf{A} - \mathbf{A}_{(\mathbf{r})}\|$ of the minimizer $\mathbf{A}_{(\mathbf{r})}$ is estimated by the bound (3.13) for RHOSVD approximation error in Theorem 3.35; see also Remark 3.43 below. For many practically interesting problems the RHOSVD approximation exhibits a good initial guess for the ALS-Tucker algorithm, which ensures fast and robust local convergence of the iterations described by Theorem 3.42.

Let us formulate the algorithm of the best rank-**r** Tucker approximation to the canonical target tensor. For the sake of computational efficiency we suppose that $r \ll R$.

Algorithm (Algorithm CP-Tucker-ALS ($\mathcal{C}_{R,\mathbf{n}} \to \mathcal{T}_{\mathcal{C}_{R},\mathbf{r}}$)). Given $\mathbf{A} \in \mathcal{C}_{R,\mathbf{n}}$, the maximal number of ALS iterations, $k_{\max}$, and the Tucker rank parameter $\mathbf{r}$:

1. For $\ell = 1, \ldots, d$, compute the truncated SVD of $U^{(\ell)} \in \mathbb{R}^{n_\ell \times R}$ to obtain the orthogonal matrices $Z_0^{(\ell)} \in \mathbb{R}^{n_\ell \times r_\ell}$, representing the rank-$r_\ell$ RHOSVD approximation of mode-ℓ dominating subspaces. The overall expense is estimated by $O(dRn \min\{R, n\})$.
2. Given the initial guess $Z_0^{(\ell)}$, $(\ell = 1, \ldots, d)$ for mode-ℓ orthogonal matrices, perform $k_{\max}$ ALS iterations as in Step 2 in the general Tucker-ALS algorithm (Section 3.4.3) applied to the problem (3.33) to obtain the maximizer $V^{(\ell)} \in \mathbb{R}^{n_\ell \times r_\ell}$, $\ell = 1, \ldots, d$, at the cost $O(dr^{d-1} n \min\{r^{d-1}, n\})$ per iteration.
3. Calculate projections of $U^{(\ell)}$ onto the basis of orthogonal vectors in $V^{(\ell)}$ defined by the coefficients represented as the matrix product $V^{(\ell)^{\mathrm{T}}} U^{(\ell)} \in \mathbb{R}^{r_\ell \times R}$ $(\ell = 1, \ldots, d)$, all at the cost $O(drRn)$.
4. Using the columns in $V^{(\ell)^{\mathrm{T}}} U^{(\ell)}$ $(\ell = 1, \ldots, d)$, calculate the rank-R core tensor $\boldsymbol{\mu} \in \mathcal{C}_{R,\mathbf{r}}$ as in (3.34), all in $O(drRn)$ operations and with $O(drR)$ storage.

Note that the *CP-Tucker-ALS* algorithm can be easily modified to the stopping criteria via fixed tolerance $\varepsilon > 0$. Furthermore, for a wide class of function related tensors the effective Tucker rank is small, hence in the case of moderate dimensions (say, $d = 3, 4$) the algorithm has a perfect linear complexity in the mode size n.

Remark 3.43. Algorithm CP-Tucker ($\mathcal{C}_{R,\mathbf{n}} \to \mathcal{T}_{\mathcal{C}_R,\mathbf{r}}$) exhibits polynomial cost in R, n,

$$O(dRn \min\{n, R\} + dr^{d-1} n \min\{r^{d-1}, n\}) ,$$

with an exponential scaling r^{d-1} in d. In the absence of Step 2 (i.e., if RHOSVD provides already satisfactory approximation), we have for any $d \geq 2$ a finite SVD based scheme with the error bound as in Theorem 3.35; see (3.13). If the R-term canonical representation of the target tensor $\mathbf{A}$ is only weakly redundant (say, in the case of partially orthogonal or 'directionally positive' input), then $\|\boldsymbol{\xi}\|$ will be of the same order as $\|\mathbf{A}\|$, and the relative error of RHOSVD becomes as good as for the HOSVD; see Theorem 3.31.

We summarize the main features of the Tucker and canonical approximations applied so far to the general algebraic tensors:
- For best orthogonal Tucker approximation of a general $\mathbb{V}_{\mathbf{n}}$-input:
  - existence of the best approximation, quadratic convergence in the 'energy';
  - reduction to the dual maximization problem to find dominating subspaces;
  - ALS iteration in $O(n^{d+1})$ complexity. (Robust, but expensive. Even for $d = 3$ the $O(n^4)$ complexity scaling is the challenge).
- For direct ALS based rank reduction methods $\mathcal{C}_{R,\mathbf{n}} \to \mathcal{C}_{R',\mathbf{n}}$, $R' < R$ exhibits polynomial cost (convergence may be slow). Complexity scaling in d depends on the favorable ALS scheme.
- For two-level versions of the best Tucker approximation to $\mathcal{C}_{R,\mathbf{n}}$-input: provides polynomial cost $O(d(R + r^q)n \min\{n, R\})$ to compute dominating subspaces for d dimensional data.

3.4.6 Multigrid Tucker approximation of function related tensors

In applications to numerical PDEs one deals with function related tensors having large mode sizes $n^{(\ell)}$, $\ell = 1, \ldots, d$, usually associated with the univariate spatial grid size in d dimensions. For such applications the above computational schemes via the Tucker-ALS and CP-Tucker-ALS algorithms can be enhanced by using the idea of *multigrid* (MG) rank-$\mathbf{r}$ Tucker approximation (MG-Tucker-ALS) introduced in [212].

The main motivation for using the multigrid approach is that in many applications (say in 3D real space quantum chemistry calculations) the univariate grid size n may be large, say of the order of 10^4–10^5. The fine tensor grids are necessary for precise resolution of functions with multiple singularities in large 3D domains. Indeed, the local mesh refinement techniques destroy the tensor structure of the data and, at the same time, make data structure extremely complicated.

The main idea of the multigrid accelerated Tucker approximation can be shortly formulated as follows:

(A) Solve the sequence of approximation problems for tensors $\mathbf{A}_n = \mathbf{A}_{n_m}$ sampled on a sequence of refined grids with $n = n_m := n_0 2^m$, $m = 0, 1, \ldots, M$. Advantage: the HOSVD step applies to a small tensor $\mathbf{A}_{n_0}$, only on the coarsest level.
(B) Use the coarse-to-fine interpolation (approximation) of vectors spanning the dominating subspaces as the initial guess to finer levels.
(C) Find positions of 'most important fibers' (MIFs) of mode-ℓ unfolding matrices on the coarse level(s) via a maximal energy principle.
(D) On fine levels of ALS iteration perform the SVD analysis on the reduced dataset by choosing only a small subset of the most representative fibers, i.e., MIFs, in the mode-ℓ unfolding matrices.

Solving the Tucker tensor approximation problem on a sequence of grids, $n = n_0, \ldots, n_M$, includes the following main ingredients:

1. Given a domain $\Omega = [-A, A]^d$, use the equidistant tensor grids $\omega_{\mathbf{d},n} := \omega_1 \times \omega_2 \cdots \times \omega_d$, where

$$\omega_\ell := \{-A + (k-1)h\colon k = 1, \ldots, n+1\} \quad (\ell = 1, \ldots, d),$$

   with the mesh size $h = 2A/n$, $n = n_0 2^m$, $m = 0, 1, \ldots, M$, and a set of collocation points $\{x_\mathbf{k}\} \in \Omega \in \mathbb{R}^d$, $\mathbf{k} \in \mathcal{I} := \{1, \ldots, n\}^d$, located at the midpoints of grid cells numbered by the multi-index $\mathbf{k} \in \mathcal{I}$.
2. Given a continuous multivariate function $f\colon \Omega \to \mathbb{R}$, the target tensor is defined by sampling f on the set of collocation points $\{x_\mathbf{k}\}$,

$$\mathbf{A}_n = [a_{n,\mathbf{k}}] \in \mathbb{R}^{\mathcal{I}}\colon \quad a_{n,\mathbf{k}} = f(x_\mathbf{k}),\ \mathbf{k} \in \mathcal{I}.$$

3. Use an 'accurate' tensor product interpolation operator $\mathbf{I}_{m-1\mapsto m}$, $m = 1, \ldots, M$, from the coarse to fine grid (we recommend interpolation by piecewise cubic splines).
4. Precompute at the coarse level the index set designating the position of most important fibers (MIFs) in the directional unfolding matrices for $\ell = 1, \ldots, d$, arising in the ALS iteration.
5. Transfer the initial guess and positions of MIFs from the coarse to the fine grids.

We comment that the index set specifying MIFs is defined on the coarse grid by selection of $O(r)$ fibers in the mode-q unfolding matrices $B_{(q)}$ of size $n \times r^{d-1}$, $q = 1, \ldots, d$, having the largest norms (*the maximum energy principle*); see illustration in Figure 3.8 for $d = 3$. For explanatory reasons, the matrix $B_{(q)}$ is presented in a tensor form.

The benefit of using the precomputed positions of MIFs is that at each step of the ALS iteration computation of the dominating subspace of dimension r (by QR or SVD algorithms) in large $n \times r^{d-1}$ unfolding matrices on fine levels, i.e., for large n, can be

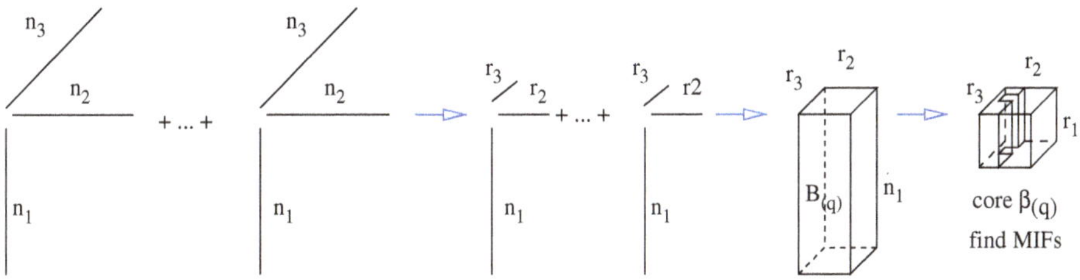

Fig. 3.8: $d = 3$: Finding MIFs in the coarse level core $\boldsymbol{\beta}_{(q)}$, $q = 1$, for the rank-R initial data on the coarse $n_1 \times n_2 \times n_3$ grid.

reduced to the SVD analysis of only a small subset of pr columns with, say, $p = 3, 4, 5$, whose indexes are predefined by the positions of MIFs. This dramatically reduces the cost of the ALS iteration on fine levels since the SVD applies only to small $n \times pr$ matrices.

Based on the above principles, we introduce the following multigrid computational scheme:

Algorithm (Algorithm MG-CP-Tucker ($\mathcal{C}_{R,\mathbf{n}_M} \to \mathcal{T}_{\mathcal{C}_R,\mathbf{r}}$)). Multigrid accelerated canonical-to-Tucker approximation:

1. Given $\mathbf{A}_m \in \mathcal{C}_{R,\mathbf{n}_m}$ in the form (3.10), corresponding to a sequence of grid parameters $n_m := n_0 2^m$, $m = 0, 1, \ldots, M$, choose the Tucker ranks $\mathbf{r} = (r, \ldots, r)$, a structural constant $p = O(1)$ (hence, $pr \ll r^{d-1}$), and the ALS iteration parameter $k_{\max}$.
2. For $m = 0$, solve the approximation problem CP-Tucker-ALS ($\mathcal{C}_{R,\mathbf{n}_0} \to \mathcal{T}_{\mathcal{C}_R,\mathbf{r}}$) and compute for $q = 1, \ldots, d$ the index set $J_{q,p}(n_0) \subset J_{\overline{r}_q}$ via identification of MIFs in the matrix unfolding $B_{(q)}$. This is done by using the maximum energy principle applied to the mode-q unfolding of the Tucker core, $\boldsymbol{\beta}_{(q)} = U^{(q)^{\mathrm{T}}} B_{(q)} \in \mathbb{R}^{r_q \times \overline{r}_q}$. Here the single hole index $\overline{r}_q$ is defined by (3.27) and $\#J_{q,p}(n_0) = pr$.
3. For $m = 1, \ldots, M$ perform the *MG accelerated Tucker approximation* by ALS iteration restricted to a small index set $J_{q,p}(n_0)$:
   3a) Compute the initial orthogonal side matrices on a level m by interpolation from the level $m-1$ (say by using cubic splines)
   $$V^{(q)} = V_m^{(q)} = \mathbf{I}_{m-1\to m}(V_{m-1}^{(q)}), \quad \text{for} \quad q = 1, \ldots, d.$$
   3b) For each $q = 1, \ldots, d$, fix $V^{(\ell)}$ ($\ell = 1, \ldots, d$, $\ell \neq q$) and perform:
   - Compute matrix products $V^{(\ell)^{\mathrm{T}}} U_m^{(\ell)} \in \mathbb{R}^{r_\ell \times R}$, $\ell = 1, \ldots, d; \ell \neq q$, and build the *restricted* mode-q matrix unfolding $B_{(q,p)}$,
     $$B_{(q,p)} = B_{(q)|J_{q,p}(n_0)} \in \mathbb{R}^{n_m \times pr},$$
     by calculating only pr fixed columns from the complete unfolding matrix $B_{(q)} \in \mathbb{R}^{n_m \times \overline{r}_q}$.

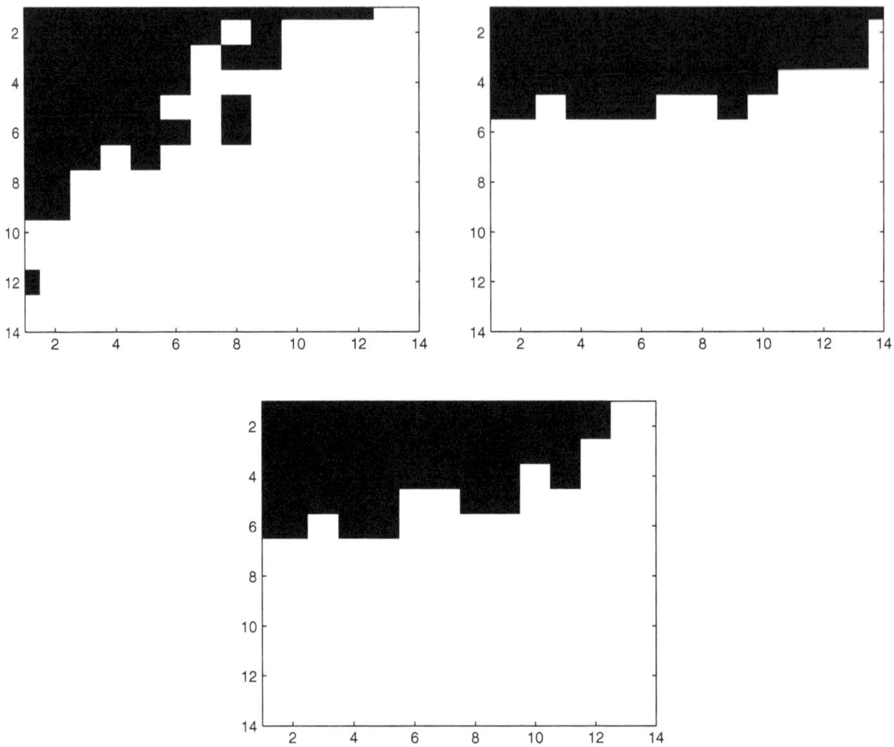

Fig. 3.9: MIFs: selected projections of fibers in the coarse level cores for computing $V^{(1)}$(left), $V^{(2)}$ (middle), and $V^{(3)}$ (right).

- Update the orthogonal matrix $V^{(q)} = V_m^{(q)} \in \mathbb{R}^{n_m \times r}$ via computing the r dimensional dominating subspace of the *restricted* matrix unfolding $B_{(q,p)}$ (truncated SVD of a small $n_m \times pr$ matrix).

4. If $m = M$, compute the rank-R core tensor $\boldsymbol{\beta} \in \mathcal{C}_{R,\mathbf{r}}$, as in Step 3 of the basic algorithm CP-Tucker-ALS($\mathcal{C}_{R,\mathbf{n}} \to \mathcal{T}_{\mathcal{C}_R,\mathbf{r}}$).

Figure 3.9 represents the index set $J_{q,p}(n_0)$ specifying selected MIFs in the coarse level cores, placed on the full $r \times r$ matrix space in a 3D case. The example corresponds to the multigrid Tucker approximation of the Hartree potential for the H_2O molecule for $d = 3$, $R = 1600$, $n = 1024$, $r = 14$, $p = 4$.

The next statement proves linear complexity of the MG-CP-Tucker-ALS algorithm.

Theorem 3.44. *Algorithm MG-CP-Tucker-ALS($\mathcal{C}_{R,\mathbf{n}_M} \to \mathcal{T}_{\mathcal{C}_R,\mathbf{r}}$) amounts to*

$$O(dRrn_M + dp^2r^2n_M)$$

operations per ALS loop, plus extra cost of the coarse mesh solver CP-Tucker-ALS ($\mathfrak{C}_{R,\mathbf{n}_0} \to \mathfrak{T}_{\mathfrak{C}_{R},\mathbf{r}}$). *The algorithm requires* $O(drn_M + drR)$ *storage to represent the result.*

Proof. Step 3a) requires $O(drn_m)$ operations and memory size. Note that for large m, we have $pr \leq n_m$, hence the complexity of the second part in Step 3b) is dominated by $O(dRrn_m + p^2r^2n_m)$ operations per iteration loop, and the same is true for the first part in Step 3b).
Rank-R representation of the core tensor $\boldsymbol{\beta} \in \mathfrak{C}_{R,\mathbf{r}}$ requires $O(drRn_m)$ operations and $O(drR)$ storage. Summing up over levels $m = 0, \ldots, M$, proves the result. □

Theorem 3.44 shows that the MG-CP-Tucker-ALS algorithm implements the fast rank reduction method that scales linearly in d, n_M, R and r on all refined levels except the coarsest level. Moreover, the complexity and approximation error of the MG Tucker approximation can be controlled by an adaptive choice of the governing parameters r, p, n_0 and $k_{\max}$.

We conclude that the multigrid accelerated scheme exhibits several favorable features:

- Robustness: fast local convergence of the ALS iteration ensured by good initial guess at all grid levels, obtained by extrapolation from the coarser level.
- Linear complexity scaling for the $\mathfrak{C}_{R,\mathbf{n}}$-input: $O(dRrn)$ up to the cost on the coarsest level.
- Complexity reduction: $O(n^{d+1}) \to O(n^d)$ on a full format target.
- Application to a rank structured tensor implementation of the convolution transform in $\mathbb{R}^d$ on large spatial $n^{\otimes d}$-grids (for example, for $d = 3$ and $n \approx 10^4 - 10^5$; see the discussion below).
- Applicability of the Richardson extrapolation on a sequence of refined grids; see for example [10, 270].

Exercise 3.45. Estimate the maximal size of $n \times n \times n$ tensors that are tractable for the HOSVD based Tucker-ALS approximation of complexity $O(n^4)$ applied to the full target.

We complete this section by presenting some numerical illustrations. Figure 3.10 demonstrates linear complexity scaling in R and in n (left), and plots singular values for the mode-1 matrix unfolding $B_{(1,p)}$, $p = 4$ (right), corresponding to the example in Figure 3.9. All data are represented on a sequence of $n \times n \times n$ grids.

Remark 3.46. The multigrid best Tucker approximation method applies to the full format $n \times n \times n$ tensors reducing the cost down to $n^{d+1} \to n^d$.

Note that in some cases the Tucker HOSVD approximation to full format $3D$ input can be based on the so called adaptive cross approximation [295].

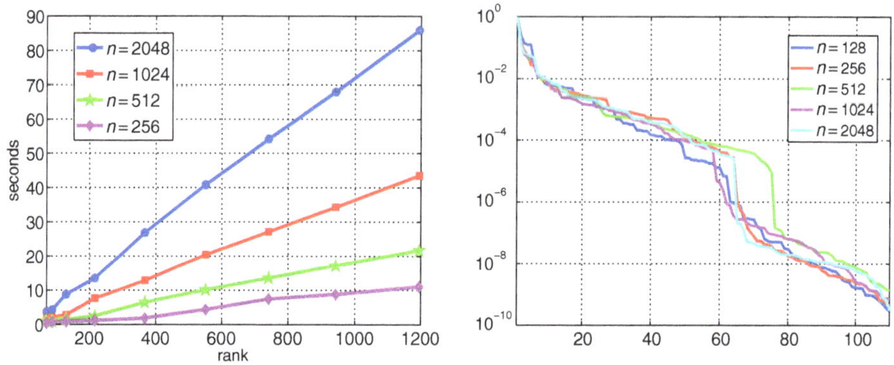

Fig. 3.10: Linear complexity scaling in R and in n (left). Singular values for the mode-1 matrix unfolding $B_{(1,p)}$, $p = 4$ (right).

3.5 Matrices in canonical and Tucker tensor formats

In this section, we consider the rank structured representation of matrices in the canonical and Tucker tensor formats. First, we define the rank-R and Tucker type matrices operating with the vectors represented as formated multidimensional tensors. Then we discuss the main properties of the Kronecker product and Kronecker sum of matrices. The important properties of a matrix exponential and an eigenvalue problem for the Kronecker sum will be addressed. We discuss how the matrix Lyapunov/Silvester equations can be solved by using the Kronecker product constructions. We introduce the Kronecker–Hadamard scalar product, which arises in the analysis of the Boltzmann equation. A special case of the Kronecker matrix rank in the case $d = 2$ will be considered. We conclude with a discussion on the complexity of the Kronecker matrix arithmetics.

3.5.1 Canonical and Tucker matrix (operator) formats

Tensor is a vector in the tensor product Hilbert space $\mathbb{V} = V_1 \otimes \cdots \otimes V_d = \mathbb{R}^{\mathcal{J}}$, $\mathcal{J} = J_1 \times \cdots \times J_d$. A rank-$R$ matrix $A \in \mathcal{M}_{R,\mathcal{I}\times\mathcal{J}}$ is supposed to represent the linear operator, which maps

$$\mathbb{R}^{\mathcal{I}\times\mathcal{J}} \ni A : \mathbb{V} \to \mathbb{W},$$

where $\mathbb{W} = W_1 \otimes \cdots \otimes W_d = \mathbb{R}^{\mathcal{I}}$, $\mathcal{I} = I_1 \times \cdots \times I_d$.

Definition 3.47. We call by $\mathcal{M}_{R,\mathcal{I}\times\mathcal{J}}$ a class of rank-R linear tensor-tensor operators (matrices) $A \in \mathbb{R}^{\mathcal{I}\times\mathcal{J}}$, $A : \mathbb{R}^{\mathcal{J}} \to \mathbb{R}^{\mathcal{I}}$,

$$A = \sum_{\nu=1}^{R} \alpha_\nu A_\nu^{(1)} \otimes \cdots \otimes A_\nu^{(d)}, \quad \alpha_\nu \in \mathbb{R}, \quad A_\nu^{(\ell)} \in \mathbb{R}^{I_\ell \times J_\ell}, \tag{3.35}$$

for which the matrix vector multiplication with the rank-1 tensor $\mathbf{V} \in \mathcal{C}_{1,\mathcal{J}}$ is defined by the rank-R canonical sum

$$\mathcal{A}\mathbf{V} := \sum_{\nu=1}^{R} \alpha_\nu A_\nu^{(1)} \mathbf{v}^{(1)} \otimes \cdots \otimes A_\nu^{(d)} \mathbf{v}^{(d)} \in \mathcal{C}_{R,\mathcal{I}}\,.$$

R is called the Kronecker/tensor rank of a matrix.

A parametrization via the canonical rank-R representation (3.35) requires a small storage size $dRn^2 \ll n^{2d}$. In turn, matrices $A_\nu^{(\ell)} : \mathbb{R}^{J_\ell} \to \mathbb{R}^{I_\ell}$ may have fully populated or data sparse structures.

Note that the Kronecker rank-R matrices in $\mathcal{M}_{R,\mathcal{I}\times\mathcal{J}}$ can be recognized as a "matricization" of canonical vectors in the rank-R CP tensor. In a similar way, the definition of rank-R matrices in $\mathcal{M}_{R,\mathcal{I}\times\mathcal{J}}$ can be extended to the Tucker format.

Definition 3.48. A matrix in the rank $\mathbf{r}$ Tucker format can be viewed as a 'matricization' of orthogonal vectors in the Tucker tensor, i.e., we say

$$\mathcal{A} = \boldsymbol{\beta} \times_1 U^{(1)} \times_2 U^{(2)} \cdots \times_d U^{(d)} \in \mathcal{M}_{\mathbf{r},\mathcal{I}\times\mathcal{J}}\,,$$

where $\boldsymbol{\beta} \in \mathbb{R}^{J_1\times\cdots\times J_d}$ is the $r_1 \times \cdots \times r_d$ core tensor, and

$$U^{(\ell)} = [\Phi_1^{(\ell)}, \ldots, \Phi_{r_\ell}^{(\ell)}] \in \mathbb{R}^{I_\ell\times J_\ell\times r_\ell}\,, \quad \Phi_k^{(\ell)} \in \mathbb{R}^{I_\ell\times J_\ell}\,.$$

By construction, the operator $\mathcal{A}$ maps $\mathcal{A} : \mathcal{C}_{1,\mathcal{J}} \to \mathcal{T}_{\mathbf{r},\mathcal{I}}$.

If tensors in $\mathbb{V}$ are vectorized (by unfolding to a vector, $\mathbf{V} \to \operatorname{vec}(\mathbf{V})$) then the respective matrix in $\mathcal{M}_{R,\mathcal{I}\times\mathcal{J}}$ can be represented by a sum of d Kronecker products of matrices of size $I_\ell \times J_\ell$ ($\ell = 1, \ldots, d$). This construction is supported by MATLAB. Likewise, the matrix in $\mathcal{M}_{\mathbf{r},\mathcal{I}\times\mathcal{J}}$ can be again represented as a sum of Kronecker products of matrices of size $I_\ell \times J_\ell$.

3.5.2 The Kronecker product of matrices revisited

We recall the definition of the Kronecker product of matrices:

Definition 3.49. The Kronecker product (KP) operation $A \otimes B$ of two matrices $A = [a_{ij}] \in \mathbb{R}^{m\times n}$, $B \in \mathbb{R}^{h\times g}$ is an $mh \times ng$ matrix that has the block representation $[a_{ij}B]$, $i = 1, \ldots, m$; $j = 1, \ldots, n$.

An equivalent representation of the Kronecker product in terms of the Khatri–Rao product reads

$$A \otimes B = [a_1 \otimes b_1\ a_1 \otimes b_2\ \ldots\ a_n \otimes b_{g-1}\ a_n \otimes b_g]\,.$$

A recursive extension of $A \otimes B$ to d-fold KP is based on the relation (Property 1 below)

$$A \otimes B \otimes C = (A \otimes B) \otimes C = A \otimes (B \otimes C)\,.$$

The following definition introduces a useful notation for the special Kronecker sum of matrices:

Definition 3.50. The Kronecker sum of $A \in \mathbb{R}^{m \times m}$ and $B \in \mathbb{R}^{n \times n}$ is defined by

$$A +^{\otimes} B = I_m \otimes B + A \otimes I_n .$$

This definition can be extended to the case of d-term Kronecker sum of matrices. The instructive particular example of the Kronecker sum is given by the finite difference (FD) Laplacian in d dimensions discretized on an $n \times \cdots \times n$ rectangular grid. Define the discrete FD discretization to the Laplacian on $H_0^1([0,1]^d)$ by

$$\Delta^{(d)} := \Delta \otimes I \otimes \cdots \otimes I + I \otimes \Delta \otimes I \otimes \cdots \otimes I + \cdots + I \otimes \cdots I \otimes \Delta \in \mathbb{R}^{n^d \times n^d} ,$$

where $I = I_n$ is the $n \times n$ identity matrix and $\Delta = \frac{1}{(n+1)^2} \operatorname{tridiag}\{-1, 2, -1\} \in \mathbb{R}^{n \times n}$ is a three-diagonal matrix. The matrix $\Delta^{(d)}$ has the Kronecker/tensor canonical rank $R = d$.

Exercise 3.51. Prove that $\Delta^{(d)} \in \mathcal{M}_{\mathbf{r}, \mathcal{I} \times \mathcal{J}}$ with the Tucker rank $\mathbf{r} = (2, 2, \ldots, 2)$.

3.5.3 General properties of the Kronecker product of matrices

The Kronecker product of matrices inherits many standard features from the matrix algebra; see [78, 127, 132, 351]. The following properties can be easily verified:

(1) Given matrices $A = [a_{ij}] \in \mathbb{R}^{m \times n}$, $B \in \mathbb{R}^{h \times n}$ and $C \in \mathbb{R}^{s \times t}$, the KP satisfies the *associative law*,

$$(A \otimes B) \otimes C = A \otimes (B \otimes C) = A \otimes B \otimes C ,$$

and therefore we skip the brackets above. The matrix $A \otimes B \otimes C := (A \otimes B) \otimes C$ has (mhs) rows and (ngt) columns.

(2) Let $C \in \mathbb{R}^{n \times r}$ and $D \in \mathbb{R}^{g \times s}$, then the standard matrix matrix product in the Kronecker format takes the form

$$(A \otimes B)(C \otimes D) = (AC) \otimes (BD) .$$

The corresponding extension to dth order products reads

$$(A_1 \otimes \cdots \otimes A_d)(B_1 \otimes \cdots \otimes B_d) = (A_1 B_1) \otimes \cdots \otimes (A_d B_d) .$$

(3) The *distributive law* reads

$$(A + B) \otimes (C + D) = A \otimes C + A \otimes D + B \otimes C + B \otimes D .$$

(4) Rank relation: $\operatorname{rank}(A \otimes B) = \operatorname{rank}(A) \operatorname{rank}(B)$, where the notation rank means the standard matrix rank.

Invariance of some matrix properties:

(5) If A and B are diagonal then $A \otimes B$ is also diagonal, and vice versa (if $A \otimes B \neq 0$).
(6) The upper/lower triangular matrices are preserved.
(7) Let A, B be Hermitian/normal/orthogonal matrices, i.e., $A^* = A$, $A^{-1} = A$, and $A^{-1} = A^T$, respectively. Then $A \otimes B$ is of the corresponding type.
(8) Let $A \in \mathbb{R}^{n \times n}$ and $B \in \mathbb{R}^{m \times m}$. Then

$$\det(A \otimes B) = (\det A)^m (\det B)^n .$$

Hint: First consider the case of diagonal matrices. Apply the induction with respect to the size of A.

Exercise 3.52. In general $A \otimes B \neq B \otimes A$. Give the condition on A and B that provides $A \otimes B = B \otimes A$.

3.5.4 Matrix operations with Kronecker products and sums

Many matrix operations with Kronecker products and sums can be reduced to those between the generating matrices.

Theorem 3.53. *Let $A \in \mathbb{R}^{n \times n}$ and $B \in \mathbb{R}^{m \times m}$ be invertible matrices. Then*

$$(A \otimes B)^{-1} = A^{-1} \otimes B^{-1} .$$

Proof. Since $\det(A) \neq 0$, $\det(B) \neq 0$ and taking into account the above property (**8**), we have $\det(A \otimes B) \neq 0$. Thus $(A \otimes B)^{-1}$ exists and we arrive at

$$(A^{-1} \otimes B^{-1})(A \otimes B) = (A^{-1}A) \otimes (B^{-1}B) = I_{nm} ,$$

which completes the proof. □

Lemma 3.54. *Let $A \in \mathbb{R}^{n \times n}$ and $B \in \mathbb{R}^{m \times m}$ be unitary matrices. Then $A \otimes B$ is a unitary matrix.*

Proof. Indeed, since $A^* = A^{-1}$, $B^* = B^{-1}$ we have

$$(A \otimes B)^* = A^* \otimes B^* = A^{-1} \otimes B^{-1} = (A \otimes B)^{-1} .$$ □

Define the commutator of matrices by $[A, B] := AB - BA$.

Lemma 3.55. *Let $A \in \mathbb{R}^{n \times n}$ and $B \in \mathbb{R}^{m \times m}$. Then*

$$[A \otimes I_n, I_m \otimes B] = 0 \in \mathbb{R}^{m^2 \times n^2} .$$

Proof. Simple calculations show that

$$\begin{aligned}[A \otimes I_n, I_m \otimes B] &= (A \otimes I_n)(I_m \otimes B) - (I_m \otimes B)(A \otimes I_n) \\ &= A \otimes B - A \otimes B = 0 .\end{aligned}$$ □

Lemma 3.56. *Let $A, B \in \mathbb{R}^{n\times n}$, $C, D \in \mathbb{R}^{m\times m}$ and $[A, B] = 0$, $[C, D] = 0$. Then*

$$[A \otimes C, B \otimes D] = 0\,.$$

Proof. Apply the identity $(A \otimes B)(C \otimes D) = (AC) \otimes (BD)$. □

Lemma 3.57. *Let $A \in \mathbb{R}^{n\times n}$ and $B \in \mathbb{R}^{m\times m}$. Then*

$$tr(A \otimes B) = tr(A)tr(B)\,.$$

Proof. Since $\operatorname{diag}(a_{ii}B) = a_{ii}\operatorname{diag}(B)$, we have

$$tr(A \otimes B) = \sum_{i=1}^{n}\sum_{j=1}^{m} a_{ii}b_{jj} = \sum_{i=1}^{n} a_{ii}\sum_{j=1}^{m} b_{jj} = tr(A)tr(B)\,.$$ □

The above properties can be combined in different ways, which may lead to more involved relations, see also [127] for further details.

3.5.5 Functions of the Kronecker products

Many properties of the matrix valued functions of the Kronecker product can be expressed in terms of functions applied to the individual arguments.

Theorem 3.58. *Let $A, B, I \in \mathbb{R}^{n\times n}$. Then*

$$\exp(A \otimes I + I \otimes B) = (\exp A) \otimes (\exp B)\,.$$

Proof. Since $[A \otimes I, I \otimes B] = 0$, we have

$$\exp(A \otimes I + I \otimes B) = \exp(A \otimes I)\exp(I \otimes B)\,.$$

Furthermore, since

$$\exp(A \otimes I) = \sum_{k=0}^{\infty}\frac{(A \otimes I)^k}{k!}\,, \qquad \exp(I \otimes B) = \sum_{m=0}^{\infty}\frac{(I \otimes B)^m}{m!}\,,$$

the arbitrary term in $\exp(A \otimes I)\exp(I \otimes B)$ can be represented by

$$\frac{1}{k!}\frac{1}{m!}(A \otimes I)^k(I \otimes B)^m\,.$$

Imposing

$$(A \otimes I)^k(I \otimes B)^m = (A^k \otimes I^k)(I^m \otimes B^m) = (A^k \otimes I)(I \otimes B^m) \equiv A^k \otimes B^m\,,$$

we finally arrive at

$$\frac{1}{k!}\frac{1}{m!}(A \otimes I)^k(I \otimes B)^m = \left(\frac{1}{k!}A^k\right) \otimes \left(\frac{1}{m!}B^m\right)\,.$$

This completes the proof. □

Theorem 3.58 can be extended to the case of many-term sums as follows:

$$\exp(A_1 \otimes I \otimes \cdots \otimes I + I \otimes A_2 \otimes \cdots \otimes I + \cdots + I \otimes \cdots \otimes I \otimes A_d) = (\mathrm{e}^{A_1}) \otimes \cdots \otimes (\mathrm{e}^{A_d}) .$$

Remark 3.59. Similar properties can be proven for other analytic functions, e.g.,

$$\sin(I_n \otimes A) = I_n \otimes \sin(A) ,$$

$$\sin(A \otimes I_m + I_n \otimes B) = \sin(A) \otimes \cos(B) + \cos(A) \otimes \sin(B) ,$$

$$\sin(A \otimes I_m + I_n \otimes B) = \frac{\sin(A) \otimes \sin(B + (b-a)I)}{\sin(b-a)} + \frac{\sin(A + (a-b)I) \otimes \sin(B)}{\sin(a-b)}$$

for all values a, b such that $\sin(a - b) \neq 0$. The latter can be extended to

$$\sin(A_1 \otimes I \otimes \cdots \otimes I + \cdots + I \otimes \cdots \otimes I \otimes A_d) ,$$

which is similar to the rank-d canonical decomposition of the function $\sin(x_1+\cdots+x_d)$.

Other simple properties can be formulated as follows:

$$(A \otimes B)^{\mathrm{T}} = A^{\mathrm{T}} \otimes B^{\mathrm{T}} , \quad (A \otimes B)^* = A^* \otimes B^* .$$

3.5.6 Eigenvalue problem for Kronecker sums

The eigenvalue problem for the Kronecker sums of matrices can be analyzed by using only the spectral information for the generating matrices.

Lemma 3.60. *Let $A \in \mathbb{R}^{m\times m}$ and $B \in \mathbb{R}^{n\times n}$ have the eigendata $\lambda_1, \ldots, \lambda_m, u_1, \ldots, u_m$, and $\mu_1, \ldots, \mu_n, v_1, \ldots, v_n$, respectively. Then $A \otimes B$ has the eigenvalues $\lambda_j \mu_k$ with the corresponding eigenvectors $u_j \otimes v_k$, $1 \leq j \leq m$, $1 \leq k \leq n$.*

Using the previous lemma the eigenvalue problem for the Kronecker sums of matrices can be reduced to the spectral problems for individual Kronecker factors.

Theorem 3.61. *Under the conditions of Lemma 3.60 the eigenvalues/eigenfunctions of $A \otimes I_n + I_m \otimes B$ can be represented by $\lambda_j + \mu_k$ and $u_j \otimes v_k$ respectively.*

Proof. Due to the previous lemma we easily calculate

$$\begin{aligned}(A \otimes I_n + I_m \otimes B)(u_j \otimes v_k) &= (A \otimes I_n)(u_j \otimes v_k) + (I_m \otimes B)(u_j \otimes v_k) \\ &= (Au_j) \otimes (I_n v_k) + (I_m u_j) \otimes (Bv_k) \\ &= (\lambda_j u_j) \otimes v_k + u_j \otimes (\mu_k v_k) \\ &= (\lambda_j + \mu_k)(u_j \otimes v_k) ,\end{aligned}$$

which proves the Kronecker structure of the eigenfunctions. □

3.5.7 Application to matrix Lyapunov/Sylvester equations

Recall that for a matrix $A \in \mathbb{R}^{m\times n}$, we use the *vector representation* $A \to \mathrm{vec}(A) \in \mathbb{R}^{mn}$, where $\mathrm{vec}(A)$ is an $nm \times 1$ vector obtained by 'stacking' A's columns

$$\mathrm{vec}(A) := [a_{11}, \ldots, a_{n1}, a_{12}, \ldots, a_{nm}]^{\mathrm{T}} .$$

In this way, $\mathrm{vec}(A)$ is a rearranged version of A.

Lemma 3.62. *Let* $A \in \mathbb{R}^{m\times m}$, $Y \in \mathbb{R}^{m\times n}$, $B \in \mathbb{R}^{n\times n}$. *Then*

$$\mathrm{vec}(AYB) = (B^{\mathrm{T}} \otimes A)\,\mathrm{vec}(Y) .$$

Proof. Applying the simple columnwise representation to the matrix matrix product,

$$AY = [A\mathbf{y}_1, \ldots, A\mathbf{y}_n] ,$$

and similar to $(AY)B$, the representation follows.

Taking into account the previous lemma, the matrix Sylvester equation for $X \in \mathbb{R}^{m\times n}$,

$$AX + XB^{\mathrm{T}} = G \in \mathbb{R}^{m\times m} , \tag{3.36}$$

with $A \in \mathbb{R}^{m\times m}$, $B \in \mathbb{R}^{n\times n}$, can be rewritten in the vector form

$$(I_n \otimes A + B \otimes I_m)\,\mathrm{vec}(X) = \mathrm{vec}(G) .$$

Now the solvability conditions and certain solution methods can be derived (compare to the results for eigenvalue problems). In particular, in view of Theorem 3.61, the matrix Sylvester equation (3.36) is uniquely solvable if

$$\lambda_j(A) + \mu_k(B) \neq 0 .$$

In the case $A = B$ we arrive at the Lyapunov equation. □

Remark 3.63. Since $I_n \otimes A$ and $B \otimes I_m$ commute, we can apply methods based on the R-term sinc quadrature approximation to the Laplace transform representation of the inverse

$$(I_n \otimes A + B \otimes I_m)^{-1} = \int_{\mathbb{R}_+} \mathrm{e}^{-t(I_n\otimes A + B\otimes I_m)}\,dt = \int_{\mathbb{R}_+} \mathrm{e}^{-tA} \otimes \mathrm{e}^{-tB}\,dt .$$

If A and B represent the discrete elliptic operators in $\mathbb{R}^d$ with separable coefficients, we obtain the low rank tensor product approximation to the Sylvester solution operator, which can be applied as the direct solver in the form

$$\mathrm{vec}(X) = (I_n \otimes A + B \otimes I_m)^{-1}\,\mathrm{vec}(G) .$$

Exercise 3.64. Approximate formally the 2D Laplacian inverse by the sinc quadrature

$$(\Delta^{(2)})^{-1} = (I_n \otimes \Delta + \Delta \otimes I_n)^{-1} \approx \sum_{k=-M}^{M} c_k \mathrm{e}^{-t_k\Delta} \otimes \mathrm{e}^{-t_k\Delta} .$$

3.5.8 Kronecker–Hadamard scalar product

In this section, we introduce the so called Kronecker–Hadamard scalar product, which is essential in the analysis of the Boltzmann equation; see [207].

Given tensors $Y \otimes U \in \mathbb{R}^{\mathcal{J}\times\mathcal{I}}$ with $U \in \mathbb{R}^{\mathcal{I}}$, $Y \in \mathbb{R}^{\mathcal{J}}$, and $B \in \mathbb{R}^{\mathcal{L}\times\mathcal{I}}$, let $\mathbf{T}: \mathbb{R}^{\mathcal{L}} \to \mathbb{R}^{\mathcal{J}}$ be the linear operator (matrix) that maps tensors defined on the index set $\mathcal{L}$ into those defined on $\mathcal{J}$. In particular, we have $\mathbf{T} \cdot B \in \mathbb{R}^{\mathcal{J}\times\mathcal{I}}$.

Definition 3.65. The Hadamard 'scalar' product $[D, C]_{\mathcal{I}} \in \mathbb{R}^{\mathcal{K}}$ of two tensors $D := [D_{\mathbf{i},\mathbf{k}}] \in \mathbb{R}^{\mathcal{I}\times\mathcal{K}}$ and $C := [C_{\mathbf{i},\mathbf{k}}] \in \mathbb{R}^{\mathcal{I}\times\mathcal{K}}$ is defined by

$$[D, C]_{\mathcal{I}} := \sum_{\mathbf{i}\in\mathcal{I}} [D_{\mathbf{i},\mathcal{K}}] \odot [C_{\mathbf{i},\mathcal{K}}] ,$$

where $\odot$ denotes the Hadamard product over the index set $\mathcal{K}$ and $[D_{\mathbf{i},\mathcal{K}}] := [D_{\mathbf{i},\mathbf{k}}]_{\mathbf{k}\in\mathcal{K}}$.

Lemma 3.66. *Let U, Y, B and $\mathbf{T}$ be given as above. Then, with $\mathcal{K} = \mathcal{J}$, the following identity is valid:*

$$[U \otimes Y, \mathbf{T}B]_{\mathcal{I}} = Y \odot (\mathbf{T}\,[U, B]_{\mathcal{I}}) \in \mathbb{R}^{\mathcal{J}} . \tag{3.37}$$

Proof. By definition of the Hadamard scalar product we have

$$\begin{aligned} [U \otimes Y, \mathbf{T}B]_{\mathcal{I}} &= \sum_{\mathbf{i}\in\mathcal{I}} [U \otimes Y]_{\mathbf{i},\mathcal{J}} \odot [\mathbf{T}B]_{\mathbf{i},\mathcal{J}} \\ &= \sum_{\mathbf{i}\in\mathcal{I}} [[U]_{\mathbf{i}} Y]_{\mathcal{J}} \odot [\mathbf{T}B]_{\mathbf{i},\mathcal{J}} \\ &= Y \odot \left(\sum_{\mathbf{i}\in\mathcal{I}} [U]_{\mathbf{i}} [\mathbf{T}B]_{\mathbf{i},\mathcal{J}} \right) \\ &= Y \odot \left(\mathbf{T} \sum_{\mathbf{i}\in\mathcal{I}} [U]_{\mathbf{i}} [B]_{\mathbf{i},\mathcal{L}} \right) , \end{aligned}$$

then the assertion follows. □

Identity (3.37) is particularly important in application to the Boltzmann equation ([207]), since in the right hand side the operator $\mathbf{T}$ is removed from the scalar product in $\mathcal{I}$ and hence it applies only once.

3.5.9 Remarks on rank structured operators (matrices)

Remark 3.67. By Definition 3.47 the matrix vector multiplication with the rank-1 tensor $\mathbf{V} \in \mathcal{C}_{1,\mathcal{J}}$ is defined by the rank-R canonical sum

$$A\mathbf{V} := \sum_{\nu=1}^{R} \alpha_\nu A_\nu^{(1)} \mathbf{v}^{(1)} \otimes \cdots \otimes A_\nu^{(d)} \mathbf{v}^{(d)} \in \mathcal{C}_{R,\mathcal{J}} .$$

Note that ⊗ is also traditionally used for the Kronecker product of matrices. The equivalent and alternative consistent notation is based on the contracted product operation,

$$A = \sum_{\nu=1}^{R} \alpha_\nu A_\nu^{(1)} \times_2 \cdots \times_d A_\nu^{(d)} , \quad \alpha_\nu \in \mathbb{R} , \quad A_\nu^{(d)} \in \mathbb{R}^{I_\ell \times J_\ell} .$$

However, if there is no confusion, we continue using ⊗ in the matrix tensor product as in Definition 3.47.

In particular, for the conventional 'long index' representation of vectors, the discrete Laplacian in $\mathbb{R}^d$ takes the form as in the example in Section 3.5.2. In the case of tensor representation of vectors, we can use the equivalent contracted product notation to write

$$\Delta^{(d)} := \Delta \times_2 I \times_3 \cdots \times_d I + I \times_2 \Delta \times_3 I \times_4 \cdots \times_d I + \cdots + I \times_2 \ldots I \times_d \Delta ,$$

with $\Delta^{(d)} \in \mathbb{R}^{(n\times n)\times\cdots\times(n\times n)}$. If there is no confusion, we continue using ⊗ in the notation for the discrete Laplacian type operators (stiffness matrices) in $\mathbb{R}^d$.

3.5.10 Comments on Kronecker matrix rank if $d = 2$

If $d = 2$, estimate of the Kronecker matrix rank can be reduced to computation of the standard matrix rank ([151, 351]).

Let $A = [a(i, j)]_{1\le i,j\le N}$, $N = n_1 n_2$. Use the bijection

$$i \leftrightarrow (i_1, i_2) , \quad j \leftrightarrow (j_1, j_2) , \quad 1 \le i_1, j_1 \le n_1 , \quad 1 \le i_2, j_2 \le n_2 ,$$

defined by FORTRAN style ordering,

$$i = i_1 + (i_2 - 1)n_1 , \quad j = j_1 + (j_2 - 1)n_1 , \quad 1 \le i_1, j_1 \le n_1 , \quad 1 \le i_2, j_2 \le n_2 .$$

The matrix A can be indexed by $a(i, j) = a(i_1, i_2, j_1, j_2)$. Introduce a new matrix $\tilde{A}$ of size $n_1^2 \times n_2^2$, indexed by the respective pairs (i_1, j_1), (i_2, j_2) (long indexes),

$$\tilde{a}(i_1, j_1; i_2, j_2) = a(i_1, i_2, j_1, j_2) .$$

New indexing also defines the bijective mapping $\mathcal{P}\colon A \to \tilde{A}$ (a rearranged version of A), which preserves the Frobenius norm.

By the way, we note that there is no permutation such that $\tilde{A} = PAP^{\mathrm{T}}$.

Applying the above construction to the rank-R Kronecker product sum,

$$A \to A_R = \sum_{k=1}^{R} U_k \otimes V_k \Longleftrightarrow \mathcal{P}(A_R) := \tilde{A}_R = \sum_{k=1}^{R} \mathbf{v}_k \mathbf{u}_k^{\mathrm{T}} ,$$

where $U_k = [u_k(i_2, j_2)]_{1\le i_2,j_2\le n_2}$, $V_k = [v_k(i_1, j_1)]_{1\le i_1,j_1\le n_1}$, $\mathbf{v}_k \in \mathbb{R}^{n_1^2}$, $\mathbf{u}_k \in \mathbb{R}^{n_2^2}$.

Finally, we note that the problem of finding a Kronecker tensor rank approximation A_R of A is identical to the problem of finding a low rank approximation $\tilde{A}_R$ of $\tilde{A}$.

3.5.11 Complexity of the Kronecker matrix arithmetics

We say that a matrix $A \in \mathcal{M}_{R,\mathcal{I}\times\mathcal{J}}$ has the $\mathcal{S}$-inherited Kronecker tensor product structure ($\mathcal{S}$ structure) if the factors $A_k^{(\ell)} \in \mathcal{S}$, where $\mathcal{S}$ is a class of data sparse matrices of complexity $O(n \log n)$, understood with respect to the storage size or the complexity of matrix-vector and matrix matrix operations. The matrix classes $\mathcal{S}$ may include the sparse matrices, $\mathcal{H}$ matrices, Toeplitz, Hankel or circulant matrices, matrix of the FFT, low rank matrices, etc.

Let $N = n^d$, and assume that $A \in \mathcal{M}_{R,\mathcal{I}\times\mathcal{J}}$ has the $\mathcal{S}$-inherited Kronecker tensor product structure. Then the following complexity issues can be specified:

- *Data compression.* The storage for A is estimated by $\mathcal{O}(dRn) = \mathcal{O}(dRN^{1/d})$. Hence, we enjoy *sublinear complexity.*
- *Matrix-by-vector complexity of* $A\mathbf{x}$, $\mathbf{x} \in \mathbb{C}^N$. For general $\mathbf{x}$ one has the linear cost $\mathcal{O}(dRN \log n)$. If $\mathbf{x} = \mathbf{x}^1 \times \cdots \times \mathbf{x}^d$, $\mathbf{x}^i \in \mathbb{C}^n$, we again arrive at a *sublinear complexity* $\mathcal{O}(dRn \log n) = \mathcal{O}(RN^{1/d} \log N^{1/d})$.
  Remarkably, in the case of general vector $\mathbf{x}$, this result is much better compared with the corresponding matrix-by-vector complexity.
- *Matrix-by matrix complexity of* AB, $A \odot B$ *and* $A \otimes \mathbf{b}$. The $\mathcal{S}$ structure of the Kronecker factors leads to $\mathcal{O}(dR^2 n \log n) = \mathcal{O}(R^2 N^{1/d} \log N^{1/d})$ arithmetic operations instead of $\mathcal{O}(N^3)$.

3.6 From additive to multiplicative dimension splitting

3.6.1 Making high dimensional functions and operators tractable

The problem of solving PDEs in higher dimensions arises in many modern applications. In particular, in the density matrix renormalization group (DMRG) for full configuration interaction (FCI) electronic structure calculations, molecular dynamics, quantum computing, and quantum information theory, as well as tensor networks in quantum chemistry. Another class of problems includes stochastic/parametric PDEs, machine learning, atmospheric modeling, FEM/BEM calculations in $\mathbb{R}^d$, financial mathematics, etc.

The main computational issue is concerned with exponential complexity scaling in d ('curse of dimensionality'): the traditional numerical methods of $O(n^d)$ linear complexity in the grid size become nontractable. In view of the approximation aspects the 'curse of dimensionality' appears in the form of a slow convergence rate in terms of the total number of degrees of freedom: for functions $u \in C^s$ the FEM approximation error decays as $O(N^{-s/d})$ with $N = n^d$.

Tensor numerical methods based on the rank structured representations in the CP or Tucker type tensor formats can be efficiently applied in the case of moderate dimensions or for the analytic sinc based approximation. The limitations of these additive type formats were already discussed in the previous chapter.

The challenging goal in solving the basic PDEs, discretized over $\underbrace{n \times n \times \cdots \times n}_{d}$ grids, by avoiding the 'curse of dimensionality' could be realized by using another concept of separation of variables, specifically by changing from the additive to multiplicative dimension splitting. The general class of such tensor formats, traditionally called the matrix product states (MPS) representations, have for a long time been applied in computational quantum chemistry and other scientific computations. The well developed version of the MPS format is called the tensor train (TT) representation.

In recent years fast and robust MPS/TT based numerical methods have been developed for representation of d-variate functions, operators, and for solving physical equations in $\mathbb{R}^d$ with linear $O(dn)$-scaling in d. Moreover, recently introduced quantics-TT (QTT) tensor approximation allows the supercompressed log-volume representation of functions and operators in $\mathbb{R}^d$ with $O(d \log n)$ storage complexity (the number of representation parameters).

3.6.2 Matrix product states and tensor train formats

The product type representation of dth order tensors, which is called in the physical literature the matrix product states (MPS) decomposition, was introduced and successfully applied in DMRG quantum computations [355, 356, 365], and, independently, in quantum molecular dynamics as the multilayer (ML) multiconfiguration time dependent Hartree (MCTDH) methods [258, 277, 360]. Representations by the MPS type formats reduce the complexity of storage to $O(dr^2N)$, where r is the maximal directional rank parameter.

In recent years various versions of the MPS type tensor format were discussed and further investigated in the mathematical literature including the hierarchical dimension splitting [206], the tensor train (TT) [289, 292], the tensor chain (TC) and combined Tucker-TT [197] and QTT-Tucker [87] formats, as well as the hierarchical Tucker representation [138] that belongs to the class of ML-MCTDH methods [360], or more generally tensor network states models. The MPS type tensor approximation was proved by extensive numerics to be efficient in high dimensional electronic/molecular structure calculations, in molecular dynamics, and in quantum information theory (see survey papers [170, 209, 298, 322, 355]).

The TT format, that is the particular case of MPS type factorization (open boundary conditions), can be defined as follows. For a given rank parameter $\mathbf{r} = (r_0, \dots, r_d)$, and the respective index sets $J_\ell = \{1, \dots, r_\ell\}$ ($\ell = 0, 1, \dots, d$), with the constraint $J_0 = J_d = \{1\}$ (i.e., $r_0 = r_d$), the rank-$\mathbf{r}$ TT format contains all elements $\mathbf{A} = [a(i_1, \dots, i_d)] \in \mathbb{W}_{\mathbf{n}}$, which can be represented as the contracted products of 3-tensors over the d-fold product index set $\mathcal{J} := \times_{\ell=1}^{d} J_\ell$, such that

$$\mathbf{A} = \sum_{\alpha \in \mathcal{J}} \mathbf{a}_{\alpha_1}^{(1)} \otimes \mathbf{a}_{\alpha_1,\alpha_2}^{(2)} \otimes \cdots \otimes \mathbf{a}_{\alpha_{d-1}}^{(d)} \equiv \mathbf{A}^{(1)} \bowtie \mathbf{A}^{(2)} \bowtie \cdots \bowtie \mathbf{A}^{(d)},$$

where $\mathbf{a}^{(\ell)}_{\alpha_\ell,\alpha_{\ell+1}} \in \mathbb{R}^{N_\ell}$, $(\ell = 1, \ldots, d)$, and $\mathbf{A}^{(\ell)} = [\mathbf{a}^{(\ell)}_{\alpha_\ell,\alpha_{\ell+1}}]$ is the vector valued $r_\ell \times r_{\ell+1}$ matrix (3-tensor). Here and in the following (e.g., in Definition 3.78) the rank product operation '$\bowtie$' is defined as a regular matrix product of the two core matrices, their fibers (blocks) being multiplied by means of a tensor product [177]. In the index notation we have

$$a(i_1, \ldots, i_d) = \sum_{\alpha_1=1}^{r_1} \cdots \sum_{\alpha_d=1}^{r_d} \mathbf{a}^{(1)}_{\alpha_1}(i_1)\mathbf{a}^{(2)}_{\alpha_1,\alpha_2}(i_2)\cdots \mathbf{a}^{(d)}_{\alpha_{d-1}}(i_d) \equiv A^{(1)}(i_1)A^{(2)}(i_2)\ldots A^{(d)}(i_d),$$

such that the latter is written in the *matrix product* form (explaining the notion MPS), where $A^{(\ell)}(i_\ell)$ is the $r_{\ell-1} \times r_\ell$ matrix.

Definition 3.68 (Tensor chain format). Given the index set $\mathcal{J} := \times_{\ell=1}^{d} J_\ell, J_\ell = \{1, \ldots, r_\ell\}$, $(\ell = 0, 1, \ldots, d)$, with the periodicity constraints $J_0 = J_d$. The rank-$\mathbf{r}$ Tensor Chain (TC) format

$$\mathrm{TC}[\mathbf{r}, \mathbf{n}, d] \equiv \mathrm{TC}[\mathbf{r}] \subset \mathbb{V}_{\mathbf{n}}, \quad \mathbf{n} = (n_1, \ldots, n_d) - d - \text{fold},$$

contains all $\mathbf{G} \in \mathbb{V}_{\mathbf{n}}$ that can be presented as the chain of contracted products of 3-tensors over the index set $\mathcal{J}$,

$$\mathbf{G} = \{\times_\ell\}_{\ell=1}^{d} \mathbf{G}^{(\ell)} \quad \text{with 3-tensors} \quad \mathbf{G}^{(\ell)} \in \mathbb{R}^{r_{\ell-1} \times n_\ell \times r_\ell}.$$

In coordinate representation we have

$$g(i_1, \ldots, i_d) = \sum_{\alpha} \mathbf{g}^{(1)}_{\alpha_1}[i_1]\mathbf{g}^{(2)}_{\alpha_1\alpha_2}[i_2]\ldots \mathbf{g}^{(d)}_{\alpha_{d-1}}[i_d] \equiv G^{(1)}[i_1]G^{(2)}[i_2]\ldots G^{(d)}[i_d].$$

If $J_0 = J_d = \{1\}$ (disconnected chain), TC format defines the tensor train representation. Denote this set of tensors by $\mathrm{TC}[\mathbf{r}] \supset \mathrm{TT}[\mathbf{r}]$.

Example 3.69. Figure 3.11 illustrates the TT representation of a fifth order tensor: each particular entry indexed by $(i_1, i_2, \ldots, i_5)$ is factorized as a product of five matrices, thus explaining the name matrix product states, MPS.

In general, the tensor chain format corresponds to the case $J_0 = J_d \neq \{1\}$. In some cases TC tensor can be represented as a sum of not more than r_* TT tensors ($r_* = \min r_\ell$), which can be converted to the TT tensor based on multilinear algebra operations like sum-and-compress. The storage cost for both TC and TT formated tensors is bounded by $O(dr^2N)$, $r = \max r_\ell$.

Clearly, one and the same tensor might have different ranks in different formats (and, hence, different numbers of representation parameters). The next example considers the Tucker and TT representations of a *function related* canonical tensor $\mathbf{F} := T(f)$ obtained by sampling of the function $f(x) = x_1 + \cdots + x_d$, $x \in [-1, 1]^d$ on the Cartesian grid of size $N^{\otimes d}$ and specified by N vectors (N is an even number) $\mathbf{x}_\ell = \{-1 - h/2 + ih\}_{i=1}^{N}$, $(h = 2/N, \ell = 1, \ldots, d)$ and all-ones vector $\mathbf{1} \in \mathbb{R}^N$,

$$\mathbf{F} = \mathbf{x}_1 \otimes \mathbf{1} \otimes \cdots \otimes \mathbf{1} + \cdots + \mathbf{1} \otimes \cdots \otimes \mathbf{1} \otimes \mathbf{x}_d.$$

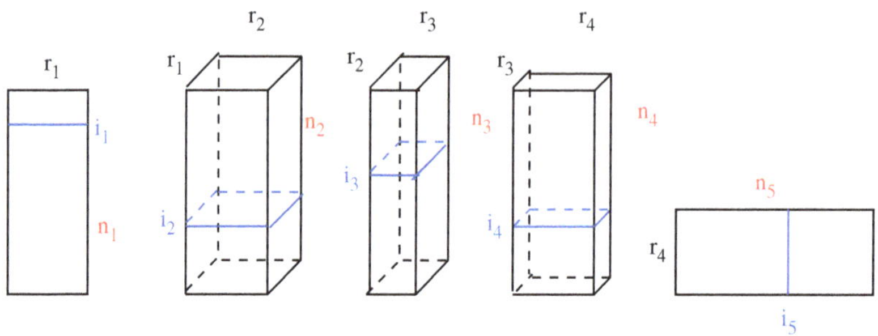

Fig. 3.11: Visualizing a fifth order TT tensor.

The canonical rank of this tensor can be proven to be exactly d, [246]. The following tensor decompositions can be verified by direct calculations.

Example 3.70. We have $\operatorname{rank}_{\mathrm{Tuck}}(\mathbf{F}) = 2$, with the explicit Tucker representation

$$\mathbf{F} = \sum_{\mathbf{k}=\mathbf{1}}^{\mathbf{2}} b_{\mathbf{k}} V_{k_1}^{(1)} \otimes \cdots \otimes V_{k_d}^{(d)}, \quad V_1^{(\ell)} = \mathbf{1},\, V_2^{(\ell)} = \mathbf{x}_\ell\,, \quad [b_{\mathbf{k}}] \in \bigotimes_{\ell=1}^{d} \mathbb{R}^2\,,$$

where the Tucker core is calculated by the contracted product $[b_{\mathbf{k}}] = \mathbf{F} \times_1 [\mathbf{1}, \mathbf{x}_1] \times_2 \cdots \times_d [\mathbf{1}, \mathbf{x}_d]$. Moreover, $\operatorname{rank}_{\mathrm{TT}}(\mathbf{F}) = 2$ holds due to the exact decomposition,

$$\mathbf{F} = \begin{bmatrix} \mathbf{x}_1 & \mathbf{1} \end{bmatrix} \bowtie \begin{bmatrix} \mathbf{1} & 0 \\ \mathbf{x}_2 & \mathbf{1} \end{bmatrix} \bowtie \cdots \bowtie \begin{bmatrix} \mathbf{1} & 0 \\ \mathbf{x}_{d-1} & \mathbf{1} \end{bmatrix} \bowtie \begin{bmatrix} \mathbf{1} \\ \mathbf{x}_d \end{bmatrix}.$$

Note that the similar decompositions remain true for a tensor generated by the function

$$f(x) = f_1(x_1) + \cdots + f_d(x_d)\,, \quad x \in [0,1]^d\,.$$

3.6.3 Specific features of the TT factorization

By definition, a tensor $\mathbf{G} \in TC[\mathbf{r}]$ is represented (approximated) by a product of matrices (matrix product states), each depending on a single ‘physical’ mode. Here $G_1(i_1)$ is a row $1 \times r_1$ vector depending on i_1, $G_\ell(i_\ell)$ is a matrix of size $r_{\ell-1} \times r_\ell$ with elements depending on i_ℓ, and $G_d(i_d)$ is a column vector of size $r_{d-1} \times 1$, depending on i_d.

It is similar to the Tucker format, but now we have localized connectivity constraints, while in the Tucker format all orthogonal subspaces are merged by the nonseparated core tensor.

In the two dimensional case, $d = 2$, all three representations, CP, Tucker, and TT tensor classes are equivalent to a rank-r matrices: CP is a skeleton sum representation of a rank-r matrix, Tucker format is the orthogonal factorization (say, SVD), and TT format is the general nonorthogonal factorization (by thin matrices).

TT/TC models have many beneficial features:
- linear in d storage,
- existence of quasioptimal SVD based rank approximation (analogy to Tucker HOSVD),
- SVD based rank truncation procedure with linear scaling in d (analogy to reduced RHOSVD for CP tensors), and
- efficient formated bilinear tensor operations and multilinear algebra.

Figure 3.12 visualizes the TC model; it shows the tensor chain format for $d = 6$.

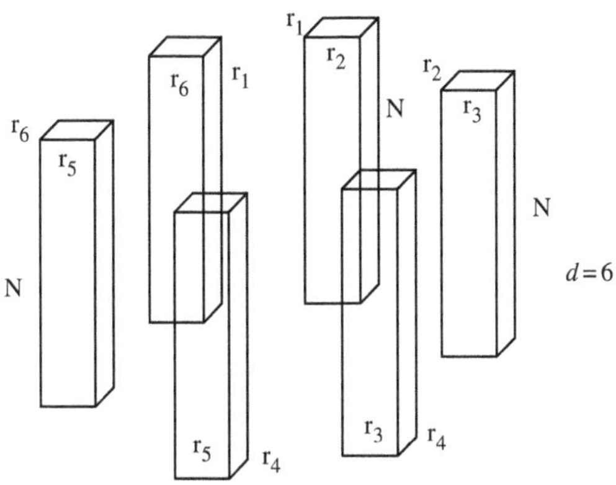

Fig. 3.12: Visualization of the TC tensor format for $d = 6$.

Recall that in the special case $r_6 = 1$ we have $\mathrm{TT}[\mathbf{r}] = \mathrm{TC}[\mathbf{r}]$, i.e., TT format can be interpreted as the disconnected chain as shown in Figure 3.11. The next definition introduces the ℓ-mode TT unfolding matrix.

Definition 3.71. Given a TT tensor $\mathbf{V} \in TT[\mathbf{r}]$, define by $V_{[\ell]}$ the ℓ-mode TT unfolding matrix, where the matrix entries are given by $V_{[\ell]}(\mathbf{i}_1, \mathbf{i}_2) := [V(i_1, \dots, i_\ell; i_{\ell+1}, \dots, i_d)]$, and the two long indexes are specified by multi-indexes $\mathbf{i}_1 = (i_1, \dots, i_\ell)$ and $\mathbf{i}_2 = (i_{\ell+1}, \dots, i_d)$.

It is interesting to compare the ℓ-mode TT unfolding matrix with the unfolding matrix in the Tucker format, $V_{(\ell)}$. For example, we have $V_{[1]} = V_{(1)}$ and $V_{[d-1]} = V_{(d)}$.

The following statement collects the important features of TT tensors concerned with the storage size, rank bounds, embedding to other tensor formats, and concatenation to higher dimensions.

Theorem 3.72. *Given a TT tensor* $\mathbf{V} \in TT[\mathbf{r}]$, *then the following properties hold:*
(A) *Storage size:* $\sum_{\ell=1}^{d} r_{\ell-1} r_\ell N \le d r^2 N$ *with* $r = \max_\ell r_\ell$.
(B) *Canonical embeddings:*

$$\mathcal{C}_{R,\mathbf{n}} \subset TT[\mathbf{r}, \mathbf{n}, d] \quad \textit{with} \quad \mathbf{r} = (R, \dots, R)\,, \qquad TT[\mathbf{r}] \subset TC[\mathbf{r}]\,.$$

(C) *Concatenation to higher dimensions:* $\mathbf{V}_1[d_1] \otimes \mathbf{V}_2[d_2] \to \widehat{\mathbf{V}}[D]$, *where* $D = d_1 + d_2$ *is the dimension of the concatenated tensor and the interface rank* $\hat{r}_{d_1}$ *of* $\widehat{\mathbf{V}}[D]$ *is the product of* r_{d_1-1} *in the first multiple with the first rank parameter* r_1 *of the second one* $\mathbf{V}_2[d_2]$.

(D) *Summary on rank bounds: There hold*

$$r_\ell \leq \operatorname{rank}_{[\ell]}(\mathbf{V}) := \operatorname{rank}(V_{[\ell]}) \leq \operatorname{rank}_{\mathrm{Can}}(\mathbf{V}) ,$$

and the boarder rank relations $r_1 = r_{1,\mathrm{Tuck}}, r_{d-1} = r_{d,\mathrm{Tuck}}$. *Furthermore, we have*

$$r_{\mathrm{Tuck}} \leq r_{TT}{}^2 .$$

Proof. Item (A) directly follows from the definition of the TT format. To verify item (B) we note that in the special case $r_d = 1$ we have $\mathrm{TT}[\mathbf{r}] = \mathrm{TC}[\mathbf{r}]$, i.e., the TT format can be interpreted as the disconnected chain. Furthermore, the rank-R canonical tensor can be considered as the rank-R TT tensor with diagonal cores. Topic (C) can be justified by the direct construction of a concatenated tensor. The rank bound $r_\ell \leq \operatorname{rank}_{[\ell]}(\mathbf{V})$ can be proven by the explicit representation of the TT matrix unfolding for the TT tensor. The last estimate can be checked by the construction of a Tucker unfolding matrix for the TT tensor. □

The formatted implementation of the scalar product of TT tensors [289]

$$\langle \mathbf{X}, \mathbf{Y} \rangle = \langle \mathbf{X} \odot \mathbf{Y}, \mathbf{1} \rangle$$

can be performed in $O(dr^3N)$ operations by using the Hadamard product in the TT format

$$\mathbf{Z} = \mathbf{X} \odot \mathbf{Y}: \quad Z_{i_k}^{(k)} = X_{i_k}^{(k)} \otimes Y_{i_k}^{(k)} .$$

The contracted product with a rank-1 tensor is given by

$$\mathbf{Z} = \mathbf{X} \times_1 U^{(1)} \cdots \times_d U^{(d)}: \quad Z^{(k)} = \sum_{i_k} X^{(k)}(i_k) U^{(k)}(i_k) .$$

For a sum of TT tensors we have

$$\mathbf{Z} = \mathbf{X} + \mathbf{Y}: \quad Z^{(k)}(i_k) = \begin{bmatrix} X^{(k)}(i_k) & 0 \\ 0 & Y^{(k)}(i_k) \end{bmatrix} .$$

In general, the above operations increase the tensor rank, hence in tensor arithmetics within fixed precision they can be complemented by the SVD based rank optimization procedures to be discussed in what follows.

Note that the canonical, Tucker, and TT ranks of a given real valued tensor may differ when considered in the fields $\mathbb{R}$ and $\mathbb{C}$. In some cases, construction of such decompositions can be done much more easily, leading to smaller separation ranks if considered over a complex field.

In what follows, we will explain how to select real and imaginary parts of the complex valued TT tensor. The result is formulated by the following theorem proven in [85] If there is no ambiguity, we use simplified notations $A^{(p)}(i_p) = A_{x_p}^{(p)}$.

Theorem 3.73. *The complex valued TT decomposition* $\mathbf{A} = A^{(1)}_{x_1} \dots A^{(d)}_{x_d}$ *with ranks* $r_0 = r_d, r_1, \dots, r_{d-1}, r_d$ *can be represented in a form* $\mathbf{A} = \hat{A}^{(1)}_{x_1} \dots \hat{A}^{(d)}_{x_d}$, *with rank parameters* $r_0 = r_d, 2r_1, \dots, 2r_{d-1}, r_d$, *and with*

$$\hat{A}^{(1)}_{x_1} = \begin{bmatrix} \operatorname{Re} A^{(1)}_{x_1} & \operatorname{Im} A^{(1)}_{x_1} \end{bmatrix}, \quad \hat{A}^{(p)}_{x_p} = \begin{bmatrix} \operatorname{Re} A^{(p)}_{x_p} & \operatorname{Im} A^{(p)}_{x_p} \\ -\operatorname{Im} A^{(p)}_{x_p} & \operatorname{Re} A^{(p)}_{x_p} \end{bmatrix}, \quad p = 2, \dots, d-1,$$
$$\hat{A}^{(d)}_{x_d} = \begin{bmatrix} \operatorname{Re} A^{(d)}_{x_d} \\ -\operatorname{Im} A^{(d)}_{x_d} \end{bmatrix} + \mathrm{i} \begin{bmatrix} \operatorname{Im} A^{(d)}_{x_d} \\ \operatorname{Re} A^{(d)}_{x_d} \end{bmatrix}, \tag{3.38}$$

where the functional cores $\hat{A}^{(p)}_{x_p}$, $p = 1, \dots, d-1$ *are all real valued.*

Proof. The proof is constructive and follows from the similar arguments as in [85]. Let $A^{(p)}_{x_p} = B^{(p)}_{x_p} + \mathrm{i}C^{(p)}_{x_p}$, with real valued $B^{(p)}_{x_p} = \operatorname{Re} A^{(p)}_{x_p}$ and $C^{(p)}_{x_p} = \operatorname{Im} A^{(p)}_{x_p}$. Then

$$A^{(1)}_{x_1} = \begin{bmatrix} B^{(1)}_{x_1} & C^{(1)}_{x_1} \end{bmatrix} \begin{bmatrix} I \\ \mathrm{i}I \end{bmatrix},$$

where I is an $r_1 \times r_1$ identity matrix. Define real valued core $\hat{A}^{(1)}_{x_1} = \begin{bmatrix} B^{(1)}_{x_1} & C^{(1)}_{x_1} \end{bmatrix}$ and multiply $\begin{bmatrix} I & \mathrm{i}I \end{bmatrix}^{\mathrm{T}}$ to the right,

$$\begin{bmatrix} I \\ \mathrm{i}I \end{bmatrix} A^{(2)}_{x_2} = \begin{bmatrix} I \\ \mathrm{i}I \end{bmatrix} \left(B^{(2)}_{x_2} + \mathrm{i}C^{(2)}_{x_2} \right) = \begin{bmatrix} B^{(2)}_{x_2} + \mathrm{i}C^{(2)}_{x_2} \\ -C^{(2)}_{x_2} + \mathrm{i}B^{(2)}_{x_2} \end{bmatrix} = \begin{bmatrix} B^{(2)}_{x_2} & C^{(2)}_{x_2} \\ -C^{(2)}_{j_2} & B^{(2)}_{x_2} \end{bmatrix} \begin{bmatrix} I \\ \mathrm{i}I \end{bmatrix}.$$

Hence, we can define

$$\begin{bmatrix} B^{(2)}_{x_2} & C^{(2)}_{x_2} \\ -C^{(2)}_{x_2} & B^{(2)}_{x_2} \end{bmatrix} \begin{bmatrix} I \\ \mathrm{i}I \end{bmatrix} =: \hat{A}^{(2)}_{x_2} \begin{bmatrix} I \\ \mathrm{i}I \end{bmatrix},$$

where in the right hand side $\hat{A}^{(2)}_{x_2}$ is a new real valued functional core and I is $r_2 \times r_2$ identity matrix. We continue the process and obtain (3.38). The last functional core reads

$$\hat{A}^{(d)}_{x_d} = \begin{bmatrix} I \\ \mathrm{i}I \end{bmatrix} A^{(d)}_{x_d} = \begin{bmatrix} I \\ \mathrm{i}I \end{bmatrix} \left(B^{(d)}_{x_d} + \mathrm{i}C^{(d)}_{x_d} \right) = \begin{bmatrix} B^{(d)}_{x_d} + \mathrm{i}C^{(d)}_{x_d} \\ -C^{(d)}_{x_d} + \mathrm{i}B^{(d)}_{x_d} \end{bmatrix} = \begin{bmatrix} B^{(d)}_{x_d} \\ -C^{(d)}_{x_d} \end{bmatrix} + \mathrm{i} \begin{bmatrix} C^{(d)}_{x_d} \\ B^{(d)}_{x_d} \end{bmatrix},$$

which completes the proof. □

The relation (3.38) allows us to represent explicitly real and imaginary parts of a matrix product function since all factors but one in the corresponding factorization are real.

Corollary 3.74. *Real and imaginary parts in a TT tensor* $\mathbf{A}$ *in* (3.38) *are given by*

$$\operatorname{Re} \mathbf{A} = \left(\prod_{p=1}^{d-1} \hat{A}^{(p)}_{x_p} \right) \operatorname{Re} \hat{A}^{(d)}_{x_d}, \quad \operatorname{Im} \mathbf{A} = \left(\prod_{p=1}^{d-1} \hat{A}^{(p)}_{x_p} \right) \operatorname{Im} \hat{A}^{(d)}_{x_d}.$$

This corollary provides a powerful tool to derive explicit TT real valued representation of a function with the minimal rank having at hand the complex valued factorization, as will be shown in the following paragraph.

3.6.4 Asymptotically optimal rank-r TT approximation

The following theorem proves the existence of the fixed rank TT approximation and provides the corresponding error estimate in terms of rank approximation to the ℓ-mode TT unfolding matrices. Here we mainly follow the constructive proof in [292], where these results were first established.

Theorem 3.75 ([292]). *For any tensor* $\mathbf{A} = [A(i_1, \dots, i_d)]$ *there exists a TT approximation* $\mathbf{T} = [T(i_1, \dots, i_d)] \in TT[\mathbf{r}]$ *with ranks* r_k *such that*

$$\|\mathbf{A} - \mathbf{T}\|_F^2 \le \sum_{k=1}^{d-1} \varepsilon_k^2 , \tag{3.39}$$

where ε_k *is the Frobenius distance from* $A_{[k]}$ *to its best rank-*r_k *approximation*

$$\varepsilon_k = \min_{\operatorname{rank} B \le r_k} \|A_{[k]} - B\|_F .$$

Proof. First, consider the case $d = 2$. Then TT decomposition reads as the rank-r_1 CP representation as follows:

$$T(i_1, i_2) = \sum_{\alpha_1=1}^{r_1} G_1(i_1, \alpha_1) G_2(\alpha_1, i_2) ,$$

which coincides with the skeleton decomposition of the matrix T. Choose T using the rank-r_1 truncated SVD of A, which guarantees that the norm $\|A - T\|_F = \varepsilon_1$ is minimal possible. Then proceed by induction. Consider the SVD of the first unfolding matrix,

$$A_{(1)} = [A(i_1; i_2 \dots i_d)] = U \Sigma V , \quad \Sigma = \operatorname{diag}\{\sigma_1, \sigma_2, \dots\} . \tag{3.40}$$

As an approximation to $A_{(1)}$, consider

$$B_1 = U_1 \Lambda V_1 , \quad \Lambda = \operatorname{diag}(\sigma_1, \dots, \sigma_{r_1}) , \tag{3.41}$$

where U_1 and V_1 contain the first r_1 columns of U and rows of V, respectively. Then B_1 is the best rank-r_1 approximation to $A_{(1)}$, i.e.,

$$A_{(1)} = B_1 + E_1 , \quad \operatorname{rank} B_1 \le r_1 , \quad \|E_1\|_F = \varepsilon_1 .$$

Obviously, B_1 can be considered as a tensor $\mathbf{B} = [B(i_1, \dots i_d)]$ and the approximation problem reduces to the one for $\mathbf{B}$.

Observe that if we take an arbitrary tensor $\mathbf{T} = [T(i_1, \dots, i_d)]$ with the first unfolding matrix $T_{(1)} = [T(i_1; i_2, \dots, i_d)]$ in the form

$$T_{(1)} = U_1 W , \tag{3.42}$$

with U_1 from (3.41) and an arbitrary matrix W with r_1 rows and as many columns as in $T_{(1)}$, then $E_1^* T_{(1)} = 0$ and this implies that

$$\|(\mathbf{A} - \mathbf{B}) + (\mathbf{B} - \mathbf{T})\|_F^2 = \|\mathbf{A} - \mathbf{B}\|_F^2 + \|\mathbf{B} - \mathbf{T}\|_F^2 . \tag{3.43}$$

However, the tensor $\mathbf{B}$ is still of dimensionality d. To reduce dimensionality, rewrite the matrix equality (3.41) in the elementwise form

$$B(i_1; i_2, \dots, i_d) = \sum_{\alpha_1=1}^{r_1} U_1(i_1; \alpha_1)\widehat{A}(\alpha_1; i_2, \dots, i_d) , \quad \text{where } \widehat{A} = \Lambda V_1 .$$

Then concatenate indexes α_1 and i_2 into one long index and consider $\widehat{A}$ as a tensor

$$\widehat{\mathbf{A}} = [\widehat{A}(\alpha_1 i_2, i_3, \dots i_d)] \quad \text{of dimensionality} \quad d - 1 .$$

By induction, $\widehat{\mathbf{A}}$ admits a TT approximately $\widehat{\mathbf{T}} = [\widehat{T}(\alpha_1 i_2, i_3, \dots i_d)]$ of the form

$$\widehat{T}(\alpha_1 i_2, i_3, \dots i_d) = \sum_{\alpha_2,\dots,\alpha_{d-1}} G_2(\alpha_1 i_2, \alpha_2) G_3(\alpha_2, i_3, \alpha_3) \dots G_d(\alpha_{d-1}, i_d) ,$$

such that

$$\|\widehat{\mathbf{A}} - \widehat{\mathbf{T}}\|_F^2 \le \sum_{k=2}^{d-1} \hat{\varepsilon}_k^2 , \quad \text{with}$$

$$\hat{\varepsilon}_k = \min_{\text{rank } C \le r_k} \|\widehat{A} - C\|_F , \quad \widehat{A}_k = [\widehat{A}(\alpha_1 i_2, \dots i_k; i_{k+1}, \dots, i_d)] .$$

Now let us set $G_1(i_1, \alpha_1) = U(i_1, \alpha_1)$, separate indexes α_1, i_2 from the long index $\alpha_1 i_2$, and define $\mathbf{T}$ by the following tensor train:

$$T(i_1, \dots, i_d) = \sum_{\alpha_1,\dots\alpha_d} G_1(i_1, \alpha_1) G_2(\alpha_1, i_2, \alpha_3) \cdots G_d(\alpha_{d-1}, i_d) .$$

It remains to estimate $\|\mathbf{A} - \mathbf{T}\|_F$. First, from (3.40) and (3.41) it stems that

$$\widehat{A} = \Lambda V_1 = U_1^* A_{(1)} ,$$

and consequently

$$\widehat{A}(\alpha_1 i_2, i_3, \dots, i_d) = \sum_{i_1} \overline{U}_1(i_1, \alpha_1) A(i_1, i_2, \dots, i_d) .$$

Let $A_k = B_k + E_k$ with rank $B_k \le r_k$ and $\|E_k\|_F = \varepsilon_k$. We can consider B_k and E_k as tensors $B_k(i_1, \dots, i_d)$ and $E_k(i_1, \dots, i_d)$. Since B_k admits a skeleton decomposition with r_k terms, we obtain

$$A(i_1, \dots i_d) = \sum_{\gamma=1}^{r_k} P(i_1, \dots i_k; \gamma) Q(\gamma; i_{k+1}, \dots, i_d) + E_k(i_1, \dots, i_d) .$$

Hence, $\widehat{A}(\alpha_1 i_2, i_3, \dots, i_d) = H_k(\alpha_1, i_2, i_3, \dots, i_d) + R_k(\alpha_1, i_2, i_3, \dots, i_d)$ with

$$H_k(\alpha_1, i_2, i_3, \dots, i_d) = \sum_{i_1} \overline{U}_1(i_1, \alpha_1) \sum_{\gamma=1}^{r_k} P(i_1, \dots, i_k; \gamma) Q(\gamma; i_{k+1}, \dots, i_d) ,$$
$$R_k(\alpha_1, i_2, i_3, \dots, i_d) = \sum_{i_1} \overline{U}_1(i_1, \alpha_1) E_k(i_1, \dots, i_d) .$$

Let us introduce a tensor L as follows:

$$L(\alpha_1, i_2, \dots, i_k, \gamma) = \sum_{i_1} \overline{U}_1(i_1, \alpha_1) P(i_1, \dots, i_k; \gamma) .$$

Then we can consider H_k as a matrix with the elements defined by a skeleton decomposition

$$H_k(\alpha_1, i_2, \dots, i_k; i_{k+1}, \dots, i_d) = L(\alpha_1, i_2, \dots, i_k; \gamma) Q(\gamma; i_{k+1}, \dots, i_d) ,$$

which makes it evident that the rank of H_k does not exceed r_k. We can also consider R_k as a matrix with the elements defined by

$$R_k(\alpha_1; i_2, i_3, \dots, i_d) = \sum_{i_1} \overline{U}_1(i_1; \alpha_1) E_k(i_1; i_2, \dots, i_d) .$$

We know that U_1 has orthonormal columns, which means that the matrix E_k is premultiplied by a matrix with orthonormal rows. Since this cannot increase its Frobenius norm, we conclude that

$$\hat{\varepsilon}_k \le \|R_k\|_F \le \|E_k\|_F = \varepsilon_k , \quad 2 \le k \le d-1 .$$

Hence, for the error tensor $\hat{\mathbf{E}}$ with the elements

$$\hat{E}(\alpha_1 i_2, i_3, \dots, i_d) = \hat{A}(\alpha_1 i_2, i_3, \dots, i_d) - \hat{T}(\alpha_1 i_2, i_3, \dots, i_d) ,$$

we obtain

$$\|\hat{\mathbf{E}}\|_F^2 \le \sum_{k=2}^{d-1} \varepsilon_k^2 .$$

Furthermore, the error tensor $\mathbf{E} = \mathbf{B} - \mathbf{T}$ can be viewed as a matrix of the form

$$E(i_1; i_2, \dots, i_d) = \sum_{\alpha_1=1}^{r_1} U_1(i_1; \alpha_1) \widehat{E}(\alpha_1; i_2, \dots, i_d) ,$$

which shows that the matrix $\widehat{E}$ is premultiplied by a matrix with the orthonormal rows, such that we have $\|\mathbf{E}\|_F^2 \le \|\widehat{\mathbf{E}}\|_F^2 \le \sum_{k=2}^{d-1} \varepsilon_k^2$.

Finally, we observe that the first unfolding matrix T_1 for $\mathbf{T}$ is exactly of the form (3.42). Thus, (3.43) is valid, completing the proof. □

Similar to the rank-**r** Tucker decomposition via HOSVD, the rank-**r** TT approximation **T** constructed in the proof of Theorem 3.75 can be viewed as the HOSVD in the TT format, see also [289]. The analysis of HOSVD for the HT format is presented in [123].

3.6.5 Comments on approximation by TT tensors

Representation of tensors in low separation rank formats is the key point in the design of fast tensor structured numerical methods for large scale higher dimensional simulations. In fact, it allows the implementation of basic linear and bilinear algebraic operations on tensors mentioned above such as addition, scalar, Hadamard, and convolution products with linear complexity in the univariate tensor size ([85, 177, 185, 201, 209]). However, the bilinear tensor operations normally increase the separation rank of the resultant tensor. Hence, the complexity control requires further 'projection' of such intermediate results to the set of tensors with smaller rank parameter (rank truncation).

To perform computation over the nonlinear set (manifold) of rank structured tensors $\mathcal{S}$ (say the canonical, Tucker, MPS/TT formats) with controllable complexity, we need to perform a 'projection' of the current iterand in $\mathcal{S}_0 \supset \mathcal{S}$ onto that manifold $\mathcal{S}$ using the 'formatted' tensor operations. This action is fulfilled by implementing the tensor truncation operator $T_{\mathcal{S}}: \mathbb{W}_{\mathbf{n},d} \to \mathcal{S}$ defined by

$$\mathbf{A}_0 \in \mathcal{S}_0 \subset \mathbb{W}_{\mathbf{n},d}: \quad T_{\mathcal{S}}\mathbf{A}_0 = \operatorname{argmin}_{\mathbf{T}\in\mathcal{S}} \|\mathbf{A}_0 - \mathbf{T}\| , \tag{3.44}$$

which reduces to a challenging nonlinear approximation problem. The replacement of $\mathbf{A}_0$ by its approximation in $\mathcal{S}$ is called the *tensor truncation* to $\mathcal{S}$ and denoted by $T_{\mathcal{S}}\mathbf{A}_0$. In practice, the computation of the minimizer $T_{\mathcal{S}}\mathbf{A}_0$ can be performed only approximately. The set $\mathcal{S}$ of rank structured tensors can be chosen adaptively in order to control the approximation error $\varepsilon > 0$,

$$\|\mathbf{A}_0 - T_{\mathcal{S}}\mathbf{A}_0\| \le \varepsilon .$$

In the case of Tucker, TT/QTT, and QTT-Tucker formats (see Chapter 4) the quasioptimal approximation can be computed by a conventional QR/SVD algorithm [87, 123, 232, 292], also known in the physical literature as the Schmidt decomposition. In particular, the Tucker tensors can be approximated by the so called HOSVD, [248]. Robust SVD based algorithms are applicable since the Tucker and TT ranks can be controlled by a certain matrix rank. Indeed, for MPS/TT format we have the equivalent definition in terms of the TT-unfolding matrix,

$$TT[\mathbf{r}] := \{\mathbf{A} \in \mathbb{V}_{\mathbf{n}}: \operatorname{rank} A_{[p]} \le r_p, p = 1, \dots, d-1\} ,$$

where according to Definition 3.71 the TT-unfolding matrix $A_{[p]}$ can be represented in the form

$$A_{[p]} := \mathbf{A}(\underbrace{j_1\, j_2\, \dots j_p}_{\text{column ind.}}; \underbrace{j_{p+1} \dots j_d}_{\text{row index}}) .$$

The definition of the Tucker rank parameters in terms of the rank of unfolding matrices was considered in Sections 3.3 and 3.4.

There are several proofs in the literature concerning the closeness of the tensor class TT[**r**] and the quasioptimality of the HOSVD approximation by the TT tensors considered in Theorem 3.75; see for example [138, 167]. We refer to [292] for the following statement:

Corollary 3.76. *Given a tensor* $\mathbf{A}$*, denote by* $\varepsilon = \inf_{\mathbf{B}\in TT[\mathbf{r}]} \|\mathbf{A}-\mathbf{B}\|_F$*. Then the minimizer for the problem* $\mathbf{B}_* = \operatorname{argmin} \min_{\mathbf{T}\in TT[\mathbf{r}]} \|\mathbf{A}-\mathbf{B}\|_F$ *exists (i.e., the infimum is in fact a minimum) and the TT approximation* $\mathbf{T}$ *constructed in the proof of Theorem 3.75 is quasioptimal in the sense that*

$$\|\mathbf{A}-\mathbf{T}\|_F \leq \sqrt{d-1}\varepsilon\,.$$

Proof. By the definition of infimum, there exists a sequence of tensor trains $\mathbf{B}^{(s)}$ ($s = 1, 2, \ldots,$) with the property $\lim_{s\to\infty} \|\mathbf{A}-\mathbf{B}^{(s)}\|_F = \varepsilon$, such that all elements of the tensor $\mathbf{B}^{(s)}$ are uniformly bounded and, hence, some subsequent $\mathbf{B}^{(s_t)}$ converges elementwise to some tensor $\mathbf{B}^{(\min)}$.

We cannot say that all elements of the corresponding tensor carriages are uniformly bounded. Nevertheless, all elements of the tensor $\mathbf{B}^{(s)}$ are uniformly bounded, and hence some subsequent $\mathbf{B}^{(s_t)}$ converges elementwise to some tensor $\mathbf{B}^{(\min)}$. The same holds true for the corresponding unfolding matrices, $B^{[s]}_{(k)} \to B^{(\min)}_{[k]}$, $1 \leq k \leq d$.

It is well known that a sequence of matrices with a common rank bound cannot converge to a matrix with a larger rank. Thus, $\operatorname{rank} B^{(s_t)}_{[k]} \leq r_k$ implies that $\operatorname{rank} B^{(\min)}_{[k]} \leq r_k$, and hence $\|\mathbf{A}-\mathbf{B}^{(\min)}\|_F = \varepsilon$, so that $\mathbf{B}^{(\min)}$ is the minimizer.

It is now sufficient to note that $\varepsilon_k \leq \varepsilon$. The reason is that ε is the approximation accuracy for every unfolding matrix A_k delivered by a special structured skeleton (dyadic) decomposition with r_k terms while ε_k stands for the best approximation accuracy without any restrictions on the vectors of skeleton decomposition. Hence, $\varepsilon_k \leq \varepsilon$, and the quasioptimality bound follows directly from (3.39). □

Theorem 3.75 converts the accuracy of the rank-**r** TT approximation in terms of the approximation errors ε_ℓ estimated over the truncated SVD of the ℓ-mode unfolding matrix of $\mathbf{A}$, $A_{(\ell)}$ ($\ell = 1, \ldots, d$). In turn, Corollary 3.76 delivers the error bound that is uniform in terms of the best approximation error for a fixed dimension parameter. Hence, this result provides an asymptotically optimal error estimate.

The embedding property (B) in Theorem 3.72 implies the following simple statement: If a tensor $\mathbf{A}$ admits an R-term canonical approximation of accuracy $\varepsilon > 0$, then there exists a TT approximation with $r_k \leq R$ and accuracy ε.

Remark 3.77. Similar to the case of the product Stiefel manifold of Tucker tensors, the set of TT tensors with fixed rank parameters, TT[**r**], can be proven to be a closed nonlinear manifold in the tensor product Hilbert space $\mathbb{V}_{\mathbf{n}}$; see [138, 167].

It it worth noting that the almost optimal approximation in TT[**r**] can be computed by stable algorithms, i.e., the accurate approximation to minimizer

$$\mathbf{T}_* = \operatorname*{argmin}_{\mathbf{T}\in TT[\mathbf{r}]} \|\mathbf{T}-\mathbf{A}\|_F$$

can be calculated by the DMRG iterative scheme based on QR/SVD decompositions. This issue will be addressed in the following sections.

We conclude with historical remarks related to the quasioptimality property via Theorem 3.75 and Corollary 3.76, which is similar to that for the rank-$\mathbf{r}$ HOSVD Tucker approximation.

First of all, the HOSVD decomposition to convert a full tensor to the Tucker form was introduced in [247, 248]. The numerical scheme based on the so called reduced HOSVD (RHOSVD) applied to the canonical target tensor with large CP rank was introduced in [212], where the multigrid version of the approach was also described. The idea of hierarchical dimension splitting based on the control of separation rank parameters was advocated in [206]. To the best of our knowledge the HOSVD type scheme for MPS tensors was first introduced in [356]. The particular HOSVD approximation scheme in the TT format applied to full format tensors was constructed in [289, 292]. The hierarchical Tucker representation in [146] is the special version of the general tensor network states models. The HOSVD for the hierarchical Tucker tensors was described in [123]. The surveys on MPS techniques can be found for example in [322, 355]. The basic properties of TT tensor manifolds were addressed in [138, 167].

Introduction (and rigorous analysis) of the quantics-TT (QTT) tensor approximation for function related vectors [197] and for operator generated matrices [286] has opened new prospects in the field of tensor numerical methods. Invention of this technique has changed the traditional outlook on numerical complexity since it has led to the new concept of logarithmic complexity scaling in the volume size of the target grid based representation of functions. It will be discussed in detail in Chapter 4.

3.6.6 Canonical, Tucker, and MPO operators (matrices)

The rank structured tensor formats like canonical, Tucker, and MPS/TT type decompositions induce the important concept of canonical, Tucker, or matrix product operators (CO/TO/MPO) acting between two tensor product Hilbert spaces, each of dimension d,

$$\mathcal{A} : \mathbb{X} = \bigotimes_{\ell=1}^{d} X^{(\ell)} \to \mathbb{Y} = \bigotimes_{\ell=1}^{d} Y^{(\ell)} .$$

For example, as was already discussed in Section 3.5, the R-term canonical operator (matrix) takes a form

$$\mathcal{A} = \sum_{\alpha=1}^{R} \bigotimes_{\ell=1}^{d} A_{\alpha}^{(\ell)} , \quad A^{(\ell)} : X^{(\ell)} \to Y^{(\ell)} .$$

The action $\mathcal{A}\mathbf{X}$ on rank-R_X canonical tensor $\mathbf{X} \in \mathbb{X}$ is defined as RR_X-term canonical sum in $\mathbb{Y}$,

$$\mathcal{A}\mathbf{X} = \sum_{\alpha=1}^{R} \sum_{\beta=1}^{R_X} \bigotimes_{\ell=1}^{d} A_{\alpha}^{(\ell)} \mathbf{x}_{\beta}^{(\ell)} \in \mathbb{Y} .$$

In the case of rank-**r** TT format the respective core matrices are defined as follows:

Definition 3.78. The rank-**r** TT operator (TTO/MPO) decomposition symbolized by a set of factorized operators, $\mathcal{A}$, is defined by

$$\mathcal{A} = \sum_{\alpha \in \mathcal{J}} A^{(1)}_{\alpha_1} \otimes A^{(2)}_{\alpha_1 \alpha_2} \otimes \cdots \otimes A^{(d)}_{\alpha_{d-1}} \equiv \mathcal{A}^{(1)} \bowtie \mathcal{A}^{(2)} \bowtie \cdots \bowtie \mathcal{A}^{(d)} ,$$

where $\mathcal{A}^{(\ell)} = [A^{(\ell)}_{\alpha_\ell \alpha_{\ell+1}}]$ denotes the operator valued $r_\ell \times r_{\ell+1}$ matrix, and where for the matrix valued entries we have $A^{(\ell)}_{\alpha_\ell \alpha_{\ell+1}} : X^{(\ell)} \to Y^{(\ell)}$, $(\ell = 1, \dots, d)$. In the index notation, we have the matrix product representation

$$\mathcal{A}(i_1, j_1, \dots, i_d, j_d) = \sum_{\alpha_1=1}^{r_1} \cdots \sum_{\alpha_{d-1}=1}^{r_{d-1}} A^{(1)}_{\alpha_1}(i_1, j_1) A^{(2)}_{\alpha_1 \alpha_2}(i_2, j_2) \cdot \dots \cdot \\ \cdot A^{(d-1)}_{\alpha_{d-2} \alpha_{d-1}}(i_{d-1}, j_{d-1}) A^{(d)}_{\alpha_{d-1}}(i_d, j_d) . \quad (3.45)$$

Given a rank-$\mathbf{r}_X$ TT tensor $\mathbf{X} = \mathbf{X}^{(1)} \bowtie \mathbf{X}^{(2)} \bowtie \cdots \bowtie \mathbf{X}^{(d)} \in \mathbb{X}$, the action $\mathcal{A}\mathbf{X} = \mathbf{Y}$ is defined as the TT element $\mathbf{Y} = \mathbf{Y}^{(1)} \bowtie \mathbf{Y}^{(2)} \bowtie \cdots \bowtie \mathbf{Y}^{(d)} \in \mathbb{Y}$,

$$\mathcal{A}\mathbf{X} = \mathbf{Y}^{(1)} \bowtie \mathbf{Y}^{(2)} \bowtie \cdots \bowtie \mathbf{Y}^{(d)} \in \mathbb{Y} , \quad \text{with} \quad \mathbf{Y}^{(\ell)} = [A^{(\ell)}_{\alpha_1 \alpha_2} \mathbf{X}^{(\ell)}_{\beta_1 \beta_2}]_{\alpha_1 \beta_1, \alpha_2 \beta_2} ,$$

where in the brackets we use the standard matrix vector multiplication. The TT rank of $\mathbf{Y}$ is bounded by $\mathbf{r}_Y \leq \mathbf{r} \odot \mathbf{r}_X$, where $\odot$ means the standard Hadamard (entrywise) product of two vectors.

To describe the index-free operator representation of the TT matrix vector product, we introduce the tensor operation denoted by $\bowtie^*$ that can be viewed as dual to $\bowtie$: it is defined as the tensor (Kronecker) product of the two corresponding core matrices, their blocks being multiplied by means of a regular matrix product operation. Now with the substitution $\mathbf{Y}^{(\ell)} = \mathcal{A}^{(\ell)} \bowtie^* \mathbf{X}^{(\ell)}$ the matrix vector product in TT format takes the operator form,

$$\mathcal{A}\mathbf{X} = (\mathcal{A}^{(1)} \bowtie^* \mathbf{X}^{(1)}) \bowtie \cdots \bowtie (\mathcal{A}^{(d)} \bowtie^* \mathbf{X}^{(d)}) .$$

As the standard example, we consider the finite difference negative d-Laplacian over the uniform tensor grid, which is known to have the Kronecker rank-d representation,

$$\Delta_d = A \otimes I_N \otimes \cdots \otimes I_N + I_N \otimes A \otimes \cdots \otimes I_N + \cdots + I_N \otimes \dots I_N \otimes \otimes A \in \mathbb{R}^{N^{\otimes d} \times N^{\otimes d}} , \quad (3.46)$$

with $A = \Delta_1 = \operatorname{tridiag}\{-1, 2, -1\} \in \mathbb{R}^{N \times N}$, and I_N being the $N \times N$ identity matrix.

For the canonical rank we have $\operatorname{rank}_{\mathrm{Can}}(\Delta_d) = d$, while the TT rank of Δ_d is equal to 2 for any dimension due to the explicit representation [177],

$$\Delta_d = \begin{bmatrix} \Delta_1 & I_N \end{bmatrix} \bowtie \begin{bmatrix} I_N & 0 \\ \Delta_1 & I_N \end{bmatrix}^{\otimes (d-2)} \bowtie \begin{bmatrix} I_N \\ \Delta_1 \end{bmatrix} ,$$

where the rank product operation $\bowtie$ in the matrix case is defined as above. A similar result is true concerning the Tucker rank, $\operatorname{rank}_{\mathrm{Tuck}}(\Delta_d) = 2$. Indeed, the explicit Tucker representation can be constructed as in Example 3.70.

3.6.7 Higher order SVD and SVD based TT rank truncation

The TT approximation scheme can be applied to both full format tensors and tensors already presented in the TT format, but having far from optimal ranks. The target tensor may be also the TTO type matrix. The proof of Theorem 3.75 offers the important constructive scheme for robust computing of a TT approximation.

First, we consider the full-to-TT approximation scheme. The following prototype algorithm is described in [292] and is implemented in the MATLAB TT Toolbox, [290].

Algorithm 3.4 (Full-to-TT approximation scheme).
Input: A tensor $\mathbf{A}$ of size $n_1 \times n_2 \cdots \times n_d$ and accuracy bound $\varepsilon > 0$.
Output: The tensor core G_k, $k = 1, \dots d$, defining a TT approximation to $\mathbf{A}$ with the relative error bound ε.

1. Compute $nrm := \|\mathbf{A}\|_F$.
2. Specify sizes of the first unfolding matrix: $N_l = n_1$, $N_r = \prod_{k=2}^{d} n_k$.
3. Temporary tensor: $\mathbf{B} = \mathbf{A}$.
4. First unfolding: $M := \mathrm{reshape}(\mathbf{B}, [N_l, N_r])$.
5. Compute the truncated SVD of $M \approx U\Lambda V^{\mathrm{T}}$, so that the approximate rank r ensures

$$\sum_{k=r+1}^{\min(N_l,N_r)} \sigma_k^2 \leq \frac{(\varepsilon \cdot nrm)^2}{d-1} .$$

6. Set $G_1 = U$, $M := \Lambda V^{\mathrm{T}}$, $r_1 = r$.
7. Process other modes.
8. *for* $k = 2$ *to* $d - 1$ *do*
9. Redefine the sizes: $N_l := n_k$, $N_r := \frac{N_r}{n_k}$.
10. Construct the next unfolding: $M := \mathrm{reshape}(M, [rN_l, N_r])$.
11. Compute the truncated SVD of $M \approx U\Lambda V^{\mathrm{T}}$, so that the approximate rank r ensures

$$\sum_{k=r+1}^{\min(N_l,N_r)} \sigma_k^2 \leq \frac{(\varepsilon \cdot nrm)^2}{d-1} .$$

12. Reshape the matrix U into a tensor: $G_k := \mathrm{reshape}(U, [r_{k-1}, n_k, r_k])$.
13. Update $M := \Lambda V^{\mathrm{T}}$.
14. *end for*
15. Set $G_d = M$.

It can be easily seen that the complexity of Algorithm 3.4 scales as $O(n^{d+1})$, which is on the same scale as the complexity of the truncated HOSVD in the Tucker format applied to $n^{\otimes d}$ input tensor. Hence, this algorithm suffers from the curse of dimensionality and it practically applies only to small size tensors and to moderate dimensions d. Note that the grid based representation of functions solving the physically relevant PDEs often requires large spacial grids with the mesh parameter of the order of $n \sim 10^4$–10^5, therefore even for 3D problems the TT-HOSVD becomes nontractable. In numerical practice the rank structured approximation of function related tensors can be realized by using the RHOSVD type algorithms as in [212].

One of the most important procedures in the rank structured tensor computations is the *recompression of formatted tensors*, that means in our case the rank optimization (reduction) in TT format by controlling the fixed approximation threshold. This task is practically important since in general the standard bilinear operations on rank structured vectors (tensors) and matrices enlarge the rank parameters.

Given a tensor $\mathbf{A} \in \mathrm{TT}[\mathbf{r}]$ with nonoptimal ranks r_k, we want to approximate it with another TT tensor $\mathbf{B}$ with smallest possible ranks $\hat{r}_k \leq r_k$, while maintaining the desired relative accuracy ε:

$$\|\mathbf{A} - \mathbf{B}\|_F \leq \varepsilon \|\mathbf{B}\|_F \,.$$

Such a 'projection' will define the ε-truncation operator,

$$\mathbf{B} = T_\varepsilon(\mathbf{A}) \,.$$

Construction of such an operator in the canonical (CP) format is a notoriously difficult task, with no best solution known. Use of the orthogonal Tucker format is limited by the curse of dimensionality in the Tucker core.

In the case of TT format the rank optimization can be implemented by using standard SVD and QR decompositions as described in Algorithm 3.5 below; see [289]. A MATLAB code for this algorithm is a part of TT Toolbox [290].

By SVD_δ in Algorithm 3.5 below, we denote truncated SVD where singular values are set to zero if smaller than δ, and by $\mathrm{QR}_{\mathrm{rows}}$ we denote QR decomposition of a matrix where Q factor has orthonormal rows. The $\mathrm{SVD}_\delta(A)$ returns three matrices U, Λ, V of the decomposition $A \approx U \Lambda V^\top$ (as MATLAB *svd* function).

Algorithm 3.5 can be considered as an extension of reduced truncated matrix SVD and it is the analogy of the RHOSVD scheme applied to the canonical target [212].

Algorithm 3.5 (TT_ε rank recompression).
Input: d dimensional tensor $\mathbf{A}$ in the TT format, required accuracy $\varepsilon > 0$.
Output: $\mathbf{B}$ in the TT format with smallest compression ranks $\hat{r}_k$, such that

$$\|\mathbf{A} - \mathbf{B}\|_F \leq \varepsilon \|\mathbf{A}\|_F \,, \quad \text{i.e.,} \quad \mathbf{B} = T_\varepsilon(\mathbf{A}) \,.$$

1. Let G_k, $k = 1, \dots, d-1$, be TT cores of $\mathbf{A}$.
2. Initialization: Compute truncation parameter $\delta = \frac{\varepsilon}{\sqrt{d-1}} \|\mathbf{A}\|_F$.
3. Right-to-left orthogonalization:
4. *for* $k = d$ *to* 2 step -1 *do*
5. $[G_k(\beta_{k-1}; i_k\beta_k), R(\alpha_{k-1}, \beta_{k-1})] := \mathrm{QR}_{\mathrm{rows}}(G_k(\alpha_{k-1}; i_k\beta_k))$.
6. $G_{k-1} := G_k \times_3 R$.
7. *end for*
8. Compression of the orthogonalized representation:
9. *for* $k = 1$ *to* $d-1$ *do* (compute δ-truncated SVD):

10. $[G_k(\beta_{k-1}i_k;\gamma_k), \Lambda, V(\beta_k,\gamma_k)] := \mathrm{SVD}_\delta[G_k(\beta_{k-1}i_k;\beta_k)]$
11. $G_{k+1} := G_{k+1} \times_1 (V\Lambda)^\top$.
12. *end for*;
13. *Return:* G_k, $k = 1,\dots,d$, as cores of **B**.

It can be verified that the complexity of Algorithm 3.5 is of the order of $\mathcal{O}(dnr^3)$; compare with Algorithm 3.4. Furthermore, this algorithm can be viewed as the RHOSVD applied to the input tensor in the TT or canonical format, and it is free from the curse of dimensionality.

We note that all basic MLA operations can be implemented in the TT format: addition, scalar product and norm calculation, matrix-by-vector product, the Hadamard product, tensor-by-vector contracted product, etc. Accomplished with the well posed recompression procedure providing quasioptimal approximation with fixed threshold ε, this gives an efficient tool for solving large scale high dimensional PDEs and other computational problems.

Exercise 3.79. Apply the rank-2 TT representation of a tensor related to the function

$$f(x) = \sin(x_1 + \cdots + x_d) = \frac{e^{ix} - e^{-ix}}{2i} = \mathrm{Im}(e^{ix}),$$

sampled on tensor product spatial grid and then approximate the exact value of the multivariate integral for $d = 5, 10, 20$,

$$I(d) = \mathrm{Im} \int\limits_{[0,1]^d} e^{i(x_1+\cdots+x_d)}dx = \mathrm{Im}\left[\left(\frac{e^i - 1}{i}\right)^d\right].$$

Hint: Use Lemma 2.18, in Chapter 2, apply simple quadrature rule on an $n \times \cdots \times n$ sampling grid, and calculate the integral as a TT scalar product. What changes if we introduce high oscillations by a substitution $f(x) \mapsto f(\omega x)$? (See also Section 5.6.)

We recall that the low rank approximation methods in the TT and HT formats are based on the HOSVD type algorithms applied to the formatted input, i.e., on the RHOSVD approximation, however this approximation scheme does not apply to the full size tensors in view of the curse of dimensionality. In the case where the target tensor can be evaluated fiberwise the heuristic adaptive cross approximation schemes can be applied [15, 16, 19, 100, 220, 228, 293, 295, 317], thus generalizing the ACA in the matrix case [20, 119, 346, 348, 361]; see also references in [20].

3.6.8 Analytic and algebraic approximation methods in tensor formats revisited

Basic tensor formats considered so far include the CP, Tucker, canonical-Tucker, and MPS/TT tensors. An important operation on multidimensional tensors is the 'projection' onto the certain nonlinear tensor manifold $\mathcal{S}$ of rank structured tensors, $\mathcal{S} \subset \mathcal{S}_0 \subset$

$\mathbb{V}_{\mathbf{n}}$, where the target input already belongs to a certain larger tensor class $\mathcal{S}_0$ characterized by larger rank parameters. Another class of approximation problems is related to rank structured representation of function generated tensors representing physical potential fields, nonlocal operators, and other physically relevant quantities discretized on a large tensor grid in $\mathbb{R}^d$.

We now summarize analytic and algebraic approximation methods in tensor analysis. Recall that the rank truncation tensor operation $T_{\mathcal{S}} : \mathcal{S}_0 \subset \mathbb{V}_{\mathbf{n}} \to \mathcal{S}$ is defined as follows. Let $\mathcal{S} \in \{\mathcal{T}_{\mathbf{r}}, \mathcal{C}_R, \mathcal{T}_{\mathcal{C}_{R,\mathbf{r}}}, \mathrm{TT}\}$. Introduce the tensor truncation operator by:

$$\text{Given } \mathbf{A} \in \mathcal{S}_0 \subset \mathbb{V}_{\mathbf{n}}: \quad \text{Find} \quad T_{\mathcal{S}}(\mathbf{A}) := \operatorname*{argmin}_{\mathbf{T} \in \mathcal{S}} \|\mathbf{T} - \mathbf{A}\| .$$

Operator $T_{\mathcal{S}}$ is an extension to $d > 2$ of the truncated SVD for matrices getting rid of the 'curse of dimensionality'. In this concern the celebrated theorem by E. Schmidt, 1907 [327], on best L^2-bilinear approximation of bivariate function in the form

$$f(x, y) \approx \sum_{k=1}^{R} a_k(x) b_k(y) ,$$

already mimics the truncated matrix SVD.

The nonlinear approximation problem on computation of the operator $T_{\mathcal{S}}$ for $\mathcal{S} \in \{\mathcal{T}_{\mathbf{r}}, TT[\mathbf{r}]\}$ allows the robust SVD based implementation providing the quasioptimal error and linear complexity scaling in d. For the special input in $\mathcal{S}_0 = \mathcal{C}_R$ the RHOSDV rank reduction algorithm is applied to the canonical target, allowing the robust SVD based implementation for a wide variety of the practically interesting classes of input tensors [212]. Remember that the approximation schemes in $\mathcal{T}_{\mathbf{r}}$ and $\mathcal{C}_R$ tensor classes have been considered in Sections 3.3 and 3.4

In many mathematical models the governing equations include some potentials and kernel functions given in the analytic form such that the combination of analytic and algebraic methods may lead to the optimal approximation strategies in tensor formats. In the case of function related tensors the analytic tensor approximation methods in the CP and Tucker formats are based on the sinc methods or tensor product polynomial/trigonometric interpolation; see the discussion in Chapters 2 and 3. After that the suboptimal analytic tensor approximations can be optimized by the RHOSVD techniques discussed above.

The TT format has one principal difference compared with the CP or Tucker approximation: there are no established analytic methods (like sinc or polynomial approximation) delivering representations of discretized functions and operators directly in the product type TT tensor format, except for a few special cases when exact explicit representations are possible (see Exercise 3.80 below). On the other hand, the HOSVD algorithm is practically not applicable to the full format tensors due to the exponential complexity $O(n^{d+1})$.

Taking this into account, the TT approximation of function generated targets can be based on the following strategy:
(a) Analytic methods delivering the moderate rank canonical or Tucker tensors.
(b) Canonical-to-TT conversion by RHOSVD, [212]. Indeed, the canonical tensor is a good starting point for further algebraic TT rank approximation based on Algorithm 3.5.
(c) SVD based recompression of Tucker tensors to the TT format.

The gainful application of the TT format for the solution of d dimensional PDEs is mainly related to efficient realization of multilinear algebra (MLA) operations on already TT formated tensors including the rank optimization. Such a situation is typical in the solution of multidimensional and/or multiparametric PDEs since the matrix vector product usually increases the tensor rank.

Exercise 3.80. In some cases a TT representation can be derived from the explicit function TT (FTT) decomposition discussed in Chapter 2; see also [85, 177, 288]. The rank-2 TT representation of the Kronecker sum tensor (compare to Example 3.70)

$$\mathbf{A} := \mathbf{x}_1 \otimes \mathbf{1} \otimes \cdots \otimes \mathbf{1} + \cdots + \mathbf{1} \otimes \cdots \otimes \mathbf{1} \otimes \mathbf{x}_d \,, \quad \mathbf{x}_\ell \in \mathbb{R}^{n_\ell}$$

is obtained as the $n_1 \times \cdots \times n_d$ grid representation of the rank-2 FTT decomposition of the function $f(x) = x_1 + x_2 + \cdots + x_d$ (Section 2.1.5),

$$f(x) = \begin{pmatrix} x_1 & 1 \end{pmatrix} \begin{pmatrix} 1 & 0 \\ x_2 & 1 \end{pmatrix} \cdots \begin{pmatrix} 1 & 0 \\ x_{d-1} & 1 \end{pmatrix} \begin{pmatrix} 1 \\ x_d \end{pmatrix}.$$

Derive the rank-2 TT decomposition for the tensor generated by the function $f(x) = \sin(x_1 + x_2 + \cdots + x_d)$.

To complete this section, we collect the main ingredients of tensor numerical methods that can be based on analytic and algebraic tools for computations with the rank structured multidimensional operators and functions:
1. Discretization of functions and operators in tensor product Hilbert space of n-d tensors, $\mathbb{V}_\mathbf{n} = \mathbb{R}^{I_1 \times \cdots \times I_d}$, $\#I_\ell = n$.
2. Explicit rank structured representation (approximation) of functions and operators.
3. Fixed precision MLA in the low separation rank tensor formats $\mathcal{S} \subset \mathbb{V}_\mathbf{n}$ with

$$\mathcal{S} = \{\mathcal{C}_R,\ \mathcal{T}_\mathbf{r},\ \mathcal{T}_{\mathcal{C}_R,\mathbf{r}},\ TT[\mathbf{r}],\ \mathrm{QTT}[\mathbf{r}]\}\,.$$

   Key point: Efficient implementation of the tensor truncation (projection), $T_\mathcal{S} : \mathcal{S}_0 \to \mathcal{S} \subset \mathcal{S}_0 \subset \mathbb{V}_\mathbf{n}$, based on the SVD and (R)HOSVD approximations, ALS/AMEn (Alternating Minimal Energy) iterations, and multigrid acceleration techniques.
4. Multilevel tensor truncated preconditioned iteration for solving linear systems.

5. Minimization on tensor manifolds via DMRG, ALS, or AMEn type iteration.
6. Quasidirect tensor solvers via efficient formatted representation of operator valued functions (matrices) such as A^{-1}, $\exp(tA)$, convolution product, etc.
7. Realization of the computational process in quantized tensor spaces (QTT approximation), which reduces the numerical complexity to the logarithmic scale in terms of full size of multidimensional data; see Chapter 4.

Note that ALS and AMEn type iterations for optimization in the TT format have been discussed in [89, 90, 168, 312, 313, 319, 320, 359].

In the next chapter we focus on the description of superfast QTT based tensor techniques.

4 Superfast computations via quantized tensor approximation

The quantized tensor train (QTT) approximation was invented by B. Khoromskij in 2009, [196, 197]. For function generated vectors (tensors), the QTT approximation was proven to provide the logarithmic data compression $O(d \log N)$ on the wide class of functions in $\mathbb{R}^d$ sampled on a tensor grid of size N^d. Specifically, it was proven that for a function generated vector of size $N = q^L$, its reshaping into a $q \times \cdots \times q$ hypercube allows a small TT rank decomposition of the resultant L dimensional tensor, [196, 197]. The low rank TT representation of the reshaped $2^L \times 2^L$ matrices, observed in numerical experiments, was reported by I. Oseledets in 2009, [286, 287]. This can be viewed as the QTT representation to the discretized operators.

Based on the QTT techniques, the asymptotic storage cost for a class of function related N–d vectors (tensors) and matrices operated on them can be reduced to the logarithmic scale $N^d \to O(d \log N)$. In recent years this breakthrough property was gainfully used for representation and numerical treatment of various complicated multivariate functions representing basic physical quantities as well as for solving d dimensional PDEs in the QTT tensor representation. The QTT decomposition subroutines, included in the TT toolbox developed in the group of I. Oseledets [290], motivated many scientists to start working in this amazing new research area.

4.1 Quantized TT approximation: TT tour of highest dimensions

In this section, we first explain on simple examples why folding (reshaping) of a function related vector to a high dimensional tensor array may lead to the highly compressed rank structured representations. We describe the quantics folding of a long vector into a high dimensional tensor and the subsequent quantized TT parametrization. Then we prove that for complex exponential, trigonometric, or polynomial functions, the vectors obtained by sampling of these functions on a uniform grid can be represented exactly with a small QTT rank that does not depend on a size of representation grid. (In some cases, it is valid also for graded grids.) Thus, for example, if we have a vector of size $2^L = 2^{20}$ obtained by the grid discretization of an exponential function, its quantized representation (that has a QTT rank equal to 1) will need only $2 \cdot 20$ numbers, that is $2 \cdot \log(2^L)$. Correspondingly, algebraic operations with the QTT images are performed with logarithmic cost.

These theoretical results provide the background for the QTT approach and establish the QTT approximation method as the new powerful tool for the solution of challenging problems in scientific computing. We discuss the analytic methods for the construction of the QTT parametrization and present several examples of explicit

https://doi.org/10.1515/9783110365917-004

low rank QTT representations of discretized functions. The numerical illustrations on the QTT approximation of some function related vectors and matrices by using the TT tensor Toolbox will be presented.

4.1.1 Main motivation for the QTT approach

In what follows, we show by some explicit examples why folding of a vector to high dimensional hypercube of data may lead to a logarithmically compressed rank structured representation. These examples were the starting point in [196, 197] for understanding and further development of the QTT tensor approximation method that can be viewed as a TT tour of higher dimensions. This approach allows us to introduce the powerful rank structured parametrization of long function generated vectors.

Proposition 4.1 (Background for the QTT tensor approximation [196, 197]).
For a given $N = 2^L$, with $L \in \mathbb{N}$ and $c, z \in \mathbb{C}$, the exponential N vector

$$\mathbf{x} := \{cz^{n-1}\}_{n=1}^{N} \in \mathbb{C}^N ,$$

can be reshaped by the dyadic folding to the rank-1, $\underbrace{2 \times 2 \times \cdots \times 2}_{L}$-tensor,

$$\mathcal{F}_{2,L}: \mathbf{x} \mapsto \mathbf{A} = c \otimes_{p=1}^{L} \begin{bmatrix} 1 \\ z^{2^{p-1}} \end{bmatrix}, \quad \mathbf{A}: \{1, 2\}^{\otimes L} \to \mathbb{C} .$$

Here the binary coding of the long index, $n = 1 + \sum_{p=1}^{L}(j_p - 1)2^{p-1}$, to the multi-index $(j_1, \ldots, j_L) \in \{1, 2\}^{\otimes L}$ is applied.

The trigonometric N vector

$$\mathbf{x} := \{\sin(h(n-1))\}_{n=1}^{N} \in \mathbb{C}^N$$

can be reshaped to the 2–L tensor of complex CP rank 2, or to a rank-2 TT tensor. A similar result remains true for the cos *N-vector.*

We postpone the proof to the forthcoming discussion. Here we only note that the second statement is the consequence of the relation $\sin z = \frac{e^{iz} - e^{-iz}}{2i} = \mathrm{Re}(e^{iz})$; see Section 2.1, where the rank-2 functional representation to $\sin(x_1 + \cdots + x_d)$ was derived. The corresponding explicit rank-2 QTT representation of the sin-vector $\mathbf{x}$ reads

$$\mathcal{F}_{2,L}: \mathbf{x} \mapsto [\sin x_1 \cos x_1] \otimes_{p=2}^{L-1} \begin{bmatrix} \cos x_p & -\sin x_p \\ \sin x_p & \cos x_p \end{bmatrix} \otimes \begin{bmatrix} \cos x_L \\ \sin x_L \end{bmatrix} \in \{0, 1\}^{\otimes L} , \tag{4.1}$$

where the parameters of the QTT representation are specified by a vector

$$\mathbf{x}_Q = [x_p = h2^{p-1} j_p]_{p=1}^{L} , \quad j_p \in \{0, 1\} , \quad n = 1 + \sum_{p=1}^{L} j_p 2^{p-1} .$$

The corresponding rank-2 decomposition for cos-vector can be justified by $\cos z = \frac{e^{iz}+e^{-iz}}{2i} = \mathrm{Im}(e^{iz})$. In Section 4.2 we describe the direct scheme to derive the above QTT representations.

We can see that in some cases the quantized function generated vectors can be represented in the form of low rank TT or CP tensors. The rank-$\mathbf{r}$ TT representation of quantized vectors is called QTT format; see [196, 197]. Similarly, we call the CP representation of quantized vectors the QCP or Q_{Can} format; see [220].

The above examples illustrate the benefits of the quantized TT or CP representations that are due to crucial reduction of the number of representation parameters to logarithmic scale $N \to 2\log_2 N$.

4.1.2 Quantics folding to higher dimension: general scheme

The QTT type approximation of an N vector with $N = q^L$, $L \in \mathbb{N}$, is defined as the tensor decomposition (approximation) in the TT or canonical formats applied to a tensor obtained by the q-adic folding (reshaping) of the target vector to an L dimensional $q \times \cdots \times q$ data array (tensor) that is thought to be an element of the L dimensional quantized tensor space. We recall the notation for the index set $I = \{1, \ldots, N\}$.

First, note that a one step folding (reshaping) of some N^2-vector leads to a $N \times N$ matrix and its subsequent low rank approximation is the known procedure to compress data in signal processing [296], and in the solution of 2D PDEs discretized on a tensor grid. In the latter case the long vector representing the FEM coefficients can be reshaped to the two-indexed data array, and it is similar in the case of the 3D problems.

Figure 4.1 visualizes the binary folding for $L = 3$, $d = 1$, $p = 3$.

In the vector case, i.e., for $d = 1$, a vector $\mathbf{x} = [x(i)]_{i\in I} \in \mathbb{V}_{N,1}$, can be reshaped to its quantized image in L dimensional tensor space,

$$\mathbb{Q}_{q,L} = \bigotimes_{\nu=1}^{L} \mathbb{K}^q\,, \quad \mathbb{K} \in \{\mathbb{R}, \mathbb{C}\}\,,$$

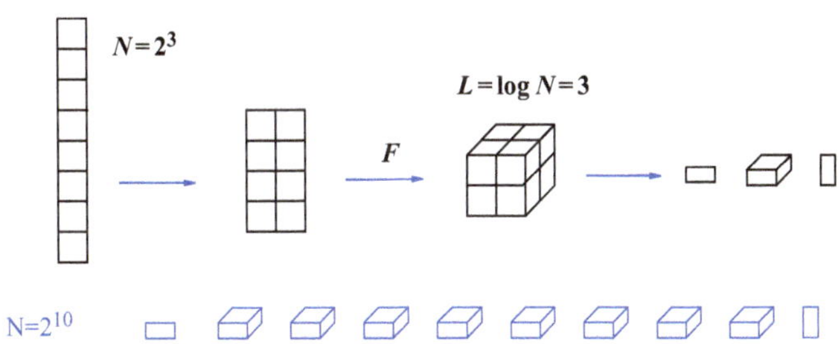

Fig. 4.1: QTT decomposition for $d = 3$ (top) and for $d = 10$ (bottom).

by q-adic folding to highest possible dimension L,

$$\mathcal{F}_{q,L}: \mathbf{x} \to \mathbf{Y} = [Y(\mathbf{j})] \in \mathbb{Q}_{q,L}, \quad \mathbf{j} = \{j_1, \dots, j_L\},$$

with $j_\nu \in \{1, 2, \dots, q\}$, $\nu = 1, \dots, L$, where for fixed i, we have $Y(\mathbf{j}) := x(i)$, and $j_\nu = j_\nu(i)$ is defined via q-coding, $j_\nu - 1 = C_{-1+\nu}$, such that the coefficients $C_{-1+\nu}$ are found from the q-adic representation of $i - 1$,

$$i - 1 = C_0 + C_1 q^1 + \cdots + C_{L-1} q^{L-1} \equiv \sum_{\nu=1}^{L} (j_\nu - 1) q^{\nu-1}.$$

Concerning the linear algebra aspect, we note that the 'reshape' command is the standard option in MATLAB.

For $d > 1$ the construction is similar [197] as described in the following definition.

Definition 4.2. The q-adic folding of degree p, $2 \le p \le L$,

$$\mathcal{F}_{q,d,p}: \mathbb{V}_{\mathbf{n},d} \to \mathbb{V}_{\mathbf{m},dp}, \quad \mathbf{m} = (\mathbf{m}_1, \dots, \mathbf{m}_\ell), \quad \mathbf{m}_\ell = (m_{\ell,1}, \dots, m_{\ell,p}),$$

$m_{\ell,1} = q^{L-p+1}$, $m_{\ell,\nu} = q$ for $\nu = 2, \dots, p$, ($\ell = 1, \dots, d$), reshapes the $\mathbf{n}$–d tensors in $\mathbb{V}_{\mathbf{n},d}$ to the elements of quantics space $\mathbb{V}_{\mathbf{m},dp}$ as follows:

(A) $d = 1$: a vector $\mathbf{x}_{(N,1)} = [x(i)]_{i \in I} \in \mathbb{V}_{N,1}$, is reshaped to $\mathbb{V}_{q^{L-p+1},p}$,

$$\mathcal{F}_{q,1,p}: \mathbf{x}_{(N,1)} \to \mathbf{Y}_{(\mathbf{m},p)} = [Y(\mathbf{j})] := [x(i)], \quad \mathbf{j} = \{j_1, \dots, j_p\},$$

$j_1 \in \{1, \dots, q^{L-p+1}\}$, and $j_\nu \in \{1, \dots, q\}$ for $\nu = 2, \dots, p$.
For fixed i, $j_\nu = j_\nu(i)$ is defined by $j_\nu = 1 + C_{L-p-1+\nu}$, ($\nu = 1, \dots, p$), where the coefficients $C_{L-p-1+\nu}$ are found from the partial radix-q representation of $i - 1$,

$$i - 1 = C_{L-p} + C_{L-p+1} q^{L-p+1} + \cdots + C_{L-1} q^{L-1}. \tag{4.2}$$

(B) For $d > 1$ a tensor $\mathbf{A}_{(\mathbf{n},d)} = [A(i_1, \dots, i_d)]$, $i_\ell \in I_\ell$, $\ell = 1, \dots, d$, is reshaped to

$$\mathcal{F}_{q,d,p}: \mathbf{A}_{(\mathbf{n},d)} \to \mathbf{B}_{(\mathbf{m},dp)} = [B(\mathbf{j}_1, \dots, \mathbf{j}_d)] := [A(i_1, \dots, i_d)],$$

$\mathbf{j}_\ell = \{j_{\ell,1}, \dots, j_{\ell,p}\}$, with $j_{\ell,1} \in \{1, \dots, q^{L-p+1}\}$, and $j_{\ell,\nu} \in \{1, \dots, q\}$, for $\nu = 2, \dots, p$, and for all $\ell = 1, \dots, d$. Now the univariate ℓ-mode index i_ℓ is reshaped into $\mathbf{j}_\ell$ as in the case $d = 1$.

(C) In the case $p = 1$, we define $\mathcal{F}_{q,d,1}$ as the identity mapping in $\mathbb{V}_{\mathbf{n},d}$.

The following examples illustrate the simple quantized representations.

Example 4.3. For $d = 1$ and $p = 2, 3$, $\mathcal{F}_{q,1,p}$ folds an N vector to a $N/q \times q$ matrix or to a $N/q^2 \times q \times q$, 3-tensor, respectively.

Example 4.4. Quantics folding of the exponential N vector. For $N = 8$, $L = 3$, $\mathbf{x} = [1\ z\ z^2\ z^3\ z^4\ z^5\ z^6\ z^7]^T \in \mathbb{C}^8$, we have $\mathcal{F}_{2,3}(\mathbf{x}) \in \mathbb{C}^{2\times2\times2}$,

$$\mathcal{F}_{2,1,3}: \mathbf{x} \mapsto \mathbf{A} = \begin{bmatrix} 1 \\ z \end{bmatrix} \otimes \begin{bmatrix} 1 \\ z^2 \end{bmatrix} \otimes \begin{bmatrix} 1 \\ z^4 \end{bmatrix} \in \mathrm{QTT}[\mathbf{1}, 3] \subset \mathbb{C}^{2\times2\times2}.$$

For fixed N, the practical choice of the model parameters q and p depends on the particular computational tasks. For the sake of higher compressibility, the *maximal degree folding* $\mathcal{F}_{q,d,L}$, corresponding to $p = L$ (for $N = 2^L$) should be applied. In this case the index $j_\nu - 1$ ($\nu = 1, \dots, L$) is the q-adic representation of $i_\ell - 1$ for $i_\ell \in I_\ell$, in radix-q system, such that $j_\nu \in \{1, \dots, q\}$. If $q = 2$, use the binary coding of $i - 1$,

$$i - 1 = \sum_{\nu=1}^{L} (j_\nu - 1) 2^{\nu-1} .$$

On the other hand, factorization of the vector size integer N is not restricted to the q-adic form. A slight generalization of the QTT format involves different *prime* dimensions of a tensor, instead of the same value q. Given initial dimension N, we decompose this integer number into smallest nontrivial prime factors, say,

$$N = q_1 \dots q_{L'},$$

such that the corresponding index factorization allows the TT or CP formats. Again, the quantization folding could be performed by the standard MATLAB command. In the general case, any other small factors like $2, 3, 5, 7, \dots,$ and so on are possible. If N is powers of 2, we end up with the classical QTT format with $2 \times \dots \times 2$ tensors.

The unfolding transform, e.g., tensor-to-matrix (matricization) or tensor-to-vector (vectorization), may be viewed as the reverse to the folding transform:

$$\mathcal{F}^{-1}_{q,d,p} : \mathbb{V}_{\mathbf{m},dp} \to \mathbb{V}_{\mathbf{n},d} .$$

The folding transform $\mathcal{F}_{q,d,p}$ exhibits many useful properties:

(F1) $\mathcal{F}_{q,d,p}$ is the linear isometry between $\mathbb{V}_{N,d}$ and $\mathbb{V}_{q^{L-p+1},dp}$ that has the inverse transform (unfolding)

$$\mathcal{F}^{-1}_{q,d,p} : \mathbb{V}_{q^{L-p+1},dp} \to \mathbb{V}_{N,d} .$$

(F2) The q-folding of a rank-1 tensor $\mathbf{X} = \mathbf{x}_1 \times \dots \times \mathbf{x}_d \in \mathbb{V}_{N,d}$, is given by the outer product of componentwise reshaping transforms of canonical vectors,

$$\mathcal{F}_{q,d,p}\mathbf{X} = \mathcal{F}_{q,1,p}\mathbf{x}_1 \otimes \dots \otimes \mathcal{F}_{q,1,p}\mathbf{x}_d .$$

The interface ranks between blocks $\mathcal{F}_{q,1,p}\mathbf{x}_\ell$ are equal to 1 for all $\ell = 1, \dots, d-1$.

(F3) Let $d = 1$, then for any $p = 2, \dots, L$ and $\mathbf{x} = [x(i)] \in \mathbb{C}^N$ we have a bound on the TT rank of a tensor $\mathcal{F}_{q,1,L}\mathbf{x}$,

$$r_{p-1} \le \operatorname{rank}(X_p) ,$$

where X_p is the reshaping of $\mathbf{x}$ to a $N/q^{p-1} \times q^{p-1}$ matrix.

4.1.3 QTT type tensor format and its hybrid versions

Recall that the rank-$\mathbf{r}$ TT representation of the quantized vector (tensor) belongs to the rank-$\mathbf{r}$ QTT format, $\mathrm{QTT}[\mathbf{r}, L] = \mathrm{QTT}[\mathbf{r}]$; see [196, 197]. The QTT parametrization requires $qr^2 \log_2 N$ numbers to store. Likewise, we call the rank-r CP representation of quantized vectors an element in QCP tensor format [220] further denoted by $QCP[r, L]$ =$QCP[r]$. Its storage size is $qr \log_2 N$. Proposition 4.1 demonstrates examples of rank-1 QCP and rank-2 QTT tensors; see also Example 4.2.

In the case of large mode size N and large r in d dimensions MLA might cause difficulties due to the polynomial complexity scaling in both r and N. In such situations the QTT and QCP tensors can be combined with basic formats leading to mixed tensor representations, which enjoy complimentary benefits of different tensor parametrizations.

To that end one may consider a combination of the Tucker and TT/QTT formats. First, we define *Tucker-TT format*, $\mathcal{T}_{\mathbf{r}}[TT[\mathbf{r}_1]]$, containing all Tucker tensors in $\mathcal{T}_{\mathbf{r},\mathbf{n}}$ such that the Tucker core is parametrized by the rank-$\mathbf{r}_1$ TT format. Now the storage complexity of representation scales linearly in r, $O(drN + d\, r_1^2\, r)$, while the representation basis is given explicitly by the 'optimal' set of orthogonal Tucker vectors. Note that for the two-level Tucker-CP format we have $\mathcal{T}_{\mathcal{C}_{R,\mathbf{r}}} \subset \mathcal{T}_{\mathbf{r}}[TT[\mathbf{r}_1]]$ with $\mathbf{r}_1 = (R, \dots, R)$.

Remark 4.5. The hierarchical Tucker (HT) format introduced in [146] is closely related to the TT model. We refer to [123, 138] for the detailed analysis on this particular version of matrix network states decomposition. The quantized images of some functions could be represented in the HT format; see [124].

The two-level Tucker-TT-QTT tensor format introduced in [87] for optimized tensor calculus has demonstrated considerably better performance compared with the direct TT or QTT approximations applied to multidimensional PDEs. In this format, designed as the subset of Tucker tensors, the Tucker core is substituted by the TT tensor (diminishing the curse of dimensionality), and each orthogonal N vector in the Tucker side matrices is represented as the QTT tensor.
The so called *canonical-QTT* format, $\mathcal{C}_{R,\mathbf{n}}[\mathrm{QTT}[\mathbf{r}, L]]$, is specified as a set of N–d tensors with $N = q^L$, represented as an R-term sum of rank-1 tensors, where each canonical skeleton N vector in rank-1 terms is given in the form of a q–L tensor in the $QTT[\mathbf{r}, L]$ format, with $L = \log_q N$. The particular representation looks like

$$\mathbf{V} = \sum_{k=1}^{R} c_k T_k^{(1)} \times_2 T_k^{(2)} \cdots \times_d T_k^{(d)} \in \mathcal{C}_{R,\mathbf{n}}[QTT[\mathbf{r}, L]] \,, \tag{4.3}$$

for $k = 1, \dots, R$, $\nu = 1, \dots, d$, such that

$$T_k^{(\nu)} := \{\times_\ell\}_{\ell=1}^{L} \mathbf{G}_{k,\nu}^{(\ell)} \in QTT[\mathbf{r}, L] \quad \text{with small size tritensors} \quad \mathbf{G}_{k,\nu}^{(\ell)} \in \mathbb{R}^{r_{\ell-1}\times q \times r_\ell} \,.$$

The storage complexity scales logarithmically in N, $O(Rr^2 d \log N)$, hence it has advantages for large mode size N.

Figure 4.2 presents the schematic view of different tensor representations including the CP, Tucker, TT, TC, tensor network states, and hierarchical tensor formats.

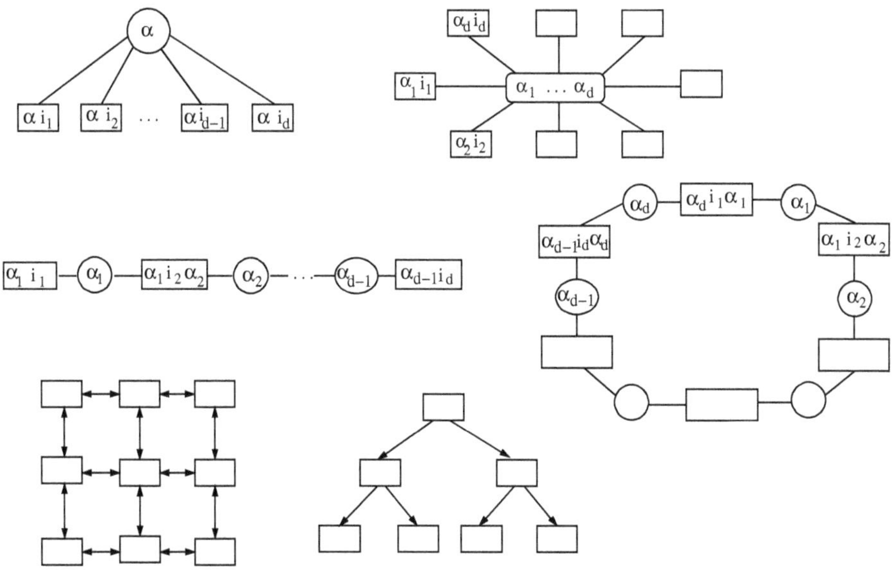

Fig. 4.2: Visualizing basic tensor models.

4.1.4 Why QTT approximation does a job

The main advantage of the rank-$\mathbf{r}$ QTT based representations is the logarithmic complexity scaling in the vector size and the quadratic scaling in the maximal QTT rank, $O(qr^2 \log_q N)$. However, the essential prerequisite for taking advantage from this favorable feature is the existence of an accurate low QTT rank representations (approximations) to the target vector (tensor). In what follows, we show that such approximations exist for a large class of function generated data.

Lemma 4.6. *For given $N = q^L$, with $q = 2, 3, \dots,$ and $L \in \mathbb{N}_+$, and for given $c_k, z_k \in \mathbb{C}$ $(k = 1, \dots, R)$, we have:*

(A) *An exponential sum N vector, $\mathbf{x} := \{x_n := \sum_{k=1}^{R} c_k z_k^{n-1}\}_{n=1}^{N}$, can be reshaped by the q-folding $\mathcal{F}_{q,1,L}$, to the rank-R, $q{-}L$ tensor in $\mathbb{V}_{\mathbf{q},L}$,*

$$\mathcal{F}_{q,1,L} : \mathbf{x} \to \mathbf{A}_{(q,L)} = \sum_{k=1}^{R} c_k \otimes_{p=1}^{L} \left[1 \; z_k^{q^{p-1}} \; \dots \; z_k^{(q-1)q^{p-1}}\right]^{\mathrm{T}} \in \mathcal{C}_{R,\mathbf{q}}[TT[\mathbf{1}]] \,.$$

(B) *A sum of trigonometric N vectors, $\mathbf{x} := \{x_n := \sum_{k=1}^{R} c_k \sin(\alpha_k(n-1))\}_{n=1}^{N}$, can be reshaped to the rank-$2R$, $q{-}L$ tensor $\mathbf{A}_{(q,L)}$, whose TT ranks do not exceed $2R$,*

$$\mathcal{F}_{q,1,L} : \mathbf{x} \to \mathbf{A}_{(q,L)} = \sum_{k=1}^{R} \mathbf{A}_k \in \mathbb{V}_{q,L} \,, \quad \text{with} \quad \mathbf{A}_k \in TT[\mathbf{2}, L] \,.$$

In both cases (A) and (B), the number of representation parameters is reduced from $(N+1)R$ to $(qL+1)R$ and to $4qLR$, respectively.

Proof.
(A) Let $R = 1$, and start the induction: $L = 2$, i.e., $N = q^2$, $\mathcal{F}_{q,1,2}: \mathbf{x}_{(q^2,1)} \to \mathbf{A}_{(q,2)}$,

$$\mathbf{A}_{(q,2)} := \begin{pmatrix} 1 & z^q & \dots & z^{(q-1)q} \\ z & \ddots & \dots & z^{(q-1)q+1} \\ \vdots & \vdots & \ddots & \vdots \\ z^{q-1} & z^{2q-1} & \dots & z^{q^2-1} \end{pmatrix} = \begin{bmatrix} 1 \\ z \\ \vdots \\ z^{q-1} \end{bmatrix} [1\ z^q\ \dots\ z^{(q-1)q}] \,.$$

Induction step: L to $L+1$, i.e., for $N = q\,q^L$. Subvectors $\mathbf{x}_1, \dots, \mathbf{x}_q \in \mathbb{R}^{q^L}$ of $\mathbf{x}_{(N,1)}$, with $x_k(i) := \mathbf{x}[i + (k-1)q^L, 1]$, $(k = 1, \dots, q,\ i = 1, \dots, q^L)$, represent the result of a one level folding ($p = 2$), by the rank-1, $N/q \times q$ matrix via rescaling of the first column $\mathbf{x}_1$,

$$\mathcal{F}_{q,1,2}: \mathbf{x}_{(N,1)} \to \mathbf{A}_{(N/q,2)} := c[\mathbf{x}_1 \mathbf{x}_2 \dots \mathbf{x}_q] = c\mathbf{x}_1 \otimes \mathbf{y} \,,$$

$\mathbf{y} := [1\ z^{q^L}\ \dots\ z^{(q-1)q^L}]^{\mathrm{T}}$.
By induction, substitute $\mathbf{x}_k$, $k = 1, \dots, q$, of size $N/q = q^L$ by rank-1 tensor,

$$\mathbf{A}_{(q,L+1)} = c\left[\otimes_{p=1}^{L} [1\ z^{q^{p-1}}\ \dots\ z^{(q-1)q^{p-1}}]^{\mathrm{T}}\right] \otimes [1\ z^{q^L}\ \dots\ z^{(q-1)q^L}]^{\mathrm{T}} \,.$$

(B) Again, we begin from the case $R = 1$. Using trigonometric identity

$$\sin z = \frac{e^{iz} - e^{-iz}}{2i} = \mathrm{Re}(e^{iz}) \,,$$

and applying item (A) with $R = 1$, we arrive at the required claim on tensor rank of $\mathbf{A}_{(q,L)}$ over field $\mathbb{C}$.
Now, the rank of each ℓ-mode TT unfolding matrix of the q–L tensor does not exceed 2, since the matrix rank does not change if we extend the field $\mathbb{R}$ to $\mathbb{C}$. Since the TT ranks do not exceed the ranks of respective directional unfolding matrices, the maximal TT rank of the q–L tensor $\mathbf{A}_{(q,L)}$ is bounded by 2.

In the case of arbitrary rank parameter $R > 1$, the result over fields $\mathbb{C}$ and $\mathbb{R}$ is obtained by summation of rank-1 (respectively rank-2) terms.

The complexity bounds then follow from the properties of CP format; see Section 3.6. To that end we recall that the existence of rank-2 QTT representation of the sin vector is proven by (4.1). □

4.1.5 QTT approximation on classes of functional vectors

Here we proceed with the description of some classes of functional vectors that allow the low rank QTT representation (approximation). First, we show that the exponential trigonometric product vector allows the $4 \log_q N$ complexity QTT representation.

Lemma 4.7. *For given $N = q^L$, with $q = 2, 3, \dots, L \in \mathbb{N}_+$, and $c, z \in \mathbb{C}, \alpha \in \mathbb{R}$, the exponential trigonometric vector*

$$\mathbf{x} := \{x_n := cz^{n-1} \sin(\alpha(n-1))\}_{n=1}^{N} ,$$

can be reshaped to

$$\mathcal{F}_{q,1,L} : \mathbf{x} \to \mathbf{A}_{(q,L)} \in QTT[2] ,$$

whose TT ranks do not exceed 2. The number of representation parameters is bounded by $4q \log_q N$.

Proof. The properties of the folding transform $\mathcal{F}_{q,1,L}$ imply that the q–L tensor $\mathbf{A}_{(q,L)}$ is obtained by the Hadamard product of the rank-1 quantics representation for a single exponential and rank-2 quantics of the trigonometric vector (compare to Lemma 4.6). Now the statement follows from the fact that the Hadamard product with rank-1 tensor does not enlarge the TT rank of the second factor that is exactly 2. Hence, the TT rank of the resultant q–L tensor $\mathbf{A}_{(q,L)}$ does not exceed 2. □

We consider the cost function that gives the storage of a rank-1 QTT tensor representing an N vector. Now we address the question of the optimal choice of the quantization base q that minimizes that cost function.

Remark 4.8. Minimization of the cost function $f_N(q) := q \log_q N$ for fixed value of $N \in \mathbb{R}_+$ leads to the optimal quantics base $q^* \in [2, 3] \approx 2.7$. This means that for large vector size N, the choice $q = 2, 3$ leads to the best compression rate.

Remark 4.8 indicates that actually the smallest possible quantization base q is the best choice with respect to the maximization of the compression rate (the function $f_N(q)$ is monotone for $q > q^*$).

Lemma 4.9. *The QTT rank of the discretized polynomial vector does not exceed $m + 1$, where m is the polynomial degree, independent of the grid size N.*

Proof. Property (F3) of the quantics folding ensures that the QTT rank of a vector obtained by equidistant sampling the polynomial of degree m, does not exceed $m+1$. In fact, the column space of the reshaped TT unfolding matrix is spanned by at most $m+1$ polynomial vectors generated by the fixed basis set $1, x, \dots, x^m$, respectively. The explicit (closed form) rank-$(m + 1)$ QTT representation is discussed in Section 4.2. □

Note that the rank-$(m + 1)$ HT representation of the polynomial vector on a uniform grid was addressed in [124].

The QTT format applies to piecewise polynomial functions, in particular to *wavelet basis functions*. For example, the QTT rank of a Haar wavelet does not exceed 2, implying that the asymptotic QTT compression properties are at least as good as for the Haar wavelets; see [219], where the superfast Haar wavelet transform of logarithmic complexity was introduced. 'Mexican hat' wavelets generated by Gaussians-times-polynomials can be shown to have low QTT ε rank as well, taking into account Lemma 4.12.

It is interesting to note that for a class of analytic functions on an interval the best approximation by polynomials of degree-m provides approximation error of the order of ε with $m = O(|\log \varepsilon|)$; see Chapter 2. Hence, in view of Lemma 4.9, the ε rank of the QTT approximation to a discretized analytic function scales as $O(|\log \varepsilon|)$. As a consequence of this observation, the QTT approximation results can be derived [85, 197, 288] for different classes of analytic (or piecewise analytic) functions discretized on a grid.

Likewise, functions with local pointwise singularities (say, x^α, e^{-x^α}, Green kernels, sinc functions, Matérn functions, etc.) can be approximated by h–p type methods using piecewise polynomials, or by sinc-quadrature techniques. To that end, the following result is useful:

Lemma 4.10. *The* sinc *vector obtained via the uniform sampling of* $\frac{\sin kx}{kx}$ *on the interval* $[-b, b]$, *allows the* q*-folding QTT approximation, whose* ε *rank scales as* $O(|\log \varepsilon|)$.

Proof. Indeed, the analytic function $\frac{\sin kx}{kx}$ allows the ε approximation via polynomials of degree $m = O(|\log \varepsilon|)$. □

In numerical tests for the sinc vector we observe that QTT rank remains bounded by a small constant (say 4, 5) almost uniformly in the vector size N (see numerics in Table 4.1).

Now we come to the following summary:

Summary 4.11. We conclude that combining the polynomial and sinc approximation theory in Chapter 2 with Lemmas 4.9 and 4.10 delivers powerful tools for the analysis of QTT approximation applied to a wide class of functions in $\mathbb{R}^d$.

The set of Gaussian functions plays an important role in approximation theory, in computational quantum chemistry and computational physics, in molecular dynamics, in signal processing, and in many other applications. The ε QTT rank estimate for the Gaussian vector, proven in [84], is presented in the following:

Lemma 4.12. *Suppose uniform grid points* $-a = x_0 < x_1 < \cdots < x_N = a$, $x_i = -a + hi$, $N = 2^L$ *are given on an interval* $[-a, a]$, *and the vector* g *is defined by its elements* $g_i = e^{-x_i^2/2p^2}$, $i = 0, \ldots, N-1$. *Assume in addition that* $\int_a^\infty e^{-x^2/2p^2} \le \frac{\varepsilon}{2} < 1$. *Then for all sufficiently small* $\varepsilon > 0$ *there exists the QTT approximation* g_r *with the ranks bounded as*

$$r(g_r) \le c\frac{a}{p}\sqrt{\log\left(\frac{1}{\varepsilon}\frac{p}{1+a}\right)},$$

and the accuracy

$$|g - g_r| \le \left(\frac{r}{2a} + 1\right)\varepsilon = \left(c\frac{1}{p}\sqrt{\log\left(\frac{1}{\varepsilon}\frac{p}{1+a}\right)} + 1\right)\varepsilon,$$

where c *does not depend on* a, p, ε, *or* N.

Proof. This lemma is based on the Fourier transform of the Gaussian function. Indeed, consider the approximation via the partial Fourier sum

$$\mathrm{e}^{-\frac{x^2}{2p^2}} = \sum_{m=0}^{M} \alpha_m \cos\left(\frac{\pi m x}{a}\right) + \eta \quad \text{on } [-a, a]\,,$$

where $|\eta| = |\sum_{m=M+1}^{\infty} \alpha_m \cos(\frac{\pi m x}{a})| < \varepsilon$. There are no sin functions in this sum, as the Gaussian function is even with respect to 0. If we discretize now this sum on a uniform grid, all vectors generated by cos functions will have exact QTT representations with all ranks equal to 2; see (4.1) and [197]. So it is enough to provide an estimate on M.

The Fourier coefficients are computed as

$$\alpha_m = \frac{\int_{-a}^{a} \mathrm{e}^{-\frac{x^2}{2p^2}} \cos\left(\frac{\pi m x}{a}\right) dx}{\int_{-a}^{a} \cos^2\left(\frac{\pi m x}{a}\right) dx}\,,$$

where all denominators are equal to a if $m > 0$, and $2a$ if $m = 0$. Let us denote them as

$$|C_m|^2 = \begin{cases} 2a\,, & \text{if } m = 0\,, \\ a\,, & \text{otherwise}\,. \end{cases}$$

In the nominator, we note that the cos function is bounded by 1, and

$$\int_{-\infty}^{\infty} \mathrm{e}^{-\frac{x^2}{2p^2}} dx = \int_{-a}^{a} \mathrm{e}^{-\frac{x^2}{2p^2}} dx + 2\int_{a}^{\infty} \mathrm{e}^{-\frac{x^2}{2p^2}} dx \le \int_{-a}^{a} \mathrm{e}^{-\frac{x^2}{2p^2}} dx + \varepsilon\,,$$

so we approximate

$$\alpha_m = \left(\int_{-\infty}^{\infty} \mathrm{e}^{-\frac{x^2}{2p^2}} \cos\left(\frac{\pi m x}{a}\right) dx - \xi_m\right) / |C_m|^2\,, \quad 0 < \xi_m < \varepsilon\,.$$

We deduce the integral over the whole axis from the continuous Fourier transform. Indeed, it is known that the Fourier image of the Gaussian function is another Gaussian function:

$$\int_{-\infty}^{\infty} \mathrm{e}^{-\frac{x^2}{2p^2}} \mathrm{e}^{\mathrm{i}\omega x} dx = \int_{-\infty}^{\infty} \mathrm{e}^{-\frac{x^2}{2p^2}} \cos(\omega x) dx = p\mathrm{e}^{-\frac{\omega^2 p^2}{2}}\,,$$

where i is the imaginary unity. So, plugging $\omega = \frac{\pi m}{a}$ in, we get the statement for α_m:

$$\alpha_m = \left(p\mathrm{e}^{-\frac{\pi^2 m^2 p^2}{2a^2}} - \xi_m\right) / |C_m|^2\,.$$

Now we truncate α_m on a value $m = M$ so that $\alpha_M \le \varepsilon$, hence for M

$$M \le \frac{\sqrt{2}}{\pi}\frac{a}{p} \log^{0.5}\left(\frac{p}{(1 + |C_M|^2)\varepsilon}\right) = \frac{\sqrt{2}}{\pi}\frac{a}{p} \log^{0.5}\left(\frac{p}{1+a}\frac{1}{\varepsilon}\right)\,,$$

which gives the first result of the lemma (up to rank 2 of each cosine function). Note that due to the very fast decay of the Fourier coefficients, the threshold $\alpha_M \le \varepsilon$ implies $\sum_{m=M+1}^{\infty} \alpha_m \le \varepsilon$ as well.

To obtain the expression for the error, recall that

$$g(x) = e^{-\frac{x^2}{2p^2}} = \sum_{m=0}^{M} \frac{1}{|C_m|^2}\left(pe^{-\frac{\pi^2 m^2 p^2}{2a^2}} - \xi_m\right)\cos\left(\frac{\pi m x}{a}\right) + \eta\,,$$

$$g_r(x) = \sum_{m=0}^{M} \frac{1}{|C_m|^2} pe^{-\frac{\pi^2 m^2 p^2}{2a^2}} \cos\left(\frac{\pi m x}{a}\right).$$

Now taking into account bounds $|\xi_m| < \varepsilon$, $|\cos(\frac{\pi m x}{a})| < 1$, $|\eta| < \varepsilon$, $r = 2M$, we arrive at the estimate for $|g - g_r|$. □

In some cases the separation rank in the QTT representation may be reduced by using special grading of sampling points. Here we show that, in fact, using *equidistant sampling points is not mandatory*. The next statement gives the uniform rank bound for polynomial vectors sampled at $N + 1$ Chebyshev–Gauss–Lobatto nodes,

$$x_j = \cos\frac{\pi j}{N} \in [-1, 1]\,, \quad j = 0, \dots, N\,,$$

and for Gaussian type function with quadratic mesh grading see [197].

Lemma 4.13.

(A) *For any* $n = 0, 1, \dots,$ *the Chebyshev polynomial* $T_n(x) = \cos(n \arccos x)$, $|x| \le 1$, *sampled over* $N + 1 = 2^L$ *Chebyshev nodes* $x_j \in [-1, 1]$, *can be represented in the quantics space of* $2 - \log N$ *tensors with both the canonical* $\mathbb{C}$ *rank and QTT rank* ≤ 2, *uniformly in* N.

(B) *The Chebyshev polynomial* $T_n(x)$, *sampled as a vector* $\mathbf{x}$ *at Chebyshev nodes*, $\theta_j = \arccos x_j$, *has the explicit rank-2 QTT representation (with* $y_p = h2^{p-1}i_p - 1$, $i_p \in \{0, 1\}$, $h = 2/N$*),*

$$\mathbf{x} \mapsto [\cos y_1 \; -\sin y_1] \otimes_{p=2}^{L-1} \begin{bmatrix} \cos y_p & -\sin y_p \\ \sin y_p & \cos y_p \end{bmatrix} \otimes \begin{bmatrix} \cos y_L \\ \sin y_L \end{bmatrix} \in \{0, 1\}^{\otimes L}\,.$$

(C) *Any polynomial of degree m sampled as an N vector over* $N + 1 = 2^L$ *Chebyshev nodes at* $[-1, 1]$ *has the QTT separation rank bounded by* $2m + 1$.

Proof.

(A) First we note that the Chebyshev polynomial $T_n(x) = \cos(n \arccos x)$, sampled at Chebyshev nodes, coincides with the cos vector sampled over uniformly graded points in variable $\theta_j = \arccos x_j$, $j = 0, \dots, N$. Then the result follows by Lemmas 4.1 and 4.6.

(B) The proof is based on the explicit FTT representation of the cos vector in Section 4.2

(C) Any polynomial of degree m can be represented in the orthogonal basis of Chebyshev polynomials by at most $m+1$ terms with $T_0 = 1$. Since the cos vector has rank 2, item (C) follows from (A). □

Lemma 4.13 can be applied to the case of polynomial interpolation over Chebyshev nodes, which are usually more preferable compared with equidistant nodes. In fact, this prevents the well known instability appearing in those interpolation process that are based on equidistant grids.

Remark 4.14. The TT rank of the q-folded discrete Gaussian type function sampled over the uniform grid, $\{e^{-\alpha(n-1)^2}\}_{n=1}^N$, appears to be greater than 2; see Lemma 4.12. (Numerical tests show that it remains to be almost uniformly bounded in the vector size N; see Table 4.1.) Lemma 4.13 implies the rank-1 QCP representation of the Gaussian vector in the case of quadratic mesh grading toward the origin, i.e., by sampling the Gaussian $e^{-\alpha t^2}$ over

$$t_n = \sqrt{h(n-1)}\,, \quad n = 1, \ldots, N\,, \quad h > 0\,.$$

We note that the previous results can be applied to R-term sums of exponential/trigonometric vectors in d dimensions, i.e., to the respective N–d tensors,

$$\mathbf{A}_{(\mathbf{n},d)} := \left\{ x_{\mathbf{n}} := \sum_{k=1}^{R} c_k \prod_{\ell=1}^{d} z_{k,\ell}^{n_\ell - 1} \right\}_{\mathbf{n}\in I^{\otimes d}}, \qquad I = \{1, \ldots, N\}\,, \tag{4.4}$$

$$\mathbf{S}_{(\mathbf{n},d)} := \left\{ x_{\mathbf{n}} := \sum_{k=1}^{R} c_k \prod_{\ell=1}^{d} \sin(\alpha_{k,\ell}(n_\ell - 1)) \right\}_{\mathbf{n}\in I^{\otimes d}}, \qquad I = \{1, \ldots, N\}\,. \tag{4.5}$$

These tensors can be reshaped to the high dimensional QCP/QTT based formats $\mathcal{C}_{R,\mathbf{q}}[QCP[\mathbf{1}, dL]]$ (complexity $dqR\log_q N$) and $\mathcal{C}_{R,\mathbf{q}}[QTT[\mathbf{2}, dL]]$ (complexity $4dqR\log_q N$), respectively.

4.1.6 QTT approximation in analytic form

Consider the error bound for the *semianalytic QTT type approximation* of function related tensors.

Lemma 4.15. *For given continuous function $f : [a, b] \to \mathbb{R}$, and $\varepsilon > 0$, let there exist an approximation such that*

$$\max_{x\in[a,b]} \left| f(x) - \sum_{k=1}^{M} c_k e^{-t_k x} \right| \le \varepsilon\,. \tag{4.6}$$

Then for any $N = q^L$, with $q = 2, 3, \ldots,$ and $L \in \mathbb{N}_+$, we have:

(A) *The function related N–d tensor $\mathbf{F} = [F_\mathbf{i}]$, defined by*

$$F_\mathbf{i} = f((i_1 + i_2 + \cdots + i_d)h)\,, \quad \mathbf{i} \in I^{\otimes d}\,, \quad h > 0\,, \quad \text{with } a \le dh \le b/N\,,$$

generated by the multivariate function $f(x_1 + \cdots + x_d)$, can be represented by the rank-M, q-dL tensor, up to the tolerance ε in the max norm.

(B) *Let $a \le dh \le b/N$, then the function related N–d tensor $\mathbf{G} = [G_\mathbf{i}]$,*

$$G_\mathbf{i} = f(x_{1,i_1}^2 + \cdots + x_{d,i_d}^2)\,, \quad x_{\ell,i_\ell} = \sqrt{hi_\ell}\,, \quad \mathbf{i} \in I^{\otimes d}\,,$$

discretizing the multivariate function $g = f(x_1^2 + \cdots + x_d^2)$ on the polynomially graded grid $\{x_{\ell,i_\ell}\}$, embedded into the region $a \le \sum_{\ell=1}^{d} x_\ell^2 \le b$, can be approximated by the rank-M, q-dL tensor with the tolerance ε in the max norm.

In both cases, the representation complexity is $O(dqM \log_q N)$.

(C) *The approximation properties by Gaussian sums via the uniform sampling, i.e., for $x_{\ell,i_\ell} = hi_\ell$, ($\ell = 1, \ldots, d$), remain essentially the same except that the rank-M, q-dL tensor should be substituted by M-term sum of rank-r QTT approximants representing each individual Gaussian vector, where $r = O(|\log \varepsilon|)$.*

Proof. We note that the previous results can be applied to R-term sums of exponential/trigonometric vectors in d dimensions, i.e., to the respective N–d tensors,

$$\mathbf{A}_{(\mathbf{n},d)} := \left\{ x_\mathbf{n} := \sum_{k=1}^{R} c_k \prod_{\ell=1}^{d} z_{k,\ell}^{n_\ell - 1} \right\}_{\mathbf{n} \in I^{\otimes d}}\,, \quad I = \{1, \ldots, N\}\,, \tag{4.7}$$

such that $\mathbf{A}_{(\mathbf{n},d)}$ can be reshaped to the hybrid QCP format $\mathcal{C}_{R,\mathbf{q}}[QCP[\mathbf{1}, dL]]$ of complexity $dqR \log_q N$. Now items (A), (B) directly follow from (4.7). Item (C) is justified by combining (4.7), Remark 4.14, and Lemma 4.12 related to representation by Gaussian sums. □

Lemma 4.15 allows us to derive the accurate $O(d \log N)$ approximations to the wide class of function related tensors in high dimension. For a class of analytic functions the basic approximability assumption (4.6) can be verified with

$$\varepsilon = O(\mathrm{e}^{-\alpha M / \log M})\,, \quad \alpha > 0\,,$$

by applying the sinc approximation. To this end, we also refer to [144, 204, 206, 208]. Note that the semianalytic QTT approximation as in Lemma 4.15 can be further optimized by applying the rank-r, $r < M$, TT approximation.

We recall that many QTT approximation results can be derived provided that accurate polynomial approximation exists,

$$\max_{x \in [a,b]} \left| f(x) - \sum_{k=1}^{M} c_k x^k \right| \le \varepsilon\,. \tag{4.8}$$

We note that in all numerical results below we use the average QTT rank, $\overline{r}$, defined by

$$\overline{r} := \sqrt{\frac{1}{d-1} \sum_{k=1}^{d-1} r_k r_{k+1}}\,.$$

Exercise 4.16. Check numerically the QTT rank of monomial of degree m, general polynomial, and Chebyshev polynomial, all over uniform and Chebyshev grids in $[-1, 1]$.

Exercise 4.17. Test the QTT rank of sin-Helmholtz and cos-Helmholtz type kernels defined by $\sin(\kappa x)/x$ and $\cos(\kappa x)/x$, respectively. (What is the scaling of the ε rank in κ?)

Exercise 4.18. Find the QTT rank of step function and Haar wavelet.

4.1.7 Examples of QTT supercompression in high dimensions

In what follows, we show with several practically interesting examples how the low rank QTT tensor approximation can be constructed in higher dimension providing the logarithmic complexity scaling in the volume size of the full representation grid. We discuss the QTT based approximation to the d dimensional Hilbert tensor that arises in the tensor representation of the d dimensional Laplacian inverse.

Example 4.19. First, we apply the sinc quadrature ε-approximation to the dth order Hilbert N–d tensor $\mathbf{A}$ of dimension $N^{\otimes d}$, to obtain the rank-$(2M+1)$ CP tensor

$$A(i_1, \ldots, i_d) = \frac{1}{i_1^2 + i_2^2 + \cdots + i_d^2} \approx \sum_{k=-M}^{M} \bigotimes_{\ell=1}^{d} c_k e^{-t_k i_\ell^2},$$

$i_1, \ldots, i_d \in \{1, \ldots, N\}$, $N = 2^L$, where the number of terms is estimated by $M = O(|\log \varepsilon|)$. Now we apply the rank-$O(|\log \varepsilon|)$ QTT approximation to each skeleton vector leading to the hybrid CP-QTT approximation to the tensor of order $D = d \log N$ and of size $2^{\otimes D}$. The storage size of this parametrization to $\mathbf{A}$ is estimated by

$$Q = O(d|\log \varepsilon|^3 \log N) \ll N^d \quad \text{reals}.$$

Numerical gain from this CP-QTT representation can be easily seen for the particular model parameters with the moderate precision ε.

Matrix case: $d = 2$, $N = 2^{15} \to Q = 30|\log \varepsilon|^3 \ll 2^{2\cdot 15}$.

High dimension: $d = 2^{10}$, $N = 2^{15} \to Q = 20 \cdot 2^{10}|\log \varepsilon|^3 \ll 2^{15\cdot 10}$.

In the following this supercompressed approximation will be applied to the efficient tensor decomposition of the d dimensional FEM-Laplacian inverse matrix.

4.1.8 Numerics on QTT and QCP approximation

Tables 4.1–4.3 represent the average QTT ranks in approximation of function related vectors and matrices up to the tolerance $\varepsilon = 10^{-5}$. All functions are defined on the square $[0, 1]^d$.

Tab. 4.1: QTT ranks of long functional N vectors, $N = 2^p$.

$N \setminus \bar{r}$	$e^{-\alpha x^2}$, $\alpha = 0.1 \div 10^2$	$\frac{\sin(\alpha x)}{x}$, $\alpha = 1 \div 10^2$	$\frac{1}{x}$	$\frac{e^{-x}}{x}$	$x, x^{10}, x^{1/10}$
2^{10}	3.2/2.8/2.8/2.2	4.0/4.7/5.5	4	3.5	1.9/2.7/3.9
2^{12}	3.1/2.9/2.9/2.6	3.8/4.8/5.6	4.2	3.8	1.9/2.6/3.9
2^{14}	2.9/2.8/2.8/2.8	3.6/4.7/5.5	4.2	3.8	1.9/2.5/3.9
2^{16}	2.8/2.7/2.8/2.8	3.6/4.5/5.4	4.2	5.3	1.9/2.4/3.9

Tab. 4.2: QTT ranks of functional $N \times N$ matrices, $N = 2^p$.

$N \setminus \bar{r}$	$1/(x_1 + x_2)$	$e^{-\|x\|}$	$e^{-\|x\|^2}$	$\mathrm{diag}[e^{-x^2}]$	$\Delta_2^{-1}\mathbf{1}$, $\varepsilon = 10^{-6}, 10^{-7}, 10^{-8}$
2^9	5.0	9.4	7.8	3.8	3.6/3.6/3.6
2^{10}	5.1	9.4	7.7	3.9	3.6/3.6/3.6
2^{11}	5.2	9.3	7.5	3.9	3.7/3.7/3.7

Tab. 4.3: QTT ranks of projected $1/\|x\|$, $x \in \mathbb{R}^3$, the Hartree potential V_H, and electron density ρ of a CH_4 molecule on $N \times N \times N$ grids in 3D.

N	128	256	512	1024
$1/\|x\|$	13.8	16.0	17.5	18.0
$\rho(x)$	32.0	40.0	45.8	48.6
V_H	32.1	34.9	20.2	28.2

Table 4.2 includes some examples of function generated matrices (vectors in the case $d = 2$). Here we denote by $\mathrm{diag}[f(x)]$ a diagonal matrix generated by the vector associated with the functional $f(x)$ sampled on the univariate grid. Here $\Delta_2^{-1}\mathbf{1}$ represents the $N \times N$ vector resulting from a 2D Laplacian inverse applied to the unit $N \times N$ vector of all ones.

Table 4.3 represents the QTT rank for functions in $\mathbb{R}^3$ arising in electronic structure calculations by the Hartree–Fock equation. This application will be considered in detail in Chapter 5.

One can observe that in all examples rank parameters are moderate, and depend very mildly on the grid size.

4.1.9 TT/QTT based tensor numerical methods: main ingredients

In this section, we outline the main ingredients that provide the basics of modern tensor numerical methods for solving multidimensional elliptic PDEs. These techniques include the standard FEM/FDM schemes for discretization of functions and operators on tensor grids in $\mathbb{R}^d$, methods of preconditioned iterations for solving the arising linear systems of equations, and the approximation error estimates. These traditional tools should be combined with the rank structured parametrization of all entities in the equations, provided that the iterative schemes and preconditioners have been properly modified. Here we collect the essential issues:

1. Discretization in tensor product Hilbert space of $N-d$ tensors, $\mathbb{V}_{\mathbf{n}} = \mathbb{R}^{I_1\times\cdots\times I_d}$, where $\#I_\ell = \{1,\ldots,N\}$, $N = 2^L$.
2. MLA in low rank TT/QTT tensor formats $\mathcal{S} \subset \mathbb{V}_{\mathbf{n}}$:

   $$\mathcal{S} = \{\mathcal{T}_{\mathbf{r}}[TT[\mathbf{r}_1]], TT[\mathbf{r}], \ QTT[\mathbf{r}], QTT_{\text{loc}}\} .$$

   The key point is the efficient implementation of tensor truncation (projection),

   $$T_{\mathcal{S}} : \mathcal{S}_0 \to \mathcal{S} \subset \mathcal{S}_0 \subset \mathbb{V}_{\mathbf{n}},$$

   based on a combination of matrix SVD, (R)HOSVD approximation, ALS/DMRG/AMEn iteration possibly combined with multigrid techniques.
3. Explicit TT/QTT representation of operators is the benefit.
4. Multilevel tensor-truncated preconditioned iteration by using the low rank preconditioners.
5. Use of quasidirect tensor solvers when available such as A^{-1}, $\exp(tA)$, convolution transforms representing the action of Green's kernels.
6. Minimization of the cost functional on rank structured tensor manifolds by using ALS, DMRG, or AMEn type algorithms.

Having at hand the efficient tensor formats and the related MLA, the tensor based solution scheme can be constructed for some particularly important classes of PDEs. The examples of such algorithms will be considered in Chapter 5.

4.2 Explicit TT/QTT representation of functional tensors

In this section we progress the discussion of the explicit TT and QTT representation of functional tensors.

First, we recall the functional TT (FTT) and the operator/matrix TT (OTT/MTT) factorizations, and then consider an explicit (closed form) FTT representation of multivariate functions in the form $f(x) = f_1(x_1) + f_2(x_2) + \cdots + f_d(x_d)$. We proceed with the decomposition of standard trigonometric functions of the form $T(x_1 + \cdots + x_d)$. Further extension to the general functions like $H(x_1 + \cdots + x_d)$ is possible, where $H(x + y)$ allows rank-r separable bilinear decomposition. In particular, we discuss the explicit QTT decomposition of polynomial vectors as well as some rational polynomials and trigonometric functions.

As a special case we consider the TT and QTT decomposition of multivariate polynomials $P(x_1,\ldots,x_d)$ and $P(x_1+\cdots+x_d)$. We focus on the practically important cases of harmonic oscillators and the Hénon–Heiles potential, which are commonly used in quantum molecular dynamics.

We conclude with the rank analysis for matrices arising in the solution of the multidimensional chemical master equation.

4.2.1 Functional TT decomposition revisited

First, we recall the functional tensor train (FTT) decomposition considered in Section 2.1; see [288] where the FTT concept was addressed.

Given the d-variate function f in the tensor product Hilbert space $\mathbb{H} = \otimes_{\ell=1}^{d} H_\ell$, and the index set $\mathcal{J} := \times_{\ell=1}^{d} J_\ell, J_\ell = \{1, \dots, r_\ell\}$, such that $J_0 = J_d$. The rank-$\mathbf{r}$ functional TT format contains products of functional tritensors over $\mathcal{J}$,

$$f(x_1, \dots, x_d) = \sum_{\mathbf{j} \in \mathcal{J}} g_1(j_d, x_1, j_1) g_2(j_1, x_2, j_2) \dots g_d(j_{d-1}, x_d, j_d)\,,$$

or in a compact form

$$\mathrm{FTT}[\mathbf{r}] := \left\{ f \in \mathbb{H} : f = \{\times_{J_\ell}\}_{\ell=1}^{d} G^{(\ell)}(x_\ell) \text{ with } G^{(\ell)} \in \mathbb{R}^{J_{\ell-1} \times H_\ell \times J_\ell} \right\}\,.$$

A function $f(x_1, \dots, x_d) \in \mathbb{H}$, $x \in [0,1]^d$ is represented (approximated) by a product of matrices (matrix product states), each depending on a single variable x_ℓ.

Clearly, efficient tensor numerical methods for multidimensional differential equations will require not only low rank FTT decomposition of functions, but also the respective multiplicative representation of operators acting on $\mathbb{H}$. FTT decomposition induces the *important concept of multiplicative formats for the operators* acting between two tensor product Hilbert spaces, $\mathbf{A}: \mathbb{X} \to \mathbb{Y}$, each of dimension d; see details in Section 4.3.

Example 4.20. The d dimensional Laplacian is a mapping between $\mathbb{X} = H_0^1([0,1]^d)$ and $\mathbb{Y} = H^{-1}([0,1]^d)$. Another pair of spaces might be $\mathbb{X} = \mathbb{Y} = L^2[0,1]^d$, which is suited for the operator of multiplication with a function. In the discrete grid based formulation one can set $\mathbb{X} = \mathbb{Y} = \mathbb{R}^{n^{\otimes d}}$.

The consideration of functional TT decompositions is motivated by the fact that in the case of continuous functions each functional decomposition can be translated to its discretized version in $\mathbb{X} = \mathbb{R}^{n^{\otimes d}}$ obtained by sampling all functional entities onto a tensor grid. In other words, the FTT can be applied to the functions of discrete arguments, i.e., to the multidimensional tensors.

4.2.2 Trigonometric functions of a sum of univariate functions

Here we discuss the instructive example of functions composed by a sum of univariate functions.

Lemma 4.21. *The function*

$$f(x) = f_1(x_1) + f_2(x_2) + \cdots + f_d(x_d)$$

allows rank-2 FTT decomposition in the form

$$f(x) = \begin{pmatrix} f_1(x_1) & 1 \end{pmatrix} \begin{pmatrix} 1 & 0 \\ f_2(x_2) & 1 \end{pmatrix} \cdots \begin{pmatrix} 1 & 0 \\ f_{d-1}(x_{d-1}) & 1 \end{pmatrix} \begin{pmatrix} 1 \\ f_d(x_d) \end{pmatrix} .$$

Proof. We apply the induction by using the simple identity for any a and b in $\mathbb{R}$,

$$\begin{pmatrix} 1 & 0 \\ a & 1 \end{pmatrix} \begin{pmatrix} 1 & 0 \\ b & 1 \end{pmatrix} = \begin{pmatrix} 1 & 0 \\ a+b & 1 \end{pmatrix} , \tag{4.9}$$

where a and b are substituted by functional values. The first induction step could be the three way decomposition. □

Applying the above decomposition (4.9) to a function of discrete variables, we obtain the rank-2 TT representation of the Kronecker sum tensor

$$\mathbf{A} := \mathbf{f}_1 \otimes I_2 \otimes \cdots \otimes I_d + \cdots + I_1 \otimes \cdots \otimes \mathbf{f}_d , \quad \mathbf{f}_\ell \in \mathbb{R}^{n_\ell} , \quad \operatorname{rank}(\mathbf{A}) = d ,$$

where vectors $\mathbf{f}_\ell$ denote the discretization of $f_\ell(x_\ell)$ by sampling on a $n_1 \times \cdots \times n_d$ tensor grid.

Now we consider the trigonometric functions of a form $f(x) = T(x_1 + \cdots + x_d)$.

Lemma 4.22. *The rank-2 FTT decomposition of*

$$f(x) := \sin\left(\sum_{j=1}^{d} x_j \right) , \quad \textit{and} \quad g(x) := \cos\left(\sum_{j=1}^{d} x_j \right) , \quad x \in \mathbb{R}^d ,$$

takes a form

$$f(x) = \begin{pmatrix} \sin x_1 & \cos x_1 \end{pmatrix} \begin{pmatrix} \cos x_2 & -\sin x_2 \\ \sin x_2 & \cos x_2 \end{pmatrix} \cdots \begin{pmatrix} \cos x_{d-1} & -\sin x_{d-1} \\ \sin x_{d-1} & \cos x_{d-1} \end{pmatrix} \begin{pmatrix} \cos x_d \\ \sin x_d \end{pmatrix} ,$$

and

$$g(x) = \begin{pmatrix} \cos x_1 & -\sin x_1 \end{pmatrix} \otimes_{p=2}^{L-1} \begin{pmatrix} \cos x_p & -\sin x_p \\ \sin x_p & \cos x_p \end{pmatrix} \begin{pmatrix} \cos x_d \\ \sin x_d \end{pmatrix} ,$$

respectively.

Proof. We prove the case $g(x)$ by induction, similar to examples in Section 2.1,

$$\begin{aligned} g(x) &= \cos x_1 \cos(x_2 + \cdots + x_d) - \sin x_1 \sin(x_2 + \cdots + x_d) \\ &= \begin{pmatrix} \cos x_1 & -\sin x_1 \end{pmatrix} \begin{pmatrix} \cos(x_2 + \cdots + x_d) \\ \sin(x_2 + \cdots + x_d) \end{pmatrix} \\ &= \begin{pmatrix} \cos x_1 & -\sin x_1 \end{pmatrix} \begin{pmatrix} \cos x_2 & -\sin x_2 \\ \sin x_2 & \cos x_2 \end{pmatrix} \begin{pmatrix} \cos(x_3 + \cdots + x_d) \\ \sin(x_3 + \cdots + x_d) \end{pmatrix} . \end{aligned}$$

We leave the proof for the function $f(x)$ as an exercise. □

It is easy to see that the analogous decompositions can be derived for the functions $\tan(\sum_{j=1}^{d} x_j)$ and $\cot(\sum_{j=1}^{d} x_j)$, etc.

Following [288], we now present the rank analysis for functions $H(x_1+x_2+\cdots+x_d)$ having special separation properties.

Theorem 4.23 ([288]). *Let f be a function that depends on a sum of arguments*

$$f(x_1,\ldots,x_d) = H(x_1+x_2+\cdots+x_d)\,,$$

where the function $H(x+y)$ has the exact separation rank r,

$$H(x+y) = \sum_{\alpha=1}^{r} u_\alpha(x)v_\alpha(y)\,,$$

and functions $\{u_\alpha(x)\}$ and $\{v_\alpha(y)\}$ form two linear independent sets. Then:

1. *All FTT ranks are bounded by r.*
2. *If, additionally, $\widehat{x}_i,\ i=1,\ldots,r$, and $\widehat{y}_j,\ j=1,\ldots,r$, are known such that a matrix with elements $H(\widehat{x}_i+\widehat{y}_j)$ is nonsingular, then FTT decomposition takes the explicit form*

   $$f = g_1(x_1)G(x_2)\cdot\cdots\cdot G(x_{d-1})g_d(x_d)\,,\quad G(x_\ell)\in\mathbb{R}^{r\times r}\,,$$

   where

   $$g_1(x_1) = (\psi_1(x_1)\,\psi_2(x_1)\ldots\psi_r(x_1))\,,$$
   $$G(x)_{ij} = \psi_i(x+\widehat{y}_j)\,,$$

   with

   $$\psi_i(z) = \sum_{j=1}^{r} M_{ij}H(\widehat{x}_i+z)\,,$$

   $$g_d(x_d) = \begin{pmatrix} H(\widehat{y}_1+x_d)\\ H(\widehat{y}_2+x_d)\\ \vdots\\ H(\widehat{y}_r+x_d)\end{pmatrix},$$

   and

   $$[M_{ij}] = [H(\widehat{x}_i+\widehat{y}_j)]^{-1}\,.$$

Proof. Due to the separability assumption, functional skeleton decomposition ensures that there exist points $\widehat{x}_i,\ i=1,\ldots,r$, and $\widehat{y}_j,\ j=1,\ldots,r$, such that

$$H(x+y) = \sum_{i,j=1}^{r} H(x+\widehat{y}_j)M_{ij}H(\widehat{x}_i+y)\,, \tag{4.10}$$

where

$$[M_{ij}] = [H(\widehat{x}_i+\widehat{y}_j)]^{-1}\,.$$

The requirement for (4.10) to be true is that nodes $\widehat{x}_i$ and $\widehat{y}_j$ are chosen so that matrix $[H(\widehat{x}_i + \widehat{y}_j)]$ is nonsingular. Let us rewrite function $H(x + y)$ in the form

$$H(x+y) = \sum_{i,j=1}^{r} H(x + \widehat{y}_j)\psi_i(y) , \tag{4.11}$$

where $\psi_i(y)$ is defined as above.

Now, let us proceed in the construction of FTT decomposition for a function f. From (4.10) and (4.11) it follows that

$$f = (\psi_1(x_1)\ \psi_2(x_1)\ \dots \psi_r(x_1)) \begin{pmatrix} H(\widehat{y}_1 + (x_2 + \dots + x_d)) \\ H(\widehat{y}_2 + (x_2 + \dots + x_d)) \\ \vdots \\ H(\widehat{y}_r + (x_2 + \dots + x_d)) \end{pmatrix} .$$

For each element of the second vector in the right hand side, x_2 can be separated:

$$H(\widehat{y}_k + (x_2 + \dots + x_d)) = \sum_{i=1}^{r} \psi_i(\widehat{y}_k + x_2) H(\widehat{y}_i + (x_3 + \dots + x_d)) ,$$

therefore

$$\begin{pmatrix} H(\widehat{y}_1 + (x_2 + \dots + x_d)) \\ H(\widehat{y}_2 + (x_2 + \dots + x_d)) \\ \vdots \\ H(\widehat{y}_r + (x_2 + \dots + x_d)) \end{pmatrix} = G_2(x_2) \begin{pmatrix} H(\widehat{y}_1 + (x_3 + \dots + x_d)) \\ H(\widehat{y}_2 + (x_3 + \dots + x_d)) \\ \vdots \\ H(\widehat{y}_r + (x_3 + \dots + x_d)) \end{pmatrix} ,$$

where G_2 is a $r \times r$ matrix with elements

$$G_2(x_2)_{ij} = \psi_i(x_2 + \widehat{y}_j) .$$

This completes the proof. □

Corollary 4.24. *For a function*

$$f(x_1, \dots, x_d) = P(x_1 + x_2 + \dots + x_d) ,$$

where function $P(z)$ is degree-p polynomial, $P(z) = \sum_{k=0}^{p} a_k z^k$, FTT decomposition takes form

$$f = g_1(x_1) G(x_2) \cdot \dots \cdot G(x_{d-1}) g_d(x_d) ,$$

where $G(x)$ is a matrix function of size $(p+1) \times (p+1)$ with entries

$$G(x)_{ij} = C_i^{i-j} x^{i-j} , \quad i \geq j , \quad \text{and} \ \ G(x)_{ij} = 0 \ \ \text{otherwise} ,$$

for $i, j = 0, \dots, p$, while $C_k^s = \frac{k!}{s!(k-s)!}$ is the binomial coefficient. Moreover,

$$g_1(x_1) = (\phi_0(x_1)\ \phi_1(x_1)\ \dots \phi_{p-1}(x_1)\ \phi_p(x_1)) ,$$

$$\phi_s(x) = \sum_{k=s}^{p} a_k C_k^s x^{k-s}, \quad s = 0, \dots, p,$$

and

$$g_d(x_d) = (1\ x_d\ x_d^2 \dots x_d^p)^{\mathrm{T}}.$$

Proof. The result follows by Theorem 4.23 due to the rank-$(p+1)$ separable representation of polynomial $P(x+y)$,

$$\sum_{k=0}^{p} a_k (x+y)^k = \sum_{k=0}^{p} a_k \sum_{s=0}^{k} C_k^s y^s x^{k-s} = \sum_{s=0}^{p} y^s \sum_{k=s}^{p} a_k C_k^s x^{k-s} = \sum_{s=0}^{p} y^s \phi_s(x),$$

with

$$\phi_s(x) = \sum_{k=s}^{p} a_k C_k^s x^{k-s}.$$

This completes the proof. □

4.2.3 QTT decomposition of rank-r separable functions

In what follows, we consider the explicit QTT decomposition of rank-r separable functions and polynomial vectors.

Theorem 4.25 ([288]). *Let $f(x)$ be a continuous function defined on an interval $[a, b]$ and let $f(x+y)$ have separation rank r. Consider a uniform grid on $[a, b]$,*

$$x_i = a + (i-1)h, \quad h = \frac{b-a}{n-1}, \quad n = 2^d, \quad i = 1, \dots, n,$$

a function generated vector $\mathbf{v} = [v(i)]$

$$v(i) = f(x_i),$$

and $2 \times 2 \times \cdots \times 2$ tensor $\mathbf{V}$ in dimension d, which is a reshape of the vector $\mathbf{v}$,

$$V(i_1, i_2, \dots, i_d) = v(i),$$

where $i_k \in \{0, 1\}$ $(k = 1, \dots, d)$ are binary digits of integer i. Then:
1. *All QTT ranks are bounded by r.*
2. *If, additionally, $\hat{x}_i$, $i = 1, \dots, r$, and $\hat{y}_j$, $j = 1, \dots, r$, are known such that matrix with elements $f(\hat{x}_i + \hat{y}_j)$ is nonsingular, then the QTT decomposition of $\mathbf{v}$ (and TT of $\mathbf{V}$) has form*

$$V(i_1, i_2, \dots, i_d) = g_1(x_1) G(x_2) \cdot \dots \cdot G(x_{d-1}) g_d(x_d),$$

where

$$g_1(x_1) = (\psi_1(x_1)\,\psi_2(x_1)\dots\psi_r(x_1))\,,$$
$$G(x)_{ij} = \psi_1(x + \widehat{y}_j)\,,$$
$$g_d(x_d) = \begin{pmatrix} f(\widehat{y}_1 + x_d) \\ f(\widehat{y}_2 + x_d) \\ \vdots \\ f(\widehat{y}_r + x_d) \end{pmatrix},$$
$$\psi_i(z) = \sum_{j=1}^{r} M_{ij} f(\widehat{x}_i + z)\,, \quad i = 1,\dots,r\,,$$
$$x_k = \frac{a}{d} + 2^{k-1} i_k h\,, \quad k = 1,\dots,d\,,$$

and $[M_{ij}] = [f(\widehat{x}_i + \widehat{y}_j)]^{-1}$.

Proof. To prove Theorem 4.25, it is sufficient to note that

$$V(i_1, i_2, \dots, i_d) = v(i) = v(i_1 + 2i_2 + 4i_3 + \dots 2^{d-1} i_d) = f(x_1 + x_2 + \dots x_d)\,,$$

where $x_k = \frac{a}{d} + 2^{k-1} i_k h$, and apply Theorem 4.23. □

4.2.4 QTT decomposition of rational polynomials and other examples

As a direct consequence of Theorem 4.25 the explicit QTT decomposition of a polynomial vector can be derived [288]. Recall that the upper bound $m + 1$ for the QTT ranks of a 'polynomial' vector is proven by Lemma 4.9. Let

$$M(x) = \sum_{k=0}^{p} a_k x^k$$

be a polynomial of degree p on an interval $[a, b]$. Consider a uniform grid on this interval,

$$x_i = a + (i-1)h\,, \quad h = \frac{b-a}{n-1}, \quad n = 2^d, \quad i = 1,\dots,n\,,$$

and vector $\mathbf{v} = [v(i)]$,

$$v(i) = M(x_i)\,, \quad i = 1,\dots,n\,,$$

and a $2 \times 2 \times \dots \times 2$ tensor $\mathbf{V}$ of dimension d, which is a reshape of $\mathbf{v}$,

$$V(i_1, i_2, \dots, i_d) = v(i)\,,$$

where $i_k \in \{0, 1\}$ ($k = 1,\dots,d$) are binary digits of integer i.

Corollary 4.26. *The QTT decomposition of* $\mathbf{v}$*, that is TT decomposition of* $\mathbf{V}$*, has form*

$$V(i_1, i_2, \dots, i_d) = g_1(i_1) G_2(i_2) \dots G_{d-1}(i_{d-1}) g_d(i_d)\,,$$

where

$$g_1(i_1) = \left(\phi_0\left(\frac{a}{d} + i_1 h\right)\ \phi_1\left(\frac{a}{d} + i_1 h\right)\ \dots \phi_{p-1}\left(\frac{a}{d} + i_1 h\right)\ \phi_p\left(\frac{a}{d} + i_1 h\right)\right),$$

$$\phi_s(x) = \sum_{k=s}^{p} a_k C_k^s x^{k-s}, \quad s = 0, \dots, p,$$

$$G_k(i_k) = G\left(\frac{a}{d} + 2^k i_k h\right),$$

$$G(x)_{ij} = \begin{cases} C_i^{i-j} x^{i-j}, & i \geq j \\ 0, & i < j \end{cases}, \quad i, j = 0, \dots p,$$

$$g_d(i_d) = g\left(\frac{a}{d} + 2^d i_d h\right),$$

with

$$g(x) = \left(1,\ x,\ x^2,\ \dots, x^p\right)^{\mathrm{T}}.$$

Proof. To prove the statement, it is sufficient to note that

$$V(i_1, i_2, \dots, i_d) = v(i) = v(i_1 + 2i_2 + 4i_3 + \dots 2^{d-1} i_d) = M(x_1 + x_2 + \dots x_d),$$

where $x_k = 2^{k-1} i_k h + \frac{a}{d}$, and apply Theorem 4.25. □

Based on Theorem 4.25 the QTT decomposition of discretized rational polynomials, trigonometric functions and logarithmic functions can be derived.

Corollary 4.27. *For a rational function $f(x) = \frac{p(x)}{q(x)}$, where p and q are polynomials defined on an interval $[a, b]$ ($|f(x)| \langle \infty,\ x \in [a, b]$) and uniform grid with $n = 2^d$ grid points, QTT ranks behave logarithmically in accuracy $\varepsilon > 0$ of the QTT approximation, and number of grid points n:*

$$r_k = O(m \log^\alpha 1/\varepsilon \log^\beta n), \quad \varepsilon, \alpha, \beta \geq 0,$$

where m is the polynomial degree of $p(x)$.

Proof. Due to Theorem 4.25, QTT rank estimates are reduced to the estimation of the ε-separation rank of the function

$$g(x, y) = f(x + y). \tag{4.12}$$

For a rational function such estimates can be obtained via constructive separable approximation schemes based on sinc quadratures (Sections 2.3, 2.4) provided that the generating function $f(x)$ is bounded $|f(x)| < \infty,\ x \in [a, b]$ and analytic. □

Recall that the closed form rank-2 QTT decomposition for trigonometric functions $f(x) = \sin(x)$, $f(x) = \cos(x)$, $f(x) = \tan(x)$, $f(x) = \cot(x)$, etc., sampled as an N vector over an equispaced grid with $N = 2^d$, has been already considered in Section 4.1. The alternative arguments can be based on Theorem 4.25. Indeed, since these functions have separation rank 2, the result follows.

Another example of well separated generating function is given by $\log(x)$, $x \in (0, b]$. In fact, the low rank separable expansion for the singular function $\log(x + y)$ was described in Chapter 2; see also [142]. This means that the 'logarithmic' vector allows the QTT decomposition with low ε rank.

4.2.5 TT ranks of multivariate polynomials

The question of low rank approximation of matrices representing multidimensional potentials $V(q_1, \dots, q_f)$ can be reduced to low rank approximation of this function sampled on a tensor grid. In particular, this problem arises in approximation of the so called potential energy surface (PES) in molecular dynamics [221].

If variables in the potential $V(q_1, \dots, q_f)$ are separated in the form

$$V(q_1, \dots, q_f) \approx \sum_{k=1}^{r} \prod_{i=1}^{f} v_i(q_i, k) ,$$

then the canonical rank of a tensor $\mathbf{V} = [V(q_1, \dots, q_f)]$ does not exceed r, and moreover TT ranks of $\mathbf{V}$ do not exceed r either. However, the TT and QTT ranks can be much smaller than the canonical rank r. For the important case of polynomial potentials, one can obtain the following estimate on TT ranks of the corresponding tensors:

Theorem 4.28 ([221]). *For a general homogeneous polynomial potential of form*

$$V(q_1, \dots, q_f) = \sum_{i_1, \dots, i_s = 1}^{f} a(i_1, \dots, i_s) \prod_{k=1}^{s} q_{i_k} ,$$

the following TT rank estimate holds:

$$\mathrm{rank}_{TT}(\mathbf{V}) = C_0 f^{[\frac{s}{2}]} + o\left(f^{[\frac{s}{2}]}\right) .$$

Proof. To fix the idea, let us first consider quadratic potential,

$$V = \sum_{i,j=1}^{f} a_{ij} q_i q_j ,$$

and estimate its TT ranks. To prove the statement, we will separate q_k one-by-one. This is exactly how numerical algorithm for computing TT decomposition works. Suppose we already have decomposition of form

$$V = G_1(q_1) \dots G_k(q_k) W(q_{k+1}, \dots, q_f) ,$$

which is a 'partial' variant of the FTT decomposition, and we want to obtain the next core. $W(q_{k+1}, \dots, q_f)$ is actually a parameter dependent vector of length r_k:

$$W(q_{k+1}, \dots, q_f) = \begin{pmatrix} W_1(q_{k+1}, \dots, q_f) \\ \vdots \\ W_{r_k}(q_{k+1}, \dots, q_f) \end{pmatrix} .$$

In each element, q_{k+1} can be separated from other variables, but we require that *the same basis functions* $R_\alpha(q_{k+2}, \dots, q_f)$, $s = 1, \dots, r_{k+1}$, are used, i.e.,

$$W_s(\dots) = \sum_{\alpha=1}^{r_{k+1}} h_{\alpha s}(q_{k+1}) R_\alpha(q_{k+2}, \dots, q_f) , \quad s = 1, \dots, r_k .$$

This can be always done: for each component W_s one can separate q_{k+1} from other variables with p_s terms, and get p_s basis functions, and no more than $\sum_{s=1}^{r_k} p_s$ basis functions are required. However, there are cases when less basis functions are needed and we can estimate their number for the polynomial PES.

At the first step, q_1 is separated from other variables. V is quadratic in q_1:

$$V = a_{11} q_1^2 + q_1 \sum_{j=2}^{f} a_{1j} q_j + \sum_{i,j=2}^{f} a_{ij} q_i q_j ,$$

hence

$$V = \begin{pmatrix} a_{11} q_1^2 & q_1 & 1 \end{pmatrix} \begin{pmatrix} 1 \\ l_1 \\ s_2 \end{pmatrix} ,$$

where l_1 is linear in $q_2, \dots, q_f$ and s_2 is quadratic in $q_2, \dots, q_f$, so $r_1 \le 3$.

Now, at the second step separation of q_2 is required. To estimate r_2, one has to bound the number of functions depending on $q_3, \dots, q_f$ required to represent each element of the vector

$$\begin{pmatrix} a_{11} \\ l_1 \\ s_1 \end{pmatrix} ,$$

as a linear combination of such functions with coefficients depending only on q_2. In order to do that, decompose l_1 as

$$l_1(q_2, \dots, q_f) = a_{12} q_2 + l_2(q_3, \dots, q_f) ,$$

where l_2 is linear in $q_3, \dots, q_f$, and

$$s_1(q_2, \dots, q_f) = a_{22} q_2^2 + q_2 l_3(q_3, \dots, q_f) + s_2(q_3, \dots, q_f) ,$$

where l_3 is linear in $q_3, \dots, q_f$ and s_2 is quadratic in $q_3, \dots, q_f$. Therefore, the following basis functions arise: $1, l_2, l_3, s_2$, i.e., there is one constant, two linear functions

Tab. 4.4: Different polynomials appearing during TT-SVD process and dimensions of the corresponding spans for order-3 polynomials.

Pol. degree	Number at kth step	Dimension of the space
0	1	1
1	k^2	$f - k$
2	k	$\mathcal{O}(f - k)^2$
3	1	$\mathcal{O}(f - k)^3$

in $q_3, \dots, q_f$, and one quadratic in $q_3, \dots, q_f$. It is easy to see what happens next: the quadratic function gives one more linear function to the basis, thus after k steps we will have one constant, k linear functions, one quadratic function in $q_{k+1}, \dots, q_f$, and the rank bound is $2 + k$.

However, dimension of the space of linear functions of $q_{k+1}, \dots, q_f$ is bounded by $(f - k)$, therefore the rank increases only until $k \le (f - k)$, i.e., $k \le [\frac{f}{2}]$ and then starts to decrease. Thus, maximal rank is $[\frac{f}{2}] + 1$.

The idea is naturally extended to higher order of polynomials. For degree three, at the first step, we will have one constant function, one linear function, one quadratic function, and one cubic function in the remaining variables. Cubic function produces one quadratic and one linear function (and one cubic remains), and each quadratic function produces one additional linear function. At the kth step there will be k quadratic functions, and the number of linear functions grows as $\mathcal{O}(k^2)$, but the dimension of linear space decreases as $(f - k)$, thus when $k \le \mathcal{O}(k^2)$ rank bound is $\mathcal{O}(k^2) + k + 2$, and the rank increases with k, whereas for $k > \mathcal{O}(k^2)$ rank bound is simply $f - k + k + 2 = f + 2$. This is depicted in Table 4.4.

Finally, the rank bound can be obtained from Table 4.4 by taking a minimum of the second and the third column. Analysis for the general case $s \ge 2$ is presented in [221]. □

4.2.6 QTT ranks of multivariate polynomials

Let us estimate the QTT ranks of discretized polynomial potentials by sampling over the uniform grid.

Theorem 4.29 ([221]). *For a general homogeneous polynomial potential of form*

$$V(q_1, \dots, q_f) = \sum_{i_1,\dots,i_s=1}^{f} a(i_1, \dots, i_s) \prod_{k=1}^{s} q_{i_k},$$

sampled over uniform grid, we have

$$\operatorname{rank}_{QTT}(\mathbf{V}) = C_0 f^{[\frac{s}{2}]} + o\left(f^{[\frac{s}{2}]}\right).$$

Proof. For continuous potentials it is sufficient to note that TT decomposition (without QTT) obtained in the proof of Theorem 4.28 gives rise to FTT decomposition (it follows from the proof directly)

$$V(q_1, \dots, q_f) = v_1(q_1)V_2(q_2) \dots V_{f-1}(q_{f-1})v_f(q_f),$$

where V_p for $p = 2, \dots, f-1$, are $r_{p-1} \times r_p$ matrices with elements that are univariate polynomials in q_p of degree at most s, and r_p are TT ranks of V.

After discretization in variable q_p with 2^L points we obtain discrete representation for V_p as $r_{p-1} \times 2 \times 2 \times \dots \times 2 \times r_p$ array, and analogously to the proof in the scalar case it can be shown that for the matrix polynomial case QTT ranks are bounded by $(r_{p-1} + r_p)(s + 1)$. □

In numerics, it is observed that constants hidden in $\mathcal{O}(\cdot)$ in estimates of Theorem 4.29 are not large. The following Hypothesis [221] summarizes the results of our numerical experiments.

Hypothesis 4.30. *Under premises of Theorem 4.29, the following rank estimates hold:*

1. *For a general quadratic potential,* $V(q_1, \dots, q_f) = \sum_{ij}^{f} a_{ij} q_i q_j$,
$$\operatorname{rank}_{\mathrm{QTT}}(\mathbf{V}) \le f + 1 .$$
2. *For a general cubic potential,* $V(q_1, \dots, q_f) = \sum_{ijk}^{f} a_{ijk} q_i q_j q_k$,
$$\operatorname{rank}_{\mathrm{QTT}}(\mathbf{V}) \le f + 1 .$$
3. *For a general quartic potential,* $V(q_1, \dots, q_f) = \sum_{ijkl}^{f} a_{ijkl} q_i q_j q_k q_l$,
$$\operatorname{rank}_{\mathrm{QTT}}(\mathbf{V}) \le f(f + 1) .$$

These upper asymptotic rank estimates remain valid for general coefficients of polynomials. However, for particular potentials ranks can be much smaller.

4.2.7 QTT ranks of special multivariate polynomials

In what follows, we consider the harmonic oscillator and the so called Henon–Heiles potential (which was used for benchmark computations in molecular dynamics).

Lemma 4.31 ([221]). *For harmonic potential,*

$$V(q_1, \dots, q_f) = \sum_{k=1}^{r} w_k q_k^2 ,$$

QTT ranks are bounded by 6, *and for the Henon–Heiles potential of form*

$$V(q_1, \dots, q_f) = \frac{1}{2} \sum_{k=1}^{f} q_k^2 + \lambda \sum_{k=1}^{f-1} \left(q_k^2 q_{k+1} - \frac{1}{3} q_k^3 \right), \tag{4.13}$$

QTT ranks are bounded by 8 *(in numerics we observe* 7*).*

Proof. Since the maximal QTT rank of discretized monomial q^2 is 3, the result for harmonic potential follows from Lemma 4.21.

To get decomposition for the Henon–Heiles potential, first separate q_1:

$$V = \begin{pmatrix} -\frac{\lambda}{3}q_1^3 + \frac{1}{2}q_1^2 & \lambda q_1 1 \end{pmatrix} \begin{pmatrix} 1 \\ q_2 \\ V_2(q_2, \dots, q_f) \end{pmatrix},$$

where $V_2(q_2, \dots, q_f)$ is the Henon–Heiles potential of $q_2, \dots, q_f$. Separation of q_2 gives

$$V = \begin{pmatrix} -\frac{\lambda}{3}q_1^3 + \frac{1}{2}q_1^2 & \lambda q_1 1 \end{pmatrix} \begin{pmatrix} 1 & 0 & 0 \\ q_2 & 0 & 0 \\ \frac{1}{2}q_2^2 - \frac{\lambda}{3}q_2^3 & q_2 & 1 \end{pmatrix} \begin{pmatrix} 1 \\ q_3 \\ V_3(q_3, \dots, q_f) \end{pmatrix},$$

which justifies the general structure of the FTT core at subsequent steps: they are 3×3 matrices,

$$G_k(q_k) = \begin{pmatrix} 1 & 0 & 0 \\ q_k & 0 & 0 \\ \frac{1}{2}q_k^2 - \frac{\lambda}{3}q_k^3 & q_k & 1 \end{pmatrix},$$

thus TT ranks are equal to 3.

To obtain the QTT decomposition on the interval $[a, b]$, one should consider binary representation of q_k,

$$q_k = a + h\left(\sum_{s=0}^{d-1} i_s 2^{s-1}\right),$$

where a is the beginning of the interval where the sampling point q_k is defined, h is a step size, and i_s take values 0 and 1 (for simplicity, index k is omitted, but of course, a, h, i_s will depend on k). Introducing new variables,

$$x_s = \frac{a}{d} + h i_s 2^{s-1},$$

we obtain $q_k = x_1 + \dots + x_s$. Estimation of QTT ranks is now reduced to the separation of indices in block parameter dependent matrix

$$G(q) = \begin{pmatrix} 1 & 0 & 0 \\ q & 0 & 0 \\ \frac{1}{2}q^2 - \frac{\lambda}{3}q^3 & q & 1 \end{pmatrix}.$$

This matrix can be split into three parts. Its element in position (3,1) is a degree-3 polynomial in q, thus

$$\operatorname{rank}_{\mathrm{QTT}}\left(\frac{1}{2}q^2 - \frac{\lambda}{3}q^3\right) \le 4,$$

and for the linear part they are bounded by $2 + 2 = 4$, therefore the overall rank estimate is 8. This completes the proof. □

Remark 4.32. Tucker ranks in all these cases are bounded by f, and lead to $\mathcal{O}(f^f)$ scaling in general, while QTT format gives polynomial storage and polynomial complexity in f, even for the most general coefficients.

Other examples of the QTT rank estimates for various functional and operator classes as well as illustrations of the efficient application of the QTT approximation in the numerical solution of multidimensional PDEs will be presented in Chapter 5.

4.3 Explicit QTT representation of multivariate matrices

In this section, we consider the QTT representation of multivariate matrices, which usually arise as the finite difference or finite element approximation of certain differential-integral operators. It is based on the concept of operator (matrix) TT (OTT/MTT) formats. We introduce two different notions of the so called vector OTT ranks and operator OTT ranks [177]. Accordingly, all these concepts will be applied also to QTT type decompositions.

First, we consider the TT and QTT structures of the Laplace type operators, which give the instructive examples of multivariate elliptic operators. Then we study the class of shift and gradient matrices that naturally arise in the QTT multilinear algebra of circulant, Toeplitz, and Hankel matrices [179, 187, 191, 201, 209, 212]. In this way, one of the most important operations is the fast matrix vector multiplication in the QTT format, which is the main ingredient of the implementation of d dimensional convolution transform with Green's convolution kernels. As the special case we discuss the 1D FEM discrete Laplacian and its multivariate version, as well as the Laplace operator inverse. General estimates on vector and operator QTT ranks for a class of discrete elliptic operators are presented. Various numerical examples illustrate the theoretical rank estimates.

4.3.1 Operator TT (OTT) decomposition

FTT decomposition induces the *important concept of multiplicative formats for the operators* acting between two tensor product Hilbert spaces $\mathbf{A}\colon \mathbb{X} \to \mathbb{Y}$, each of dimension d; see examples in Section 4.2.

Definition 4.33. (Operator TT decomposition). Introduce the rank-$\mathbf{r}$ operator TT (OTT) decomposition symbolized by a set of factorized operators $\mathbf{A}$ in the form

$$\mathbf{A} = \sum_{\mathbf{j}\in\mathcal{J}} G^{(1)}(j_1)G^{(2)}(j_1, j_2)\dots G^{(d)}(j_{d-1}),$$

with $G^{(\ell)} = [G^{(\ell)}(j_\ell, j_{\ell+1})]$ being the operator valued $r_\ell \times r_{\ell+1}$ matrix, where

$$G^{(\ell)}(j_\ell, j_{\ell+1})\colon X_\ell \to Y_\ell\,, \quad \ell = 1, \dots, d\,,$$

such that the action $\mathbf{A}f$ on rank-1 function $f = f_1 \otimes \cdots \otimes f_d \in \mathbb{X}$ is defined as the rank-$\mathbf{r}$ TT element in $\mathbb{Y}$,

$$(\mathbf{A}f)(y_1, \ldots, y_d) := \sum_{\mathbf{j} \in \mathcal{J}} g_1(y_1, j_1) g_2(j_1, y_2, j_2) \ldots g_d(j_{d-1}, y_d),$$

with

$$g_\ell(j_{\ell-1}, y_\ell, j_\ell) = [G^{(\ell)}(j_{\ell-1}, j_\ell) f_\ell](y_\ell).$$

In the case $X_\ell = \mathbb{R}^{n_\ell}$, $Y_\ell = \mathbb{R}^{m_\ell}$ we arrive at the definition for matrix TT decomposition.

4.3.2 Vector TT and QTT ranks of a multiway matrix

Our presentation here is mainly based on [177]. TT representation may be applied to an operator $\boldsymbol{A}$ over a vector space (as well as to a vector from it), which could be recognized as a 'matricization' of TT cores of a 'vectorized' matrix, e.g.,

$$\boldsymbol{A}(i_1, j_1, \ldots, i_D, j_D) = \sum_{\alpha_1=1}^{r_1} \cdots \sum_{\alpha_{D-1}=1}^{r_{D-1}} U_1(i_1, j_1, \alpha_1)\, U_2(\alpha_1, i_2, j_2, \alpha_2) \cdot \cdots \cdot U_{D-1}(\alpha_{D-2}, i_{D-1}, j_{D-1}, \alpha_{D-1})\, U_D(\alpha_{D-1}, i_D, j_D). \tag{4.14}$$

Let us now recall equation (4.14) in view of the basic results obtained on the minimal possible kth rank of an exact rather than approximate TT decomposition of a tensor $\boldsymbol{A}$ (the kth TT rank of $\boldsymbol{A}$). We recall the definition of TT rank:

Definition 4.34. ([177]) Given a multiway $n_1 \times \cdots \times n_D$ vector

$$\boldsymbol{A} \in \mathbb{R}^{n_1} \times \cdots \times \mathbb{R}^{n_D},$$

its *kth TT rank* is the rank of its matrix unfolding $A^{(k)}$ with the elements

$$A^{(k)}(i_1 \ldots i_k\,;\, i_{k+1} \ldots i_D) = \boldsymbol{A}(i_1 \ldots i_D), \quad 1 \le k \le D-1.$$

Once we apply this to a multiway matrix rather than vector, TT decomposition, which is given by (4.14), we arrive at the same concept of matrix TT rank. This implies application of TT to a 'vectorization' of the matrix, i.e., a matrix is considered merely as a *vector* in (4.14), and its ability to neither map vectors to vectors nor related properties are taken into consideration. To emphasize this, we refer to this ranks as *vector TT ranks*.

Definition 4.35. ([177]) (Vector TT rank of a matrix) A multiway $(m_1 \times n_1) \times \cdots \times (m_D \times n_D)$ matrix

$$\boldsymbol{A} : \mathbb{R}^{n_1} \times \cdots \times \mathbb{R}^{n_D} \mapsto \mathbb{R}^{m_1} \times \cdots \times \mathbb{R}^{m_D}$$

is given, its *kth vector TT rank* is the rank of its unfolding $A^{(k)}$ $(1 \le k \le D-1)$ with the elements

$$\boldsymbol{A}^{(k)}(i_1 j_1 \ldots i_k j_k\,;\, i_{k+1} j_{k+1} \ldots i_D j_D) = A(i_1 j_1 \ldots i_D j_D),$$

$1 \le k \le D-1$.

In particular, this means that the minimal vector ranks in the TT decomposition of a certain matrix are somewhat independent from one other, depending on the matrix in the aggregate. So we may consider a minimal rank decomposition, which holds that no one of $D-1$ ranks can be reduced without introducing an error in (4.14) even if we allow the others to grow. This makes it reasonable to compare ranks elementwise.

Hence, we define bounds on vector TT rank of a matrix as follows: We say that a multiway matrix (vector) is of ranks not greater than $r_1 \dots r_{D-1}$ if and only if for any k: $1 \le k \le D-1$ its kth vector TT rank is not greater than r_k.

4.3.3 Operator TT and QTT ranks of a matrix

Vector TT rank of a matrix is of great importance in view of storage costs and complexity of such basic operations as dot product, multidimensional contraction, matrix-by-vector multiplication, rank reduction, and orthogonalization of a tensor train. Their complexity upper bound is linear with respect to vector TT rank upper bound raised to the power 2 or 3.

Even if we manage to perform a matrix-by-vector multiplication, this may not be enough for solution of the problem involved. For example, when developing iterative solvers we are likely to be concerned with vector TT ranks of a matrix-by-vector product.

Formally, ranks of TT decompositions are multiplied when two matrices or a matrix and a vector are multiplied. Often this obvious estimate of ranks of the product leads to nonaffordable complexity estimates, but fortunately it is not sharp, so that low rank approximation is possible.

A reasonable a priori estimate of ranks would allow one to rely upon such an approximation procedure, the complexity of which is cubic in respect of ranks. Below we introduce the concept of *operator TT rank*.

Definition 4.36. ([177]) (Operator TT rank of a matrix). Given a multiway matrix $\boldsymbol{A}\colon \mathbb{R}^{n_1} \times \dots \times \mathbb{R}^{n_D} \mapsto \mathbb{R}^{m_1} \times \dots \times \mathbb{R}^{m_D}$, for any vector $\boldsymbol{X} \in \mathbb{R}^{n_1} \times \dots \times \mathbb{R}^{n_D}$ let us denote vector TT ranks of the matrix-by-vector product $\boldsymbol{AX}$ by $r_1 \dots r_{D-1}$. Then let us refer to

$$\operatorname{rank}_o(\boldsymbol{A}) = \max_{\substack{k=1\dots D-1,\\ \boldsymbol{X} \text{ is of vector TT rank } 1\dots1}} r_k$$

as the *operator TT rank* of $\boldsymbol{A}$.

The following statement gives obvious inequality between the two ranks introduced by Definitions 4.35 and 4.36:

Proposition 4.37. *Operator TT rank does not exceed the maximum component of vector TT rank.*

This estimate is essentially not sharp. For example, consider two vectors $\boldsymbol{X}, \boldsymbol{Y} \in \mathbb{R}^{n_1} \times \cdots \times \mathbb{R}^{n_D}$ such that $\boldsymbol{X}$ is of vector TT rank $1 \dots 1$. Then for any vector $\boldsymbol{Z} \in \mathbb{R}^{n_1} \times \cdots \times \mathbb{R}^{n_D}$ of vector TT rank $1 \dots 1$ the tensor $(\boldsymbol{X}\boldsymbol{Y}')\boldsymbol{Z} = \langle \boldsymbol{Y}, \boldsymbol{Z} \rangle \boldsymbol{X}$ is of vector TT rank $1 \dots 1$, while $(\boldsymbol{Y}\boldsymbol{X}')\boldsymbol{Z} = \langle \boldsymbol{X}, \boldsymbol{Z} \rangle \boldsymbol{Y}$ is of the same vector TT rank as $\boldsymbol{Y}$. Consequently, operator TT rank of $\boldsymbol{X}\boldsymbol{Y}'$ is equal to 1, while that of $\boldsymbol{Y}\boldsymbol{X}'$ is as high as the maximum rank of TT cores of $\boldsymbol{Y}$, which can be random and have a very bad QTT structure resulting in a high vector TT rank of $\boldsymbol{X}\boldsymbol{Y}'$.

4.3.4 Notations to explicit QTT decomposition of matrices

First, we discuss the class of discrete elliptic type operators of consideration. We focus on QTT structure of finite difference discretization $\boldsymbol{\Delta}^{(d_1 \dots d_D)}$ of Laplace operator, considered over a D dimensional cube on tensor uniform grids. In the one dimensional case we also consider its inverse as well.

The grids in question are tensor products of D one dimensional uniform grids, each kth of them comprising 2^{d_k} points. By discrete Laplace operator we mean a matrix

$$\boldsymbol{\Delta}^{(d_1 \dots d_D)} = a_1 \boldsymbol{\Delta}_1^{(d_1)} \otimes I_{2^{d_2}} \cdots \otimes I_{2^{d_D}} + \cdots + I_{2^{d_1}} \otimes \cdots \otimes I_{2^{d_{D-1}}} \otimes a_D \boldsymbol{\Delta}_D^{(d_D)} , \tag{4.15}$$

summed by D terms with I_m being an $m \times m$ identity matrix. The weights a_k are there to take into consideration both the difference in grid steps and anisotropy. For the sake of brevity these weights are 1 below unless otherwise stated.

Each of univariate discrete Laplacians $\boldsymbol{\Delta}_k^{(d_k)}$ may be any of the following $2^{d_k} \times 2^{d_k}$ matrices, depending on the boundary conditions imposed:

$$\boldsymbol{\Delta}_{DD}^{(d_k)} = \begin{pmatrix} 2 & -1 & & & \\ -1 & 2 & \ddots & & \\ & \ddots & \ddots & -1 & \\ & & -1 & 2 & -1 \\ & & & -1 & 2 \end{pmatrix}, \quad \boldsymbol{\Delta}_{NN}^{(d_k)} = \begin{pmatrix} 1 & -1 & & & \\ -1 & 2 & \ddots & & \\ & \ddots & \ddots & -1 & \\ & & -1 & 2 & -1 \\ & & & -1 & 1 \end{pmatrix} \tag{4.16}$$

are the ones for Dirichlet and Neumann boundary conditions respectively, while

$$\boldsymbol{\Delta}_{DN}^{(d_k)} = \begin{pmatrix} 2 & -1 & & & \\ -1 & 2 & \ddots & & \\ & \ddots & \ddots & -1 & \\ & & -1 & 2 & -1 \\ & & & -1 & 1 \end{pmatrix}, \quad \boldsymbol{\Delta}_{ND}^{(d_k)} = \begin{pmatrix} 1 & -1 & & & \\ -1 & 2 & \ddots & & \\ & \ddots & \ddots & -1 & \\ & & -1 & 2 & -1 \\ & & & -1 & 2 \end{pmatrix} \tag{4.17}$$

correspond to Dirichlet–Neumann and Neumann–Dirichlet boundary conditions, respectively. Finally, the matrix

$$\Delta_P^{(d_k)} = \begin{pmatrix} 2 & -1 & & & -1 \\ -1 & 2 & \ddots & & \\ & \ddots & \ddots & -1 & \\ & & -1 & 2 & -1 \\ -1 & & & -1 & 2 \end{pmatrix} \tag{4.18}$$

corresponds to periodic boundary conditions.

We introduce the set of 2×2 matrices (Pauli type matrices), which are useful for the description of QTT structured matrices:

$$\begin{gathered} I = \begin{pmatrix} 1 & 0 \\ 0 & 1 \end{pmatrix}, \quad J = \begin{pmatrix} 0 & 1 \\ 0 & 0 \end{pmatrix}, \quad I_1 = \begin{pmatrix} 1 & 0 \\ 0 & 0 \end{pmatrix}, \quad I_2 = \begin{pmatrix} 0 & 0 \\ 0 & 1 \end{pmatrix}, \quad P = \begin{pmatrix} 0 & 1 \\ 1 & 0 \end{pmatrix}, \\ E = \begin{pmatrix} 1 & 1 \\ 1 & 1 \end{pmatrix}, \quad F = \begin{pmatrix} 1 & -1 \\ -1 & 1 \end{pmatrix}, \quad K = \begin{pmatrix} -1 & 0 \\ 0 & 1 \end{pmatrix}, \quad L = \begin{pmatrix} 0 & -1 \\ 1 & 0 \end{pmatrix}. \end{gathered} \tag{4.19}$$

To deal with 3 and 4 dimensional TT cores efficiently, we use matrix notation for them and their convolutions. For instance, if $n \times m$ matrices $A_{\alpha\beta}$, $\alpha = 1 \dots r_1$, $\beta = 1 \dots r_2$ are the TT blocks of a TT core U of mode sizes n and m, left rank r_1 and right rank r_2, so that $U(\alpha, i, j, \beta) = (A_{\alpha\beta})_{ij}$ for all the values of the indices, then we write it just as a *core matrix* (in square brackets)

$$U = \begin{bmatrix} A_{11} & \dots & A_{1r_2} \\ \vdots & \vdots & \vdots \\ A_{r_1 1} & \dots & A_{r_1 r_2} \end{bmatrix}.$$

As long as we aim to present TT structure in terms of a narrow set of TT blocks, we need to focus on rank structure of the cores, and that is why such a notation is convenient in handling the cores of TT decomposition.

For every two TT cores U and V of proper sizes we define an *inner core product* of them, $U \bowtie V$, as a regular product of the two core matrices, their elements (TT blocks) being multiplied by means of tensor product, e.g.,

$$U \bowtie V = \begin{bmatrix} A_{11} & A_{12} \\ A_{21} & A_{22} \end{bmatrix} \begin{bmatrix} B_{11} & B_{12} \\ B_{21} & B_{22} \end{bmatrix} = \begin{bmatrix} A_{11} \otimes B_{11} + A_{12} \otimes B_{21} & A_{11} \otimes B_{12} + A_{12} \otimes B_{22} \\ A_{21} \otimes B_{11} + A_{22} \otimes B_{21} & A_{21} \otimes B_{12} + A_{22} \otimes B_{22} \end{bmatrix}.$$

Furthermore, we define an *outer core product* of these matrices, $U \bullet V$, as a tensor product of the two core matrices, their elements (TT blocks) being multiplied by means of regular matrix product, e.g.,

$$U \bullet V = \begin{bmatrix} A_{11} & A_{12} \\ A_{21} & A_{22} \end{bmatrix} \bullet \begin{bmatrix} B_{11} & B_{12} \\ B_{21} & B_{22} \end{bmatrix} = \begin{bmatrix} A_{11}B_{11} & A_{11}B_{12} & A_{12}B_{11} & A_{12}B_{12} \\ A_{11}B_{21} & A_{11}B_{22} & A_{12}B_{21} & A_{12}B_{22} \\ A_{21}B_{11} & A_{21}B_{12} & A_{22}B_{11} & A_{22}B_{12} \\ A_{21}B_{21} & A_{21}B_{22} & A_{22}B_{21} & A_{22}B_{22} \end{bmatrix} .$$

In order to avoid confusion we use square brackets for TT cores, which are to be multiplied by means of inner or outer core product, and round brackets for regular matrices, which are to be multiplied as usual.

We comment that both core products introduced above arise naturally from the multilinear representations in the TT and QTT formats. For instance, (4.14) could be recast as

$$\boldsymbol{A} = U_1 \bowtie U_2 \bowtie \ldots \bowtie U_{D-1} \bowtie U_D ,$$

and a matrix product of $\boldsymbol{A}$ and $\boldsymbol{B} = V_1 \bowtie V_2 \bowtie \ldots \bowtie V_{D-1} \bowtie V_D$ could be put down as

$$\boldsymbol{AB} = (U_1 \bullet V_1) \bowtie (U_2 \bullet V_2) \bowtie \ldots \bowtie (U_{D-1} \bullet V_{D-1}) \bowtie (U_D \bullet V_D) ,$$

letting $\boldsymbol{B}$ be either a matrix or a vector.

As usual, by $A^{\otimes k}$ we mean a kth tensor power of A. For example, $I^{\otimes 3} = I \otimes I \otimes I$, and the same way for the core product operations '$\bowtie$' and '$\bullet$', i.e., for $A^{\bowtie k}$ and $A^{\bullet k}$.

4.3.5 'One dimensional' shift and gradient matrices in QTT format

Let us introduce the QTT structure of the two such recognizable 'one dimensional' operators as shift and gradient matrices:

$$\boldsymbol{S}^{(d)} = \begin{pmatrix} 0 & 1 & 0 & & & \\ & \ddots & \ddots & \ddots & & \\ & & 0 & 1 & 0 & \\ & & & 0 & 1 & \\ & & & & 0 & \end{pmatrix}, \quad \boldsymbol{G}^{(d)} = \begin{pmatrix} 1 & -1 & 0 & & \\ & \ddots & \ddots & \ddots & \\ & & 1 & -1 & 0 \\ & & & 1 & -1 \\ & & & & 1 \end{pmatrix},$$

the size of both being equal 2^d. These matrices are known to provide the building blocks for structured representation of circulant and Toeplitz matrices, as well as for the div-grad representation of FEM Laplacian.

A simple recursive block structure of $\boldsymbol{G}^{(k)}$

$$\boldsymbol{G}^{(k)} = \left(\begin{array}{c|c} \boldsymbol{G}^{(k-1)} & -J'^{\otimes(k-1)} \\ \hline & \boldsymbol{G}^{(k-1)} \end{array}\right) = I \otimes \boldsymbol{G}^{(k-1)} - J \otimes J'^{\otimes(k-1)} ,$$

in our core product notation leads straightforwardly to

$$\boldsymbol{G}^{(d)} = \begin{bmatrix} I & J \end{bmatrix} \bowtie \begin{bmatrix} \boldsymbol{G}^{(d-1)} \\ -J'^{\otimes(d-1)} \end{bmatrix} = \begin{bmatrix} I & J \end{bmatrix} \bowtie \begin{bmatrix} I & J & \\ & & J' \end{bmatrix} \bowtie \begin{bmatrix} \boldsymbol{G}^{(d-2)} \\ -J'^{\otimes(d-2)} \\ -J'^{\otimes(d-2)} \end{bmatrix}$$

$$= \begin{bmatrix} I & J \end{bmatrix} \bowtie \begin{bmatrix} I & J \\ & J' \end{bmatrix} \bowtie \begin{bmatrix} \mathbf{G}^{(d-2)} \\ -J'^{\otimes(d-2)} \end{bmatrix} = \cdots = \begin{bmatrix} I & J \end{bmatrix} \bowtie \begin{bmatrix} I & J \\ & J' \end{bmatrix}^{\bowtie(d-2)} \bowtie \begin{bmatrix} \mathbf{G}^{(1)} \\ -J' \end{bmatrix}$$

$$= \begin{bmatrix} I & J \end{bmatrix} \bowtie \begin{bmatrix} I & J \\ & J' \end{bmatrix}^{\bowtie(d-2)} \bowtie \begin{bmatrix} I-J \\ -J' \end{bmatrix}.$$

Decomposition of a shift matrix is obtained by the same token:

$$\mathbf{S}^{(d)} = \begin{bmatrix} I & J \end{bmatrix} \bowtie \begin{bmatrix} I & J \\ & J' \end{bmatrix}^{\bowtie(d-2)} \bowtie \begin{bmatrix} \mathbf{S}^{(1)} \\ J' \end{bmatrix} = \begin{bmatrix} I & J \end{bmatrix} \bowtie \begin{bmatrix} I & J \\ & J' \end{bmatrix}^{\bowtie(d-2)} \bowtie \begin{bmatrix} J \\ J' \end{bmatrix}.$$

Based on the explicit tensor decompositions of shift and gradient matrices we shall consider the QTT form of the one dimensional FEM Laplacian.

4.3.6 QTT representation of the one dimensional Laplacian

Let us consider 'one dimensional' Laplace operator $\boldsymbol{\Delta}_{DD}^{(d)}$. Similar to the shift and gradient matrices considered above, there is a low rank QTT structure of $\boldsymbol{\Delta}_{DD}^{(d)}$ described in the next lemma.

Lemma 4.38 ([177]). *For any $d \geq 2$ the rank-3 QTT representation holds*

$$\boldsymbol{\Delta}_{DD}^{(d)} = \begin{bmatrix} I & J' & J \end{bmatrix} \bowtie \begin{bmatrix} I & J' & J \\ & J & \\ & & J' \end{bmatrix}^{\bowtie(d-2)} \bowtie \begin{bmatrix} 2I-J-J' \\ -J \\ -J' \end{bmatrix}.$$

Proof. Similar to a gradient matrix, $\boldsymbol{\Delta}_{DD}^{(d)}$ exhibits a recursive block structure:

$$\boldsymbol{\Delta}_{DD}^{(k)} = \left(\begin{array}{c|c} \boldsymbol{\Delta}_{DD}^{(k-1)} & -J'^{\otimes(k-1)} \\ \hline -J^{\otimes(k-1)} & \boldsymbol{\Delta}_{DD}^{(k-1)} \end{array} \right) = I \otimes \boldsymbol{\Delta}_{DD}^{(k-1)} - J' \otimes J^{\otimes(k-1)} - J \otimes J'^{\otimes(k-1)},$$

which yields us its low rank QTT representation:

$$\boldsymbol{\Delta}_{DD}^{(d)} = \begin{bmatrix} I & J' & J \end{bmatrix} \bowtie \begin{bmatrix} \boldsymbol{\Delta}^{(d-1)} \\ -J^{\otimes(d-1)} \\ -J'^{\otimes(d-1)} \end{bmatrix} = \begin{bmatrix} I & J' & J \end{bmatrix} \bowtie \begin{bmatrix} I & J' & J & & \\ & & & J & \\ & & & & J' \end{bmatrix} \bowtie \begin{bmatrix} \boldsymbol{\Delta}^{(d-2)} \\ -J^{\otimes(d-2)} \\ -J'^{\otimes(d-2)} \\ -J^{\otimes(d-2)} \\ -J'^{\otimes(d-2)} \end{bmatrix}$$

$$= \begin{bmatrix} I & J' & J \end{bmatrix} \bowtie \begin{bmatrix} I & J' & J \\ & J & \\ & & J' \end{bmatrix} \bowtie \begin{bmatrix} \boldsymbol{\Delta}^{(d-2)} \\ -J^{\otimes(d-2)} \\ -J'^{\otimes(d-2)} \end{bmatrix} = \cdots =$$

$$= \begin{bmatrix} I & J' & J \end{bmatrix} \bowtie \begin{bmatrix} I & J' & J \\ & J & \\ & & J' \end{bmatrix}^{\bowtie(d-2)} \bowtie \begin{bmatrix} 2I-J-J' \\ -J \\ -J' \end{bmatrix},$$

which completes the constructive proof. □

The explicit representation of one dimensional Laplacian can be considered as the main building block for tensor representation of this operator in many dimensions.

4.3.7 TT and QTT decomposition of the D dimensional Laplacian

Now we are in a position to describe the TT and QTT structure of a 'D dimensional' Laplace like operator. Below we will also need a Laplace like operator $\mathcal{L}^{(D)}$, $D \geq 2$, with a slightly more general structure:

$$\begin{aligned}\mathcal{L}^{(D)} = {} & M_1 \otimes R_2 \otimes R_3 \otimes \ldots \otimes R_{D-2} \otimes R_{D-1} \otimes R_D \\ & + L_1 \otimes M_2 \otimes R_3 \otimes \ldots \otimes R_{D-2} \otimes R_{D-1} \otimes R_D + \ldots \\ & + L_1 \otimes L_2 \otimes L_3 \otimes \ldots \otimes L_{D-2} \otimes M_{D-1} \otimes R_D \\ & + L_1 \otimes L_2 \otimes L_3 \otimes \ldots \otimes L_{D-2} \otimes L_{D-1} \otimes M_D \,, \end{aligned} \tag{4.20}$$

with matrices L_k, M_k and R_k being of size $m_k \times n_k$, $1 \leq k \leq D$.

Lemma 4.39 ([177]). *For any $D \geq 2$ the Laplace like operator $\mathcal{L}^{(D)}$ allows the following rank-$2 \ldots 2$ TT representation in terms of the blocks L_k, M_k and R_k:*

$$\mathcal{L}^{(D)} = \begin{bmatrix} L_1 & M_1 \end{bmatrix} \bowtie \begin{bmatrix} L_2 & M_2 \\ & R_2 \end{bmatrix} \bowtie \ldots \bowtie \begin{bmatrix} L_{D-1} & M_{D-1} \\ & R_{D-1} \end{bmatrix} \bowtie \begin{bmatrix} M_D \\ R_D \end{bmatrix} .$$

This lemma can be verified by induction using the definition of the inner core product $\bowtie$.

Remark 4.40. Once QTT decompositions of each of the 'one dimensional' operators L_k, M_k and R_k, $1 \leq k \leq D$, are known, they can easily be merged into a QTT decomposition of the 'D dimensional' operator $\mathcal{L}^{(D)}$ according to Lemma 4.39.

As soon as $M_k = a_k \boldsymbol{\Delta}_k^{(d_k)}$ and $L_k = R_k = I^{\otimes d_k}$ for any $k = 1 \ldots D$ are given, the operator $\mathcal{L}^{(D)}$ in (4.20) represents the Laplace operator $\boldsymbol{\Delta}^{(d_1 \ldots d_D)}$ (4.15), and hence the following corollary to Lemma 4.39 holds:

Corollary 4.41 ([177]). *For any $D \geq 2$ the 'D dimensional' Laplace operator defined by (4.15) has the following QTT structure in terms of the 'one dimensional' Laplace operators $a_k \boldsymbol{\Delta}_k^{(d_k)}$, $1 \leq k \leq D$:*

$$\begin{aligned}\boldsymbol{\Delta}^{(d_1 \ldots d_D)} = {} & \begin{bmatrix} I^{\otimes d_1} & a_1 \boldsymbol{\Delta}_1^{(d_1)} \end{bmatrix} \\ & \bowtie \begin{bmatrix} I^{\otimes d_2} & a_2 \boldsymbol{\Delta}_2^{(d_2)} \\ & I^{\otimes d_2} \end{bmatrix} \bowtie \ldots \bowtie \begin{bmatrix} I^{\otimes d_{D-1}} & a_{D-1} \boldsymbol{\Delta}_{D-1}^{(d_{D-1})} \\ & I^{\otimes d_{D-1}} \end{bmatrix} \bowtie \begin{bmatrix} a_D \boldsymbol{\Delta}_D^{(d_D)} \\ I^{\otimes d_D} \end{bmatrix} .\end{aligned}$$

Corollary 4.42 ([177]). *For any $d \geq 3$ the following QTT representations hold:*

$$\begin{bmatrix} I^{\otimes d_k} & a_k \boldsymbol{\Delta}_{DD}^{(d_k)} \\ & I^{\otimes d_k} \end{bmatrix} = \begin{bmatrix} I & J' & J & \\ & & & I \end{bmatrix} \bowtie \begin{bmatrix} I & J' & J & \\ & J & & \\ & & J' & \\ & & & I \end{bmatrix}^{\bowtie(d-2)} \bowtie \begin{bmatrix} I & a_k(2I - J - J') \\ & -a_k J \\ & -a_k J' \\ & I \end{bmatrix},$$

$$\begin{bmatrix} I^{\otimes d_k} & a_k \boldsymbol{\Delta}_{DD}^{(d_k)} \end{bmatrix} = \begin{bmatrix} I & J' & J \end{bmatrix} \bowtie \begin{bmatrix} I & J' & J \\ & J & \\ & & J' \end{bmatrix}^{\bowtie(d-2)} \bowtie \begin{bmatrix} I & a_k(2I - J - J') \\ & -a_k J \\ & -a_k J' \end{bmatrix},$$

$$\begin{bmatrix} a_k \boldsymbol{\Delta}_{DD}^{(d_k)} \\ I^{\otimes d_k} \end{bmatrix} = \begin{bmatrix} I & J' & J & \\ & & & I \end{bmatrix} \bowtie \begin{bmatrix} I & J' & J & \\ & J & & \\ & & J' & \\ & & & I \end{bmatrix}^{\bowtie(d-3)}$$

$$\bowtie \begin{bmatrix} a_k I & a_k J' & a_k J \\ & a_k J & \\ & & a_k J' \\ \frac{1}{2} I & -\frac{1}{2} I & -\frac{1}{2} I \end{bmatrix} \bowtie \begin{bmatrix} 2I - J - J' \\ -J \\ -J' \end{bmatrix}.$$

Proof. We match results of Lemma 4.39, Lemma 4.38, and Remark 4.40. As soon as we derive low rank QTT representations of the supercores involved, we will have one of the D dimensional Laplace operators comprising these supercores at once.

In case of Dirichlet boundary conditions we put QTT cores into the supercores involved and do the same thing as before: reduce ranks as possible by elimination of dependent QTT blocks, which could be conceived as sweeping column (in regard to the left core) or row (in regard to the right core) transformation matrices through the 'tensor train' just as was done in the proof of Lemma 4.38,

$$\begin{bmatrix} I^{\otimes d_k} & a_k \boldsymbol{\Delta}_{DD}^{(d_k)} \\ & I^{\otimes d_k} \end{bmatrix} = \begin{bmatrix} I & I & J' & J & \\ & & & & I \end{bmatrix}$$

$$\bowtie \begin{bmatrix} I & & & & \\ & I & J' & J & \\ & & J & & \\ & & & J' & \\ & & & & I \end{bmatrix}^{\bowtie(d-2)} \bowtie \begin{bmatrix} I & \\ & a_k(2I - J - J') \\ & -a_k J \\ & -a_k J' \\ & I \end{bmatrix}$$

$$= \begin{bmatrix} I & J' & J & \\ & & & I \end{bmatrix} \bowtie \begin{bmatrix} I & J' & J & \\ & J & & \\ & & J' & \\ & & & I \end{bmatrix}^{\bowtie(d-2)} \bowtie \begin{bmatrix} I & a_k(2I - J - J') \\ & -a_k J \\ & -a_k J' \\ & I \end{bmatrix},$$

for a middle supercore. The terminal supercores are subcores, which allow us to reduce ranks of them similarly to how it was done in the proof of Lemma 4.38.

In cases of other boundary conditions QTT decompositions may be derived by the same scheme and the alterations to the Dirichlet boundary conditions case required are quite straightforward. □

4.3.8 Laplace operator inverse for $d = 1$

Next we derive low rank QTT decompositions of inverse to the discretized 1D Laplace operator, where Dirichlet–Neumann or Dirichlet–Dirichlet boundary conditions are imposed. We will proceed from explicit representation of $\boldsymbol{\Delta}_{DD}^{(d)^{-1}}$ and $\boldsymbol{\Delta}_{DN}^{(d)^{-1}}$. The next statement follows by a direct check.

Proposition 4.43. *Let Δ_{DD}, Δ_{DN}, be $n \times n$ matrices. Then*

$$\Delta_{DD}{}^{-1}{}_{ij} = \frac{1}{n+1} \begin{cases} i(n+1-j)\,, & 1 \le i \le j \le n \\ (n+1-i)j\,, & 1 \le j < i \le n \end{cases},$$

$$\Delta_{DN}{}^{-1}{}_{ij} = \frac{1}{n+1} \begin{cases} i(n+1)\,, & 1 \le i \le j \le n \\ (n+1)j\,, & 1 \le j < i \le n \end{cases}$$

follows from either explicit expressions of Green's functions of the corresponding Sturm–Liouville problems or by a direct check (see, for example, [357]).

The next lemma proves the explicit QTT decomposition for mixed Dirichlet–Neumann Laplacian inverse. For technical reasons, we introduce matrices

$$\boldsymbol{K}^{(k)} = \begin{pmatrix} 1 & 2 & 3 & \dots & 2^k \\ \vdots & \dots & \dots & \dots & \vdots \\ \vdots & \dots & \dots & \dots & \vdots \\ \vdots & \dots & \dots & \dots & \vdots \\ 1 & 2 & 3 & \dots & 2^k \end{pmatrix},$$

for $1 \le k \le D - 1$.

Lemma 4.44. *For any $d \ge 2$ it holds that*

$$\boldsymbol{\Delta}_{DN}^{(d)^{-1}} = \begin{bmatrix} I & I_2 J & J' \end{bmatrix} \bowtie \begin{bmatrix} I & I_2 & J & J' \\ & 2E & & \\ & I_2 + J' & E & \\ & I_2 + J & & E \end{bmatrix}^{\bowtie (d-2)} \bowtie \begin{bmatrix} E + I_2 \\ 2E \\ E + I_2 + J' \\ E + I_2 + J \end{bmatrix}.$$

Proof. According to Proposition 4.43, the inverse of the matrix $\boldsymbol{\Delta}_{DN}^{(d)}$ has the following form:

$$\boldsymbol{\Delta}_{DN}^{(d)^{-1}} = \begin{pmatrix} 1 & \dots & \dots & \dots & 1 \\ \vdots & 2 & \dots & \dots & 2 \\ \vdots & \vdots & 3 & \dots & 3 \\ \vdots & \vdots & \vdots & \ddots & \vdots \\ 1 & 2 & 3 & \dots & 2^d \end{pmatrix},$$

and hence, using matrices $\boldsymbol{K}^{(k)}$ allows the explicit representation

$$\boldsymbol{K}^{(k)} = \left(\begin{array}{c|c} \boldsymbol{K}^{(k-1)} & 2^{k-1}E^{\otimes(k-1)} + \boldsymbol{K}^{(k-1)} \\ \hline \boldsymbol{K}^{(k-1)} & 2^{k-1}E^{\otimes(k-1)} + \boldsymbol{K}^{(k-1)} \end{array}\right) = \begin{bmatrix} I_2 + J & E \end{bmatrix} \bowtie \begin{bmatrix} 2^{k-1}E^{\otimes(k-1)} \\ \boldsymbol{K}^{(k-1)} \end{bmatrix},$$

and we draw up the following factorization:

$$\boldsymbol{\Delta}_{DN}^{(d)^{-1}} = \begin{bmatrix} I & I_2 & J & J' \end{bmatrix} \bowtie \begin{bmatrix} \boldsymbol{\Delta}_{DN}^{(d-1)^{-1}} \\ 2^{d-1}E^{\otimes(d-1)} \\ \boldsymbol{K}^{(d-1)'} \\ \boldsymbol{K}^{(d-1)} \end{bmatrix}$$

$$= \begin{bmatrix} I & I_2 & J & J' \end{bmatrix} \bowtie \begin{bmatrix} I & I_2 & J & J' & & & & & & \\ & & & & 2E & & & & & \\ & & & & & I_2 + J' & E & & & \\ & & & & & & & & I_2 + J & E \end{bmatrix}$$

$$\bowtie \begin{bmatrix} \boldsymbol{\Delta}_{DN}^{(d-2)^{-1}} \\ 2^{d-2}E^{\otimes(d-2)} \\ \boldsymbol{K}^{(d-2)'} \\ \boldsymbol{K}^{(d-2)} \\ 2^{d-2}E^{\otimes(d-2)} \\ 2^{d-2}E^{\otimes(d-2)} \\ \boldsymbol{K}^{(d-2)'} \\ 2^{d-2}E^{\otimes(d-2)} \\ \boldsymbol{K}^{(d-2)} \end{bmatrix} = \begin{bmatrix} I & I_2 & J & J' \end{bmatrix} \bowtie \begin{bmatrix} I & I_2 & J & J' \\ & 2E & & \\ & I_2 + J' & E & \\ & I_2 + J & & E \end{bmatrix} \bowtie \begin{bmatrix} \boldsymbol{\Delta}_{DN}^{(d-2)^{-1}} \\ 2^{d-2}E^{\otimes(d-2)} \\ \boldsymbol{K}^{(d-2)'} \\ \boldsymbol{K}^{(d-2)} \end{bmatrix}$$

$$= \dots = \begin{bmatrix} I & I_2 & J & J' \end{bmatrix} \bowtie \begin{bmatrix} I & I_2 & J & J' \\ & 2E & & \\ & I_2 + J' & E & \\ & I_2 + J & & E \end{bmatrix}^{\bowtie(d-2)} \bowtie \begin{bmatrix} E + I_2 \\ 2E \\ E + I_2 + J' \\ E + I_2 + J \end{bmatrix}.$$

This completes the proof. □

The following lemma presents the explicit rank-5 QTT decomposition to the 1D Laplacian inverse with the Dirichlet–Dirichlet boundary conditions. The proof is based on the application of the Sherman–Morrison–Woodbury formula and recursive representations for $k = 2, \ldots, d$; see [177] for the detailed considerations.

Lemma 4.45. *Let $d \geq 2$ and*

$$\lambda_k = -\frac{2^{k-1}+1}{2}, \quad \mu_k = \frac{(2^{k-1}+1)^2}{2^k},$$

$$\xi_k = \frac{2^{k-1}+1}{2^k+1}, \quad \eta_k = \frac{2^{k-2}}{2^k+1}, \quad C^{(k)} = \begin{pmatrix} \lambda_k & \mu_k \\ \mu_k & \lambda_k \end{pmatrix}$$

for $1 \leq k \leq d$. Then ${\boldsymbol{\Delta}_{DD}^{(d)}}^{-1}$ has a rank-$5\ldots5$ QTT representation

$${\boldsymbol{\Delta}_{DD}^{(d)}}^{-1} = W_d \bowtie W_{d-1} \bowtie \ldots \bowtie W_2 \bowtie W_1 ,$$

which consists of the TT cores

$$W_d = \begin{bmatrix} I & \frac{1}{4}C^{(d)} & C^{(d)} & -\lambda_d K & -\mu_d L \end{bmatrix}, \quad W_1 = \begin{bmatrix} \frac{1}{3}(I+E) \\ -E \\ -\frac{1}{36}F \\ -\frac{1}{3}K \\ \frac{1}{3}L \end{bmatrix},$$

$$W_k = \begin{bmatrix} I & \frac{1}{4}C^{(k)} & C^{(k)} & -\lambda_k K & -\mu_k L \\ & E & & & \\ & \eta_k^2 F & \xi_k^2 E & \xi_k\eta_k K & \xi_k\eta_k L \\ & \eta_k^2 K & & \xi_k\eta_k E & \\ & -\eta_k^2 L & & & \xi_k\eta_k E \end{bmatrix}, \quad 2 \leq k \leq d-1 .$$

Now we collect the results on the QTT representation to the Laplacian type operators that follow from the lemmas above presenting explicit QTT representation for a class of discrete elliptic operators; see also [177] for the detailed discussion. Here by bold font we denote the 'interface' rank parameters, which correspond to the separation between different spacial variables.

Summary 4.46. The following upper bounds of vector QTT ranks for the corresponding matrices hold:

$$\begin{aligned}
\boldsymbol{\Delta}_{DD}^{(d)} &: 3\ldots3 \quad \text{(Lemma 4.38)} \\
\boldsymbol{\Delta}_{DN}^{(d)}, \boldsymbol{\Delta}_{ND}^{(d)} &: 4\ldots4 \\
\boldsymbol{\Delta}_{NN}^{(d)} &: 4, 5\ldots5, 4 \\
\boldsymbol{\Delta}_{P}^{(d)} &: 2, 3\ldots3 \quad [177] \\
{\boldsymbol{\Delta}_{DD}^{(d)}}^{-1} &: 4, 5\ldots5, 4 \quad \text{(Lemma 4.45)} \\
{\boldsymbol{\Delta}_{DN}^{(d)}}^{-1}, {\boldsymbol{\Delta}_{ND}^{(d)}}^{-1} &: 4\ldots4 \quad \text{(Lemma 4.44)}
\end{aligned}$$

$$\Delta_{DD}^{(d_1 \dots d_d)} : 3 \dots 3, \mathbf{2}, 4 \dots 4, \mathbf{2} \dots\dots \mathbf{2}, 4 \dots 4, \mathbf{2}, 4 \dots 4, 3$$

(Corollary 4.44, 4.42)

$$\Delta_{DN}^{(d_1 \dots d_d)}, \Delta_{ND}^{(d_1 \dots d_d)} : 4 \dots 4, \mathbf{2}, 5 \dots 5, \mathbf{2} \dots\dots \mathbf{2}, 5 \dots 5, \mathbf{2}, 5 \dots 5, 4$$

$$\Delta_{NN}^{(d_1 \dots d_d)} : 4, 5 \dots 5, \mathbf{2}, 5, 6 \dots 6, 5, \mathbf{2} \dots\dots \mathbf{2}, 5, 6 \dots 6, 5, \mathbf{2}, 5, 6 \dots 6, 4$$

$$\Delta_{P}^{(d_1 \dots d_d)} : 2, 3 \dots 3, \mathbf{2}, 3, 4 \dots 4, \mathbf{2} \dots\dots \mathbf{2}, 3, 4 \dots 4, \mathbf{2}, 3, 4 \dots 4, 3 \ [177] .$$

In the case of many spacial dimensions there are two basic approaches to the rank structured tensor representation of Laplacian inverse to be discussed in what follows.

4.3.9 Laplace operator inverse for $d \geq 2$

In the following d means the spatial dimension. Consider the problem: Find $u \in H_0^1(\Omega)$, such that

$$-\Delta u = f \quad \text{in} \quad \Omega = (0,1)^d .$$

The finite difference discretization on the uniform grid leads to

$$\Delta_d U = F \quad \text{on} \quad N \times N \times \dots \times N - \text{grid in} \quad \mathbb{R}^d ,$$

where the discrete Laplacian reads

$$\Delta_d = \Delta_1 \otimes I_N \otimes \dots \otimes I_N + \dots + I_N \otimes I_N \dots \otimes \Delta_1 \in \mathbb{R}^{N^{\otimes d} \times N^{\otimes d}} .$$

The first approach for approximation of the Laplacian inverse is based on the diagonalization via d dimensional FFT_d further denoted by F_d, which has the canonical rank-1 decomposition, that is

$$-\Delta_d = F_d^* \operatorname{diag}\{A\} F_d ,$$

where A depends on λ_ℓ^2 the eigenvalues of univariate Laplacian in variable ℓ, and $F_d = \bigotimes_{\ell=1}^d F_\ell$, where F_ℓ is the 1D sin FFT.

Here the diagonal matrix is treated by sinc quadrature as in Example 4.19, Subsection 4.1.7. We apply the sinc quadrature ε-approximation to the dth order Hilbert type N–d tensor $\mathbf{A}$ of dimension $N^{\otimes d}$, to obtain the rank-$(2M+1)$ CP tensor decomposition

$$A(i_1, i_1; \dots ; i_d, i_d) = \frac{1}{\lambda_{i_1}^2 + \lambda_{i_2}^2 + \dots + \lambda_{i_d}^2} \approx \sum_{k=-M}^{M} \bigotimes_{\ell=1}^{d} c_k e^{-t_k \lambda_{i_\ell}^2} ,$$

$i_1, \dots, i_d \in \{1, \dots, N\}$, $N = 2^L$, where the number of terms, i.e., the CP rank, is estimated by $M = O(|\log \varepsilon|^2)$.

Now we apply the rank-r QTT approximation to each skeleton vector leading to the hybrid CP-QTT approximation to the tensor of order $D = d \log N$ and of size $2^{\otimes D}$. The storage size of this parametrization to $\mathbf{A}$ is estimated by

$$Q = dr^2 |\log \varepsilon|^2 \log N \ll N^d \quad \text{reals} .$$

The second approach is based on the direct sinc quadrature ε-approximation to the d-Laplacian inverse. It was already shown that the matrix exponential of the d-Laplacian is rank-1 separable, that means

$$e^{-t\Delta_d} = \bigotimes_{\ell=1}^{d} e^{-t\Delta_1} ,$$

where $\Delta_1 = F_1^* \Lambda_1 F_1$ and F_1 is the 1D sin FFT. Moreover, the univariate matrix exponential is diagonalizable in the Fourier basis $e^{-t\Delta_1} = F_1^* e^{-t\Lambda_1} F_1$.

The sinc quadrature approximation $G_M \approx G =: \Delta_d{}^{-1}$ in the rank-R canonical format takes the form

$$\Delta_d^{-1} = \int_0^\infty e^{-t\Delta_d} dt \approx \sum_{k=-M}^{M} c_k \bigotimes_{\ell=1}^{d} \exp(-t_k \Delta_1) := \sum_{k=-M}^{M} c_k \bigotimes_{\ell=1}^{d} F_1^* e^{-t_k \Lambda_1} F_1 := G_M , \qquad (4.21)$$

$$t_k = e^{k\mathfrak{h}} , \quad c_k = \mathfrak{h} t_k , \quad \mathfrak{h} = \pi/\sqrt{M} ,$$

with the exponential convergence rate in $R = 2M + 1$ ([108, 111]),

$$\left\| \Delta_d^{-1} - G_M \right\|_\infty \leq C e^{-\pi\sqrt{M}} , \quad (\text{or } \leq C e^{-\pi M/\log M}) .$$

The rank structured approximation in (4.21) already separates the spatial variables. For the superfast implementation of this decomposition in the QTT format the QTT compression of the family of matrix exponentials $\{\exp(-t_k \Delta_1)\}$ is required. To that end we consider the simplified approach for the choice of sinc quadrature nodes ([222]) based on the well known scaling and squaring scheme.

Let us describe the recursive quadrature for matrix exponential family that allows the collective treatment of the whole set of matrix exponentials involved in the sinc quadrature. We discuss a simple recursion that connects previously computed exponents $\exp(-t_p \Delta_1)$, $p < k$, with the new one for the number k. Denote these matrices by Φ_k,

$$\Phi_k = \exp(-t_k \Delta_1) .$$

The simplest possible recursion is

$$\Phi_k = \Phi_{k-1}^2 , \quad \text{corresponding to} \quad t_k = 2t_{k-1} .$$

This is possible by choosing M such that $e^{\mathfrak{h}} = 2$, or equivalently, $\mathfrak{h} = \log 2 = \frac{\pi}{\sqrt{M}}$, therefore

$$M = \left(\frac{\pi}{\log 2} \right)^2 \approx 20.54 .$$

Since M should be integer, we select $M = 21$ or $M = 20$ and slightly modify $\mathfrak{h}$ ($\mathfrak{h} = \log 2$) to make the recursion exact. This yields a new sinc type quadrature formula with

$$t_k = 2^k, \; c_k = 2^k \log 2, \; k = -M, \dots, M . \qquad (4.22)$$

This quadrature will be called 2^k-*quadrature.*

The accuracy of quadrature formula (4.22) depends on the interval where it is considered (i.e., on the spectrum of Δ_1), but our numerical tests indicate that it always gives an excellent approximate inverse to serve as a preconditioner (with relative accuracy not smaller than 10^{-3}). The special structure of the quadrature nodes allows the computation of all exponents fast – in $2M+1$ multiplications – in exact arithmetic.

However, in the approximate case, the error may accumulate during the squaring (since for $k = -M$ the exponents are close to the identity matrix), and a mixed scheme is more preferable. Up to some k_0 the exponents are computed by a scaling and squaring method, and after that they are just the squares of the previously computed exponents.

A similar approach can be adopted to obtain a more accurate quadrature. For example, another possible recurrence relation is given by

$$\Phi_k = \Phi_{k-1}\Phi_{k-2}\,, \quad \text{or} \quad t_k = t_{k-1} + t_{k-2}\,.$$

Denoting $a = e^{\mathfrak{h}}$ implies that a satisfies the quadratic equation

$$a^2 - a - 1 = 0\,, \quad \text{hence } a \text{ is a } \textit{golden ratio}\text{:} \quad a = \rho = \frac{1+\sqrt{5}}{2} \approx 1.6180\,.$$

The corresponding M is larger:

$$M = \left(\frac{\pi}{\log \rho}\right)^2 \approx 42.62\,,$$

so one can choose $M = 42$ or $M = 43$.

The respective quadrature weights and nodes are given by (ρ^k-quadrature)

$$t_k = \rho^k, \; c_k = \rho^k \log \rho\,.$$

This quadrature formula is around 1000 times more accurate than 2^k-quadrature (indication on the exponentially fast convergence of the sinc quadratures). The number of quadrature points can be slightly decreased, since the norm of several first and last summands is negligible.

There are other possible recursions:

$$\Phi_k = \Phi_{k-2}^2\,, \quad \text{i.e.,} \quad t_k = 2t_{k-2}\,, \quad \text{and} \quad \Phi_k = \Phi_{k-2}\Phi_{k-3}\,, \quad \text{i.e.,} \quad t_k = t_{k-2} + t_{k-3}\,,$$

and so on, which lead to increased M and increased accuracy, yielding a whole family of 'cheap' quadrature rules.

This approach can be applied to calculation of the complex matrix exponential (Section 4.1.7), as well in the implementation of preconditioners for parametric elliptic PDEs (Section 5.4).

In the numerical experiment, we consider the d dimensional Poisson equation for $d \le 100$ in the C-QTT format with the right hand side $F = 1$. The results are presented in the following table.

N	Precomp	Time for sol	Residue	Relative L_2 error
2^8	6.14	2.98	6.6e-06	7.0e-06
2^9	8.37	3.52	8.7e-06	7.0e-06
2^{10}	10.81	4.02	9.4e-06	7.0e-06

The numerical complexity is estimated by $W = O(d|\log \varepsilon|^2 \log N)$.

Let us consider the example from [150]: Find the minimal λ of the eigenvalue problem $-\Delta_d u = \lambda u$ in $(0, \pi)^d$. We apply the truncated inverse power iteration

$$u_{n+1} = -\Delta_d^{-1} u_n , \quad u_{n+1} = u_{n+1}/\|u_{n+1}\| ,$$

where the rank-$(2M + 1)$, sinc approximation of Δ_d^{-1} with $M = 49$ for $d = 3, 10, 50$ is used as described above.

CPU time (sec.) per iteration indicates the linear complexity scaling in d, see the next table.

d	Time/it	Δ_λ error	Δ_u error
3	0.9	$3.1 \cdot 10^{-6}$	$4.5 \cdot 10^{-4}$
10	2.9	$3.1 \cdot 10^{-6}$	$3.8 \cdot 10^{-4}$
50	14.7	$3.1 \cdot 10^{-6}$	$3.1 \cdot 10^{-4}$

In all cases we have $N_{\text{iter}} \le 6$. Here the finite difference approximation is defined on $n^{\otimes d}$ grid points with $n = 2^9$. In what follows, we consider the QTT matrix decompositions to some multi-indexed matrices, which are not directly related to the Laplacian type examples.

4.3.10 Stiffness matrix for elliptic operators with separable coefficients

We consider the FEM-Galerkin piecewise linear approximation in $\Omega = (0, 1)^d$ to the elliptic boundary value problem

$$\text{Find } u \in H_0^1(\Omega): \quad -\operatorname{div}(a_\epsilon \nabla u) = f \text{ in } \Omega , \quad f \in L^2(\Omega) , \tag{4.23}$$

where the small parameter $\epsilon > 0$ can be associated with the size of the unit cell in the quasiperiodic structure of the coefficients; see Section 5.4 for further details.

First, we consider the case $d = 1$. Let equation (4.23) be discretized on the uniform grid with $N = 2^L$. The entries of the stiffness matrix $A[a_\epsilon]$ read

$$(A[a_\epsilon])_{i,i'} = \left(a_\epsilon(x)\nabla\phi_i(x), \nabla\phi_{i'}(x)\right)_{L_2(\Omega)} , \quad i, i' = 1, \dots, N ,$$

such that the full matrix

$$A[a_\epsilon] = \frac{1}{h}\begin{bmatrix} a_1 + a_2 & -a_2 & & & \\ -a_2 & a_2 + a_3 & -a_3 & & \\ & \ddots & \ddots & \ddots & \\ & & -a_{N-1} & a_{N-1} + a_N & -a_N \\ & & & -a_N & 2a_N \end{bmatrix}$$

is specified by the only coefficient vector

$$\mathbf{a} = [a_i] \in \mathbb{R}^N , \quad a_i = a_\epsilon(x_{i-1/2}) , \quad i = 1, \dots, N .$$

The next theorem was proven in [79].

Theorem 4.47. *Assume that the coefficient vector* $\mathbf{a}$ *is given in a QTT form*

$$i \mapsto (j_1, \dots, j_L): \; a(i) = \sum_{\gamma_1,\dots,\gamma_{L-1}=1}^{r_1,\dots,r_{L-1}} a^{(1)}_{\gamma_1}(j_1) \dots a^{(L)}_{\gamma_{L-1}}(j_L) \, ,$$

with the following QTT rank bounds: $r_p \leq R$, *for* $p = 1, \dots, L-1$. *Then the QTT ranks of the matrix* $A[a_\epsilon]$ *are bounded by*

$$r_{\mathrm{QTT}}(A[a_\epsilon]) \leq 7R \, .$$

Proof. Let us represent the stiffness matrix in the form

$$A\,[a_\epsilon] = S\,\mathrm{diag}(\mathbf{a}) + \mathrm{diag}(\mathbf{a} + S\mathbf{a}) + \mathrm{diag}(\mathbf{a})S^\top \, ,$$

where $S \in \mathbb{R}^{N \times N}$ means the upper shift matrix

$$S_{i,i'} = \begin{cases} 1, & i' = i + 1 \, , \\ 0, & \text{else} \, , \end{cases}$$

for which we have $r_{\mathrm{QTT}}(S) = 2$. We take into account that $r_{\mathrm{QTT}}(\mathrm{diag}(\mathbf{a})) = r_{\mathrm{QTT}}(\mathbf{a})$, then the rank estimate follows. □

Some QTT rank estimates for a special classes of 1D elliptic problems can be found in [178, 180, 181, 224]. The rank estimates in multidimensional case were considered in [84, 86, 87, 177].

The results for $d = 1$ can be extended to multidimensional cases as follows.

We apply the FEM-Galerkin discretization of equation (4.23) by means of tensor product piecewise affine basis functions (instead of 'linear finite elements')

$$\{\varphi_{\mathbf{i}}(x) := \varphi_{i_1}(x_1) \dots \varphi_{i_d}(x_d)\} \, , \quad \mathbf{i} = (i_1, \dots, i_d) \, , \; i_\ell \in \mathcal{I}_\ell = \{1, \dots, n_\ell\} \, ,$$

for $\ell = 1, \dots, d$, where φ_{i_k} are 1D finite element basis functions (say, piecewise linear hat functions).

We associate the univariate basis functions with the uniform grid $\{v_j\}$, $j = 1, \dots, n_\ell$, on $[0, 1]$ with the mesh size $h = 1/(n_\ell + 1)$. In this construction we have $N = n_1 n_2 \dots n_d$ basis functions $\varphi_{\mathbf{i}}$. Note that the univariate grid size n_ℓ is of the order of $n_\ell = O(1/\epsilon)$ designating the total problem size $N = O(1/\epsilon^d)$.

For ease of exposition we first consider the case $d = 2$, and further assume that the scalar diffusion coefficient $a(x_1, x_2)$ can be represented in the form

$$a(x_1, x_2) = \sum_{k=1}^{R} a^{(1)}_k(x_1) a^{(2)}_k(x_2) > 0$$

with a small rank parameter R.

The $N \times N$ stiffness matrix is constructed by the standard mapping of the multi-index $\mathbf{i}$ into the N-long univariate index i representing all degrees of freedom. For instance, we use the so called big-endian convention for $d = 3$ and $d = 2$

$$\mathbf{i} \mapsto i := i_3 + (i_2 - 1)n_3 + (i_1 - 1)n_2 n_3 \,, \quad \mathbf{i} \mapsto i := i_2 + (i_1 - 1)n_2 \,,$$

respectively. Hence all matrices and vectors are defined on the long index i as usual, however, the special Kronecker structure allows the low storage and low complexity matrix vector multiplications when appropriate, i.e., when a vector also admits the low rank Kronecker form representation. In particular, the basis function $\varphi_{\mathbf{i}}$ is designated via the long index, i.e., $\varphi_i = \varphi_{\mathbf{i}}$.

First, we consider the simplest case $R = 1$ and let $d = 2$. We construct the Galerkin stiffness matrix $A = [a_{ij}] \in \mathbb{R}^{N \times N}$ in the form of a sum of Kronecker products of small 'univariate' matrices. Recall that given $p_1 \times q_1$ matrix A and $p_2 \times q_2$ matrix B, their Kronecker product is defined as a $p_1 p_2 \times q_1 q_2$ matrix C via the block representation

$$C = A \otimes B = [a_{ij} B] \,, \quad i = 1, \ldots, p_1 \,, \; j = 1, \ldots, q_1 \,.$$

We say that the Kronecker rank of the matrix A in the representation above equals to 1. Now the elements of Galerkin stiffness matrix take a form

$$\begin{aligned} a_{ij} = \langle A\psi_{\mathbf{i}}, \psi_{\mathbf{j}} \rangle &= \int_{\Omega} a^{(1)}(x_1) a^{(2)}(x_2) \nabla\psi_i \cdot \nabla\psi_j dx \\ &= \int_{(0.1)} a^{(1)}(x_1) \frac{\partial\psi_{i_1}(x_1)}{\partial x_1} \frac{\partial\psi_{j_1}(x_1)}{\partial x_1} dx_1 \int_{(0.1)} a^{(2)}(x_2) \psi_{i_2}(x_2) \psi_{j_2}(x_2) dx_2 \\ &+ \int_{(0.1)} a^{(1)}(x_1) \psi_{i_1}(x_1) \psi_{j_1}(x_1) dx_1 \int_{(0.1)} a^{(2)}(x_2) \frac{\partial\psi_{i_2}(x_2)}{\partial x_2} \frac{\partial\psi_{j_2}(x_2)}{\partial x_2} dx_2 \,, \end{aligned}$$

which leads to the rank-2 Kronecker product representation

$$A = [a_{ij}] = A_1 \otimes M_2 + M_1 \otimes A_2 \,.$$

Here $A_1 = [a_{i_1 j_1}] \in \mathbb{R}^{n_1 \times n_1}$ and $A_2 = [a_{i_2 j_2}] \in \mathbb{R}^{n_2 \times n_2}$ denote the univariate stiffness matrices and $M_1 = [m_{i_1 j_1}] \in \mathbb{R}^{n_1 \times n_1}$ and $M_2 = [m_{i_2 j_2}] \in \mathbb{R}^{n_2 \times n_2}$ define the corresponding weighted mass matrices, e.g.,

$$a_{i_1 j_1} = \int_0^1 a^{(1)}(x_1) \frac{\partial\varphi_{i_1}(x_1)}{\partial x_1} \frac{\partial\varphi_{j_1}(x_1)}{\partial x_1} dx_1 \,, \quad m_{i_1 j_1} = \int_0^1 a^{(1)}(x_1) \varphi_{i_1}(x_1) \varphi_{j_1}(x_1) dx_1 \,.$$

By simple algebraic transformations (e.g., by lumping of the tridiagonal mass matrices, which does not effect the approximation order of the FEM discretization) the matrix A can be simplified to the form

$$A \mapsto A = A_1 \otimes D_2 + D_1 \otimes A_2 \,, \tag{4.24}$$

where D_1, D_2 are the diagonal matrices. The matrix A corresponds to the FEM discretization of the initial elliptic PDE with complicated highly oscillating coefficients.

In simplest case the representation in (4.24) is simplified to the discrete Laplacian matrix in the form of rank-2 Kronecker sum

$$A \mapsto A = A_1 \otimes I_2 + I_1 \otimes A_2 , \tag{4.25}$$

where I_1 and I_2 denote the identity matrices of the corresponding size. Here the simple treediagonal matrices A_1 and A_2 represent the FEM/FDM Laplacian like operators in 1D.

Now we apply the QTT approximation to 1D stiffness matrices A_1 and A_2 as in Theorem 4.47 to obtain the QTT rank estimates for the 2D discrete elliptic operator. Here we assume that diagonal matrices allows the low rank QTT representation.

The matrix A is constructed in general for the R-term separable coefficient $a(x_1, x_2)$ with $R \geq 1$, which leads to the rank-$2R$ Kronecker sum representation

$$A = \sum_{k=1}^{R} [A_{1,k} \otimes D_{2,k} + D_{1,k} \otimes A_{2,k}] ,$$

with matrices of the respective size. In this case the QTT rank will be multiplied by the factor R.

The extension to many dimensions, i.e., for $d \geq 3$, can be made in a similar way; see representations in Lemma 4.34 and Corollary 4.36.

4.3.11 Multidimensional bilinear forms

This section presents some results on the TT/QTT decomposition of the multidimensional bilinear forms, considered in [87] in full detail.

In many applications we are to compute the following bilinear form:

$$V(X', X) = \sum_{i,j=1}^{d} B_{ij} X'_i \circledast X_j , \quad \text{where} \tag{4.26}$$

X'_i, X_j are the following rank-1 tensor product objects:

$$X'_i = e_1 \otimes e_2 \otimes \cdots \otimes x'_i \otimes \cdots \otimes e_d , \quad X_j = e_1 \otimes e_2 \otimes \cdots \otimes x_j \otimes \cdots \otimes e_d ,$$
$$e_k, x_k, x'_k \in W \sim \mathbb{R}^n , \quad \text{which is a certain Euclidean vector space.}$$

Here $\circledast\colon W \times W \to W$ is a multiplicative bilinear operation, distributive with '$\cdot$' (e.g., Hadamard or matrix product), e_k are the unities with respect to the operation $\circledast$, and $B = B_{ij}$ is a matrix of scalar numbers. Note that the mode indices are omitted in this section for brevity.

The quadratic *Lyapunov* function[1] is such a form, with e_k being vectors of all ones, $x_k = x'_k$ are the vectors of grid points, and $\circledast$ is the Hadamard product. For a general

1 For example, it arises in calculation of the typical d dimensional probability density $\psi = \exp(-V)$.

polynomial, it was investigated in [221], where the rank bound $d + 2$ was proven for the polynomial degree 2 see Subsections 4.2.5–4.2.7. Here, we give a refined result. An explicit construction of a low rank representation of V, provided that the basis elements e_k, x_k, x'_k, $(x'_k \circledast x_k) \equiv x_k^2$ are defined, gives the following:

Theorem 4.48 ([87]). *Introduce the off-diagonal lower $C^k = B_{k+1:\,d,\,1:\,k}$, and upper $\hat{C}^k = B_{1:\,k,\,k+1:\,d}$ submatrices. The bilinear form (4.26) possesses an exact TT decomposition $V = V_1 \dots V_d$ with the ranks:*

$$r_{TT,k} = \operatorname{rank}(C^k) + \operatorname{rank}(\hat{C}^k) + 2 \le d + 2\,. \tag{4.27}$$

Moreover, if $x_k = x'_k$ (symmetric case), the ranks reduce to

$$r_{TT,k} = \operatorname{rank}\left((C^{k\top} + \hat{C}^k)/2\right) + 2 \le d/2 + 2\,.$$

In the QTT-Tucker format, the estimates above correspond to the TT ranks of the core, $r_{c,k} = r_{TT,k}$. The Tucker ranks $r_{T,k}$ equal 4, in the general case, and 3 in the symmetric case. The QTT ranks of the Tucker factors depend on the vectors e_k, x_k, x'_k, x_k^2:

$$r_{f,k} \le r_{qtt}(e_k) + r_{qtt}(x_k) + r_{qtt}(x'_k) + r_{qtt}(x_k^2)\,.$$

Proof. The proof is rather technical. The full details can be found in [87]. First of all, we compute the off-diagonal skeleton decompositions

$$C^k = W^k U^k\,, \qquad W^k \in \mathbb{R}^{d-k \times r_k}\,, \quad U^k \in \mathbb{R}^{r_k \times k}\,,$$
$$\hat{C}^k = \hat{U}^{k\top} \hat{W}^{k\top}\,, \quad \hat{W}^k \in \mathbb{R}^{d-k \times \hat{r}_k}\,, \quad \hat{U}^k \in \mathbb{R}^{\hat{r}_k \times k}\,,$$

with orthonormal U^k, $\hat{U}^k$. Now, we separate the TT blocks recursively.

In the first step we have $V = B_{11}X_1^2 + \sum_{i=2}^d B_{i1}X'_i \circledast X_1 + \sum_{j=2}^d B_{1j}X'_1 \circledast X_j + V^{[2]}$, where $V^{[k]} = \sum_{i,j=k}^d B_{ij}\, X'_i \circledast X_j$. So,

$$V = \begin{bmatrix} x_1^2 B_{11} & x'_1 & x_1 & e_1 \end{bmatrix} \begin{bmatrix} E^{[2]} & \sum_{j=2}^{d} X_j^{[2]} \hat{B}_{1,j} & \sum_{i=2}^{d} X_i'^{[2]} B_{i,1} & V^{[2]} \end{bmatrix}^\top\,, \tag{4.28}$$

where $E^{[k]} = e_k \dots e_d$, $X_j^{[k]} = e_k \dots e_{j-1} \cdot x_j \cdot e_{j+1} \dots e_d$, and $X_i'^{[k]} = e_k \dots e_{i-1} \cdot x'_i \cdot e_{i+1} \dots e_d$, for $i, j \ge k$. The first term in (4.28) is the first TT block V_1. On the other hand, we can represent it using the skeleton factors for the first row and column:

$$V = \begin{bmatrix} x_1^2 B_{11} & x'_1 \hat{U}^1 & x_1 U^1 & e_1 \end{bmatrix} \begin{bmatrix} E^{[2]} & \sum_{j=2}^{d} X_j^{[2]} \hat{W}_{j,1}^1 & \sum_{i=2}^{d} X_i'^{[2]} W_{i,1}^1 & V^{[2]} \end{bmatrix}^\top\,.$$

Now, we need to derive the recursive representation for the second term.

Suppose we have

$$\tilde{V} = \begin{bmatrix} E^{[k]} & \sum_{j=k}^{d} X_j^{[k]} \hat{W}_{j,1:\,\hat{r}_{k-1}}^{k-1} & \sum_{i=k}^{d} X_i'^{[k]} W_{i,1:\,r_{k-1}}^{k-1} & V^{[k]} \end{bmatrix}^\top\,. \tag{4.29}$$

Applying the first step to $V^{[k]}$, we obtain ($I_{r_{k-1}} = \operatorname{diag}[e_k, \dots, e_k]$):

$$\tilde V = \begin{bmatrix} e_k & & & & & \\ x_k \hat W^{k-1\top}_{k,:} & I_{\hat r_{k-1}} & & & & \\ x'_k W^{k-1\top}_{k,:} & & I_{r_{k-1}} & & & \\ x_k^2 B_{kk} & & & x'_k & x_k & e_k \end{bmatrix} \begin{bmatrix} E^{[k+1]} \\ \sum\limits_{j=k+1}^{d} X_j^{[k+1]} \hat W^{k-1\top}_{j,:} \\ \sum\limits_{i=k+1}^{d} X_i'^{[k+1]} W^{k-1\top}_{i,:} \\ \sum\limits_{j=k+1}^{d} X_j^{[k+1]} B_{k,j} \\ \sum\limits_{i=k+1}^{d} X_i'^{[k+1]} B_{i,k} \\ V^{[k+1]} \end{bmatrix},$$

or, separating the scalar coefficients from X related data in the last column,

$$\tilde V = \begin{bmatrix} e_k & & & & & \\ x_k \hat W^{k-1\top}_{k,:} & I_{\hat r_{k-1}} & & & & \\ x'_k W^{k-1\top}_{k,:} & & I_{r_{k-1}} & & & \\ x_k^2 B_{kk} & & & x'_k & x_k & e_k \end{bmatrix} \begin{bmatrix} 1 & & & \\ & \hat W^{k-1\top}_{k+1:\,d,:} & & \\ & & W^{k-1\top}_{k+1:\,d,:} & \\ & B_{k,k+1:\,d} & & \\ & & B^\top_{k+1:\,d,k} & \\ & & & 1 \end{bmatrix} \begin{bmatrix} E^{[k+1]} \\ X^{[k+1]}_{k+1:\,d} \\ X'^{[k+1]}_{k+1:\,d} \\ V^{[k+1]} \end{bmatrix}. \tag{4.30}$$

For brevity, denote the second matrix in this statement R_k. Now, recall that $\hat W^{k-1\top} = \hat U^{k-1} B_{1:\,k-1,k+1:\,d}$, $W^{k-1} = B_{k+1:\,d,1:\,k-1} U^{k-1\top}$:

$$R_k = \begin{bmatrix} 1 & & & & & \\ & \hat U^{k-1} & & & & \\ & & U^{k-1} & & & \\ & & & 1 & & \\ & & & & 1 & \\ & & & & & 1 \end{bmatrix} \begin{bmatrix} 1 & & & \\ & B_{1:\,k-1,k+1:\,d} & & \\ & & B^\top_{k+1:\,d,1:\,k-1} & \\ & B_{k,k+1:\,d} & & \\ & & B^\top_{k+1:\,d,k} & \\ & & & 1 \end{bmatrix} = R_k^U R_k^B .$$

For the second term we use the kth skeleton decomposition:

$$R_k^B = \begin{bmatrix} 1 & & & \\ & \hat U^{k\top}_{1:\,\hat r_k,1:\,k-1} & & \\ & & U^{k\top}_{1:\,r_k,1:\,k-1} & \\ & \hat U^{k\top}_{1:\,\hat r_k,k} & & \\ & & U^{k\top}_{1:\,r_k,k} & \\ & & & 1 \end{bmatrix} \begin{bmatrix} 1 & & & \\ & \hat W^{k\top}_{k+1:\,d,1:\,\hat r_k} & & \\ & & W^{k\top}_{k+1:\,d,1:\,r_k} & \\ & & & 1 \end{bmatrix}.$$

The last term contains only W coefficients, i.e., yields the form of $\tilde V$ (4.29). Multiplying the first two terms, and plugging into (4.30), we get

$$\tilde{V} = \begin{bmatrix} e_k & & & \\ x_k \hat{W}^{k-1\top}_{k,:} & e_k \hat{U}^{k-1} \hat{U}^{k\top}_{1:\hat{r}_k, 1:k-1} & & \\ x'_k W^{k-1\top}_{k,:} & & e_k U^{k-1} U^{k\top}_{1:r_k,1:k-1} & \\ x_k^2 B_{kk} & x'_k \hat{U}^{k\top}_{1:\hat{r}_k,k} & x_k U^{k\top}_{1:r_k,k} & e_k \end{bmatrix} \begin{bmatrix} E^{[k+1]} \\ \sum_{j=k+1}^{d} X_j^{[k+1]} \hat{W}^{k\top}_{j,1:\hat{r}_k} \\ \sum_{j=k+1}^{d} X_j'^{[k+1]} W^{k\top}_{j,1:r_k} \\ V^{[k+1]} \end{bmatrix}. \tag{4.31}$$

The first term is nothing else than the kth TT block V_k, since it contains only x_k, e_k. It has the sizes $(1 + \hat{r}_{k-1} + r_{k-1} + 1) \times (1 + \hat{r}_k + r_k + 1)$, which confirms the first statement of the theorem. With the second term we can proceed by recursion, and the last TT block reads (since $V^{[d]}$ is just a one dimensional $x_d^2 B_{dd}$)

$$V_d = \begin{bmatrix} e_d & x_d \hat{W}^d_{d,1:\hat{r}_{d-1}} & x'_d W^d_{d,1:r_{d-1}} & x_d^2 B_{dd} \end{bmatrix}^\top . \tag{4.32}$$

If the first-order term is presented with a unique object $x_k = x'_k$, the ranks are reduced as follows. First,

$$V_1 = \begin{bmatrix} x_1^2 B_{11} & 2x_1 & e_1 \end{bmatrix} \begin{bmatrix} 1 & & & \\ & 0.5 & 0.5 & \\ & & & 1 \end{bmatrix}.$$

Reassigning $\tilde{V}_1 = \begin{bmatrix} x_1^2 B_{11} & 2x_1 & e_1 \end{bmatrix}$, we contract the rest of the matrix with the middle blocks, obtaining

$$\tilde{V}_k = \begin{bmatrix} e_k & & & \\ x_k W^{k-1\top}_{k,:} & \frac{1}{2} e_k U^{k-1} U^{k\top}_{1:r_k,1:k-1} & \frac{1}{2} e_k U^{k-1} U^{k\top}_{1:r_k,1:k-1} & \\ x_k^2 B_{kk} & x_k U^{k\top}_{1:r_k,k} & x_k U^{k\top}_{1:r_k,k} & e_k \end{bmatrix}.$$

We note here that the constraint $x_k = x'_k$ yields $V \equiv 0$ if $B = -B^\top$, so that only the symmetric part $B := 0.5(B + B^\top)$ is relevant. Thus, $\hat{W}^k = W^k$, $\hat{U}^k = U^k$. The rest of the TT blocks are simplified as follows:

$$\begin{bmatrix} E^{[k+1]} \\ \sum_{j=k+1}^{d} X_j^{[k+1]} \hat{W}^{k\top}_{j,1:\hat{r}_k} \\ \sum_{j=k+1}^{d} X_j'^{[k+1]} W^{k\top}_{j,1:r_k} \\ V^{[k+1]} \end{bmatrix} = \begin{bmatrix} 1 & & \\ & 1 & \\ & 1 & \\ & & 1 \end{bmatrix} \begin{bmatrix} E^{[k+1]} \\ \sum_{j=k+1}^{d} X_j^{[k+1]} W^{k\top}_{j,1:r_k} \\ V^{[k+1]} \end{bmatrix}.$$

Multiplying $\tilde{V}_k$ with the scalar matrix appearing here, we obtain the reduced block

$$\tilde{V}_k = \begin{bmatrix} e_k & & \\ x_k W^{k-1\top}_{k,:} & e_k U^{k-1} U^{k\top}_{1:r_k,1:k-1} & \\ x_k^2 B_{kk} & 2x_k U^{k\top}_{1:r_k,k} & e_k \end{bmatrix} \quad \text{of size } (r_{k-1} + 2) \times (r_k + 2)\,. \tag{4.33}$$

Applying the TT-to-Tucker conversion technique (Section 3.6) we obtain that the TT ranks of the Tucker core are equal to the ranks obtained above, and the Tucker ranks equal 4 (3 in the symmetric case), since there are 4 (respectively 3) linearly independent elements in each block, e_k, x_k, x'_k and x^2_k. □

Remark 4.49. The proof gives a constructive routine for fast assembly of a bilinear form in TT format. Indeed, performing the SVDs of submatrices in B (which are in the order of tens) and building the blocks from (4.28), (4.31) (or (4.33) in symmetric case), (4.32) we get the analytical TT representation.

Note that in the case of a degree-2 polynomial on a uniform grid, the QTT ranks of the Tucker factors are equal to 3 [196, 197, 288].

Corollary 4.50. *The estimate* (4.27) *is applicable to the d dimensional anisotropic elliptic operator* $\sum_{i,j} \nabla_i^\top B_{i,j} \nabla_j$ *with a constant matrix B, by setting* $e_k = I_k$, $x_k = \nabla_k$, $x'_k = \nabla_k^\top$, $x^2_k = \Delta_k$, *and* $\circledast$ *being the operator composition. A discussion on such structures was started in [84, 177].*

Remark 4.51. The ranks of the off-diagonal blocks are the so called *semiseparable* ranks. Theorem 4.48 establishes a connection between the matrix semiseparability and the TT structure. The rank estimate is numerically sharp. If the matrix is diagonal, i.e., $C^{k\top} = \hat{C}^k = 0$, then the TT ranks equal 2 (Laplace operator, harmonic oscillator potential).

There are similar results on the explicit TT/QTT representation to the discrete multidimensional diffusion operator with variable coefficients, and the sharp rank estimates in terms of a rank bound for the accompanying semiseparable coefficient matrix have been considered in [181].

We comment that results in Theorem 4.48 will be applied in Chapter 5 within the numerical scheme for the TT/QTT tensor approximation to the chemical master equation (CME).

4.3.12 Toward numerical issues

We note that numerical experiments carried out with TT Toolbox prove all the upper bounds for vector TT/QTT ranks given in Summary 4.46 are sharp, i.e., the corresponding explicit representations are of minimal rank.

We recommend several exercises that demonstrate the explicit QTT matrix decompositions presented above.

Exercise 4.52. $\Delta_{DD}^{(d)}$ has a rank-5 . . . 5 explicit QTT representation in Lemma 4.38. Confirm it with the numerical QTT decomposition. Compare the CPU time for explicit and algebraic decomposition for large $n = 2^d$.

Exercise 4.53. Recall that eigenvectors of the matrix $\Delta_{DD}^{(d_k)}$ have the explicit rank-2 QTT decomposition; see Section 4.1. Check it by 'analytic' calculation of the matrix vectors product of the respective QTT matrix and QTT vector decompositions.

Exercise 4.54. Compare vector and operator ranks of ${\Delta_{DD}^{(d)}}^{-1}$ by QTT calculations.

Tables 4.5 and 4.6 represent the average QTT ranks in ε-approximation of $N \times N$ function related matrices up to tolerance $\varepsilon = 10^{-5}$, where N is the univariate grid size. In particular, Table 4.5 includes the example of matrix exponential; see [222]. One can observe that rank parameters are small, and depend very mildly on the grid size.

Here we used the same notations as in Section 4.1 and Section 4.3.

Tab. 4.5: QTT matrix ranks of $N \times N$ matrices for $N = 2^p$.

$N \backslash \bar{r}$	$e^{-\alpha\Delta_1}$, $\alpha = 0.1, 1, 10, 10^2$	diag$(1/x^2)$	diag(e^{-x^2})
2^9	6.2/6.8/9.7/11.2	5.1	4.0
2^{10}	6.3/6.8/9.5/10.8	5.3	4.0
2^{11}	6.4/6.8/9.0/10.4	5.5	4.1

Tab. 4.6: QTT ranks of functional $N \times N$ matrices, $N = 2^p$.

$N \backslash \bar{r}$	$1/(x_1 + x_2)$	$e^{-\|x\|}$	$e^{-\|x\|^2}$	diag(e^{-x^2})	$\Delta_2^{-1}1$, $\varepsilon = 10^{-6}, 10^{-7}, 10^{-8}$
2^9	5.0	9.4	7.8	3.8	3.6/3.6/3.6
2^{10}	5.1	9.4	7.7	3.9	3.6/3.6/3.6
2^{11}	5.2	9.3	7.5	3.9	3.7/3.7/3.7

4.4 QTT-FFT and convolution transform in logarithmic time

In this section we describe the superfast FFT and convolution transforms of vectors, based on the QTT tensor approximation.

4.4.1 Diagonalizing circulant matrix revisited

Let us recall the definition of circulant and block circulant matrices.

Definition 4.55. A one level block circulant matrix $A \in \mathcal{BC}(L, m_0)$ is defined by

$$A = \mathrm{bcirc}\{A_0, A_1, \dots, A_{L-1}\} = \begin{bmatrix} A_0 & A_{L-1} & \dots & A_2 & A_1 \\ A_1 & A_0 & \dots & \vdots & A_2 \\ \vdots & \vdots & \ddots & A_0 & \vdots \\ A_{L-1} & A_{L-2} & \dots & A_1 & A_0 \end{bmatrix} \in \mathbb{R}^{Lm_0 \times Lm_0}, \quad (4.34)$$

where $A_k \in \mathbb{R}^{m_0 \times m_0}$ for $k = 0, 1, \dots, L-1$, are matrices of general structure.

The equivalent Kronecker product representation is defined by the associated matrix polynomial,

$$A = \sum_{k=0}^{L-1} \pi^k \otimes A_k =: p_A(\pi), \quad (4.35)$$

where $\pi = \pi_L \in \mathbb{R}^{L \times L}$ is the periodic downward shift (cycling permutation) matrix,

$$\pi_L := \begin{bmatrix} 0 & 0 & \dots & 0 & 1 \\ 1 & 0 & \dots & 0 & 0 \\ \vdots & \vdots & \ddots & \vdots & \vdots \\ 0 & \dots & 1 & 0 & 0 \\ 0 & 0 & \dots & 1 & 0 \end{bmatrix}. \quad (4.36)$$

In the case $m_0 = 1$ a matrix $A \in \mathcal{BC}(L, 1)$ defines a circulant matrix generated by its first column vector $\mathbf{a} = (a_0, \dots, a_{L-1})^{\mathrm{T}}$. The associated scalar polynomial then reads

$$p_A(z) := a_0 + a_1 z + \dots + a_{L-1} z^{L-1},$$

so that (4.35) simplifies to

$$A = p_A(\pi_L).$$

Let $\omega = \omega_L = \exp(-\frac{2\pi i}{L})$, and denote the unitary matrix of the Fourier transform by

$$F_L = \{f_{k\ell}\} \in \mathbb{R}^{L \times L}, \quad \text{with} \quad f_{k\ell} = \frac{1}{\sqrt{L}} \omega_L^{(k-1)(\ell-1)}, \quad k, l = 1, \dots, L.$$

Since the shift matrix π_L is diagonalizable in the Fourier basis,

$$\pi_L = F_L^* D_L F_L, \quad D_L = \mathrm{diag}\{1, \omega, \dots, \omega^{L-1}\}, \quad (4.37)$$

the same holds for any circulant matrix,

$$A = p_A(\pi_L) = F_L^* p_A(D_L) F_L, \quad (4.38)$$

where

$$p_A(D_L) = \mathrm{diag}\{p_A(1), p_A(\omega), \dots, p_A(\omega^{L-1})\} = \mathrm{diag}\{F_L \mathbf{a}\}.$$

Matrix vector product can be implemented in $O(L \log L)$ operations as follows:

$$A\mathbf{x} = F_L^* p_A(D_L) F_L \mathbf{x} = F_L^* \left(\mathrm{diag}\{F_L \mathbf{a}\}(F_L \mathbf{x})\right).$$

4.4.2 Discrete circulant/Toeplitz convolution

Definition 4.56. A vector $\mathbf{g} = [g_n]$ is the discrete convolution of signals $\mathbf{f}$ and $\mathbf{h}$ supported by the indices $0 \leq n \leq M-1$,

$$g_n = (\mathbf{f} * \mathbf{h})_n = \sum_{k=-\infty}^{\infty} f_k h_{n-k} .$$

The naive implementation requires $M(M+1)$ operations.

It can be represented as a matrix-by-vector product with the Toeplitz matrix

$$\mathbf{g} = \mathbf{f} * \mathbf{h} = T\mathbf{f}: \quad T = \{h_{n-k}\}_{0 \leq n,k < M} \in \mathbb{R}^{M \times M} .$$

Extending $\mathbf{f}$ and $\mathbf{h}$ with over M samples by

$$\tilde{h}_M = 0 , \quad \tilde{h}_{2M-i} = h_i , \quad i = 1, \dots, M-1 ,$$

$$\tilde{f}_n = 0 , \quad n = M, \dots, 2M-1 ,$$

we reduce the problem to the matrix vector product with a circulant matrix $\mathcal{C} \in \mathbb{R}^{2M \times 2M}$ specified by the first row $\tilde{h} \in \mathbb{R}^{2M}$. The latter can be multiplied with a vector by FFT algorithm (diagonalization by DFT).

Note that Toeplitz/circulant type matrices apply to quasiperiodic systems [187], and in geometric homogenization methods [224]. The detailed discussion will be postponed to Chapter 5.

4.4.3 QTT decomposition of 1D shift matrices of size $2^d \times 2^d$

In this and the next section we discuss the fast matrix vector multiplication with circulant/Toeplitz matrices in the QTT tensor format. This presentation follows [177, 179].

Rank-2 QTT representation of periodic downward shift ($n \times n$ cycling permutation) takes a form:

$$P_1^d \equiv \pi_n := \begin{pmatrix} 0 & & & & 1 \\ 1 & \ddots & & & \\ 0 & \ddots & \ddots & & \\ & \ddots & \ddots & \ddots & \\ & & 0 & 1 & 0 \end{pmatrix} , \quad \mathbf{P}_1^d = \begin{bmatrix} I & P \end{bmatrix} \bowtie \begin{bmatrix} I & J' \\ & J \end{bmatrix}^{\bowtie (d-2)} \bowtie \begin{bmatrix} J' \\ J \end{bmatrix} ,$$

$$I = \begin{pmatrix} 1 & 0 \\ 0 & 1 \end{pmatrix} , \quad J = \begin{pmatrix} 0 & 1 \\ 0 & 0 \end{pmatrix} , \quad J' = \begin{pmatrix} 0 & 0 \\ 1 & 0 \end{pmatrix} \quad \text{and} \quad P = \begin{pmatrix} 0 & 1 \\ 1 & 0 \end{pmatrix} .$$

Define

$$P_m^d = (P_1^d)^m ,$$

and stack the matrices P_m^d, $1 \le m \le 2^d$, into $2^d \times 2^d \times 2^d$ tensors $\mathbf{P}^{(d)}$,

$$\mathbf{P}^d_{\cdot,\cdot,m} = P^d_{m-1}(\cdot,\cdot)\,, \quad 1 \le m \le 2^d\,.$$

In view of (4.38) we summarize that given an n component vector $\mathbf{a} = (a_1, \dots, a_n)^{\mathrm{T}}$, defining a *circulant matrix* as in (4.34) for $m_0 = 1$, then we have

$$A = \sum_{k=1}^{n} a_k (P_1^d)^{k-1} = F_n^* \operatorname{diag}\{F_n \mathbf{a}\} F_n\,.$$

Hence, as the consequence of the above representation, the fast QTT convolution reduces to QTT-FFT (but the latter has full QTT rank).

4.4.4 QTT based circulant/Toeplitz convolution in $O(\log N)$ cost

The next structural lemma allows us to construct the efficient QTT arithmetics with circulant matrices; see [179] for the proof.

Lemma 4.57. *Let $d \ge 2$. Then the tensor $\boldsymbol{P}^{(d)}$ has the following rank-2 QTT representation:*

$$\boldsymbol{P}^{(d)} = P \bowtie W^{\bowtie(d-2)} \bowtie V\,,$$

where TT cores are

$$P = \begin{bmatrix} I|P & P|I \end{bmatrix}\,, \quad W = \begin{bmatrix} I|J' & J'|O \\ O|J & J|I \end{bmatrix}\,, \quad V = \begin{bmatrix} I|J' \\ O|J \end{bmatrix}\,.$$

Lemma 4.57 allows us to derive the algorithm for superfast QTT based circulant convolution taking into account the important property that a circulant is expressed in terms of contracted product involving shift matrices and **a**:

$$A = \mathbf{P}^d \times_3 \mathbf{a}\,.$$

Theorem 4.58 ([179]). *Let $\tilde{\mathbf{x}}$ be a 2^d-component vector, $d \ge 2$, given in a QTT representation*

$$\tilde{\mathbf{x}} = \tilde{X}_d \bowtie \tilde{X}_{d-1} \bowtie \cdots \bowtie \tilde{X}_2 \bowtie \tilde{X}_1$$

of ranks $r_{d-1}, \dots, r_1$. Then a circulant, $2^d \times 2^d$ matrix $\mathbf{C}$, generated by the vector $\tilde{\mathbf{x}}$ has the explicit QTT decomposition of ranks $2r_{d-1}, \dots, 2r_1$:

$$\mathbf{C} = \left(P \bullet^3 \tilde{X}_d\right) \bowtie \left(W \bullet^3 \tilde{X}_{d-1}\right) \bowtie \cdots \bowtie \left(W \bullet^3 \tilde{X}_2\right) \bowtie \left(V \bullet^3 \tilde{X}_1\right)\,,$$

with TT cores P, Q, R, W, and V being the same as in Lemma 4.57.

As the direct consequence of the constitution Theorem 4.58 one derives the superfast algorithm for matrix vector multiplication with the complexity $O(\log N)$.

Theorem 4.59 ([179]). *Assume that $\tilde{\mathbf{x}}$ and $\mathbf{y}$ are 2^d-component vectors such that a $2^d \times 2^d$ matrix $\tilde{\mathbf{x}}\mathbf{y}'$ has a QTT representation at the cost $O(dR^2)$,*

$$\tilde{\mathbf{x}}\mathbf{y}' = \tilde{G}_{d+1} \bowtie \tilde{G}_d \bowtie \tilde{G}_{d-1} \bowtie \dots \bowtie \tilde{G}_2 \bowtie \tilde{G}_1$$

of ranks $t_{d-1}, \dots, t_1$. Then the matrix vector product of a circulant $2^d \times 2^d$ matrix $\mathbf{C} = \mathbf{C}(\tilde{\mathbf{x}})$, is generated by the vector $\tilde{\mathbf{x}}$ in the sense of (4.34), and the vector $\mathbf{y}$ has the QTT representation

$$\mathbf{C}\mathbf{y} = A_d \bowtie A_{d-1} \bowtie \dots \bowtie A_2 \bowtie A_1 \quad \text{of ranks} \quad 2t_{d-1}, \dots, 2t_1 ,$$

where $A_d = P \bullet^{2,3} \tilde{G}_d$, $A_k = W \bullet^{2,3} \tilde{G}_k$ for $d-1 \geq k \geq 2$ and $A_1 = V \bullet^{2,3} \tilde{G}_1$, with TT cores P, Q, R, W, and V being the same as in Lemma 4.57.

The full proof of above statements concerning the explicit QTT representation of circulant matrices and the related matrix vector arithmetics can be found in [179].

Table 4.7 (see the full data presented in [179]) demonstrates the $O(\log N)$ scaling of the superfast QTT based convolution transform and its comparison with the QTT-FFT algorithm to be presented in the next section.

We summarize that the basic matrix vector transforms such as the circulant/ Toeplitz convolution, FTT, and fast wavelet transform ([219]), can be realized in the QTT arithmetics with the logarithmic complexity in the vector size.

4.4.5 QTT decomposition of FFT matrix has irreducible ε rank

In what follows, we describe the superfast QTT based fast Fourier transform (FFT). Our presentation is based on [85].

Recall that the unitary FTT matrix of size $N = 2^d$ is defined by

$$F_d = \frac{1}{2^{d/2}} \left[\omega_d^{jk}\right]_{j,k=0}^{2^d-1} , \qquad \omega_d = \exp\left(-\frac{2\pi \mathrm{i}}{2^d}\right), \quad \mathrm{i}^2 = -1 .$$

We recall that QTT format for a matrix $A = [a(i,j)]$ is defined in terms of the d-fold quantized multi-index by

$$a(i,j) = a(\overline{j_1 \dots j_d}, \overline{k_1 \dots k_d}) = \mathbf{A}(j_1 k_1, j_2 k_2, \dots, j_d k_d) = A^{(1)}_{j_1 k_1} A^{(2)}_{j_2 k_2} \dots A^{(d)}_{j_d k_d} .$$

with the QTT rank parameters

$$r_p = \operatorname{rank} \dot{a}(j_1 k_1\, j_2 k_2 \dots j_p k_p; j_{p+1} k_{p+1} \dots j_d k_d) .$$

There are commonly used matrix transforms that allow low rank QTT representation. As an example, we mention the Hadamard (Walsh) transform that has QTT ranks equal to one,

$$H_d = H_1^{\otimes d} , \quad H_1 = \frac{1}{\sqrt{2}} \begin{bmatrix} 1 & 1 \\ 1 & -1 \end{bmatrix} .$$

Tab. 4.7: Time (sec.) of convolution of random 2^d-vectors. The two columns marked with * present time of three FFTs and one QTT-FFT of a sum of r plane waves with random amplitudes and frequencies.

	in the full format				in the QTT format		
d	FFT conv.	3 × FFT*	r	exact	exact+tr.	DMRG	3 × QTT-FFT*
20	$2.6 \cdot 10^{-1}$	$1.1 \cdot 10^{-1}$	5	$6.5 \cdot 10^{-3}$	$2.8 \cdot 10^{-2}$	$2.8 \cdot 10^{-2}$	$1.3 \cdot 10^{-1}$
			15	$1.8 \cdot 10^{-1}$	$4.5 \cdot 10^{+0}$	$6.9 \cdot 10^{-2}$	$8.4 \cdot 10^{-1}$
			40	$1.7 \cdot 10^{+1}$	$1.1 \cdot 10^{+3}$	$1.1 \cdot 10^{+0}$	$4.8 \cdot 10^{+0}$
21	$6.3 \cdot 10^{-1}$	$2.4 \cdot 10^{-1}$	5	$6.9 \cdot 10^{-3}$	$3.0 \cdot 10^{-2}$	$4.4 \cdot 10^{-2}$	$1.3 \cdot 10^{-1}$
			15	$2.0 \cdot 10^{-1}$	$4.9 \cdot 10^{+0}$	$9.9 \cdot 10^{-2}$	$8.7 \cdot 10^{-1}$
			40	$1.7 \cdot 10^{+1}$	$1.2 \cdot 10^{+3}$	$1.1 \cdot 10^{+0}$	$5.4 \cdot 10^{+0}$
22	$1.5 \cdot 10^{+0}$	$5.0 \cdot 10^{-1}$	5	$7.4 \cdot 10^{-3}$	$3.2 \cdot 10^{-2}$	$4.6 \cdot 10^{-2}$	$1.5 \cdot 10^{-1}$
			15	$2.3 \cdot 10^{-1}$	$5.5 \cdot 10^{+0}$	$8.3 \cdot 10^{-2}$	$1.0 \cdot 10^{+0}$
			40	$1.8 \cdot 10^{+1}$	$1.3 \cdot 10^{+3}$	$1.3 \cdot 10^{+0}$	$6.1 \cdot 10^{+0}$
23	$2.9 \cdot 10^{+0}$	$1.1 \cdot 10^{+0}$	5	$7.6 \cdot 10^{-3}$	$3.4 \cdot 10^{-2}$	$3.2 \cdot 10^{-2}$	$1.7 \cdot 10^{-1}$
			15	$2.4 \cdot 10^{-1}$	$5.8 \cdot 10^{+0}$	$8.3 \cdot 10^{-2}$	$1.1 \cdot 10^{+0}$
			40	$1.9 \cdot 10^{+1}$	$1.4 \cdot 10^{+3}$	$1.3 \cdot 10^{+0}$	$7.2 \cdot 10^{+0}$
24	$6.5 \cdot 10^{+0}$	$2.4 \cdot 10^{+0}$	5	$8.1 \cdot 10^{-3}$	$3.5 \cdot 10^{-2}$	$5.1 \cdot 10^{-2}$	$1.9 \cdot 10^{-1}$
			15	$2.6 \cdot 10^{-1}$	$6.2 \cdot 10^{+0}$	$9.0 \cdot 10^{-2}$	$1.2 \cdot 10^{+0}$
			40	$2.0 \cdot 10^{+1}$	$1.5 \cdot 10^{+3}$	$1.4 \cdot 10^{+0}$	$7.6 \cdot 10^{+0}$
25	$1.4 \cdot 10^{+1}$	$5.4 \cdot 10^{+0}$	5	$8.3 \cdot 10^{-3}$	$3.8 \cdot 10^{-2}$	$5.4 \cdot 10^{-2}$	$2.1 \cdot 10^{-1}$
			15	$2.7 \cdot 10^{-1}$	$6.5 \cdot 10^{+0}$	$9.5 \cdot 10^{-2}$	$1.3 \cdot 10^{+0}$
			40	$2.2 \cdot 10^{+1}$	$1.7 \cdot 10^{+3}$	$1.5 \cdot 10^{+0}$	$8.5 \cdot 10^{+0}$
26	$3.0 \cdot 10^{+1}$	$1.2 \cdot 10^{+1}$	5	$8.8 \cdot 10^{-3}$	$3.9 \cdot 10^{-2}$	$3.7 \cdot 10^{-2}$	$2.1 \cdot 10^{-1}$
			15	$2.9 \cdot 10^{-1}$	$6.9 \cdot 10^{+0}$	$1.3 \cdot 10^{-1}$	$1.4 \cdot 10^{+0}$
			40	$2.2 \cdot 10^{+1}$	$1.7 \cdot 10^{+3}$	$1.5 \cdot 10^{+0}$	$9.2 \cdot 10^{+0}$
27	$6.6 \cdot 10^{+1}$	$3.6 \cdot 10^{+1}$	5	$9.0 \cdot 10^{-3}$	$4.1 \cdot 10^{-2}$	$5.9 \cdot 10^{-2}$	$2.3 \cdot 10^{-1}$
			15	$3.5 \cdot 10^{-1}$	$7.5 \cdot 10^{+0}$	$1.0 \cdot 10^{-1}$	$1.5 \cdot 10^{+0}$
			40	$2.4 \cdot 10^{+1}$	$1.9 \cdot 10^{+3}$	$1.7 \cdot 10^{+0}$	$1.0 \cdot 10^{+1}$

Remark 4.60. As a matter of fact, it was shown in [85] that the QTT decomposition of unitary FFT matrix has full rank. Moreover, it was shown numerically that the QTT-FFT matrix has full ε rank, which means that the low rank ε-approximation of this matrix is not possible.

In spite of the pessimistic observation in Remark 4.60 it is still possible to construct the fast FFT-QTT algorithm by using the QTT rank truncated operations at all steps of FFT transform as discussed in what follows.

4.4.6 Fast QTT-FFT based on Cooley–Tuckey recursion

The Fourier transform of a vector $\mathbf{x}$ reads $\mathbf{y} = \mathrm{FFT}_n\,\mathbf{x}$

$$y = \frac{1}{\sqrt{n}} F_n x \quad \text{where} \quad y_k = \frac{1}{\sqrt{n}} \sum_{j=0}^{n-1} x_j \exp\left(-\frac{2\pi \mathrm{i}}{n} jk\right), \quad j, k = 0, \dots, n-1 .$$

The FFT for dense vectors costs $O(n \log n)$. It is thanks to the idea of Cooley and Tuckey [73] based on the recurrence factorization of the action of FFT matrix on a vector

$$P_d F_d \mathbf{x} = \frac{1}{\sqrt{2}} \begin{bmatrix} F_{d-1} & \\ & F_{d-1} \end{bmatrix} \begin{bmatrix} I & \\ & \Omega_{d-1} \end{bmatrix} \begin{bmatrix} I & I \\ I & -I \end{bmatrix} \begin{bmatrix} x_- \\ x_+ \end{bmatrix}, \tag{4.39}$$

where P_d is the so called *bit-shift* permutation, agglomerating even and odd elements of a vector. Here the ‘twiddle factors’ are specified by

$$\Omega = \operatorname{diag}\left\{\exp\left(-\frac{2\pi \mathrm{i}}{2^d} j\right)\right\}_{j=0}^{2^{d-1}-1} = \operatorname{diag}\left\{\exp\left(-\frac{2\pi \mathrm{i}}{2^d} j_1\right)\right\} \dots \operatorname{diag}\left\{\exp\left(-\frac{2\pi \mathrm{i}}{2} j_{d-1}\right)\right\} .$$

The recurrence (4.39) is the key point for fast FFT transform in the QTT format: Given QTT vector $\mathbf{x}$ in the QTT format, the vector $\mathbf{y} = \frac{1}{\sqrt{n}} F_d \mathbf{x}$ can be computed in QTT format at the cost $O(d^2 R^3)$ with the storage demand $O(dR^2)$. In what follows, we describe the main steps of the algorithm, as presented in [84].

Given half vectors $\mathbf{x}_-$ and $\mathbf{x}_+$ in the QTT format (here $j_d = 0$ or $j_d = 1$)

$$\begin{aligned} x(\overline{j_1 \dots j_{d-1} j_d}) &= X_{j_1}^{(1)} \dots X_{j_{d-1}}^{(d-1)} X_{j_d}^{(d)} \\ x_-(\overline{j_1 \dots j_{d-1} 0}) &= X_{j_1}^{(1)} \dots X_{j_{d-1}}^{(d-1)} X_{j_d=0}^{(d)} \\ x_+(\overline{j_1 \dots j_{d-1} 1}) &= X_{j_1}^{(1)} \dots X_{j_{d-1}}^{(d-1)} X_{j_d=1}^{(d)} . \end{aligned}$$

Then the one-core operations take a form

$$\frac{1}{\sqrt{2}} \begin{bmatrix} I & I \\ I & -I \end{bmatrix} x(\overline{j_1 \dots j_{d-1} j_d}) = X_{j_1}^{(1)} \dots X_{j_{d-1}}^{(d-1)} \begin{Bmatrix} X_0^{(d)} + X_1^{(d)} \\ X_0^{(d)} - X_1^{(d)} \end{Bmatrix}_{j_d} \frac{1}{\sqrt{2}},$$

which can be rewritten as

$$\hat{\mathbf{x}} := \frac{1}{\sqrt{2}} \begin{bmatrix} I & I \\ I & -I \end{bmatrix} x(\overline{j_1 \dots j_{d-1} j_d}) = X_{j_1}^{(1)} \dots X_{j_{d-1}}^{(d-1)} \hat{X}_{j_d}^{(d)} .$$

It is worth to note that the QTT ranks did not change yet.

In the next step the multiplication by the diagonal matrix of twiddle factors is accomplished with the rank truncation, that is the Hadamard product operation in the QTT format

$$\mathbf{z} = \begin{bmatrix} I & \\ & \Omega_{d-1} \end{bmatrix} \frac{1}{\sqrt{2}} \begin{bmatrix} I & I \\ I & -I \end{bmatrix} \mathbf{x} = \begin{bmatrix} I & \\ & \Omega_{d-1} \end{bmatrix} \hat{\mathbf{x}} = \mathbf{w}_d \odot \hat{\mathbf{x}} ,$$

where the vector $\mathbf{w}_d$ is defined by

$$\mathbf{w}_d^{\mathrm{T}} \overset{\text{def}}{=} \Bigg[\underbrace{1 \ldots 1}_{2^{d-1} \text{ elements}} \; \underbrace{1 \, \omega_d \, \omega_d^2 \ldots \omega_d^{2^{d-1}-1}}_{2^{d-1} \text{ elements}} \Bigg] .$$

Here the vector $\mathbf{w}_d = [w_d(k)]$ has the rank-2 QTT decomposition (i.e., it doubles the QTT ranks),

$$w_d(k) = w_d(\overline{k_1 \ldots k_d}) = \begin{bmatrix} 1 & \omega_d^{k_1} \end{bmatrix} \begin{bmatrix} 1 & \\ & \omega_d^{2k_2} \end{bmatrix} \cdots \begin{bmatrix} 1 & \\ & \omega_d^{2^{d-2}k_{d-1}} \end{bmatrix} \begin{bmatrix} 1 - k_d \\ k_d \end{bmatrix} .$$

As result, the vector $\mathbf{z} = \operatorname{diag}(w_d)\hat{\mathbf{x}} = \mathbf{w}_d \odot \hat{\mathbf{x}}$ has the following QTT decomposition:

$$z(k) = w_d(k)\hat{x}(k) = Z_{k_1}^{(1)} Z_{k_2}^{(2)} \ldots Z_{k_d}^{(d)} , \quad \text{with} \quad Z_{k_1}^{(1)} = \begin{bmatrix} X_{k_1}^{(1)} & \omega_d^{k_1} X_{k_1}^{(1)} \end{bmatrix} ,$$

$$Z_{k_p}^{(p)} = \begin{bmatrix} X_{k_p}^{(p)} & \\ & \omega_{d-p+1}^{k_p} X_{k_p}^{(p)} \end{bmatrix} , \quad p = 2, \ldots, d-1 , \quad Z_{k_d}^{(d)} = \begin{bmatrix} (1 - k_d)\hat{X}_{k_d}^{(d)} \\ k_d \hat{X}_{k_d}^{(d)} \end{bmatrix} .$$

Now the half-size Fourier transform applies to the 'top' and 'bottom' parts of the vector $\mathbf{z}$,

$$x(\overline{j_1 \ldots j_{d-1} j_d}) = X_{j_1}^{(1)} \ldots X_{j_{d-1}}^{(d-1)} X_{j_d}^{(d)} ,$$

$$\begin{aligned} z(j) &= \begin{bmatrix} X_{j_1}^{(1)} & \omega_d^{j_1} X_{j_1}^{(1)} \end{bmatrix} \cdots \begin{bmatrix} X_{j_p}^{(p)} & \\ & \omega_d^{2^{p-1} j_p} X_{j_p}^{(p)} \end{bmatrix} \cdots \begin{bmatrix} (1 - j_d)\hat{X}_{j_d}^{(d)} \\ j_d \hat{X}_{j_d}^{(d)} \end{bmatrix} \\ &\approx \underbrace{\tilde{Z}_{j_1}^{(1)} \ldots \tilde{Z}_{j_{d-1}}^{(d-1)}}_{\text{apply } F_{d-1}} \tilde{Z}_{j_d}^{(d)} . \end{aligned}$$

Here each radix-2 step applies the bit-shift permutation to the result:

$$P_1 P_2 \ldots P_d = R_d ,$$

implying bit-reverse permutation

$$(R_d y)(\overline{j_d j_{d-1} \ldots j_1}) = y(\overline{j_1 \ldots j_d}) .$$

In turn, the bit-reverse permutation R_d can be nicely implemented in the QTT format without any computations just by reversing the order of cores in the tensor train as follows:

$$x(k) = x(\overline{k_1 \ldots k_d}) = X_{k_1}^{(1)} \ldots X_{k_d}^{(d)} , \quad (R_d x)(\overline{k_d \ldots k_1}) = \left(X_{k_d}^{(d)}\right)^{\mathrm{T}} \ldots \left(X_{k_1}^{(1)}\right)^{\mathrm{T}} .$$

Finally, combining exact QTT-FFT recurrence scheme with the rank truncation leads to the approximate algorithm.

Figure 4.3 illustrates the flow chart of the QTT-FFT algorithm (ε-approximation).

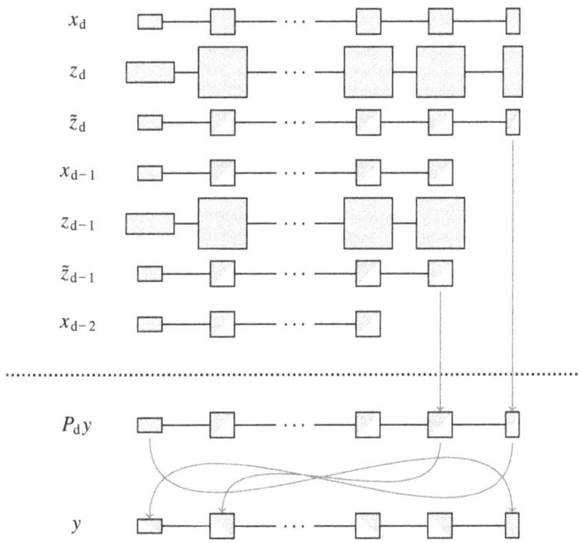

Fig. 4.3: Visualization of the QTT-FFT algorithm (ϵ-approximation).

4.4.7 QTT-FFT versus sparse FFT: numerical comparison

In this section, we demonstrate by numerical examples that combining the exact QTT-FFT with rank truncation leads to enormous data compression even in comparison with the so called sparse FFT. All these numerical experiments have been performed and presented in [85]. The schematic illustration of the data transforms by the QTT-FFT and the sparse FFT [97] is presented in Figure 4.4.

For the comparisons with fast sparse FFT transform the FFTW (Fastest Fourier Transform in the West library) has been used. The numerical results are presented in Table 4.8.

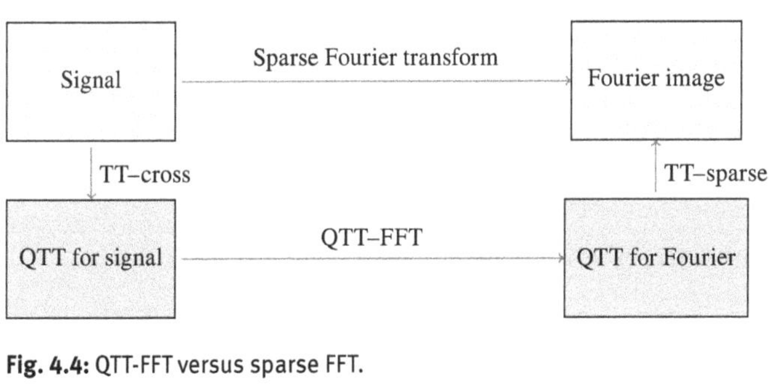

Fig. 4.4: QTT-FFT versus sparse FFT.

Tab. 4.8: Time for QTT-FFT (in milliseconds) versus size $n = 2^d$ and accuracy ε. time_{QTT} is the runtime of QTT-FFT algorithm, $\text{time}_{\text{FFTW}}$ is the runtime of the FFT from the FFTW library, and rank $\hat{f}$ is the effective QTT rank of the Fourier image.

	$f = \Pi(t)$	$\varepsilon = 10^{-4}$		$\varepsilon = 10^{-8}$		$\varepsilon = 10^{-12}$	
d	$\text{time}_{\text{FFTW}}$	rank $\hat{f}$	time_{QTT}	rank $\hat{f}$	time_{QTT}	rank $\hat{f}$	time_{QTT}
16	1.7	4.66	7.9	6.85	13.8	8.85	20.0
18	8.9	4.70	9.7	6.86	16.7	8.82	23.4
20	42.5	4.75	11.3	6.85	19.8	8.86	30.6
22	180	4.77	13.1	6.83	23.3	8.89	36.4
24	810	4.74	15.0	6.72	26.3	8.94	41.7
26	4 100	4.62	17.0	6.76	30.0	8.89	46.5
28	26 300	4.57	18.9	6.80	33.0	8.88	51.2
30	–	4.72	20.3	6.78	36.2	8.84	57.0
40	–	4.20	29.1	6.59	53.6	8.78	83.2
50	–	3.96	39.3	6.45	70.5	8.48	109
60	–	3.69	50.0	6.25	87.6	8.32	133

We consider the *rectangle pulse* function, for which the Fourier transform is known,

$$\Pi(t) = \begin{cases} 0\,, & \text{if } \; |t| > 1/2\,, \\ 1/2\,, & \text{if } \; |t| = 1/2\,, \\ 1\,, & \text{if } \; |t| < 1/2\,, \end{cases} \qquad \hat{\Pi}(\xi) = \operatorname{sinc}(\xi) \overset{\text{def}}{=} \frac{\sin \pi\xi}{\pi\xi}\,.$$

The Fourier integral is approximated by rectangular rule

$$\hat{f}(\xi) = \int_{-\infty}^{+\infty} f(t) \exp(-2\pi i t\xi) dt\,.$$

Here $f(t) = \Pi(t)$ is real and even, hence we write for $k, j = 0, \dots, n-1$, $n = 2^d$,

$$\hat{f}(\xi_j) = 2\,\mathrm{Re} \int_{0}^{+\infty} f(t) \exp(-2\pi i t\xi_j) dt \approx 2\,\mathrm{Re} \sum_{k=0}^{n-1} f(t_k) \exp(-2\pi i t_k \xi_j) h_t\,,$$

$t_k = (k + 1/2)h_t$, $\xi_j = (j + 1/2)h_\xi$, and then use discrete FT for $h_t = h_\xi = \frac{1}{2^{d/2}}$ and d even. The results are presented in Table 4.8.

The QTT representation of the rectangular pulse has QTT ranks one, i.e.,

$$\Pi(t_k) = \Pi\left(\frac{h}{2} + \overline{k_1 \dots k_{d/2-1}}h + \overline{k_{d/2} \dots k_d}/2\right) = (1 - k_{d/2}) \dots (1 - k_d)\,.$$

Consider a sum of p plane waves in m dimensional space,

$$f(k) = \sum_{p=1}^{R} a_p \exp\left(\frac{2\pi i}{n} f_p \cdot k\right), \qquad f_p \in \mathbb{R}^m\,, \quad k = \{0, \dots, n-1\}^m\,, \quad n = 2^d\,.$$

Each plane wave has QTT ranks one, independent of the accuracy and vector size.

The QTT ranks of $f = [f(k)]$ are not larger than R.

A signal consisting of three components with limited bandwidths is given by

$$\hat{f}(\xi) = \exp\left(-\frac{\xi^2}{2\sigma^2}\right) + \frac{1}{10}\exp\left(-\frac{(\xi-\xi_*)^2}{2\sigma^2}\right) + \frac{1}{10}\exp\left(-\frac{(\xi+\xi_*)^2}{2\sigma^2}\right), \qquad (4.40)$$

where $\sigma = 0.02$ and $\xi_* = 0.4$.

Table 4.9 presents a comparison of FFTW and 'TT-ACA + QTT-FFT' for limited bandwidth signal of length $n = 2^d$; see (4.40). 'FFTW' columns: R_* is minimal number of Fourier terms that allow us to detect the sideband, `time` and `nval` are the corresponding runtime (in milliseconds) and the number of samples, and ε is the relative accuracy of the Fourier sparse representation with R_* terms. 'TT-ACA + QTT-FFT' columns: ε is the desired accuracy, R is the maximum effective QTT rank seen in the computation, and `time` and `nval` are the corresponding runtime (in milliseconds) and the number of samples.

Tab. 4.9: Comparison of FFTW and 'TT-ACA + QTT-FFT' for limited bandwidth signal of length $n = 2^d$.

	FFTW			TT-ACA + QTT-FFT					
	$\varepsilon = 10^{-1}$			$\varepsilon = 10^{-2}$			$\varepsilon = 10^{-10}$		
d	R_*	time	nval	R	time	nval	R	time	nval
10	46	< 190	$7.36 \cdot 10^4$	3.00	5.00	$3.59 \cdot 10^3$	7.29	10.9	$4.74 \cdot 10^3$
12	183	$1.2 \cdot 10^3$	$3.29 \cdot 10^5$	2.78	6.72	$2.61 \cdot 10^3$	7.30	19.8	$4.92 \cdot 10^3$
14	710	$6.4 \cdot 10^3$	$1.42 \cdot 10^6$	2.58	9.00	$3.60 \cdot 10^3$	7.13	28.1	$5.00 \cdot 10^3$
16	2800	$8.4 \cdot 10^4$	$6.16 \cdot 10^6$	2.42	9.81	$3.83 \cdot 10^3$	6.78	36.1	$6.96 \cdot 10^3$
18	11 100	$4.2 \cdot 10^5$	$2.67 \cdot 10^7$	2.30	11.2	$3.43 \cdot 10^3$	6.44	43.8	$5.52 \cdot 10^3$
20	–	–	–	2.19	12.9	$1.56 \cdot 10^4$	6.15	49.5	$5.88 \cdot 10^3$

5 Tensor approach to multidimensional integrodifferential equations

The novel tensor numerical methods for the solution of multidimensional PDEs discretized on large tensor grids can be viewed as a bridging of the nonlinear approximation theory for multivariate functions and operators with the traditional and recently developed multilinear algebra. The construction of the particular discretization schemes and the iterative solution methods oriented on the rank structured data naturally requires the classical FEM/FDM techniques on the analysis, approximation, and error control for numerical solution of PDEs. We refer, for example, to [35, 53, 54, 65, 103, 117, 136, 229, 270, 282, 323, 335], and [11, 169, 258, 264, 316, 344, 372] among others.

In this chapter we present several examples of successful applications of tensor numerical methods in the solution of PDEs in many dimensions.

5.1 Tensor approximation of multivariate convolution

5.1.1 Problem setting

The multidimensional convolution arises in a wide range of mathematical models, which include multivariate correlation functions, Green's functions of an elliptic operator, or some other translation invariant integral transforms (filtering). As examples in scientific computing, we mention many-particle modeling based on the Hartree–Fock, Kohn–Sham, and Boltzmann equations as well as the Lippmann–Schwinger formulation of the Schrödinger equation. Further applications appear in image/signal processing, population modeling, and financial mathematics.

The multidimensional convolution in $L^2(\mathbb{R}^d)$ is defined by the integral transform

$$w(x) := (f \star g)(x) := \int_{\mathbb{R}^d} f(y)g(x-y)dy \quad f, g \in L^2(\mathbb{R}^d), \quad x \in \mathbb{R}^d . \tag{5.1}$$

We are interested in approximate computation of $f \star g$ in some fixed box $\Omega = [-A, A]^d$, assuming that the convolving function f has a support in $\Omega' := [-B, B]^d \subset \Omega$ $(B < A)$, i.e., $\operatorname{supp} f \subset \Omega'$. In electronic structure calculations the convolving function f may represent electron orbitals or electron densities, which normally have an exponential decay in $\mathbb{R}^3$.

The common example of the convolving kernel g is given by the restriction of the fundamental solution of an elliptic operator in $\mathbb{R}^d$. For example, in the case of the Laplacian in $\mathbb{R}^d$, $d \geq 3$, we have

$$g(x) = c(d)/\|x\|^{d-2}, \quad x = (x_1, \dots, x_d) \in \mathbb{R}^d, \quad \|x\| = \sqrt{x_1^2 + \dots + x_d^2},$$

where $c(d) = -2^{\frac{d}{4-d}}/\Gamma(d/2-1)$. This example will be considered in more detail.

https://doi.org/10.1515/9783110365917-005

Our presentation follows papers [201, 212]. Note that [212] presents the first fast $O(N \log N)$ tensor based convolution algorithm supported by numerical tests for 3D Newton potentials discretized on a $N \times N \times N$ grid. The cross approximation approach for low rank computations of the convolution transform was discussed in [303]. Note that in some cases the quasi-Monte Carlo algorithms can be applied for calculation of multidimensional integrals [329]. The QTT approximation for a class of integral operators in complex geometries was considered in [75].

5.1.2 Discretization of translation invariant integral operators

There are three commonly used discretization methods for the integral operators: the so called Nyström, collocation, and Galerkin type schemes. For the sake of simplicity, first we consider the case of uniform grids.

Introduce the equidistant tensor product lattice $\omega_{\mathbf{d}} := \omega_1 \times \cdots \times \omega_d$ of size $h = 2A/n$ by setting $\omega_\ell := \{-A + (k-1)h \colon k = 1, \ldots, n+1\}$, where for the sake of convenience $n = 2p$, $p \in \mathbb{N}$, and define the tensor product index set $\mathcal{I} := \{1, \ldots, n\}^d$. Hence $\Omega = \cup_{\mathbf{i} \in \mathcal{I}} \Omega_{\mathbf{i}}$ becomes the union of closed boxes $\Omega_{\mathbf{i}} = \bigotimes_{\ell=1}^{d} \Omega_{i_\ell}$ with intervals

$$\Omega_{i_\ell} := \{x_\ell : x_\ell \in [-A + (i_\ell - 1)h, -A + i_\ell h]\} \subset \mathbb{R}, \quad (\ell = 1, \ldots, d). \tag{5.2}$$

The *Nyström type scheme* leads to simple discretization

$$(f \star g)(x_{\mathbf{j}}) \approx h^d \sum_{\mathbf{i} \in \mathcal{I}} f(y_{\mathbf{i}}) g(x_{\mathbf{j}} - y_{\mathbf{i}}), \quad \mathbf{j} \in \mathcal{I},$$

where, for the ease of presentation, the evaluation points $x_{\mathbf{j}}$ and the collocation points $y_{\mathbf{i}}$, $\mathbf{i}, \mathbf{j} \in \mathcal{I}$, are assumed to be located on the same cell-centered tensor product grid corresponding to $\omega_{\mathbf{d}}$. The Nyström type scheme applies to the continuous functions f, g, which leads to certain limitations in the case of singular kernels g.

The *collocation-projection discretization* can be applied to a much more general class of integral operators than the Nyström methods including Green's kernels with the diagonal singularity, say to the Newton potential $g(x) = 1/\|x\|$. We consider the case of tensor product piecewise constant basis functions $\{\phi_{\mathbf{i}}\}$ associated with $\omega_{\mathbf{d}}$, so that $\phi_{\mathbf{i}} = \chi_{\Omega_{\mathbf{i}}}$ is the characteristic function of $\Omega_{\mathbf{i}}$,

$$\phi_{\mathbf{i}}(x) = \prod_{\ell=1}^{d} \phi_{i_\ell}(x_\ell), \quad \text{where} \quad \phi_{i_\ell} = \chi_{\Omega_{i_\ell}}. \tag{5.3}$$

Let $x_{\mathbf{m}} \in \omega_{\mathbf{d}}$ be the set of collocation points with $\mathbf{m} \in \mathcal{M}_n := \{1, \ldots, n+1\}^d$ (we use the notation $\mathcal{M}_n = \mathcal{M}$ if there is no confusion), and let $f_{\mathbf{i}}$ be the representation coefficients of f in $\{\phi_{\mathbf{i}}\}$,

$$f(y) \approx \tilde{f}(y) := \sum_{\mathbf{i} \in \mathcal{I}} f_{\mathbf{i}} \phi_{\mathbf{i}}(y).$$

In the following we specify the coefficients as $f_{\mathbf{i}} = f(y_{\mathbf{i}})$, where $y_{\mathbf{i}}$ is the midpoint of $\Omega_{\mathbf{i}}$, $\mathbf{i} \in \mathcal{I}$. We consider the following *discrete collocation-projection scheme*:

$$f \star g \approx \{w_{\mathbf{m}}\}, \quad w_{\mathbf{m}} := \sum_{\mathbf{i} \in \mathcal{I}} f_{\mathbf{i}} \int_{\mathbb{R}^d} \phi_{\mathbf{i}}(y) g(x_{\mathbf{m}} - y)\, dy, \quad x_{\mathbf{m}} \in \omega_{\mathbf{d}}, \quad \mathbf{m} \in \mathcal{M}. \tag{5.4}$$

Pointwise evaluation of this scheme requires $O(n^{2d})$ operations. In the case of equidistant grids the computational complexity can be reduced to $O(n^d \log n)$ by applying the multidimensional FFT.

To transform the collocation scheme (5.4) to the discrete convolution, we precompute the collocation coefficients

$$g_{\mathbf{i}} = \int_{\mathbb{R}^d} \phi_{\mathbf{i}}(y) g(-y) dy \,, \quad \mathbf{i} \in \mathcal{I} \,, \tag{5.5}$$

define the dth order tensors $\mathbf{F} = \{f_{\mathbf{i}}\}$, $\mathbf{G} = \{g_{\mathbf{i}}\} \in \mathbb{R}^{\mathcal{I}}$, and introduce the d dimensional *discrete convolution*

$$\mathbf{F} \star \mathbf{G} := \{z_{\mathbf{j}}\} \,, \quad z_{\mathbf{j}} := \sum_{\mathbf{i}} f_{\mathbf{i}} g_{\mathbf{j}-\mathbf{i}+\mathbf{1}} \,, \quad \mathbf{j} \in \mathcal{J} := \{1, \ldots, 2n-1\}^d \,, \tag{5.6}$$

where the sum is over all $\mathbf{i} \in \mathcal{I}$, which leads to legal subscripts for $g_{\mathbf{j}-\mathbf{i}+\mathbf{1}}$, $\mathbf{j} - \mathbf{i} + \mathbf{1} \in \mathcal{I}$. Specifically, for $j_\ell = 1, \ldots, 2n-1$,

$$i_\ell \in [\max(1, j_\ell + 1 - n), \min(j_\ell, n)] \,, \quad \ell = 1, \ldots, d \,.$$

The discrete convolution can be gainfully applied to fast calculation of $\{w_{\mathbf{m}}\}_{\mathbf{m} \in \mathcal{M}}$ in the collocation scheme (5.4) as shown in the following statement:

Proposition 5.1 ([201]). *The discrete collocation scheme* $\{w_{\mathbf{m}}\}$, $\mathbf{m} \in \mathcal{M}$, *is obtained by copying the corresponding portion of* $\{z_{\mathbf{j}}\}$ *from* (5.6), *centered at* $\mathbf{j} = \mathbf{n} = n^{\otimes d}$,

$$\{w_{\mathbf{m}}\} = \{z_{\mathbf{j}}\}_{|\mathbf{j}=\mathbf{j}_0+\mathbf{m}} \,, \quad \mathbf{m} \in \mathcal{M} \,, \quad \mathbf{j}_0 = \mathbf{n}/2 \,.$$

Proof. In the 1D case we have

$$z(1) = f(1) \cdot g(1) \,, \quad z(2) = f(1) \cdot g(2) + f(2) \cdot g(1), \ldots \,,$$
$$z(n) = f(1) \cdot g(n) + f(2) \cdot g(n-1) + \cdots + f(n) \cdot g(1), \ldots, z(2n-1) = f(n) \cdot g(n) \,.$$

Then we find that elements $\{w_m\}$ coincide with $\{z_j\}_{|j=j_0+m}$, $m \in \mathcal{M}$, $j_0 = n/2$. The general case $d \geq 1$ can be justified by applying the above argument to each spatial variable. □

The *Galerkin method* of discretization reads as follows:

$$f \star g \approx \sum_{\mathbf{i}, \mathbf{j}-\mathbf{i}+\mathbf{1} \in \mathcal{I}, \, \mathbf{j} \in \mathbf{j}_0 + \mathcal{M}} f_{\mathbf{i}} g_{\mathbf{j}-\mathbf{i}+\mathbf{1}} \quad \text{with} \quad g_{\mathbf{j}-\mathbf{i}+\mathbf{1}} := \int_{\mathbb{R}^d} \phi_{\mathbf{j}}(x) \phi_{\mathbf{i}}(y) g(x-y) dx dy$$

and with the choice $f_{\mathbf{i}} = \langle f, \phi_{\mathbf{i}} \rangle_{L^2}$. The Galerkin scheme is known as the most convenient for theoretical error analysis. However, compared with the collocation method, it has higher implementation cost because of the presence of double integration. Hence classical discretization methods mentioned above may differ from each other by construction of the tensor product decompositions. To keep a reasonable compromise between the numerical complexity of the scheme and its generality, in the following we focus on the collocation method by simple low order finite elements.

5.1.3 $O(h^2)$ and $O(h^3)$ error bounds

In the case of piecewise constant basis functions we prove the error bound $O(h^2)$ for the collocation scheme and then present a more refined error analysis, which justifies the Richardson extrapolation method on a sequence of grids providing the better approximation error $O(h^3)$. Such an extrapolation, when available, allows a substantial reduction of the approximation error without extra cost. It is worth noting that the Richardson extrapolation can also be applied to some functionals of the convolution product, say to eigenvalues of the operator that include the discrete convolution.

We use the multivariate Taylor expansion to find a local polynomial approximation of order m for a function with certain smoothness. Let us suppose that $f \in C^m(\mathbb{R}^d)$. The *Taylor polynomial* of order m evaluated at y is given by

$$T_y^m f(x) := \sum_{|\alpha|<m} \frac{1}{\alpha!} D^\alpha f(y)(x-y)^\alpha , \quad x, y \in \mathbb{R}^d ,$$

where $\alpha = (\alpha_1, \dots, \alpha_d)$ is an d-tuple of nonnegative integers, $x^\alpha = \prod_{\ell=1}^d x_\ell^{\alpha_\ell}$, $\alpha! = \prod_{\ell=1}^d \alpha_\ell !$, and $|\alpha| = \sum_{\ell=1}^d \alpha_\ell$. We restrict to the case of m-times continuously differentiable functions. For a given hypercube $B \in \mathbb{R}^d$ of size H, let $f \in C^m(B)$. We apply the Taylor expansion at the point $y \in B$ in the form

$$f(x) = T_y^m f(x) + R_y^{(m)}(x) , \quad x \in B \tag{5.7}$$

with

$$R_y^{(m)}(x) := m \sum_{|\alpha|=m} (x-y)^\alpha \int_0^1 \frac{1}{\alpha!} s^{m-1} D^\alpha f(x + s(y-x)) ds .$$

In the following we need the standard error estimate

$$\|f(x) - T_y^m f(x)\|_{L^\infty(B)} \le C_{m,d} H^m \|f\|_{C^m(B)} . \tag{5.8}$$

We recall that continuous Fourier transform in $\mathbb{R}^d$ is given by

$$\mathcal{F}(f)(\kappa) := \int_{\mathbb{R}^d} f(x) e^{-i\langle \kappa, x\rangle} dx , \quad \kappa \in \mathbb{R}^d .$$

Theorem 5.2 ([201]). *Let $f \in C^2(\Omega)$ and let $g \in L^1(\Omega)$. Furthermore, we assume that there exist $\mu \ge 1$ and $\beta > 0$, such that*

$$|\mathcal{F}(g)(\kappa)| \le C/\|\kappa\|^\mu \quad \text{as } \|\kappa\| \to \infty, \quad \kappa \in \mathbb{R}^d \tag{5.9}$$

and

$$|\nabla_y g(x-y)| \le C/\|x-y\|^\beta \quad \text{for } x, y \in \Omega, \quad x \ne y . \tag{5.10}$$

Then there is a constant $C > 0$ independent of h such that for w defined in (5.1), *and for $w_{\mathbf{m}}$ defined in* (5.4), *we have*

$$|w(x_{\mathbf{m}}) - w_{\mathbf{m}}| \le C h^2 , \quad \mathbf{m} \in \mathcal{M} . \tag{5.11}$$

Proof. Introduce the 'local' interpolation error by

$$\Delta_{\mathbf{i}}(y) = (f(y) - f(y_{\mathbf{i}}))\phi_{\mathbf{i}}(y)\,, \quad y \in \Omega \quad \text{with } \operatorname{supp}(\Delta_{\mathbf{i}}) = \Omega_{\mathbf{i}}\,.$$

Define the error function as

$$E(x) := w(x) - \widetilde{f} \star g(x) = \sum_{\mathbf{i}\in\mathcal{I}} \Delta_{\mathbf{i}} \star g(x) \quad \text{with } \widetilde{f} = \sum_{\mathbf{i}\in\mathcal{I}} f(y_{\mathbf{i}})\phi_{\mathbf{i}}\,.$$

For any fixed $\mathbf{i} \in \mathcal{I}$, we will estimate the individual term of the total error, $E_{\mathbf{i}}(x) = \Delta_{\mathbf{i}} \star g(x)$. To that end let us apply the Taylor expansion (5.7) on $B = \Omega_{\mathbf{i}}$ with $m = 2$ to obtain

$$\Delta_{\mathbf{i}}(y) = \langle \nabla f(y_{\mathbf{i}}), y - y_{\mathbf{i}} \rangle + R^{(2)}_{y_{\mathbf{i}}}(y)\,, \quad y \in B\,.$$

Step 1. It is easy to see that (5.8) implies

$$\|R^{(2)}_{y_{\mathbf{i}}}(\cdot)\|_{L^\infty(B)} \le Ch^2\,,$$

hence the condition $g \in L^1(\Omega)$ leads to

$$\left\| \sum_{\mathbf{i}\in\mathcal{I}} R^{(2)}_{y_{\mathbf{i}}} \star g \right\|_{L^\infty(\Omega)} \le Ch^2 \|g\|_{L^1(\Omega)} = O(h^2)\,. \tag{5.12}$$

Next we analyze the remaining part of $E(x)$ at some fixed collocation point $x_{\mathbf{m}}$, $\mathbf{m} \in \mathcal{M}$.

Step 2. Let us consider the contribution to the error from the individual terms $\langle \nabla f(y_{\mathbf{i}}), \cdot - y_{\mathbf{i}} \rangle \star g(\cdot)$ for all

$$\mathbf{i} \in \Sigma_{\mathbf{m}} := \{\mathbf{j} \in \mathcal{I} : x_{\mathbf{m}} \in \Omega_{\mathbf{j}}\}\,.$$

To that end we estimate the Fourier transform of such terms,

$$\mathcal{F}\left(\langle \nabla f(y_{\mathbf{i}}), \cdot - y_{\mathbf{i}} \rangle \star g(\cdot)\right) = \mathcal{F}\left(\langle \nabla f(y_{\mathbf{i}}), \cdot - y_{\mathbf{i}} \rangle\right) \cdot \mathcal{F}(g)\,, \tag{5.13}$$

where $\mathcal{F}(g)$ is understood as a temporary distribution. Since $g \in L^1(\Omega)$, we have

$$\|\mathcal{F}(g)\|_{L^\infty(\mathbb{R}^d)} \le C\|g\|_{L^1}\,.$$

Furthermore, we will need a 'directional' estimate on $|\mathcal{F}(g)|$. At this point we apply the classical inequality of the harmonic and geometric mean: let $a_1, \dots, a_d$ be the positive real numbers, then

$$\frac{d}{\frac{1}{a_1} + \cdots + \frac{1}{a_d}} \le \sqrt[d]{a_1 a_2 \dots a_d}\,.$$

Let us set $a_k = 1/x_k^2$ for $x \in \mathbb{R}^d$, which leads to

$$\frac{1}{\|x\|} = \frac{1}{\sqrt{x_1^2 + \cdots + x_d^2}} \le \frac{1}{\sqrt{d}} \prod_{\ell=1}^{d} \frac{1}{\sqrt[d]{|x_\ell|}}\,.$$

Hence, the assumption on the decay property (5.9) implies the desired 'directional' bound

$$|\mathcal{F}(g)(\kappa)| \leq \frac{C}{\|\kappa\|^{\mu}} \leq \frac{C}{\sqrt{d^{\mu}}} \prod_{\ell=1}^{d} \frac{1}{|\kappa_\ell|^{\mu/d}} . \tag{5.14}$$

Furthermore, for the first factor in the right hand side of (5.13) we are able to prove

$$|\mathcal{F}(\langle \nabla f(y_{\mathbf{i}}), \cdot - y_{\mathbf{i}} \rangle)| \leq C h^{d+2} P_{\mathbf{i}}, \quad P_{\mathbf{i}} > 0 \tag{5.15}$$

with the uniformly bounded sum $\sum_{\mathbf{i}\in\mathcal{J}} P_{\mathbf{i}} \leq C$. In fact, due to separability of $\mathcal{F}$ in $\mathbb{R}^d$ with respect to the one dimensional Fourier transforms $\mathcal{F}_k$ in variable y_k ($k = 1, \ldots, d$), one can represent

$$\mathcal{F}(\langle \nabla f(y_{\mathbf{i}}), y - y_{\mathbf{i}} \rangle) = \langle \nabla f(y_{\mathbf{i}}), U_{\mathbf{i}} \rangle , \quad U_{\mathbf{i}} \in (L^{\infty}(\mathbb{R}))^d$$

with

$$(U_{\mathbf{i}})_k(\kappa) = \mathcal{F}_k(\chi_{\Omega_{i_k}}(\cdot - y_{\mathbf{i},k}))(\kappa_k) \prod_{\ell=1,\ \ell\neq k}^{d} \mathcal{F}_\ell(\chi_{\Omega_{i_\ell}})(\kappa_\ell) .$$

For each fixed $k = 1, \ldots, d$, consider the individual term

$$\mathcal{F}_k(\chi_{\Omega_{i_k}}(t - y_{\mathbf{i},k}))(\tau) = \mathrm{e}^{-i y_{\mathbf{i},k}\tau} \mathcal{F}(\chi_{[-h/2,h/2]} t)(\tau)$$

with $\mathcal{F}$ being the Fourier transform in $\mathbb{R}$, and derive

$$\begin{aligned} e_h(\tau) := \mathcal{F}(\chi_{[-h/2,h/2]} t)(\tau) &= \left[\frac{\sin(\tau y)}{\tau^2} - \frac{y\cos(\tau y)}{\tau} \right]_{-h/2}^{h/2} \\ &= \frac{2\sin(\tau h/2)}{\tau^2} - \frac{h\cos(\tau h/2)}{\tau} . \end{aligned}$$

Hence we have the asymptotic expansions

$$e_h(\tau) = \frac{\tau h^3}{12} + O(h^5) \quad \text{as } |\tau h| \leq O(1)$$

and

$$|e_h(\tau)| \leq C\left(\frac{h}{\tau} + \frac{1}{\tau^2}\right) \quad \text{as } |\tau| \to \infty .$$

We apply (5.14) with $\mu > 0$, take the directional factor

$$g_1(\tau) = \min\{1, 1/\tau^{\mu/d}\} ,$$

and then consider the parametric function

$$p_h(\tau) := g_1(\tau)\left(\frac{2\sin(\tau h/2)}{\tau^2} - \frac{h\cos(\tau h/2)}{\tau}\right) .$$

We can prove by the scaling argument that

$$|p_h(\tau)| = C h^{2+\mu/d} P(u) , \quad \text{with } C = C(\mu, d) ,$$

where, with $u = \tau h/2$,

$$P(u) = \left[\frac{\sin u}{u^2} - \frac{\cos u}{u}\right] \min\{h^{-\mu/d}, u^{-\mu/d}\} \in L^1(\mathbb{R}) .$$

The standard scaling argument leads to the relation

$$\|p_h(\tau)\|_{L^1} \le C h^{1+\mu/d} \|P(u)\|_{L^1} . \tag{5.16}$$

Likewise, we have

$$\mathcal{F}(\chi_{[-h/2,h/2]})(\tau) = \left[\frac{\sin(\tau y)}{\tau}\right]_{-h/2}^{h/2} = \frac{2\sin(\tau h/2)}{\tau} ,$$

$$q_h(\tau) := g_1(\tau)\mathcal{F}(\chi_{[-h/2,h/2]})(\tau) = C h^{1+\mu/d} \operatorname{sinc}(u) \min\{h^{-\mu/d}, u^{-\mu/d}\} , \tag{5.17}$$

which implies

$$\|q_h(\tau)\|_{L^1} \le C h^{\mu/d} \| \operatorname{sinc}(u) \min\{h^{-\mu/d}, u^{-\mu/d}\}\|_{L^1} .$$

With fixed index $\mathbf{i} \in \Sigma_{\mathbf{m}}$, we apply the inverse Fourier transform $\mathcal{F}^{-1}$ to (5.13), then make use of the bounds (5.14), (5.16), and (5.17) to obtain

$$\begin{aligned} |\langle \nabla f(y_{\mathbf{i}}), \cdot - y_{\mathbf{i}}\rangle \star g(\cdot)| &\le \|\mathcal{F}(\langle \nabla f(y_{\mathbf{i}}), \cdot - y_{\mathbf{i}}\rangle) \cdot \mathcal{F}(g)\|_{L^1} \\ &\le \| \, |\mathcal{F}(\langle \nabla f(y_{\mathbf{i}}), \cdot - y_{\mathbf{i}}\rangle)| \cdot |\mathcal{F}(g)| \|_{L^1} \\ &\le C d \|p_h(\tau)\|_{L^1} \prod_{\ell=2,\dots,d} \|q_h(\tau_\ell)\|_{L^1} \\ &\le C d h^{1+\mu/d} \prod_{\ell=2,\dots,d} h^{\mu/d} = C d h^{1+\mu} . \end{aligned} \tag{5.18}$$

Summing over $\mathbf{i} \in \Sigma_{\mathbf{m}}$ leads to the desired 'local' estimate of order $Cd\, 2^d\, h^{1+\mu}$.

Step 3. In the final step, we estimate the contribution from 'nondiagonal' terms corresponding to $\mathbf{i} \in \mathcal{I} \setminus \Sigma_{\mathbf{m}}$. For such terms we just apply the Taylor expansion around $y_{\mathbf{i}}$ with $m = 2$ to the convolving kernel $g(x_{\mathbf{m}} - y)$, $y \in B = \Omega_{\mathbf{i}}$, and take into account (5.10), which leads to the bound (with $n = 1/h$, $\mathbf{n} = n^{\otimes d}$ and $\beta \ne d$)

$$\begin{aligned} &\left| \sum_{\mathbf{i} \in \mathcal{I}\setminus\Sigma_{\mathbf{m}}} (\langle \nabla f(y_{\mathbf{i}}), \cdot - y_{\mathbf{i}}\rangle \star g)(x_{\mathbf{m}}) \right| \\ &= \left| \sum_{\mathbf{i} \in \mathcal{I}\setminus\Sigma_{\mathbf{m}}} \int_{\Omega_{\mathbf{i}}} \langle \nabla f(y_{\mathbf{i}}), y - y_{\mathbf{i}}\rangle g(x_{\mathbf{m}} - y) dy \right| \\ &= \left| \sum_{\mathbf{i} \in \mathcal{I}\setminus\Sigma_{\mathbf{m}}} \int_{\Omega_{\mathbf{i}}} \langle \nabla f(y_{\mathbf{i}}), y - y_{\mathbf{i}}\rangle \left(\langle \nabla_y g(x_{\mathbf{m}} - y_{\mathbf{i}}), y - y_{\mathbf{i}}\rangle + R^{(2)}_{y_{\mathbf{i}}}(y) \right) dy \right| \\ &\le \sum_{\mathbf{i} \in \mathcal{I}\setminus\Sigma_{\mathbf{m}}} \int_{\Omega_{\mathbf{i}}} |\langle \nabla f(y_{\mathbf{i}}), y - y_{\mathbf{i}}\rangle| \cdot |y - y_{\mathbf{i}}| / \|x_{\mathbf{m}} - y_{\mathbf{i}}\|^{\beta} dy + O(h^3) \end{aligned}$$

$$= Cd \sum_{\mathbf{k}=1}^{\mathbf{n}} \frac{h^{d+2}}{|\mathbf{k}h|^{\beta}} = C\frac{h^{2+d}}{h^{\beta}} \sum_{\mathbf{k}=1}^{\mathbf{n}} \frac{1}{|\mathbf{k}|^{\beta}}$$
$$\leq C\frac{h^{2+d}}{h^{\beta}} \cdot h^{\beta-d} = Ch^2 .$$

Combining this result with (5.12) completes the proof. □

Theorem 5.2 indicates the 'superconvergence property' for low order elements in the case of smooth enough convolving functions. To illustrate the applicability of Theorem 5.2 we note that the fundamental solution of the Laplace operator in $\mathbb{R}^d$ is given by $g(x) = c(d)/\|x\|^{d-2}$ with the Fourier transform $\mathcal{F}(g) = C/\|\kappa\|^2$. Hence Theorem 5.2 applies with $\beta = d-1$, $\mu = 2$. It also applies to the Yukawa potential

$$g(x) = \mathrm{e}^{-\lambda\|x\|}/\|x\| \quad \text{for} x \in \mathbb{R}^3$$

with any $\mu \geq 1$ and with $\beta = 2$.

The approximation error $O(h^2)$ can be improved up to $O(h^3)$ using the Richardson extrapolation scheme on a sequence of grids. We show that the linear combination of solutions $w_{\mathbf{m}}^{(n)}$, $\mathbf{m} \in \mathcal{M}_n$, and $w_{\mathbf{m}}^{(2n)}$, $\mathbf{m} \in \mathcal{M}_{2n}$, corresponding to the grid sizes n and $2n$, respectively, ensures the expected high order approximation.

Theorem 5.3 ([201]). *Let $f \in C^3(\Omega)$, and assume that the conditions of Theorem 5.2 are satisfied with $\mu \geq 2$ and $\beta \neq d$ (technical condition). Moreover, suppose that*

$$|\nabla_y^2 g(x-y)| \leq C/\|x-y\|^{\gamma} \quad \textit{with} \quad \gamma > 0 . \tag{5.19}$$

Then for $\mathbf{m} \in \mathcal{M}_n$, there is a constant $C > 0$ independent of h such that

$$(4w_{\mathbf{m}}^{(2n)} - w_{\mathbf{m}}^{(n)})/3 = w(x_{\mathbf{m}}) + \eta_{\mathbf{m},n} , \quad \eta_{\mathbf{m},n} \in \mathbb{R} \quad \textit{with} \ |\eta_{\mathbf{m},n}| \leq Ch^3 . \tag{5.20}$$

Remark 5.4. The Newton potential in 3D, $g(x) = 1/\|x\|$, $x \in \mathbb{R}^3$, satisfies the conditions of Theorem 5.3 with $\mu = 2$ and $\beta = 2$.

Note that in the case $\beta = d$ some logarithmic terms in the error estimate may arise.

Below we give numerical examples for the Newton kernel with $d = 3$.

5.1.4 Rank structured tensor approximation to discrete convolution

We note that the multidimensional convolution product appears to be one of the most computationally elaborate MLA operations. In the present approach, the key idea is to calculate the d dimensional convolution approximately using rank structured tensor approximations. Recall that for given dth order tensors $\mathbf{F}, \mathbf{G} \in \mathcal{T}_{\mathbf{r}}$, represented by

$$\mathbf{F} = \boldsymbol{\beta} \times_1 F^{(1)} \times_2 F^{(2)} \cdots \times_d F^{(d)} , \quad \text{and} \quad \mathbf{G} = \boldsymbol{\gamma} \times_1 G^{(1)} \times_2 G^{(2)} \cdots \times_d G^{(d)} ,$$

the convolution product can be 'separated' via ([211])

$$\mathbf{F} \star \mathbf{G} := \sum_{\mathbf{k}=1}^{\mathbf{r}} \sum_{\mathbf{m}=1}^{\mathbf{r}} \beta_{k_1 \ldots k_d} \gamma_{m_1 \ldots m_d} \left(\mathbf{f}_1^{k_1} \star \mathbf{g}_1^{m_1} \right) \otimes \cdots \otimes \left(\mathbf{f}_d^{k_d} \star \mathbf{g}_d^{m_d} \right) . \tag{5.21}$$

Computing 1D convolution $\mathbf{f}_\ell^{k_\ell} \star \mathbf{g}_\ell^{m_\ell} \in \mathbb{R}^{2n-1}$ in $O(n \log n)$ operations leads to the overall linear-logarithmic complexity in n,

$$\mathcal{N}_{T \star T} = O(dr^2 n \log n + \#\boldsymbol{\beta} \cdot \#\boldsymbol{\gamma}) .$$

In general, one might have $\#\boldsymbol{\beta} \cdot \#\boldsymbol{\gamma} = O(r^{2d})$, which may be restrictive even for moderate d.

Significant complexity reduction is observed if at least one of the convolving tensors can be represented by the canonical model. Letting $\mathbf{F} \in \mathcal{T}_{\mathbf{r}}$, $\mathbf{G} \in \mathcal{C}_R$, i.e., $\boldsymbol{\gamma} = \operatorname{diag}\{\gamma_1, \ldots, \gamma_R\}$, we tensorize the convolution product as follows:

$$\mathbf{F} \star \mathbf{G} = \sum_{\mathbf{k}=1}^{\mathbf{r}} \sum_{m=1}^{R} \beta_{k_1 \ldots k_d} \gamma_m \left(\mathbf{f}_1^{k_1} \star \mathbf{g}_1^{m} \right) \otimes \cdots \otimes \left(\mathbf{f}_d^{k_d} \star \mathbf{g}_d^{m} \right) . \tag{5.22}$$

However, the calculation by (5.22) still scales exponentially in d, which leads to certain limitations in the case of higher dimensions.

To get rid of this exponential scaling, we propose to perform the convolution transform using the two-level tensor format, i.e., $\mathbf{F} \in \mathcal{T}_{\mathcal{C}_{R_1,\mathbf{r}}}$ (Sections 3.2 and 3.4) in such a way that the result $\mathbf{U} = \mathbf{F} \star \mathbf{G}$ with $\mathbf{G} \in \mathcal{C}_{R_G}$ is represented in the two-level Tucker format $\mathcal{T}_{\mathcal{C}_{R_1 R_G, \mathbf{r} R_G}}$. Recall that an explicit representation for $\mathbf{F} \in \mathcal{T}_{\mathcal{C}_{R_1,\mathbf{r}}}$ is given by

$$\mathbf{F} = \left(\sum_{\nu=1}^{R_1} \beta_\nu \, \mathbf{z}_1^\nu \otimes \cdots \otimes \mathbf{z}_d^\nu \right) \times_1 F^{(1)} \times_2 F^{(2)} \cdots \times_d F^{(d)} , \tag{5.23}$$

so that we have the embedding $\mathcal{T}_{\mathcal{C}_{R_1,\mathbf{r}}} \subset \mathcal{C}_{R_1,\mathbf{n}}$ with the corresponding (nonorthogonal) side matrices $S^{(\ell)} = [F^{(\ell)} z_\ell^1 \ldots F^{(\ell)} z_\ell^{R_1}] \in \mathbb{R}^{n \times R_1}$, and scaling factors β_ν ($\nu = 1, \ldots, R_1$). Now we represent the tensor product convolution in the two-level format

$$\mathbf{F} \star \mathbf{G} = \sum_{m=1}^{R_G} \gamma_m \left(\sum_{\nu=1}^{R_1} \beta_\nu \, \mathbf{z}_1^\nu \otimes \cdots \otimes \mathbf{z}_d^\nu \right) \times_1 \left(F^{(1)} \star \mathbf{g}_1^m \right) \times_2 \cdots \times_d \left(F^{(d)} \star \mathbf{g}_d^m \right) , \tag{5.24}$$

such that the above expansion can be evaluated by the following algorithm.

Algorithm 5.1 (d dimensional tensor convolution of type $\mathcal{T}_{\mathcal{C}_{R_1,\mathbf{r}}} \star \mathcal{C}_{R_G,\mathbf{n}} \to \mathcal{T}_{\mathcal{C}_{R_1 R_G, \mathbf{r} R_G}}$).

1. Given $\mathbf{F} \in \mathcal{T}_{\mathcal{C}_{R_1,\mathbf{r}}}$ with the core $\boldsymbol{\beta} = \sum_{\nu=1}^{R_1} \beta_\nu \mathbf{z}_1^\nu \otimes \cdots \otimes \mathbf{z}_d^\nu \in \mathcal{C}_{R_1,\mathbf{r}}$, and $\mathbf{G} \in \mathcal{C}_{R_G,\mathbf{n}}$.
2. For $\ell = 1, \ldots, d$, compute the set of 1D convolutions $\mathbf{u}_\ell^{k,m} = \mathbf{f}_\ell^k \star \mathbf{g}_\ell^m$ ($k = 1, \ldots, r$, $m = 1, \ldots, R_G$) of size $2n - 1$, restrict the results to the index set I_ℓ, and form the $n \times rR_G$ side matrices $U^{(\ell)} = [U_1^{(\ell)} \ldots U_{R_G}^{(\ell)}]$, composed of the blocks $U_m^{(\ell)}$ with columns $\mathbf{u}_\ell^{k,m}$ as $U_m^{(\ell)} = [\mathbf{f}_\ell^1 \star \mathbf{g}_\ell^m \ldots \mathbf{f}_\ell^r \star \mathbf{g}_\ell^m]$, all at the cost $O(drR_G n \log n)$.

3. Build the core tensor $\boldsymbol{\omega} = \operatorname{blockdiag}\{\gamma_1\boldsymbol{\beta}, \dots, \gamma_R\boldsymbol{\beta}\}$ and represent the resultant two-level Tucker tensor in the form (storage demand is $R_G + R_1 + drR_1 + drR_G n$)

$$\mathbf{U} = \boldsymbol{\omega} \times_1 U^{(1)} \times_2 \cdots \times_d U^{(d)} \in \mathcal{T}_{\mathcal{C}_{R_1 R_G, \mathbf{r} R_G}}.$$

In some cases one may require the consequent rank reduction of the target tensor $\mathbf{U}$ to the two-level format $\mathcal{T}_{\mathcal{C}_{R_0,\mathbf{r}_0}}$ with moderate rank parameters R_0 and $\mathbf{r}_0 = (r_0, \dots, r_0)$. This can be accomplished by the heuristic Algorithm 5.1′.

Recall that the higher order SVD (HOSVD, [247]) tensor approximation is defined by truncated SVD of the mode-ℓ unfolding matrices.

Algorithm 5.1′ (Rank reduction for Algorithm 5.1).
1. Given tensor $\mathbf{U}$ defined by Algorithm 5.1, and the rank parameters $r_0, R_0 \in \mathbb{N}$ (suppose that $R_0 \ll R_1 R_G$, $r_0 < r$).
2. For $\ell = 1, \dots, d$, compute the ℓ-mode r_0-dimensional dominating subspace for the matrix $U^{(\ell)}$, specified by the rank-r_0 truncated SVD, given by $Z_0^{(\ell)} D_0^{(\ell)} V_0^{(\ell)^{\mathrm{T}}}$ (cost $O(dnrR_G \min\{n, rR_G\})$).
3. Project the target tensor $\mathbf{U}$ onto orthogonal basis defined by columns of $Z_0^{(\ell)}$ by calculating the core tensor of size $\mathbf{r}_0 = (r_0, \dots, r_0)$ in the product canonical format (the so called reduced HOSVD, or shortly RHOSVD, [212]),

$$\widetilde{\boldsymbol{\beta}}_0 = \sum_{m=1}^{R_G} \gamma_m \left(\sum_{\nu=1}^{R_1} \beta_\nu \bigotimes_{\ell=1}^{d} D_0^{(\ell)} M_{m,0}^{(\ell)^{\mathrm{T}}} \mathbf{z}_\ell^\nu \right) \in \mathcal{C}_{R_1 R_G, \mathbf{r}_0},$$

and represent the RHOSVD approximation in the form

$$\mathbf{U}_{(\mathbf{r}_0)} = \widetilde{\boldsymbol{\beta}}_0 \times_1 Z_0^{(1)} \times_2 \cdots \times_d Z_0^{(d)} \in \mathcal{T}_{\mathcal{C}_{R_1 R_G, \mathbf{r}_0}}.$$

The related cost is $O(dR_1 R_G r r_0)$.
4. Recompress the core $\widetilde{\boldsymbol{\beta}}_0$ to the rank-R_0 canonical tensor $\boldsymbol{\beta}_0$ and constitute the result in the contracted product form

$$\mathbf{W}_0 = \boldsymbol{\beta}_0 \times_1 Z_0^{(1)} \times_2 \cdots \times_d Z_0^{(d)} \in \mathcal{T}_{\mathcal{C}_{R_0,\mathbf{r}_0}}.$$

5. (Optional.) Use tensor $\mathbf{W}_0$ as the initial guess for few nonlinear (say ALS) iterations to approximate the target tensor $\mathbf{U}$ in the $\mathcal{T}_{\mathcal{C}_{R_0,\mathbf{r}_0}}$ format.

Note that the iterative Step 5 in Algorithm 5.1′ is not mandatory. In our applications the approximation W_0 usually provides sufficiently good accuracy (see Theorem 3.42 on the RHOSVD). The justification of Algorithm 5.1′ is based on the effective error control of the RHOSVD for $\mathbf{U}$ (compare to [212, Theorem 2.5] for the case of canonical input tensor).

If $\mathbf{F} \in \mathcal{C}_{R_F}$ with $\boldsymbol{\beta} = \operatorname{diag}\{\beta_1, \dots, \beta_{R_F}\}$, and $\mathbf{G} \in \mathcal{C}_{R_G}$ as above, then

$$\mathbf{F} * \mathbf{G} = \sum_{k=1}^{R_F} \sum_{m=1}^{R_G} \beta_k \gamma_m \left(\mathbf{f}_1^k * \mathbf{g}_1^m\right) \otimes \cdots \otimes \left(\mathbf{f}_d^k * \mathbf{g}_d^m\right), \tag{5.25}$$

leading to the reduced cost that scales linearly in dimensionality parameter d and linear logarithmically in n,

$$\mathcal{N}_{C\star C\to C} = O(dR_F R_G n \log n) .$$

Algorithm 5.2 (Multidimensional tensor product convolution of type $C \star C \to C$).
1. Given $\mathbf{F} \in \mathcal{C}_{R_F,\mathbf{n}}$, $\mathbf{G} \in \mathcal{C}_{R_G,\mathbf{n}}$.
2. For $\ell = 1, \dots, d$, compute the set of 1D convolutions $\mathbf{f}_\ell^k \star \mathbf{g}_\ell^m$ ($k = 1, \dots, R_F$, $m = 1, \dots, R_G$) of size $2n - 1$, restrict the results to the index set I_ℓ, and form the $n \times R_F R_G$ side matrix $U^{(\ell)}$ (cost $dR_F R_G n \log n$).
3. Compute the set of scaling factors $\beta_k \gamma_m$ as in (5.25).

We have proven the following complexity bounds:

Lemma 5.5. *Algorithm 5.1 scales log linearly in n and linearly in d,*

$$\mathcal{N}_{\mathcal{T}_\mathcal{C}\star C\to \mathcal{T}_\mathcal{C}} = O(drR_G n \log n + drR_F + drR_G n) .$$

Algorithm 5.2 exhibits the complexity bound $O(dR_F R_G n \log n)$.

The resultant convolution product $\mathbf{F} \star \mathbf{G}$ in (5.25) may be approximated in either Tucker or canonical formats, depending on further MLA operations applied to this tensor. In the framework of approximate iterations with structured matrices and vectors, we can fix the $\mathcal{C}_{R_0}$ format for the output tensors, hence the rank-R_0 canonical approximation (with $R_0 < R_F R_G$) would be the proper choice to represent $\mathbf{F} \star \mathbf{G}$. The tensor truncation of the rank-$R_F R_G$ intermediate result to rank-R_0 tensor can be accomplished by fast multigrid accelerated tensor approximation at the cost $O(dR_F R_G R_0 n \log n)$ ([212]), and then the result can be stored by $O(dR_0 n)$ reals.

Based on our experience with Algorithms 5.1 and 5.2, applied in electronic structure calculations in 3D, we note that Algorithm 5.2 is preferable in the case of moderate grid size (say, $n \le 10^4$), while Algorithm 5.1 is faster for large grids. For example, Algorithm 5.2 works perfectly in electronic structure calculations by the Hartree–Fock model for $d = 3$ [212, 213]. In particular, the Hartree potential of simple molecules can be calculated on the $n \times n \times n$ grid up to $n \le 1.6 \cdot 10^4$ in a few minutes providing the relative accuracy of about 10^{-7} already with $n = 8192$. Further numerical illustrations will be given in the next Section 5.2.

5.1.5 Tensor product convolution on generic nonuniform grids in $\mathbb{R}^d$

In this section, we present a few remarks concerning the design of multidimensional fast convolution transform (FCT) on nonuniform grids. Again, our key principle is the low rank tensor approximation of the multidimensional convolution transform described above. We stress the following issues:

- As soon as the multidimensional convolution is represented in the tensor product form as in (5.21)–(5.25), the computation is reduced to the fast 1D convolution transforms of ℓ-mode univariate components for $\ell = 1, \dots, d$ on equidistant grids, leading to practically negligible cost $O(n \log n)$ in the large range of a grid size n.
- The 1D convolution on the hierarchically structured refined grids can be effectively computed in almost linear cost as discussed in [134, 135].
- In the case of general tensor product grids with adaptive grid refinement one can apply the *embedding strategy* ([211]) to reduce the computation to 1D FFT on a uniform grid. Specifically, assume that a 1D refined grid of size n is obtained by agglomeration of subintervals of the auxiliary fine grid of size N (usually $n \ll N$, say $n = O(\log N)$), so that there is a natural extension operator $P_{n \to N}$ from adaptive to fine uniform grid. Further, assuming that the convolving tensors $\mathbf{F}$ and $\mathbf{G}$ in the collocation scheme (5.4) are represented in some structured tensor formats as above, the summation over $\mathbf{i}$ and for all $\mathbf{m} \in \mathcal{M}$ can be reduced to the tensor product convolution on the auxiliary uniform grid with the cost $O(dr^2 N \log N + \#\boldsymbol{\beta} \cdot \#\boldsymbol{\gamma})$. Taking the proper subvectors of size N from the corresponding ℓ-mode components given on the grid of size $2N - 1$ ($\ell = 1, \dots, d$) and interpolating the results to the initial 'small' grid, we obtain the approximate convolution on the adaptive grid and in the tensor product form. Detailed discussion of this issue is beyond the scope of our paper.

Numerical examples illustrating the efficiency of the convolution product in the Tucker/canonical formats are given in [211]. These results indicate that 1D FFT on the auxiliary equidistant fine grid has negligible cost compared with the summation in (5.21), (5.22), at least in the parameter domain $N \le 10^4$. Hence, the embedding strategy can be successfully applied in the case of moderate mesh refinement. To reduce the FFT cost $O(N \log N)$ on the auxiliary uniform grid to the linear-logarithmic complexity in n, we will describe a multidimensional FCT on the two-level composite grids (for ease of presentation we discuss the case of piecewise constant basis functions).

5.1.6 $O(n \log n)$ convolution on 1D composite grid

Based on results in [201], let us describe the FCT on a two-level composite grid defined by the coarse level lattice with mesh size $H = 2A/n_0$. We introduce the coarse space $V_{n_0} = x\,\mathrm{span}\{\phi_{\mathbf{i}_0}\}$, $\mathbf{i}_0 \in \mathbb{R}^{n_0^{\otimes d}}$, of piecewise constant basic functions (it is only for the ease of exposition) supported by the domain Ω. Assume that p intervals Ω_i, $i = 1, \dots, p$ with $p \ll n_0$ are further decomposed by fine uniform grid of size $h = H/n$ (5.2). The union of subdomains will be called $\Omega^{(p)} = \cup_{i=1}^{p} \Omega_i \subset \Omega$. We define by V_n the corresponding fine space of piecewise constant basic functions supported by $\Omega^{(p)}$

having zero mean value at each subinterval $\Omega_1, \dots, \Omega_p$, then introduce the composite space $V = V_{n_0} + V_n$. Our goal is the fast evaluation of the convolution product

$$\mathbf{w} = (\mathbf{x}_0 + \mathbf{x}_h) \star (\mathbf{y}_0 + \mathbf{y}_h) \quad \text{with } \mathbf{x}_0, \mathbf{y}_0 \in V_{n_0}, \ \mathbf{x}_h, \mathbf{y}_h \in V_n$$

at the cost $O(n_0 \log n_0 + n \log n)$ assuming that the result is projected to the initial composite space V. The corresponding numerical scheme can be implemented in four steps as follows (for the ease of presentation we further simplify and set $p = 1$).

Algorithm 5.3 (FCT on two-level composite grid).
1. Given $\mathbf{x}_0, \mathbf{y}_0 \in V_{n_0}, \ \mathbf{x}_h, \mathbf{y}_h \in V_n$.
2. Compute $\mathbf{w}_h = \mathbf{x}_h \star \mathbf{y}_h$ and project the result to the coarse and fine spaces. This includes one convolution product of size n whose result will be defined on the union of intervals $\Omega_1 \cup \Omega_2$. The coarse components supported by Ω_1 and Ω_2 will be calculated using the mean values of $\mathbf{w}_{h|\Omega_1}$ and $\mathbf{w}_{h|\Omega_2}$, respectively. Consequently, the fine projection onto V_n has zero mean value. Altogether, this amounts to $O(n \log n)$ operations.
3. Compute $\mathbf{w}_{0h} = \mathbf{x}_0 \star \mathbf{y}_h + \mathbf{x}_h \star \mathbf{y}_0$ and project the result to the coarse and fine spaces. The computational scheme is clear from the representation

   $$\mathbf{x}_h \star \mathbf{y}_0 = \mathbf{x}_h \star \sum_{i=1}^{n_0} a_i \chi_i = \sum_{i=1}^{n_0} a_i \left(\mathbf{x}_h \star \chi_i\right)_{|\Omega_{i-1} \cup \Omega_i \cup \Omega_{i+1}},$$

   where χ_i is the indicator function of the interval Ω_i. This includes one convolution product and three scalar products of size n, plus calculation of the coarse grid projection. The numerical cost is estimated by $O(n \log n + n_0)$.
4. Compute $\mathbf{w}_0 = \mathbf{x}_0 \star \mathbf{y}_0$ and project the result to the coarse and fine spaces. The corresponding computational ansatz

   $$\mathbf{x}_0 \star \mathbf{y}_0 = \left(\sum_{i=1}^{n_0} a_i \chi_i\right) \star \left(\sum_{j=1}^{n_0} b_j \chi_j\right)$$

   is evaluated at the coarse level by FFT of size n_0. To obtain the projection onto V_n we compute the weighted convolution $a_1 b_1 (\chi_1 \star \chi_1)$ for the vectors of size n supported by Ω_1. Hence, the total cost of Step 4 is estimated by $O(n_0 \log n_0 + n \log n)$.
5. Collect the contributions from Steps 1–4 in the coarse and fine spaces, which amounts to $O(n_0 + n)$ operations.

This proves the following result:

Lemma 5.6. *The numerical complexity of Algorithm 5.3 is estimated by* $O(n_0 \log n_0 + n \log n)$.

Algorithm 5.3 applies to two-level composite grids. However, it can be easily extended in a recursive manner to the case of multilevel composite grids.

Tab. 5.1: CPU times (s) for FFT on full grids and vs. FCT on the two-level composite grid.

ℓ_0	4	5	6	7	8	9	9	9	9	9	9
ℓ_n	8	8	8	8	8	8	7	6	5	4	3
FFT	0.28	0.77	2.7	10.8	45.9	40.1	45.7	10.4	2.7	0.75	0.27
FCT	0.05	0.05	0.04	0.04	0.05	0.08	0.06	0.06	0.05	0.05	0.06

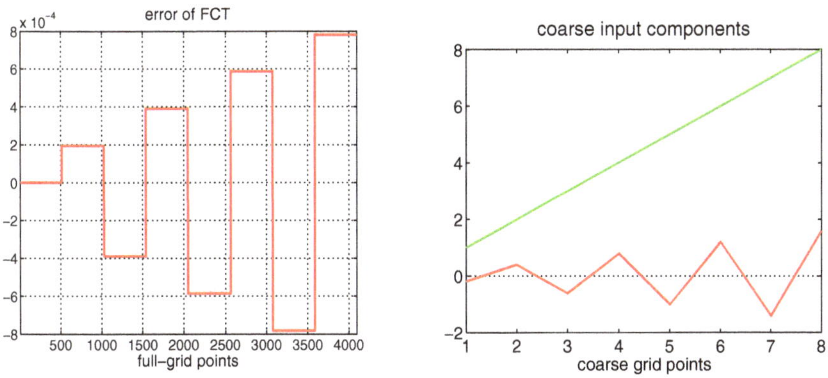

Fig. 5.1: The error of the FCT (left) and coarse components $\mathbf{x}_0$, $\mathbf{y}_0$.

In what follows, we present numerical results for FCT on the two-level composite grid with $n_0 = 2^{\ell_0}$, $n = 2^{\ell_n}$, where n_0 and n are the dimensions of the coarse and fine spaces respectively. The full grid size is given by $n_f = n_0 \cdot n$, which might be very large in our numerical examples (say, $n_f = 2^{17}$ with $n_0 = 2^9$, $n = 2^8$). Algorithm 5.3 is implemented in MATLAB 7.3. Table 5.1 presents CPU times (in s.) for FFT on the corresponding full grid and for FCT on the composite two-level grid.

In this example the finest auxiliary 1D grid attains the size 2^{17}, which is more than enough to resolve arising singularities. The corresponding FCT appears to be at about $4 \cdot 10^4$ times faster than 1D FFT. Numerics clearly demonstrate the advantage of FCT on large composite grids.

Figure 5.1 represents the error of the FCT (left) and coarse components

$$\mathbf{x}_0 = 0.2\,\{(-1)^k\,k\}_{k=1}^{n_0}\,, \quad \mathbf{y}_0 = \{k\}_{k=1}^{n_0}$$

of the input vectors $\mathbf{x} = \mathbf{x}_0 + \mathbf{x}_h$ and $\mathbf{y} = \mathbf{y}_0 + \mathbf{y}_h$ defined on the coarse grid with $n_0 = 8$, and with step size $H = 1$. Fine components of the input vectors are given by $\mathbf{x}_h = \{\sin(2\pi i \cdot h\cdot)\}_{i=1}^{n}$ and $\mathbf{y}_h = \{(-1)^i\}_{i=1}^{n}$ with fine grid size $n = 2^8$.

Figure 5.2 represents the fine (left) and coarse (middle) components of the output vector, as well as the graph of the resultant convolution product.

In applications related to electronic structure calculations the number of refined zones may correspond to the number of atoms in the molecule requiring high resolution.

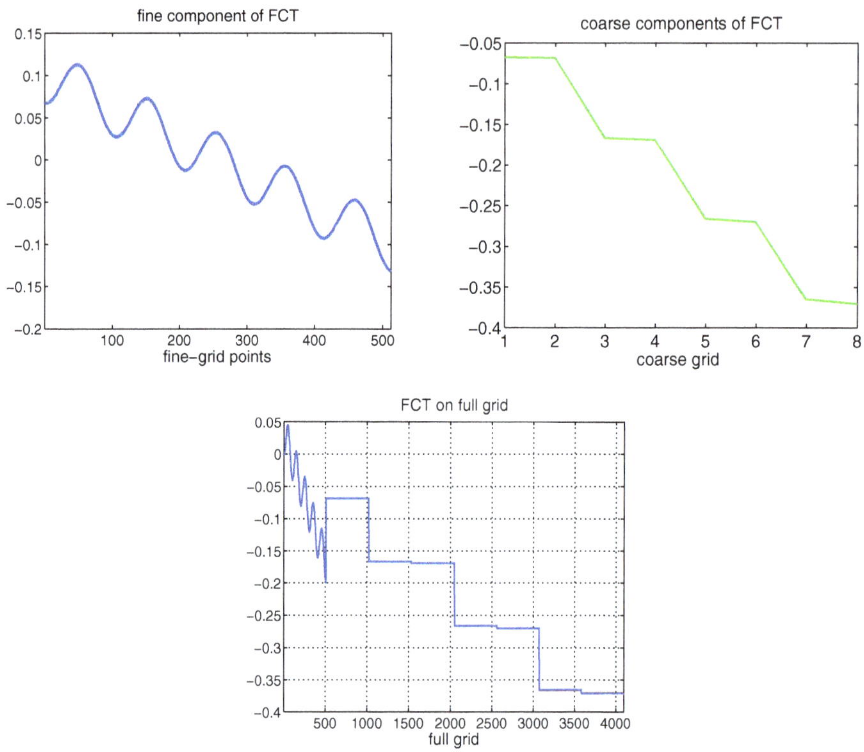

Fig. 5.2: Fine and coarse components of the output vector (upper left, upper right) and resultant convolution vector (bottom).

5.1.7 Low rank sinc approximation of convolving tensors, algebraic rank reduction

In applications related to electronic structure calculations, the function related collocation coefficient tensor $\mathbf{F} = [f_{\mathbf{i}}]_{\mathbf{i}\in\mathcal{J}}$ can be generated by the electron density $\rho(x)$, by the product of the interaction potential $V(x)$ with the electron orbitals, $V(x)\psi(x)$, or by some related terms. In this way we make an a priori assumption on the existence of low rank approximation to the corresponding tensors. This assumption is not easy to analyze, however it works well in practice.

Example 5.7. In the case of a hydrogen atom we have

$$\rho(x) = \mathrm{e}^{-2\|x\|}, \quad \text{and} \quad V(x)\psi(x) = \frac{\mathrm{e}^{-\|x\|}}{\|x\|} \quad \text{with} \quad V(x) = \frac{1}{\|x\|}, \quad x \in \mathbb{R}^3,$$

hence the existence of corresponding low rank tensor approximations can be proven along the lines of [206, Lemma 4.3] and [204, Theorem 3].

To construct a low rank approximation of the tensor $\mathbf{G}$, we consider a class of multivariate spherically symmetric convolving kernels $g\colon \mathbb{R}^d \to \mathbb{R}$ parametrized by

$$g = g(\rho(y)) \quad \text{with} \quad \rho \equiv \rho(y) = y_1^2 + \cdots + y_d^2 ,$$

where the univariate function $g\colon \mathbb{R}_+ \to \mathbb{R}$ can be represented via a generalized Laplace transform

$$g(\rho) = \int_{\mathbb{R}_+} \widehat{g}(\tau^2) \mathrm{e}^{-\rho\tau^2} d\tau . \tag{5.26}$$

Without loss of generality, we introduce one and the same scaling function

$$\phi_i(\cdot) = \phi(\cdot + (i-1)h) , \quad i \in I_n ,$$

for all spatial dimensions $\ell = 1, \dots, d$, where $h > 0$ is the mesh parameter, so that the corresponding tensor product basis function $\phi_{\mathbf{i}}$ is defined in (5.3).

Using sinc quadrature methods, we approximate the collocation coefficient tensor $\mathbf{G} = [g_{\mathbf{i}}]_{\mathbf{i} \in \mathcal{J}}$ in (5.5) via rank-$(2M + 1)$ canonical decomposition

$$g \approx \sum_{k=-M}^{M} w_k \,\mathcal{E}(\tau_k) \quad \text{with} \quad \mathcal{E} = [e_{\mathbf{i}}(\tau_k)] , \quad \mathbf{i} \in \mathcal{J} , \tag{5.27}$$

with suitably chosen coefficients $w_k \in \mathbb{R}$ and quadrature points $\tau_k \in \mathbb{R}_+$, and with the rank-1 components $\mathcal{E}(\tau_k) \in \mathbb{R}^{\mathcal{J}}$ given by

$$e_{\mathbf{i}}(\tau_k) = \widehat{g}(\tau_k^2) \prod_{\ell=1}^{d} \int_{\mathbb{R}} \mathrm{e}^{-y_\ell^2 \tau_k^2} \phi_{i_\ell}(y_\ell) dy_\ell . \tag{5.28}$$

For a class of analytic functions the exponentially fast convergence of above quadrature in M can be proven ([141, 204]). Note that the quadrature points τ_k can be chosen symmetrically, i.e., $\tau_k = \tau_{-k}$, hence reducing the number of terms in (5.27) to $r = M+1$.

In our particular applications in electronic structure calculations we are interested in fast convolution with the Newton or Yukawa kernels. In the case of the Newton kernel, $g(x) = 1/\|x\|$, the approximation theory can be found in [141]. In the case of the Yukawa potential $\mathrm{e}^{-\kappa\|x\|}/\|x\|$ for $\kappa \in [0, \infty)$, we apply the generalized Laplace transform (5.62)

$$g(\rho) = \frac{\mathrm{e}^{-\kappa\sqrt{\rho}}}{\sqrt{\rho}} = \frac{2}{\sqrt{\pi}} \int_{\mathbb{R}_+} \exp(-\rho\tau^2 - \kappa^2/4\tau^2) d\tau , \tag{5.29}$$

corresponding to the choice

$$\widehat{g}(\tau^2) = \frac{2}{\sqrt{\pi}} \mathrm{e}^{-\kappa^2/4\tau^2} .$$

Approximation theory in the case of Yukawa potential is presented in [204], see also Section 5.2.10.

In our numerical experiments below the collocation coefficient tensor $\mathbf{G} \in \mathbb{R}^{\mathcal{J}}$ for the Newton kernel is approximated in the rank-R canonical format with $R \in [20, 30]$ providing an accuracy of about $10^{-7} \div 10^{-5}$ for the grid size up to $n = 10^4$.

In the case of large computational grids the tensor rank of the (problem independent) convolving kernel g can be reduced by an algebraic recompression procedure. For ease of presentation let us consider the case $d = 3$. The idea of our recompression algorithm is based on the observation that a typical feature of the analytic tensor approximation by the sinc quadratures as in (5.27)–(5.28) (for symmetric quadrature points it is agglomerated to the sequence with $k = 0, 1, \ldots, M$) is the presence of many terms all supported only by a few grid points from $p \times p \times p$ grid (domain $\Omega^{(p)}$) in the vicinity of the point type singularity (say at $x = 0$). Assume that this group of rank-1 tensors is numbered by $k = 0, \ldots, K < M$. The sum of these tensors, further called $\mathbf{A}_p$, effectively belongs to the low dimensional space of trilinear $p \times p \times p$ tensors, hence the maximal tensor rank of $\mathbf{A}_p$ does not exceed $r = p^2 \le K$. Furthermore, we can perform the rank-R_0 canonical approximation of this small tensor with $R_0 \ll K$ using the ALS or gradient type optimization.

Algorithm 5.4 (Rank recompression for the canonical sinc based approximation).
1. Given the canonical tensor $\mathbf{A}$ of rank $R = M + 1$.
2. Agglomerate all rank-1 terms supported by only one point, say by $\Omega^{(1)}$, into one rank-1 tensor, further called $\mathbf{A}_1$.
3. Agglomerate by a summation all terms supported by $\Omega^{(2)} \setminus \Omega^{(1)}$ in one tensor $\mathbf{A}_2$ (with maximal rank 3), and approximate with the tensor rank $r_2 \le 3$, and so on until we end up with tensor $\mathbf{A}_p$ supported by $\Omega^{(p)} \setminus \Omega^{(p-1)} \setminus \cdots \setminus \Omega^{(1)}$.
4. Approximate the canonical sum $\mathbf{A}_1 + \cdots + \mathbf{A}_p$ by low rank tensor.

Note that in our sinc quadrature approximations most of the 'local' terms are supported by only one point, say by Ω_1, hence they are all agglomerated in a rank-1 tensor. In approximation of the classical potentials like $1/\|x\|$ or $\mathrm{e}^{-\|x\|}/\|x\|$ the usual choice is $p = 1, 2$.

The simple rank recompression procedure described above allows us to noticeably reduce the initial rank $R = M + 1$ appearing in the (symmetric) sinc quadratures. Numerical examples on the corresponding rank reduction by Algorithm 5.4 are depicted in [204], Figure 2.

Figure 5.3 presents the rank parameters obtained from the sinc approximations of $g(x) = 1/\|x\|$ up to the threshold $\varepsilon = 0.5 \cdot 10^{-6}$ in max norm, computed on $n \times n \times n$ grids with $n = 2^{L+3}$ for the level number $L = 1, \ldots, 8$ (upper curve), and the corresponding values obtained by Algorithm 5.4 with $p = 1$ (lower curve). One observes the significant reduction of the tensor rank.

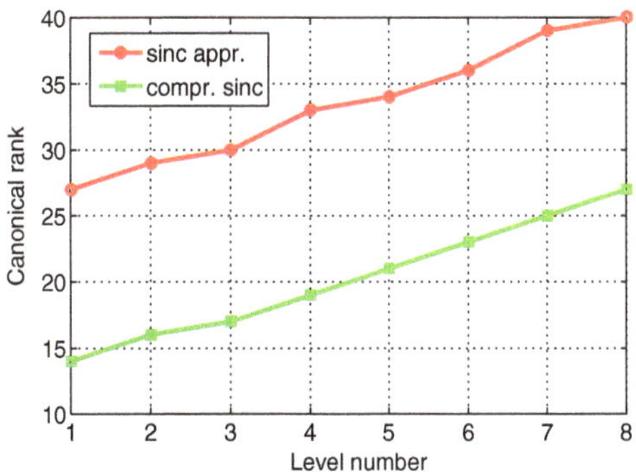

Fig. 5.3: Tensor rank of the sinc and recompressed sinc approximation for $1/\|x\|$.

5.1.8 Numerical verification on quantum chemistry data

We test the approximation error of the tensor product collocation convolution scheme on practically interesting data arising in electronic structure calculations using the Hartree–Fock equation (see [213] for more details). We consider the (pseudo)electron density of a simple CH_4-molecule represented by the exponential sum

$$f(x) := \sum_{\nu=1}^{M} \left(\sum_{k=1}^{R_0} c_{\nu,k}(x - x_k)^{\beta_k} e^{-\lambda_k (x-x_k)^2} \right)^2 , \quad x \in \mathbb{R}^3 , \quad R_0 = 50, \; M = 4 \qquad (5.30)$$

with x_k corresponding to the locations of the C and H atoms. We extract the 'principal exponential' approximation of the electron density, f_0, obtained by setting $\beta_k = 0$ ($k = 1, \dots, R_0$) in (5.30). Using the fast tensor product convolution method, the Hartree potential of f_0,

$$V_{\mathrm{H}}(x) = \int_{\Omega} \frac{f_0(y)}{\|x-y\|} dy , \quad x \in \Omega = [-A, A]^3 ,$$

is computed with high accuracy on a sequence of uniform $(n+1) \times (n+1) \times (n+1)$ grids with $n = 2^p$, $p = 5, 5, \dots, 12$, and with $A = 9.6$. The initial rank of the input tensor $\mathbf{F} = [f_0(y_{\mathbf{i}})]_{\mathbf{i} \in \mathcal{I}}$, presented in the canonical format, is bounded by $R \leq \frac{R_0(R_0+1)}{2}$ (even for simple molecules it normally takes about several thousands). The collocation coefficients tensor $\mathbf{G}$ in (5.5) for the Newton kernel is approximated by the sinc method with the algebraic rank recompression described in Algorithm 5.4. Figure 5.4 represents the shape of the functions f_0 and V_{H}.

Note that the Hartree potential has slow polynomial decay, i.e.,

$$V_{\mathrm{H}}(x) = O\left(\frac{1}{\|x\|}\right) \quad \text{as} \quad \|x\| \to \infty ,$$

however, the density f_0 decays exponentially. Hence the accurate tensor approximation is computed in some smaller box $\Omega' = [-B, B]^3 \subset \Omega$, $B < A$.

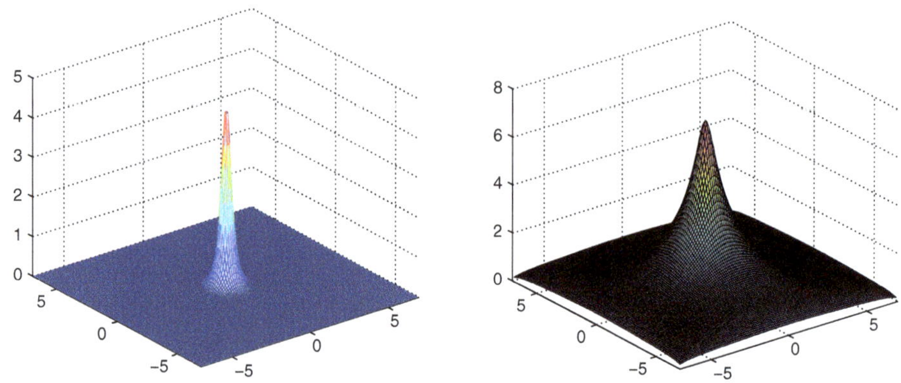

Fig. 5.4: The density $f_0(x_1, x_2, 0)$ (left) and its Hartree potential $V_H(x_1, x_2, 0)$ (right).

In this numerical example the resultant convolution product with the Newton convolving kernel can be calculated exactly by using the analytic representation for each individual Gaussian,

$$\left(e^{-\alpha\|\cdot\|^2} \star \frac{1}{\|\cdot\|}\right)(x) = \left(\frac{\alpha}{\pi}\right)^{-3/2} \frac{1}{\|x\|} \operatorname{erf}(\sqrt{\alpha}\|x\|) ,$$

where the erf function is defined by

$$\operatorname{erf}(t) := \frac{2}{\sqrt{\pi}} \int_0^t \exp(-\tau) d\tau , \quad t \geq 0 .$$

The Hartree potential $V_H = f_0 \star 1/\|\cdot\|$ attains its maximum value at the origin $x = 0$, that is $V_H(0) = 7.19$.

Figure 5.5 (upper left) demonstrates the accuracy $O(h^2)$ of our tensor approximation and of the corresponding improved values, $O(h^3)$, due to the Richardson extrapolation. Here the grid size is given by $n = n_\ell = 2^{\ell+4}$ for the level number $\ell = 1, \ldots, 7$, with the finest grid size $n_7 = 2048$. It is seen that beginning from the level number $\ell = 5$ ($n_5 = 512$) the extrapolated scheme already achieves the saturation error 10^{-6} of the tensor approximation related to the chosen Tucker rank $r = 22$. This demonstrates high accuracy of the Richardson extrapolation.

Absolute errors for the Hartree potential of a CH_4 molecule are given in Figure 5.5 (upper right) compared with the commonly used MOLPRO [363] calculations (here we have $\max|V_H| = 8.6$).

Figure 5.5 (bottom) presents the CPU times (min) to compute the 3D FCT on $n\times n\times n$ grid for a sequence of grid sizes $n \in [64, 128, \ldots, 8192]$, and with the input rank $R_2 = 256, 484$. It confirms the theoretical result on linear logarithmic complexity scaling in n, and linear scaling in R_2.

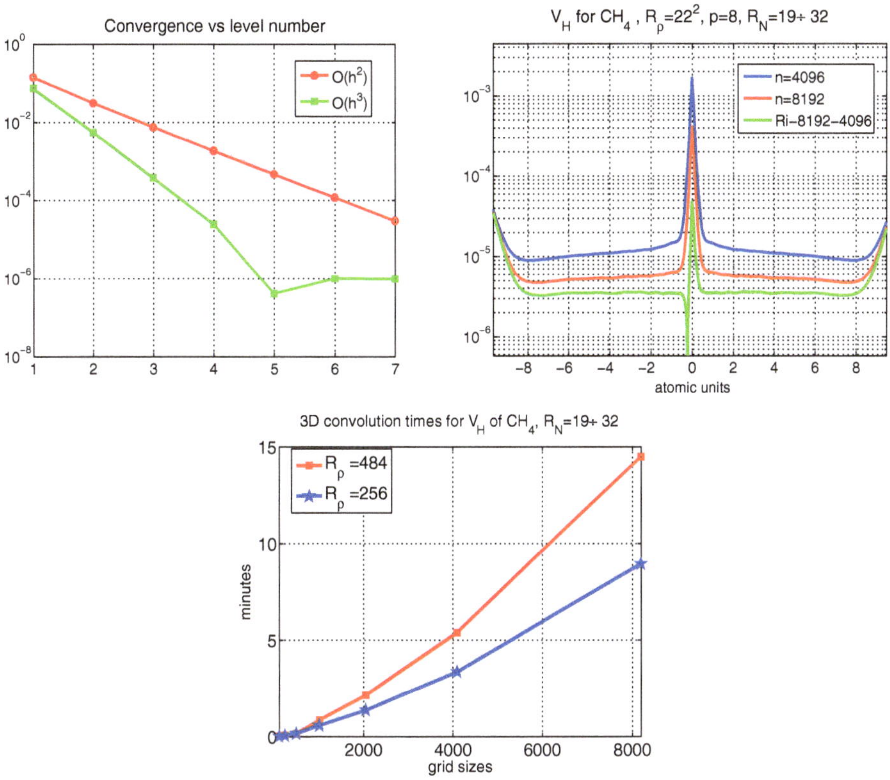

Fig. 5.5: Convergence history for the $O(h^2)$ and $O(h^3)$ extrapolated convolution schemes (upper left); absolute error for the Hartree potential of CH_4 molecule (upper right); CPU time versus the grid size and the input rank R_2 (bottom).

5.2 Tensor numerical methods in Hartree–Fock calculations

The Hartree–Fock equation is the basic model reduction approach for the approximation of the general Schrödinger equation. A number of approximation and solution methods in quantum chemistry and in particular for the solution of the Hartree–Fock equation have long been developed in the community of computational quantum chemistry. We refer to papers and monographs [17, 46, 48, 51, 57, 91, 92, 99, 104, 157, 158, 161, 183, 185, 191, 195, 214, 250], and [41, 42, 47, 243, 249, 267, 300–302, 309, 314, 338, 353, 362, 368, 371], which include citations on many other interesting works on the topic. In what follows, we briefly discuss the recently developed grid based tensor numerical approach to the solution of the Hartree–Fock equation. The more detailed presentation can be found in [185, 191, 192, 214].

5.2.1 Nonlinear eigenvalue problem

The Hartree–Fock (HF) equation for determination of the ground state energy of a molecular system consisting of M nuclei and N_{orb} electrons (closed shell case) is given by the following nonlinear eigenvalue problem in $H^1(\mathbb{R}^3)$ [249]:

$$(\mathcal{F}\phi_i)(x) = \lambda_i\,\phi_i(x)\,, \quad \int_{\mathbb{R}^3} \phi_i(x)\phi_j(x)\,dx = \Delta_{ij},\ i,j = 1,\dots,N_{\text{orb}}\,, \tag{5.31}$$

where the nonlinear integrodifferential Fock operator $\mathcal{F}$ is given by

$$\mathcal{F} := -\frac{1}{2}\Delta - V_c + V_{\text{H}} + \mathcal{V}_E\,, \quad V_c = \sum_{\nu=1}^{M} \frac{Z_\nu}{\|x - A_\nu\|}\,, \tag{5.32}$$

with the Hartree potential, $V_{\text{H}}(x)$, and the nonlocal exchange operator, $\mathcal{V}_E$, defined by

$$V_{\text{H}}(x) := \int_{\mathbb{R}^3} \frac{\tau(y,y)}{\|x-y\|}\,d\mathbf{y}\,, \quad \text{and} \quad \mathcal{V}_E\phi := -\frac{1}{2}\int_{\mathbb{R}^3} \frac{\tau(x,y)}{\|x-y\|}\phi(y)\,dy\,,$$

respectively. Here, $1/\|\cdot\|: \mathbb{R}^3 \to \mathbb{R}$ corresponds to the Newton (Coulomb) potential, and $Z_\nu \in \mathbb{R}_+$, $A_\nu \in \mathbb{R}^3$ $(\nu = 1,\dots,M)$ specify charges and positions of M nuclei. The electron density matrix

$$\tau: \mathbb{R}^3 \times \mathbb{R}^3 \to \mathbb{R}\,,$$

is given by $\tau(x,y) = 2\sum_{i=1}^{N_{\text{orb}}} \phi_i(x)\phi_i^*(y)$, specifying the electron density $\rho(x) = \tau(x,x)$.

Note that the Hartree–Fock equation is a nonlinear eigenvalue problem in the sense that one should solve it when the nonlinear part $V_{\text{H}} + \mathcal{V}_E$ of the governing operator depends on the unknown eigenvectors. This dependence is expressed by the 3D convolution integral transform with the Newton convolving kernel, while the electron density $\rho(x)$, $x \in \mathbb{R}^3$, contains multiple strong singularities corresponding to each nuclei location. Therefore solution of the HF equation requires iterative solvers, with multiply repeated recalculation of these convolution integrals.

Usually, the Hartree–Fock equation is approximated by the standard Galerkin projection of the initial problem (5.31) posed in $H^1(\mathbb{R}^3)$ (see [249] for more details). For a given finite Galerkin basis set $\{g_\mu\}_{1\le\mu\le N_b}$, $g_\mu \in H^1(\mathbb{R}^3)$, the molecular orbitals ϕ_i are expanded (approximately) by

$$\phi_i = \sum_{\mu=1}^{N_b} C_{\mu i} g_\mu\,, \quad i = 1,\dots,N_{\text{orb}}\,, \tag{5.33}$$

[190] yielding the Galerkin system of nonlinear equations for the coefficients matrix $C = \{c_{\mu i}\} \in \mathbb{R}^{N_b \times N_{\text{orb}}}$, concatenating the eigenvectors $C_i \in \mathbb{R}^{N_b}$ (and the density matrix $D = 2CC^* \in \mathbb{R}^{N_b\times N_b}$)

$$F(D)C = SC\Lambda\,, \quad \Lambda = \operatorname{diag}(\lambda_1,\dots,\lambda_{N_b})\,, \quad C^{\text{T}}SC = I_{N_b}\,. \tag{5.34}$$

Here S is the overlap (stiffness) matrix for $\{g_\mu\}_{1\le\mu\le N_b}$, and the Fock matrix

$$F(D) = H + J(D) + K(D), \tag{5.35}$$

is a sum of the stiffness matrix $H = \{h_{\mu\nu}\}$ of the core Hamiltonian $\mathcal{H} = -\frac{1}{2}\Delta + V_c$ (the single-electron integrals),

$$h_{\mu\nu} = \frac{1}{2}\int_{\mathbb{R}^3} \nabla g_\mu \cdot \nabla g_\nu dx + \int_{\mathbb{R}^3} V_c(x) g_\mu g_\nu dx\,, \quad 1 \le \mu, \nu \le N_b\,,$$

and the two nonlinear terms $J(D) + K(D)$, representing the Galerkin approximation to the Hartree and exchange operators. This is the main computational task, which is traditionally treated by using the two-electron integrals tensor $\mathbf{B} = [b_{\mu\nu\kappa\lambda}]$, defined as follows: Given the finite basis set $\{g_\mu\}_{1\le\mu\le N_b}$, $g_\mu \in H^1(\mathbb{R}^3)$, the associated fourth order two-electron integrals (TEI) tensor, $\mathbf{B} = [b_{\mu\nu\kappa\lambda}]$, is defined entrywise by

$$b_{\mu\nu\kappa\lambda} = \int_{\mathbb{R}^3}\int_{\mathbb{R}^3} \frac{g_\mu(x) g_\nu(x) g_\kappa(y) g_\lambda(y)}{\|x-y\|} dxdy\,, \quad \mu, \nu, \kappa, \lambda \in \{1, \dots, N_b\}\,. \tag{5.36}$$

In the straightforward implementation based on the analytically precomputed integrals, the computational and storage complexity for the TEI tensor is of the order of $O(N_b^4)$, or even $O(N_b^5)$, which becomes computationally nontractable already for N_b of the order of several hundred.

Given TEI tensor, the $N_b \times N_b$ matrices $J(D)$ and $K(D)$ can be represented by

$$J(D)_{\mu\nu} = \sum_{\kappa,\lambda=1}^{N_b} b_{\mu\nu,\kappa\lambda} D_{\kappa\lambda}\,, \quad K(D)_{\mu\nu} = -\frac{1}{2}\sum_{\kappa,\lambda=1}^{N_b} b_{\mu\lambda,\nu\kappa} D_{\kappa\lambda}\,, \tag{5.37}$$

with the low rank symmetric density matrix, $D = 2CC^T \in \mathbb{R}^{N_b\times N_b}$, such that

$$\operatorname{rank}(D) = N_{\mathrm{orb}} \ll N_b\,.$$

Equations (5.37) can be rewritten in terms of the TEI matrix $B = [b_{\mu\nu,\kappa\lambda}] \in \mathbb{R}^{N_b^2\times N_b^2}$.

The total energy is computed as

$$E_{\mathrm{HF}} = 2\sum_{i=1}^{N_{\mathrm{orb}}} \lambda_i - \sum_{i=1}^{N_{\mathrm{orb}}} \left(\widetilde{J}_i - \widetilde{K}_i\right)\,,$$

where $\widetilde{J}_i = (\phi_i, V_{\mathrm{H}}\phi_i)_{L^2} = \langle C_i, JC_i\rangle$ and $\widetilde{K}_i = (\phi_i, K\phi_i)_{L^2} = \langle C_i, KC_i\rangle$, $i = 1, \dots, N_{\mathrm{orb}}$, are the Coulomb and exchange integrals in the basis of orbitals ϕ_i.

Given geometry of nuclei, the resulting ground state energy of a molecule, E_0, includes the nuclei repulsion energy E_{nuc},

$$E_0 = E_{\mathrm{HF}} + E_{\mathrm{nuc}}\,, \quad \text{where} \quad E_{\mathrm{nuc}} = \sum_{k=1}^{M}\sum_{m<k}^{M} \frac{Z_k Z_m}{\|x_k - x_m\|}\,. \tag{5.38}$$

The standard quantum chemical implementations are based on the analytically precomputed set of the two-electron integrals (5.36) in a naturally separable Gaussian basis with the computational and storage complexity for the TEI tensor of the order of $O(N_b^4) - O(N_b^5)$, which becomes nontractable already for N_b of the order of several hundred.

5.2.2 Grid based rank structured approximation in Hartree–Fock calculus

The tensor structured numerical methods, both the name and the concept, appeared during the work on the 3D grid based tensor approach to the solution of the Hartree–Fock equation. They led to 'black box' numerical treatment of the Hartree–Fock problem based on the low rank representation of the basis functions in a volume box, using the $n \times n \times n$ $3D$ Cartesian grids positioned arbitrarily with respect to the atomic centers [185, 186, 195, 214]. In 2008 [184, 212] it was shown that within the tensor structured paradigm the core Hamiltonian and the $3D$ convolutions with the Newton kernel, involved in the Coulomb and exchange operators, can be calculated in $1D$ complexity by the rank structured tensor operations reduced to $1D$ convolutions and Hadamard and scalar products [183, 201, 212]. Due to elimination of the analytical integrability requirements it gives us a choice to use rather general physically relevant basis sets represented on the grid. High accuracy is achieved due to grid sizes up to the order of $n \simeq 10^6$, yielding a volume box of size $n^3 \simeq 10^{18}$. It corresponds to mesh resolution up to $h = 10^{-5}$ Å (close to sizes of atomic radii) in the volume box with the equal sizes of 20 Å for each spatial variable. MATLAB on a laptop can be used for all algorithms, without parallelizations and supercomputing.

Chronologically, two different approaches have been developed for the 3D grid based tensor structured solution of the Hartree–Fock equation. Both use the rank structured calculation of the core Hamiltonian [183, 186] for a given grid based basis set as described in what follows.

- The tensor solver I does not use the two-electron integrals. Instead, the Coulomb and exchange operators are recomputed 'on the fly' using the refined 3D grids and rank structured (1D) operations; see [184, 185, 214]. This approach has low storage demand, but might be time consuming. It can be applicable to the Kohn–Sham type models as well.
- The 'black box' solver II is based on calculation of the TEI matrix B by the truncated Cholesky decomposition and the redundancy free factorization by the algebraically reduced product basis, yielding the reduced storage consumption $O(N_b^3)$ [183, 190, 194]. Its performance in time and accuracy are both compatible with the benchmark packages in quantum chemistry based on analytical precalculation of involved multidimensional integrals.

It is worth noting that the grid based numerical methods in computational quantum chemistry are traditionally based on the concept of multiresolution analysis ([157]), which require the storage of rather unstructured 3D discrete data, which are complicated for implementation of the convolution transform. The discussion on Kronecker product approximation to some functions arising in quantum chemistry can be found in [66, 98]. The recent attempt at the direct treatment of the Hartree–Fock problem in the TT/QTT tensor format via the discretization on the 3D tensor grid was reported in [304].

5.2.3 Rank structured representation of the two-electron integrals tensor

In the following, we briefly discuss the tensor algorithm for computation of TEIs used in the solver II. We suppose that all basis functions $\{g_\mu\}_{1\le\mu\le N_b}$ are supported by the finite volume box $\Omega = [-b, b]^3 \in \mathbb{R}^3$, and assume for ease of presentation that $\operatorname{rank}(g_\mu) = 1$. Introducing the $n \times n \times n$ Cartesian grid over Ω and using the standard collocation discretization in the volume by piecewise constant basis functions, we define a grid based tensor representation of the initial basis set $g_\mu(x) \in \mathbb{R}^3$, $\mu = 1, \dots N_b$,

$$g_\mu(x) = g_\mu^{(1)}(x_1) g_\mu^{(2)}(x_2) g_\mu^{(3)}(x_3) \approx \mathbf{G}_\mu = \mathbf{g}_\mu^{(1)} \otimes \mathbf{g}_\mu^{(2)} \otimes \mathbf{g}_\mu^{(3)} \in \mathbb{R}^{n\times n\times n} .$$

Define the respective product-basis tensor

$$\mathbf{G} = [\mathbf{G}_{\mu\nu}] \in \mathbb{R}^{N_b\times N_b\times n^{\otimes 3}} \quad \text{with} \quad \mathbf{G}_{\mu\nu} = \mathbf{G}_\mu \odot \mathbf{G}_\nu \in \mathbb{R}^{n^{\otimes 3}} ,$$

where $\mu, \nu \in \{1, \dots, N_b\}$, then both the TEI tensor and TEI matrix are represented by tensor operations,

$$\mathbf{B} = \mathbf{G} \times_{n^{\otimes 3}} \mathbf{P} *_{n^{\otimes 3}} \mathbf{G} , \quad b_{\mu\nu,\kappa\lambda} = \langle \mathbf{G}_{\mu\nu}, \mathbf{P} * \mathbf{G}_{\kappa\lambda} \rangle_{n^{\otimes 3}} . \tag{5.39}$$

Here the rank-$R_{\mathcal{N}}$ canonical tensor

$$\mathbf{P} = \sum_{k=1}^{R_{\mathcal{N}}} \mathbf{p}_k^{(1)} \otimes \mathbf{p}_k^{(2)} \otimes \mathbf{p}_k^{(3)} \in \mathbb{R}^{n^{\otimes 3}}$$

approximates the Newton potential $\frac{1}{\|x\|}$ (see [38, 212] for more details), $*$ stands for the 3D tensor convolution, and $\odot$ denotes the Hadamard product of tensors. Sinc quadrature approximation of the Newton kernel is discussed in Sections 2.4.4, 5.1.7 and 5.5.1.

Though tensor methods reduce the multidimensional integration to 1D complexity operations, the direct tensor structured evaluation of (5.39) needs a storage size of at least $O(R_{\mathcal{N}} N_b^2 n)$, which can be prohibitive for large $N_b \sim 10^2$ and $n \approx 10^5$. We apply the RHOSVD type factorization [212] to the fourth order tensor $\mathbf{G}$ by approximating its side matrices, $G^{(\ell)} \in \mathbb{R}^{n\times N_b^2}$, ($\ell = 1, 2, 3$) in a 'squeezed' factorized form,

$$G^{(\ell)} \cong U^{(\ell)} V^{(\ell)\mathrm{T}} ,$$

according to the chosen ε-truncation. This step can be implemented by the truncated SVD in combination with incomplete truncated Cholesky decomposition.

This provides the construction of dominating subspaces in the x, y, and z components in the product basis set defined by an $n \times R_\ell$ matrix $U^{(\ell)}$ (left orthogonal basis) and $N_b^2 \times R_\ell$ matrix $V^{(\ell)}$ (right basis). Then for the TEI matrix $B \in \mathbb{R}^{N_b^2 \times N_b^2}$, we obtain a factorization [190, 195],

$$B \cong B_\varepsilon := \sum_{k=1}^{R_\mathcal{N}} \odot_{\ell=1}^{3} V^{(\ell)} M_k^{(\ell)} V^{(\ell)^{\mathrm{T}}} , \tag{5.40}$$

where $V^{(\ell)}$ is the corresponding right redundancy free basis, $\odot$ denotes the pointwise (Hadamard) product of matrices, and

$$M_k^{(\ell)} = U^{(\ell)^{\mathrm{T}}} (\mathbf{p}_k^{(\ell)} *_n U^{(\ell)}) \in \mathbb{R}^{R_\ell \times R_\ell} , \quad k = 1, \ldots, R_\mathcal{N} . \tag{5.41}$$

Ultimately, the TEI matrix B is approximated in a form of the truncated Cholesky factorization,

$$B \approx LL^{\mathrm{T}} , \quad L \in \mathbb{R}^{N_b^2 \times R_B} , \quad R_B = O(N_b) ,$$

such that the required columns of the matrix B are easily computed by using (5.40) at the low cost.

Note that the redundancy free factorization (5.40) can be viewed as the algebraic tensor structured counterpart of the *density fitting scheme* commonly used in quantum chemistry [3]. In our approach the 'one dimensional density fitting' independently for each space dimension reduces the ε ranks of the dominating directional bases to the lowest possible value. The robust error control in the proposed basis optimization method is based on the low rank approximation by purely algebraic SVD like procedure that allows us to eliminate the redundancy in the product basis set up to given precision $\varepsilon > 0$.

Vectorizing matrices

$$\overline{J} = \mathrm{vec}(J(D)) , \quad \overline{K} = \mathrm{vec}(K(D)) , \quad \overline{D} = \mathrm{vec}(D) ,$$

we arrive at the simple matrix representations

$$\overline{J} = B\overline{D} \approx L(L^{\mathrm{T}}\overline{D}) , \quad \mathrm{vec}(K) = \overline{K} = \widetilde{B}\overline{D} , \tag{5.42}$$

where $\widetilde{B} = \mathrm{mat}(\widetilde{\mathbf{B}})$ is the matrix unfolding of the permuted tensor $\widetilde{\mathbf{B}} = [\widetilde{b}_{\mu\nu\kappa\lambda}]$ such that $\widetilde{b}_{\mu\nu\kappa\lambda} = b_{\mu\kappa\nu\lambda}$.

The nonlinear eigenvalue problem (5.34) is solved by the commonly used DIIS self-consistent iteration, which requires the update of both Hartree and exchange operators at each iterative step; see [183, 185, 214] for more details.

5.2.4 Calculating multidimensional integrals by using tensor formats

Calculation of the multidimensional convolution integral operators with the Newton kernel in tensor formats has been introduced and implemented in [184, 212, 214], where in the examples of the Hartree and exchange operators in the HF equation it was shown that calculation of the 3D and 6D convolution integrals can be reduced to a combination of 1D Hadamard products, 1D convolutions, and 1D scalar products.

The molecule is embedded in a certain fixed computational box $\Omega = [-b, b]^3 \in \mathbb{R}^3$, as shown in Figure 5.6. (Usually for small to medium size molecules we use the computational box of size 40^3 bohr.) For a given discretization parameter $n \in \mathbb{N}$, we use the equidistant $n \times n \times n$ tensor grid $\omega_{\mathbf{3},n} = \{x_{\mathbf{i}}\}$, $\mathbf{i} \in \mathcal{I} := \{1, \dots, n\}^3$, with the mesh size $h = 2b/(n+1)$. In calculations of integral terms, the Gaussian basis functions $g_k(x)$, $x \in \mathbb{R}^3$ are approximated by sampling their values at the centers of discretization intervals, as in Figure 5.7, using univariate piecewise constant basis functions, such that $g_k(x) \approx \overline{g}_k(x) = \prod_{\ell=1}^{3} \overline{g}_k^{(\ell)}(x_\ell)$, $\ell = 1, 2, 3$, yielding their rank-1 tensor representation,

$$\mathbf{G}_k = \mathbf{g}_k^{(1)} \otimes \mathbf{g}_k^{(2)} \otimes \mathbf{g}_k^{(3)} \in \mathbb{R}^{n\times n\times n}, \quad k = 1, \dots, N_b \,. \tag{5.43}$$

Let us consider the numerical scheme for the rank structured tensor calculation of the Hartree potential

$$V_{\mathrm{H}}(x) := \int_{\mathbb{R}^3} \frac{\rho(y)}{\|x-y\|}\, dy\,,$$

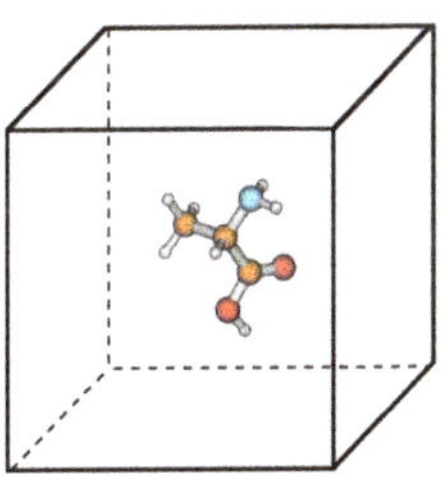

Fig. 5.6: Glycine amino acid in a computational box.

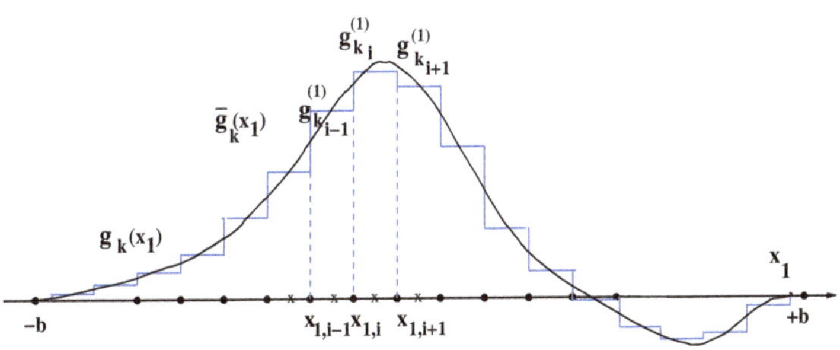

Fig. 5.7: Piecewise constant approximation of the Gaussian type basis function.

and the corresponding Coulomb matrix,

$$J_{km} := \int_{\mathbb{R}^3} g_k(x) g_m(x) V_{\mathrm{H}}(x) dx\,, \quad k,m = 1,\dots N_b \quad x \in \mathbb{R}^3\,,$$

where the electron density, $\rho(x) = 2\sum_{a=1}^{N_{\mathrm{orb}}} (\varphi_a)^2$, is represented in terms of molecular orbitals $\varphi_a(x) = \sum_{k=1}^{N_b} c_{a,k} g_k(x)$. Given the discrete tensor representation of basis functions in (5.43), the electron density is approximated by using 1D Hadamard products of rank-1 tensors (instead of product of 3D Gaussians):

$$\rho \approx \Theta = 2 \sum_{a=1}^{N_{\mathrm{orb}}} \sum_{k=1}^{N_b} \sum_{m=1}^{N_b} c_{a,m} c_{a,k} \left(\mathbf{g}_k^{(1)} \odot \mathbf{g}_m^{(1)}\right) \otimes \left(\mathbf{g}_k^{(2)} \odot \mathbf{g}_m^{(2)}\right) \otimes \left(\mathbf{g}_k^{(3)} \odot \mathbf{g}_m^{(3)}\right) \in \mathbb{R}^{n\times n\times n}\,.$$

Furthermore, the representation of the Newton kernel $\frac{1}{\|x-y\|}$ by a canonical rank-R_N tensor via the optimized sinc approximation [38] is used:

$$\mathbf{P}_R = \sum_{q=1}^{R_N} \mathbf{p}_q^{(1)} \otimes \mathbf{p}_q^{(2)} \otimes \mathbf{p}_q^{(3)} \in \mathbb{R}^{n\times n\times n}\,. \tag{5.44}$$

Since large ranks make tensor operations inefficient, the multigrid canonical-to-Tucker and Tucker-to-canonical algorithms should be applied to reduce the initial rank of $\Theta \mapsto \Theta'$ by several orders of magnitude, from $N_b^2/2$ to $R_\rho \ll N_b^2/2$. Then the 3D tensor representation of the Hartree potential is calculated by using the 3D tensor product convolution, which is a sum of tensor products of 1D convolutions,

$$V_{\mathrm{H}} \approx \mathbf{V}_H = \Theta' * \mathbf{P}_R = \sum_{m=1}^{R_\rho} \sum_{q=1}^{R_N} c_m \left(\mathbf{u}_m^{(1)} * \mathbf{p}_q^{(1)}\right) \otimes \left(\mathbf{u}_m^{(2)} * \mathbf{p}_q^{(2)}\right) \otimes \left(\mathbf{u}_m^{(3)} * \mathbf{p}_q^{(3)}\right)\,.$$

The entries of the Coulomb matrix, J_{km}, are obtained by 1D scalar products of $\mathbf{V}_H$ with the Galerkin basis consisting of rank-1 tensors,

$$J_{km} \approx \langle \mathbf{G}_k \odot \mathbf{G}_m, \mathbf{V}_H \rangle\,, \quad k,m = 1,\dots N_b\,.$$

The cost of 3D tensor product convolution is $O(n \log n)$ instead of $O(n^3 \log n)$ for the standard benchmark 3D convolution using the 3D FFT. Table 5.2 shows CPU times (s) for the MATLAB computation of V_{H} for a H_2O molecule [212] on a small workstation. The tensor based computation is two orders of magnitude faster than the corresponding 3D FFT based algorithm. C2T shows the time for the canonical-to-Tucker rank reduction.

In a similar way, the algorithm for 3D grid based tensor structured calculation of 6D integrals in the exchange potential operator was introduced in [184], $K_{km} = \sum_{a=1}^{N_{\mathrm{orb}}} K_{km,a}$ with

$$K_{km,a} := \int_{\mathbb{R}^3}\int_{\mathbb{R}^3} g_k(x) \frac{\varphi_a(x)\varphi_a(y)}{|x-y|} g_m(y) dx dy\,, \quad k,m = 1,\dots N_b\,.$$

Tab. 5.2: Times (s) for the 3D tensor product convolution.

n^3	1024^3	2048^3	4096^3	8192^3	$16\,384^3$
$C * C$	8.8	20.0	61.0	157.5	299.2
C2T	6.9	10.9	20.0	37.9	86.0

The contributions from the ath orbital are approximated by tensor ansatz,

$$K_{km,a} := \left\langle \mathbf{G}_k \odot \left[\sum_{\mu=1}^{N_b} c_{\mu a} \mathbf{G}_\mu \right], \left[\mathbf{G}_m \odot \sum_{\nu=1}^{N_b} c_{\nu a} \mathbf{G}_\nu \right] * \mathbf{P}_R \right\rangle .$$

Here, the tensor product convolution is first calculated for each ath orbital, and then scalar products in canonical format yield the contributions to entries of the exchange Galerkin matrix from the ath orbital.

These algorithms were employed in the tensor structured solver I using 3D grid based evaluation of the Coulomb and exchange matrices in 1D complexity at every step of the SCF iteration for solving the eigenvalue problem [185, 214]. A sequence of dyadic refined 3D Cartesian grids was used for reducing time in first iterations, with an ε convergence criterion for switching to larger grids. This is a direct grid based computational scheme avoiding calculation of the two-electron integrals. The accuracy for small molecules like H_2O and CH_4 was of the order of 10^{-4} Hartree. Although time performance of this solver was not compatible with the standard Hartree–Fock packages it was the first proof of concept for the tensor numerical methods.

5.2.5 Core Hamiltonian on tensor grid

The Galerkin representation of the 3D Laplace operator in the nonlocal Gaussian basis $\{g_k(x)\}_{1\le k\le N_b}$, $x \in \mathbb{R}^3$, leads to the fully populated matrix $A_g = [a_{km}] \in \mathbb{R}^{N_b \times N_b}$. Tensor calculation of the matrix entries a_{km} for the discrete Laplacian A_g in the separable Gaussian basis is reduced to 1D matrix operations [183] involving the FEM Laplacian Δ_3, defined on $n \times n \times n$ grid,

$$\Delta_3 = \Delta_1 \otimes I \otimes I + I \otimes \Delta_1 \otimes I + I \otimes I \otimes \Delta_1 ,$$

where $\Delta_1 = \frac{1}{h}$ tridiag$\{-1, 2, -1\}$. Specifically, we have

$$a_{km} = \langle \Delta_3 \mathbf{G}_k, \mathbf{G}_m \rangle ,$$

where $\mathbf{G}_k$ is the tensor representation of Gaussian basis functions using the piecewise linear finite elements (Figure 5.8). In the case of large $n\times n\times n$ grids, this calculation can be implemented with logarithmic cost in n by using the low rank QTT representation of the large matrix Δ_3; see [177, 184, 186].

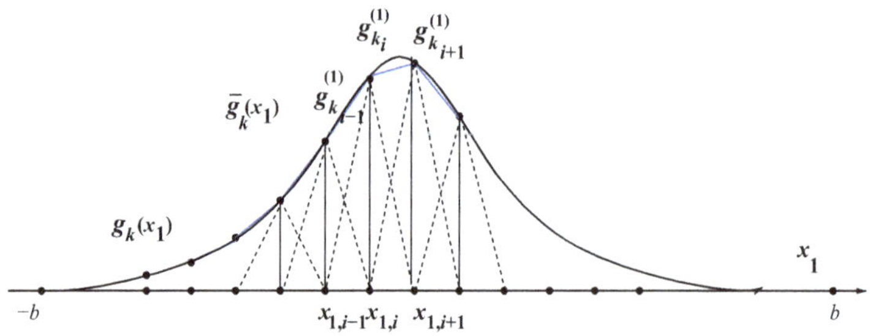

Fig. 5.8: Discretization of a Gaussian by piecewise linear finite elements.

For tensor calculation of the nuclear potential operator

$$V_c(x) = -\sum_{\alpha=1}^{M} \frac{Z_\alpha}{\|x - a_\alpha\|}, \quad Z_\alpha > 0, \quad x, a_\alpha \in \mathbb{R}^3 ,$$

we apply the rank-1 windowing operator, $\mathcal{W}_\alpha = \mathcal{W}_\alpha^{(1)} \otimes \mathcal{W}_\alpha^{(2)} \otimes \mathcal{W}_\alpha^{(3)}$, for shifting the reference Newton kernel $\widetilde{\mathbf{P}}_R \in \mathbb{R}^{2n\times 2n\times 2n}$ according to the coordinates of nuclei in a molecule. Then the resulting nuclear potential, $\mathbf{P}_c \in \mathbb{R}^{n\times n\times n}$, is obtained as a direct tensor sum of shifted potentials [183],

$$\mathbf{P}_c = \sum_{\alpha=1}^{M} Z_\alpha \mathcal{W}_\alpha \widetilde{\mathbf{P}}_R = \sum_{\alpha=1}^{M} Z_\alpha \sum_{q=1}^{R} \mathcal{W}_\alpha^{(1)} \widetilde{\mathbf{p}}_q^{(1)} \otimes \mathcal{W}_\alpha^{(2)} \widetilde{\mathbf{p}}_q^{(2)} \otimes \mathcal{W}_\alpha^{(3)} \widetilde{\mathbf{p}}_q^{(3)} .$$

This leads to the following representation of the Galerkin matrix, $V_c = [\overline{v}_{km}]$, by tensor operations

$$\overline{v}_{km} = \int_{\mathbb{R}^3} V_c(x)\overline{g}_k(x)\overline{g}_m(x)dx \approx \langle \mathbf{G}_k \odot \mathbf{G}_m, \mathbf{P}_c \rangle , \quad 1 \le k, m \le N_b .$$

Figure 5.9, left, shows several vectors of the canonical representation of the Coulomb kernel along one of the variables. Figure 5.9, right, represents the cross section of the resulting nuclear potential $\mathbf{P}_c$ for a C_2H_5OH molecule.

5.2.6 Numerical illustrations to the Hartree–Fock solver

The tensor structured Hartree–Fock solver [183] based on factorized calculation of the two-electron integrals [195] includes efficient tensor implementation of the MP2 energy correction [190] scheme. Though it is yet to be implemented in MATLAB, its performance in time and accuracy is compatible with the standard packages based on analytical evaluation of the two-electron integrals. Due to 1D complexity of all calculations, it enables 3D grids of the size 10^{15}, yielding mesh size of the order of atomic radii, 10^{-4} Å. This ensures high accuracy of calculations, which is controlled by the ε ranks of tensor truncation.

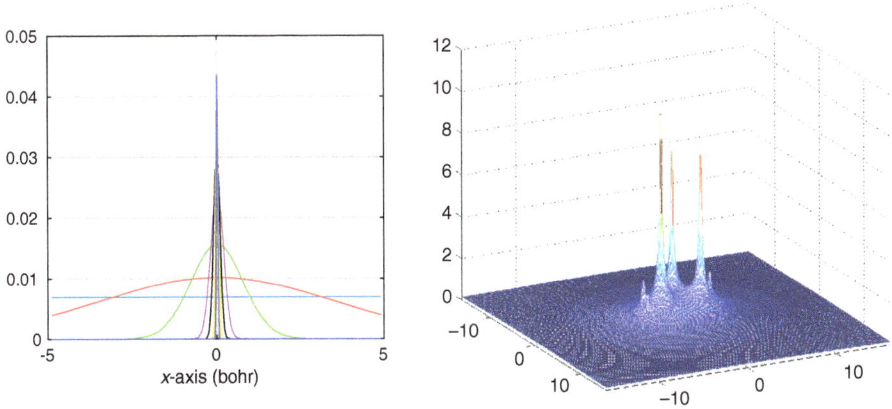

Fig. 5.9: Several skeleton vectors of the canonical representation to the Newton kernel along one of the variables (left); a sum of the nuclear potentials for an ethanol molecule (right).

Solver II works in a black box way: input the grid based basis functions and coordinates of nuclei in a molecule and start the program. Calculation of TEI for H_2O on grids $32\,768^3$ takes two minutes on a laptop. The time for TEI with $n^3 = 13\,1072^3$ for the alanine amino acid takes approximately one hour in MATLAB, including incorporated density fitting.

Figure 5.10, upper left, illustrates the convergence history for self-consistent iterations by tensor solver II compared with the output of a standard quantum chemical package MOLPRO based on analytical calculations [362] for the glycine amino acid, $C_2H_5NO_2$. The basis set cc-pVDZ of 170 Gaussians is used for both analytical and 3D grid based calculations. Here TEI is calculated on the grids $n^3 = 131\,072^3$. For this iteration the core Hamiltonian is taken from MOLPRO. Time for one iteration is about six seconds in MATLAB.

Figure 5.10, upper right, shows the convergence in ground state energy $E_{0,it,h}$ while the right side displays the zoomed difference of ground state energy for last iterations. Here the energy $E_{0,it,h}$ is computed by (5.38) at each iteration step, it, representing the convergence history. The stagnation in the energy on lower than MOLPRO level (relative error $7 \cdot 10^{-7}$) may indicate the actual accuracy in computation of 3D convolution integrals in that code, and additionally, some possible instabilities of the grid based algorithms applied for huge spatial grids ($n^3 = 131\,072^3$). This topic needs further analysis to be done elsewhere.

Figures 5.11 (left) shows the absolute error with respect to MOLPRO output of the tensor calculation of the density matrix for glycine amino acid using TEI matrix computed on the grid of size $n^3 = 131\,072^3$. Figure 5.11 (right) shows the error of tensor calculations for the alanine amino acid, $C_3H_7NO_2$, with $N_b^2 \times N_b^2$ TEI matrix computed on the grid of size $n^3 = 32\,768^3$, with $N_b = 211$ basis functions.

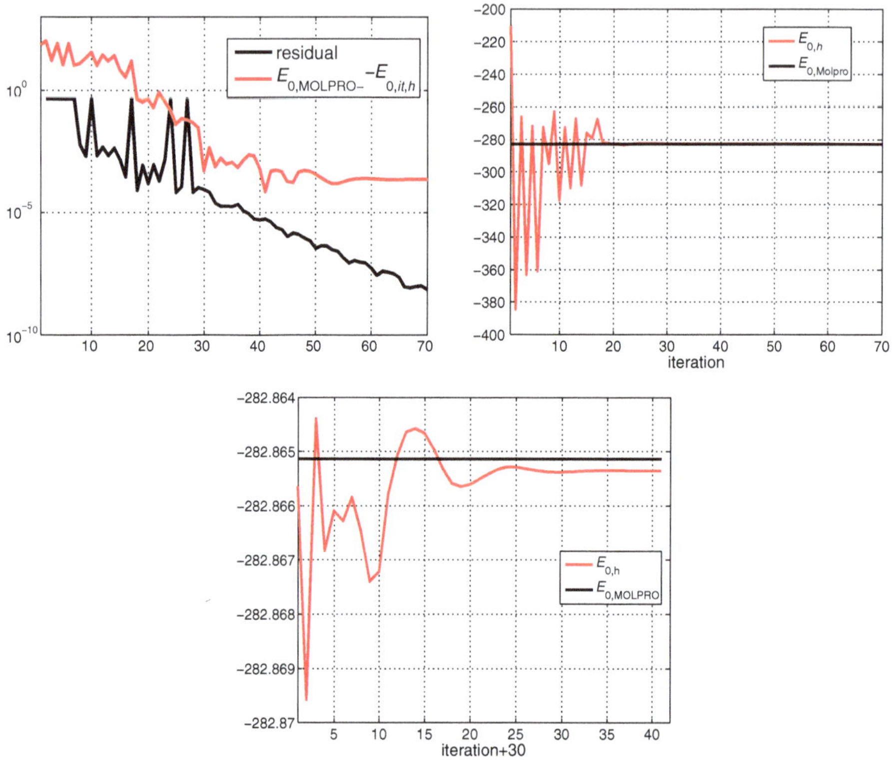

Fig. 5.10: Upper left: iterations history for a glycine molecule with TEI calculated on the grids $n^3 = 131\,072^3$. Upper right: convergence in energy. Bottom: a zoom for energy difference at last iteration.

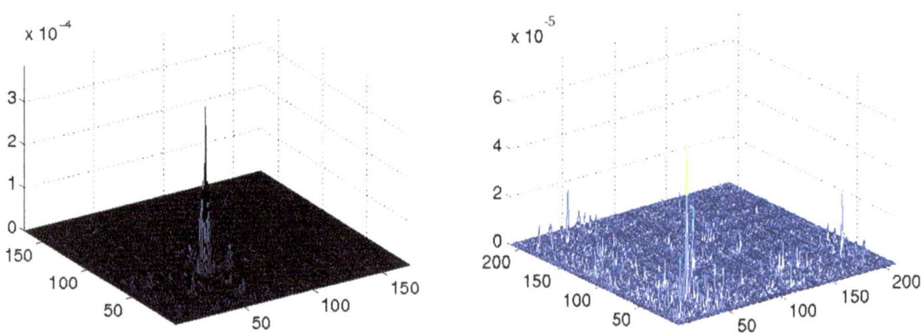

Fig. 5.11: Left: the error in density matrix resulting from calculations of TEI on a grid of size $n^3 = 131\,072^3$. Right: the error in density matrix for the alanine amino acid, TEI computed with $n^3 = 32\,768^3$.

The dominating part in the above tensor calculus resulting in rather large mode size n, can be reduced to the logarithmic scale $O(\log n)$ by applying the QTT approximation. Numerical illustrations on the QTT approximation of functions and operators arising in the solution of the Hartree–Fock equation are given in Table 5.3.

Tab. 5.3: Average QTT ranks for canonical vectors of the tensor $\mathbf{P}_R$ and for the product basis $\{g_\mu g_\nu\}$ designed for the CH_4 molecule.

n^3	$2^{14\cdot 3}$	$2^{15\cdot 3}$	$2^{16\cdot 3}$	$2^{17\cdot 3}$
$1/\|x\|$	3.15	3.13	3.13	3.11
$g_\mu g_\nu$	3.77	3.78	3.76	3.76

This indicates the low rank QTT approximation of (a) the canonical skeleton vectors in low rank decomposition $\mathbf{P}_R$ to the Newton kernel $1/\|x\|$, in $\mathbb{R}^3$; see [212], and (b) the product basis set designed for the CH_4 molecule, both discretized over large $n \times n \times n$ spatial grids. In all cases the QTT approximation accuracy $\varepsilon = 10^{-6}$ is achieved. We observe that QTT ranks of the canonical vectors for both the Newton potential and the product basis remain practically constant in n, ensuring $O(\log \varepsilon^{-1} \log n)$ complexity scaling.

Tab. 5.4: Time (s) for one ab initio SCF EVP iteration and accuracy of the solution with respect to grid size in TEI calculations. MATLAB on an Intel Xeon X5650.

	H_2O	H_2O_2	$C_2H_5NO_2$
N_{orb}, N_b	5; 41	9; 68	20; 170
Time	0.35	0.55	6.0
$\Delta E_{0,g}$, $(65\,536^3)$	$3.0 \cdot 10^{-7}$	$8.0 \cdot 10^{-8}$	$9.1 \cdot 10^{-7}$
$\Delta E_{0,g}$, $(131\,072^3)$	$1.4 \cdot 10^{-7}$	$3.9 \cdot 10^{-8}$	$8.0 \cdot 10^{-7}$

The first row in Table 5.4 shows the number of orbitals and basis functions used in Hartree–Fock calculations for H_2O (cc-pVDZ-41), H_2O_2 (cc-pVDZ-68), and $C_2H_5NO_2$ (cc-pVDZ-170) molecules. The second row represents times (s) for one step of SCF iteration in ab initio solution of the Hartree Fock spectral problem. The next two rows show the relative difference in energy (5.38) for different grid sizes used in TEI calculations. This numeric demonstrates that in the case of fine enough spacial $n \times n \times n$ grids the relative accuracy of about $10^{-7} - 10^{-8}$ can be achieved for moderate size molecules up to small amino acids.

All calculations are performed in the computational box of size $[-20, 20]^3$ bohr^3. This tensor based solver can be considered as the computational tool capable of providing the alternatives to Gaussian type basis sets. Indeed, it applies to any basis set of well separable functions given on a tensor grid.

These numerical results show that the grid based tensor structured Hartree–Fock solver II exhibits the accuracy and computation time comparable with the analytical calculations from MOLPRO. Further numerical results, including the complete grid based calculations with the grid based core Hamiltonian and their comparison with MOLPRO are given in [183]. The Laplacian matrix within the core Hamiltonian is calculated on large 3D grids by using the QTT format.

We summarize that tensor numerical methods described above are implemented in the MATLAB program package Tensor based Electronic Structure Calculation (TESC) by V. Khoromskaia and B. Khoromskij. The TESC package allows the efficient grid based solution of the 3D nonlinear Hartree–Fock equation discretized in a general set of basis functions characterized by the existence of low rank separable representation. All 3D and 6D integrals involved are approximated on large $n \times n \times n$ grids and computed by the black box algorithms in the 1D complexity, $O(n)$, or even in $O(\log n)$ operations, which allows us the high resolution with the large univariate grid size up to $n = 10^6$.

5.2.7 MP2 correction scheme by low rank tensor decompositions of two-electron integrals

The Møller–Plesset perturbation theory (MP2) provides an efficient tool for a correction to the Hartree–Fock energy by relatively modest numerical efforts [3, 161, 280]. Since the straightforward calculation of the MP2 correction scales as $O(N_b^5)$ flops with respect to the number of basis functions, efficient methods are consistently developed to make the problem tractable for larger molecular systems.

The direct method for evaluating the MP2 energy contribution and the energy gradient, which reduces the storage needs to $O(N_b^2)$ at the expense of calculation time, were introduced in [159]. The advantageous technique using the Cholesky factorization of the two-electron integrals introduced in [23] was efficiently applied for MP2 calculations [6]. A linear scaling MP2 scheme for extended systems is considered in [9]. Recently, the MP2 scheme attracted huge interest due to efficient algorithms for the multielectron integrals [309, 367], the density fitting approach exhibiting a low cost when considering extended molecular systems [266, 305, 364], and owing to application of tensor factorization methods [369].

The tensor structured method to calculate the Møller–Plesset (MP2) correction with reduced computational consumptions [190] originates from the $3D$ grid based low rank factorization of the two-electron integrals performed by the purely algebraic optimization.

Given the set of Hartree–Fock molecular orbitals $\{C_p\}$ and the corresponding energies $\{\varepsilon_p\}$, $p = 1, 2, \ldots, N_b$, where $\{C_i\}$ and $\{C_a\}$ denote the occupied and virtual orbitals, respectively. First, one has to transform the TEI matrix $B = [b_{\mu\nu,\lambda\sigma}]$, corresponding to the initial atomic orbitals basis set, to those represented in the molecular orbital (MO) basis,

$$V = [v_{ia,jb}] : \quad v_{iajb} = \sum_{\mu,\nu,\lambda,\sigma=1}^{N_b} C_{\mu i} C_{\nu a} C_{\lambda j} C_{\sigma b} b_{\mu\nu,\lambda\sigma} , \tag{5.45}$$

where $a, b \in \mathcal{I}_v$, $i, j \in \mathcal{I}_o$, and $\mathcal{I}_o := \{1, \ldots, N_{\rm orb}\}$, $\mathcal{I}_v := \{N_{\rm orb} + 1, \ldots, N_b\}$, with $N_{\rm orb}$ denoting the number of occupied orbitals. In the following, we shall use the notation

$$N_v = N_b - N_{\rm orb} , \quad N_{ov} = N_{\rm orb} N_v .$$

The straightforward computation of the matrix V by the above representation makes the dominating impact to the overall numerical cost of order $O(N_b^5)$. The method of complexity $O(N_b^4)$ based on the low rank tensor decomposition of the matrix V was introduced in [190]. Indeed, it can be shown that the rank $R_B = O(N_b)$ approximation to the TEI matrix

$$B \approx LL^{\rm T} ,$$

with the $N \times R_B$ Cholesky factor L, allows us to introduce the low rank representation of the matrix V ([190] and [29]),

$$V = L_V L_V^{\rm T} , \quad L_V \in \mathbb{R}^{N_{ov} \times R_B} ,$$

and then reduce the asymptotic complexity of calculations in (5.45) to $O(N_b^4)$.

Given the tensor $\mathbf{V} = [v_{iajb}]$, the second order MP2 perturbation to the HF energy is calculated by

$$E_{MP2} = - \sum_{a,b \in \mathcal{I}_v} \sum_{i,j \in \mathcal{I}_o} \frac{v_{iajb}(2v_{iajb} - v_{ibja})}{\varepsilon_a + \varepsilon_b - \varepsilon_i - \varepsilon_j} , \tag{5.46}$$

where the real numbers ε_k, $k = 1, \ldots, N_b$ represent the Hartree–Fock eigenvalues.

Introducing the so called doubles amplitude tensor $\mathbf{T}$,

$$\mathbf{T} = [t_{iajb}] : \ t_{iajb} = \frac{(2v_{iajb} - v_{ibja})}{\varepsilon_a + \varepsilon_b - \varepsilon_i - \varepsilon_j} , \quad a, b \in \mathcal{I}_v;\ i, j \in \mathcal{I}_o ,$$

the MP2 perturbation takes the form of a simple scalar product of tensors,

$$E_{MP2} = -\langle \mathbf{V}, \mathbf{T} \rangle = -\langle \mathbf{V} \odot \mathbf{T}, \mathbf{1} \rangle ,$$

where $\mathbf{1}$ denotes the rank-1 all-ones tensor. Now we introduce the low ε rank reciprocal 'energy' tensor

$$\mathbf{E} = [e_{abij}] := \left[\frac{1}{\varepsilon_a + \varepsilon_b - \varepsilon_i - \varepsilon_j} \right] , \quad a, b \in \mathcal{I}_v;\ i, j \in \mathcal{I}_o \tag{5.47}$$

and the partly transposed tensor (transposition in indices a and b)

$$\mathbf{V}' = [v'_{iajb}] := [v_{ibja}] ,$$

which allows us to decompose the doubles amplitude tensor **T** as follows:

$$\mathbf{T} = \mathbf{T}^{(1)} + \mathbf{T}^{(2)} = 2\mathbf{V} \odot \mathbf{E} - \mathbf{V}' \odot \mathbf{E}\,. \tag{5.48}$$

Note that the denominator in (5.46) remains strongly positive if $\varepsilon_a > 0$ for $a \in \mathcal{I}_v$ and $\varepsilon_i < 0$ for $i \in \mathcal{I}_o$. The latter condition (nonzero HOMO-LUMO gap) allows us to prove the low ε rank decomposition of the tensor **E**; see [190, 195].

Each term in the right hand side in (5.48) can be treated separately by using ranks structured tensor decompositions of **V** and **E**, possibly combined with various symmetries and data sparsity. Numerical tests illustrating the tensor approach to the MP2 energy correction are presented in [190].

5.2.8 Toward calculation of excited states: reduced basis approach by low rank approximation to the Bethe–Salpeter Hamiltonian

One of the commonly used approaches for calculation of the excited states in molecules and solids, along with the time dependent DFT, is based on the solution of the Bethe–Salpeter equation (BSE); see for example [61, 160, 284, 307, 315].

The BSE approach leads to the challenging computational task of the solution of the eigenvalue problem for determining the excitation energies ω_n, governed by a large fully populated matrix of size $O(N_{ov}^2) \approx O(N_b^2)$,

$$\begin{pmatrix} A & B \\ B^* & A^* \end{pmatrix} \begin{pmatrix} \mathbf{x}_n \\ \mathbf{y}_n \end{pmatrix} = \omega_n \begin{pmatrix} I & 0 \\ 0 & -I \end{pmatrix} \begin{pmatrix} \mathbf{x}_n \\ \mathbf{y}_n \end{pmatrix}, \tag{5.49}$$

so that the computation of the entire spectrum is prohibitively expensive. Here the large matrix blocks of size $N_{ov} \times N_{ov}$ take a form

$$A = \boldsymbol{\Delta\varepsilon} + V - \overline{W}\,, \quad B = V - \widetilde{W}\,,$$

where the diagonal 'energy' matrix is defined by

$$\boldsymbol{\Delta\varepsilon} = [\Delta\varepsilon_{ia,jb}] \in \mathbb{R}^{N_{ov}\times N_{ov}}: \quad \Delta\varepsilon_{ia,jb} = (\varepsilon_a - \varepsilon_i)\Delta_{ij}\Delta_{ab}\,,$$

while the matrices $\overline{W} = [\overline{w}_{ia,jb}]$ and $\widetilde{W} = [\widetilde{w}_{ia,jb}]$ are determined by permutation of the so called static screened interaction matrix $W = [w_{ia,jb}]$, via $\overline{w}_{ia,jb} = w_{ij,ab}$ and $[\widetilde{w}_{ia,jb}] = [w_{ib,aj}]$, respectively. In turn, the forth order tensor $\mathbf{W} = [w_{iajb}]$ is constructed by certain linear transformations of the tensor $\mathbf{V} = [v_{iajb}]$; see [61, 307] for more details. Here V denotes the TEI matrix in the molecular orbital basis, described in the previous section.

A number of numerical methods for structured eigenvalue problems have been discussed in the literature [28, 33, 81, 237–239, 253]. The general aspects of low rank approximation in the model order reduction were addressed in [283].

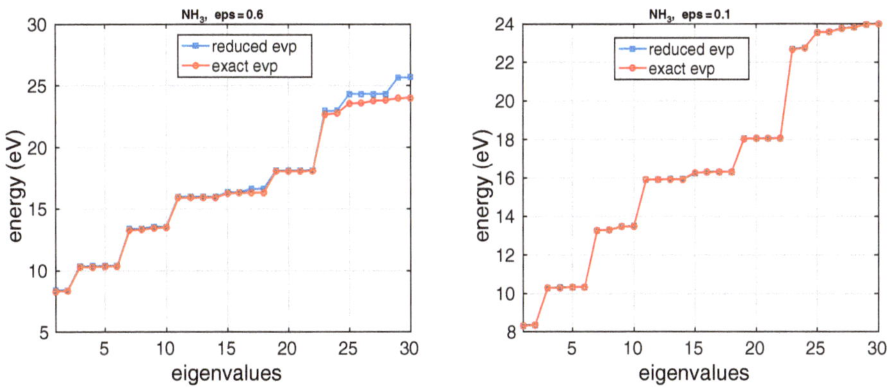

Fig. 5.12: Comparison of $m_0 = 30$ lower eigenvalues for the reduced and exact BSE systems for NH_3 molecule: $\varepsilon = 0.6$, left; $\varepsilon = 0.1$, right.

In what follows, we present the new approach based on the use of low rank sparsity pattern in the BSE system matrix. The tensor approach to the solution of the partial BSE eigenvalue problem for equation (5.49) proposed in [29] suggests to compute the reduced basis set by solving the simplified eigenvalue problem via the low rank plus diagonal approximation to the matrix blocks A and B, and then solve the spectral problem for the subsequent Galerkin projection of the initial system (5.49) to this reduced basis. This procedure relies entirely on multiplication of the simplified BSE matrix with vectors, which reduces dramatically both the storage size and the numerical complexity.

Note that the crucial point for the inventing of this new approach for computation of excitation spectrum was the low rank representation of the TEI matrix developed in [190].

It was demonstrated on the examples of moderate size molecules[29] that a small reduced basis set, obtained by separable approximation with the rank parameters of about several tens, allows us to reveal several lowest excitation energies and respective excited states with the accuracy of about 0.1–0.02eV.

Figure 5.12 illustrates the BSE energy spectrum of the NH_3 molecule (based on HF calculations with cc-pDVZ-48 GTO basis) for the lowest $N_{\mathrm{red}} = 30$ eigenvalues versus the rank truncation parameter $\varepsilon = 0.6$ and 0.1, where the ranks of V and the BSE matrix block W are 4, 5 and 28, 30, respectively. For the above choice of ε, the error in the first (lowest) eigenvalue for the solution of the problem in reduced basis is about 0.11eV and 0.025eV correspondingly. The CPU time in the laptop MATLAB implementation of each example is about five seconds.

Paper [27] presents efficient iterative solution techniques for solving the Bethe–Salpeter large scale eigenvalue problem using the reduced basis approach via low rank factorizations introduced in [29].

For the statically screened interaction part $\overline{W} = [\overline{w}_{ia,jb}]$ of the BSE submatrix, a beneficial substitution was found by a small subblock, which reduces the approximation error by an order of magnitude. Moreover, it provides two-sided error estimates for the exact BSE excitation energies in the case of compact and chain type molecular systems. It was show that the structured inverse iterations (by using matrix inverse) provide fast convergence for calculation of the required central part of the BSE spectrum. For both BSE and TDA models (i.e., $B = 0$), the inverse matrix can be represented in the same diagonal plus low rank plus reduced block format by using the Sherman–Morrison–Woodbury scheme. The estimates on the complexity of algorithms for diagonal plus low rank plus reduced block inverse iterations have been presented, see [27].

Furthermore, the solution of the BSE spectral problem in the QTT format is discussed in [27]. The QTT tensor transform of the initial BSE system to the higher dimensional setting allows us to construct a structural solver of complexity $\mathcal{O}(\log(N_o)N_o^2)$. This complexity is determined by only the number of occupied orbitals, N_o, in the molecular system (i.e., on physical characteristics of the molecule), but it is almost independent of the system size determined by the number of atomic orbitals basis functions, N_b. The results are confirmed by a number of numerical tests where we observe dramatic reduction of solution time. For example, TDA calculations in QTT format for the C_2H_5OH molecule with matrix size 1430^2 take 0.14 s, while for the $C_3H_7NO_2$ (alanine amino acid) with TDA matrix size 4488^2, the CPU time increases only to 0.63 s. Note that the solution of the eigenvalue problem with the rank structured representation of the BSE matrix reduces calculation times for large enough molecules at least by two orders of magnitude.

5.2.9 Sketch on the Green function iteration for the Kohn–Sham equation

Recall that the Hartree–Fock equation for the N-electron system is a self-consistent eigenvalue problem in $L^2(\mathbb{R}^3)$

$$\mathcal{F}_\Phi \phi_i(x) = \lambda_i \, \phi_i(x) \, , \quad \int_{\mathbb{R}^3} \phi_i \phi_j = \Delta_{ij} \, , \quad \text{for} \quad i,j = 1, \ldots, N \, , \tag{5.50}$$

where $\mathcal{F}_\Phi$ is the nonlinear Fock operator

$$\mathcal{F}_\Phi := -\frac{1}{2}\Delta - \sum_{\nu=1}^{M} \frac{Z_\nu}{|x - a_\nu|} + \left(\rho \star \frac{1}{|x|}\right) - \sum_{j=1}^{N} \left(\cdot \, \phi_j \star \frac{1}{|x|}\right) \phi_j \, ,$$

where $\star$ denotes the convolution product in $L^2(\mathbb{R}^3)$.

In the Kohn–Sham model the operator $\mathcal{F}_\Phi$ is simplified to

$$\mathcal{F}_\Phi \mapsto \mathcal{K}_\rho := -\frac{1}{2}\Delta - \sum_{\nu=1}^{M} \frac{Z_\nu}{|x - a_\nu|} + \left(\rho \star \frac{1}{|x|}\right) + v_{xc}(\rho) \, , \tag{5.51}$$

where $v_{xc}(\rho)$ is a scalar function that depends only upon $\rho(x)$ (one of the possible choices is $v_{xc}(\rho) = \rho^{1/3}$).

Standard methods to solve the nonlinear equation (5.50) or (5.51) are based on the two-level iteration, which include: (a) the so called *self-consistent field* (SCF) algorithms, i.e., the iterations on the nonlinearity, and (b) at each SCF cycle, computation of the spectral projection to build the new electron density/density matrix, using the current (frozen) discretization to the Fock or Kohn–Sham operators.

In the present notes we focus on the solution of the spectral problem in step (b). Here we will follow [204] and consider the eigenvalue problem of the form

$$H\phi := \left[-\frac{1}{2}\Delta + V\right]\phi = \lambda\phi\,, \tag{5.52}$$

where the 'interaction' potential V ensures the existence of the discrete spectrum that belongs to $(-\infty, 0)$, and also allows the low separation rank representation in a sense that will be specified later on. To avoid unessential technicalities, below we consider the Kohn–Sham model, which is represented by the local multiplication operator with the potential

$$V := -\sum_{\nu=1}^{M} \frac{Z_\nu}{|\cdot - a_\nu|} + \left(\rho \star \frac{1}{|x|}\right) + v_{xc}(\rho)\,. \tag{5.53}$$

However, the case of the Hartree–Fock equation can be treated with minor modifications.

For the ease of presentation, we consider the computation of the minimal eigenvalue

$$\lambda^* = \min\left(\sigma\left(-\frac{1}{2}\Delta + V\right)\right)$$

and the corresponding eigenfunction ϕ^* of the problem (5.52). Introducing the Green's function operator (the elliptic resolvent)

$$R_z = (-\Delta + zI)^{-1}$$

with the kernel function defined by the Yukawa potential

$$G_z(x) := \frac{e^{-z|x|}}{4\pi|x|}\,, \qquad z \in (0, \infty)\,,$$

and setting $z = \sqrt{-2\lambda}$, we obtain the equivalent formulation to the eigenvalue problem (5.52)

$$\phi = \mathcal{G}_z\phi \quad \text{with} \quad \mathcal{G}_z := -2(V\cdot) \star G_z \equiv -2R_z V \tag{5.54}$$

with the compact operator $\mathcal{G}_z$. The important feature of this formulation is that any eigenvalue-eigenfunction pair, (λ, ϕ), (λ in the discrete spectrum), for the operator H is a fixed point solution of the problem (5.54) ([157]).

Numerical algorithms for solving the integral equation (5.54) are based on considering the eigenvalue problem for the Lippmann–Schwinger type parametric integral operator

$$\mu_z\phi_z = \mathcal{G}_z\phi_z \quad \text{for} \quad z > 0\,, \tag{5.55}$$

where both $\mu_z = \mu(z)$ and ϕ_z depend on the parameter z.

Lemma 5.8. *Suppose that the 'exchange part' $v_{xc}(\rho)$ in the Kohn–Sham potential* (5.53) *satisfies the following assumption:*

$$v_{xc}(\rho)(x) = V_1(x) + V_2(x)$$

with $V_1 \in L^2(\mathbb{R}^3)$ and $V_2 \in L^\infty(\mathbb{R}^3)$ where V_2 can be taken as arbitrarily small in the L^∞ sense. Then $\mu_{\lambda^*} = 1$ is the largest eigenvalue of $\mathcal{G}_{\lambda^*}$, and the corresponding single eigenfunction $\phi_{\lambda^*} = \phi_*$ is the desired ground state eigenfunction of (5.52).

Proof. The proof is a slight modification of respective arguments in the proof of Theorem 1.1 in [279]. □

Lemma 5.8 implies that finding the eigenvalue λ^* can be reduced to an iterative solution of the scalar nonlinear equation

$$\mu(z) = 1\,, \quad z \in \mathbb{R}_+ \tag{5.56}$$

with an initial guess $z_0 \in \mathbb{R}_+$ that belongs to the attraction basin of λ^* (we never start electronic structure calculations from scratch).

Other eigenvalues/eigenfunctions may be obtained by deflation, which is used to recast the integral equation for each orbital as a ground state problem (see [157] for more details). Here we briefly repeat the argument from [157]. Let $\mathcal{P}_m$ be the orthoprojection onto the space of eigenfunctions of lower energy than orbital m. Then the mth occupied orbital ψ_m will be the lowest energy solution of

$$(1 - \mathcal{P}_m)H(1 - \mathcal{P}_m)\psi_m = \lambda_m \psi_m\,,$$

which leads to the modified integral formulation

$$\psi_m = -2R_z(V + \mathcal{P}_m H(1 - \mathcal{P}_m))\psi_m\,.$$

Several iteration schemes were considered in the literature to solve the nonlinear problem (5.54). In particular, in the case of the Schrödinger equation, the convergence of the power method in the form

$$\phi_{n+1} = \frac{\mathcal{G}_{z_n}\phi_n}{\|\mathcal{G}_{z_n}\phi_n\|}\,, \quad z_{n+1} = \langle H\phi_{n+1}, \phi_{n+1}\rangle$$

was analyzed in [279]. Under the corresponding assumptions, the power method can be applied to the Kohn–Sham equation as well. Newton's type iteration in the form

$$\widetilde{\phi}_n = \mathcal{G}_{z_n}\phi_n\,, \quad \phi_{n+1} = \widetilde{\phi}_n/\|\widetilde{\phi}_n\|\,,$$

$$z_{n+1} = z_n - \langle V\phi_n, \phi_n - \widetilde{\phi}_n\rangle/\|\widetilde{\phi}_n\|^2\,,$$

was applied in [157] to the Kohn–Sham equation defined by the potential (5.53).

The following result provides sufficient conditions for the quadratic (local) convergence of the Newton iteration applied to equation (5.56),

$$\text{given } z_0: \quad z_{n+1} = z_n - \frac{\mu(z_n)}{\frac{\partial \mu}{\partial z}(z_n)}, \quad n = 0, 1, \ldots \tag{5.57}$$

Particular realization of the iteration (5.57) is determined by discretization chosen for $\mu(z_n)$ and $\frac{\partial \mu}{\partial z}(z_n)$.

Theorem 5.9 ([204]).

(I) *Under assumptions in Lemma 5.8 the power method converges geometrically.*

(II) *Let the exact eigenvalue-eigenfunction pair of* (5.52), (λ_*, ϕ_*), *satisfy*

$$\langle (V\phi_*) \star e^{-\lambda_*|x|}, \phi_* \rangle \neq 0\,. \tag{5.58}$$

Then (a) the Newton iteration (5.57) *converges (locally) quadratically, and (b) the quasi-Newton iteration defined by the approximation*

$$\frac{\partial \mu}{\partial z}(z_n) \approx \langle (V\phi_n) \star e^{-z_n|x|}, \phi_n \rangle\,,$$

converges (locally) quadratically as well.

Proof. First, we note that G_z is an analytic family in z, hence $(V\,\cdot) \star G_z$ is an analytic family and its eigenvalue μ_z and eigenfunction ϕ_z depend analytically on z. Differentiating equation (5.55) at $z = \lambda_*$, we obtain

$$\frac{\partial \mu_z}{\partial z}\phi_* + \lambda_* \frac{\partial \phi_z}{\partial z} = \frac{\partial \mathcal{G}_z}{\partial z}\phi_* + \mathcal{G}_{\lambda_*}\frac{\partial \phi_z}{\partial z}\,.$$

Scalar multiplication of this equation with ϕ_* and taking into account that

$$\mathcal{G}_{\lambda_*}\phi_* = \lambda_*\phi_*\,,$$

leads to

$$\frac{\partial \mu_z}{\partial z}_{|z=\lambda_*} = \left\langle \frac{\partial \mathcal{G}_z}{\partial z}_{|z=\lambda_*}\phi_*, \phi_* \right\rangle\,. \tag{5.59}$$

Moreover, direct computation shows that

$$\frac{\partial \mathcal{G}_z}{\partial z}_{|z=\lambda_*} = (V\,\cdot) \star e^{-\lambda_*|x|}\,. \tag{5.60}$$

This proves (a).

Furthermore, due to analyticity of G_z in z, and in view of (5.59) and (5.60), we conclude that the approximation (in general noncomputable)

$$\frac{\partial \mu_z}{\partial z}(z_n) \approx \frac{\partial \mu_z}{\partial z}(\lambda_*) = \langle (V\phi_*) \star e^{-\lambda_*|x|}, \phi_* \rangle$$

ensures the local quadratic convergence. The desired computable approximation can be proven by perturbation argument. Indeed, the functional $f\colon \mathbb{R}_+ \times L^2(\mathbb{R}^3) \to \mathbb{R}$,

$$f(z, \phi) := 1/\langle (V\phi) \star e^{-z|x|}, \phi \rangle, \quad z > 0,$$

is continuously differentiable at (λ_*, ϕ_*), hence in the small vicinity of (λ_*, ϕ_*) we have

$$f(z_n, \phi_n) = 1/\langle (V\phi_*) \star e^{-\lambda_*|x|}, \phi_* \rangle + \zeta_n, \quad \zeta_n \in \mathbb{R},$$

where

$$|\zeta_n| \le C(|z_* - z_n| + \|\phi_* - \phi_n\|).$$

This completes the proof of part (b). □

In what follows, we focus on the efficient implementation of the Lippmann–Schwinger type operator $\mathcal{G}_z$ for $z > 0$, which constitutes the computational kernel of the Green iterations in both power method and Newton's scheme [204].

Example 5.10. The nondegeneracy condition (5.58) can be justified in the case of a hydrogen atom/ion with $V(x) = C/\|x\|$ and $\phi_*(x) = e^{-\mu_*\|x\|}$ with $\mu_* > 0$. In fact, it is easy to check (using representation in spherical coordinates) that

$$\left\langle \frac{e^{-\mu_*\|x\|}}{\|x\|} \star e^{-\mu_*\|x\|}, e^{-\mu_*\|x\|} \right\rangle \neq 0.$$

Low tensor rank representation of operators. We are interested in the low rank tensor approximation of the operator $\mathcal{G}_z$ in (5.54), which can be presented (up to the scaling factor) in one of the following forms:

$$\mathcal{G}_z = (V\,\cdot) \star G_z \equiv R_z V. \tag{5.61}$$

The first one is well suited for the integral representation of equation (5.54), while the second form, which contains the elliptic resolvent, can be applied to the FE/FD discretization of the initial equation (5.52) posed on the bounded domain. We refer to [110, 142, 143] concerning the low rank tensor approximation methods for the elliptic resolvent operators.

To fix the idea, in what follows we will discuss the method of fast tensor product convolution of the Yukawa kernel and the action of interaction potential, $(V\,\phi) \star G_z$, involved in the integral representation of equation (5.54). The approach is based on the ideas from [134, 201] applied to the collocation convolution schemes. The main advantages of the integral representation $\mathcal{G}_z = (V\,\cdot) \star G_z$ are the following:

- Compactness of the operator $\mathcal{G}_z$.
- Applicability of simple collocation schemes with discontinuous basis functions (L^2 setting).
- Possibility to compute the total energy using the only integral operators (avoid the application of the Laplace operator Δ that requires at least H^1 basis functions).

- Existence of low separation rank approximations to the operators V and G_Z.
- Nonlinear (say Newton's) quadratically convergent iteration, which is well suited for the numerical multilinear algebra via truncation to fixed tensor format.

Collocation discretization of the convolution operator. Recall that the multidimensional convolution transform in $L^2(\mathbb{R}^d)$ is given by

$$w(x) := (f * g)(x) := \int_{\mathbb{R}^d} f(y)g(x-y)dy \quad f, g \in L^2(\mathbb{R}^d), \quad x \in \mathbb{R}^d.$$

We are interested in approximately computing $f * g$ in some fixed box $\Omega = [-A, A]^d$; see Section 5.1 further details. We suppose that the convolving function f has support in $\Omega' = [-B, B]^d \subset \Omega$ ($B < A$, i.e., supp $f, \subset \Omega'$). Our particular choice will correspond to $g(x) = e^{-z|x|}/|x|$ and $f(x) = V(x)\phi(x)$, where ϕ is an exponentially decaying function. Note that for both Hartree–Fock and Kohn–Sham equations the interaction potential V already contains the term $\rho \star \frac{1}{|x|}$ (ρ is the electron density), which can be efficiently approximated by the low tensor rank convolution algorithms, to be described below, see also Section 5.1. The numerical results and theoretical analysis can be found in [201].

In the following, we focus on the collocation methods. In this case the approximation properties have been discussed in Section 5.1.

For the collocation method with piecewise constant basis functions and for C^2 input data f, we are able to prove the error bound $O(h^2)$ (superconvergence); see Section 5.1. More refined error analysis justifies the Richardson extrapolation method on a sequence of grids providing the better approximation error $O(h^3)$. Such an extrapolation allows substantial improvement of the approximation accuracy without an extra cost.

5.2.10 On separable approximation to the convolving functions

In our applications, the function related collocation coefficients tensor $\mathbf{F} = [f_{\mathbf{i}}]_{\mathbf{i} \in \mathcal{J}}$ is generated by either density function $\rho(x)$ or by the product $V(x)\psi(x)$. In this way we make a priori assumptions on the existence of low rank approximation to both tensors. This assumption is not easy to analyze, however it works well in practice.

Example 5.11. In the case of a hydrogen atom we have

$$\rho(x) = e^{-2\|x\|}, \quad \text{and} \quad V(x)\psi(x) = \frac{e^{-\|x\|}}{\|x\|},$$

hence the corresponding low rank tensor approximations can be proven along the line of Lemma 4.3 [206] and Theorem 5.12 below.

To construct low rank approximation of function generated tensor **G**, we consider a class of multivariate spherically symmetric convolving kernels $g\colon \mathbb{R}^d \to \mathbb{R}$ parametrized by

$$g(y) = G(\rho(y)) \equiv G(\rho) \quad \text{with} \quad \rho \equiv \rho(y) = y_1^2 + \cdots + y_d^2 ,$$

where the univariate function $G\colon \mathbb{R}_+ \to \mathbb{R}$ can be represented via the generalized Laplace transform

$$G(\rho) = \int_{\mathbb{R}_+} \widehat{G}(\tau^2) e^{-\rho\tau^2} d\tau . \tag{5.62}$$

Without loss of generality, we introduce one and the same scaling function

$$\phi_i(\cdot) = \phi(\cdot + (i-1)h) , \quad i \in I_n ,$$

for all spatial dimensions $\ell = 1, \ldots, d$, where $h > 0$ is the mesh parameter, so that the corresponding tensor product basis function $\phi_{\mathbf{i}}$ is defined by (5.3).

Using sinc quadrature methods we approximate the collocation coefficients tensor $\mathbf{G} = [g_{\mathbf{i}}]_{\mathbf{i}\in\mathcal{I}}$ in (5.5) via rank-$(2M+1)$ canonical decomposition

$$\mathcal{G} \approx \sum_{k=-M}^{M} w_k \, \mathcal{E}(\tau_k) \quad \text{with} \quad \mathcal{E} = [e_{\mathbf{i}}] , \quad \mathbf{i} \in \mathcal{I} ,$$

with suitably chosen coefficients $w_k \in \mathbb{R}$ and $\tau_k \in \mathbb{R}_+$, and with the rank-1 components $\mathcal{E}(\tau_k) \in \mathbb{R}^{\mathcal{I}}$ given by

$$e_{\mathbf{i}}(\tau_k) = \widehat{G}(\tau_k^2) \prod_{\ell=1}^{d} \int_{\mathbb{R}} e^{-y_\ell^2 \tau_k^2} \phi_{i_\ell}(y_\ell) dy_\ell . \tag{5.63}$$

Following the standard description of the sinc methods, we introduce the Hardy space $H^1(D_\Delta)$. Given $f \in H^1(D_\Delta)$, $\mathfrak{h} > 0$, and $M \in \mathbb{N}_0$, the corresponding sinc quadrature reads as

$$T_M(f, \mathfrak{h}) := \mathfrak{h} \sum_{k=-M}^{M} f(k\mathfrak{h}) \approx \int_{\mathbb{R}} f(\xi) d\xi . \tag{5.64}$$

We apply Theorem 2.46 from Section 2.3 on the the hyperexponential decay of the sinc approximation method.

For a class of analytic functions the exponential convergence of the above quadrature in M can be proven ([141]). In our particular case of the Yukawa potential for $\kappa \in [0, \infty)$, we apply the Gauss transform (5.62) to obtain

$$G(\rho) = \frac{e^{-\kappa\sqrt{\rho}}}{\sqrt{\rho}} = \frac{2}{\sqrt{\pi}} \int_{\mathbb{R}_+} \exp(-\rho\tau^2 - \kappa^2/\tau^2) d\tau , \tag{5.65}$$

corresponding to the choice

$$\widehat{G}(\tau^2) = \frac{2}{\sqrt{\pi}} e^{-\kappa^2/\tau^2} .$$

Theorem 5.12. *For given $G(\rho)$ in (5.65) with fixed $\kappa > 0$, we set*

$$w_k = h_M \widehat{G}(\tau_k^2) \quad \text{and} \quad \tau_k = e^{t_k} ,$$

where $t_k = kh_M$ with $h_M = C_0 \log(M)/M$ for some $C_0 > 0$. Then we have the exponentially convergent rank-$(2M+1)$ CP approximation to the collocation coefficients tensor $\mathbf{G} = [g_{\mathbf{i}}]_{\mathbf{i}\in\mathcal{I}}$ in (5.5) generated by the Yukawa potential $G(\rho)$,

$$\left\| \mathcal{G} - \sum_{k=-M}^{M} w_k \, \mathcal{E}(\tau_k) \right\| \le C e^{-\pi^2 M/(C+\log(M))} , \tag{5.66}$$

where $\mathcal{E} = [e_{\mathbf{i}}]$ with $e_{\mathbf{i}}(\tau_k)$ given by (5.63).

$$e_{\mathbf{i}}(\tau_k) = \widehat{G}(\tau_k^2) \prod_{\ell=1}^{d} \int_{\mathbb{R}} e^{-y_\ell^2 \tau_k^2} \phi_{i_\ell}(y_\ell) dy_\ell .$$

The detailed proof is presented in Section 2.4.7.

Theorem 5.12 applies to the case with fixed $\kappa > 0$. To construct the approximation, which is uniform for $\kappa \in [0, \infty)$ (i.e., including the case of Coulomb potential corresponding to $\kappa = 0$) we modify the above quadrature using a variable transformation $t = \sinh(u)$ to obtain the quadrature

$$\int_{\mathbb{R}} p_{\mathbf{i}}(t)\, dt = \int_{\mathbb{R}_+} 2\cosh(u)\, p_{\mathbf{i}}(\sinh(u))\, du \approx \sum_{k=0}^{M} w_k \, p_{\mathbf{i}}(t_k)$$

with $t_k := \sinh(k\mathfrak{h}_M)$ and

$$w_k := \begin{cases} \mathfrak{h}_M & \text{for } k = 0 \\ 2\mathfrak{h}_M \cosh(k\mathfrak{h}_M) & \text{for } k > 0 \end{cases} \tag{5.67}$$

with the choice $\mathfrak{h}_M = C_0 \frac{\ln(M)}{M}$ for some $C_0 > 0$. The analysis is similar to those in Lemma 4.3 [141], taking into account the symmetry of the integrand; see also Section 2.4.

In what follows, we consider the discretization and complexity issues, and present some numerical illustrations. Our primal interest is concerned with integral formulation in a L^2-setting. Hence, we consider the collocation type approximation of the operator $\mathcal{G}_z = (V\cdot) \star G_z$ with respect to the certain ansatz space $W = \operatorname{span}\{\phi_{\mathbf{i}}\} \subset L^2(\mathbb{R}^3)$ of discontinuous functions; see (5.3). Letting

$$\psi = \sum_{\mathbf{i}\in\mathcal{I}} a_{\mathbf{i}} \phi_{\mathbf{i}} ,$$

we define the stiffness matrix $\mathbb{V} = \{\langle V\phi_{\mathbf{i}}, \phi_{\mathbf{j}}\rangle\}_{\mathbf{i},\mathbf{j}\in\mathcal{I}}$. Now we calculate the L^2-projection of $V\psi$ onto W,

$$\mathcal{P}_W(V\psi) = \sum_{\mathbf{j}\in\mathcal{I}} \sum_{\mathbf{i}\in\mathcal{I}} a_{\mathbf{i}} \langle V\phi_{\mathbf{i}}, \phi_{\mathbf{j}}\rangle \phi_{\mathbf{j}} = \sum_{\mathbf{j}\in\mathcal{I}} b_{\mathbf{j}} \phi_{\mathbf{j}} ,$$

with $b_{\mathbf{j}} = (\mathbb{V}\mathbf{A})_{\mathbf{j}}$, $\mathbf{A} = [a_{\mathbf{i}}]$, and introduce the coefficients tensor

$$\mathbf{F} = [b_{\mathbf{j}}]_{\mathbf{j}\in\mathcal{J}} \in \mathbb{R}^{n\times n\times n}, \quad \text{i.e., } \mathbf{F} = \mathbb{V}\mathbf{A}\,.$$

The discretization of $\mathcal{G}_z\psi$ is then defined by the tensor structured convolution

$$\mathcal{G}_z\psi \approx \mathbf{G} \star \mathbf{F}\,,$$

where tensor **G** was described in Section 5.1.7. In the case of piecewise constant basis functions (5.3) the stiffness matrix $\mathbb{V}$ becomes diagonal, $\mathbb{V} = \operatorname{diag}\{\langle V\phi_{\mathbf{i}}, \phi_{\mathbf{i}}\rangle\}_{\mathbf{i}\in\mathcal{J}}$.

Finite element or finite difference approximations result in the representation $\mathcal{G}_z\psi = R_z V\psi$, where R_z is the discrete elliptic resolvent and $V\psi$ is the action of the discrete interaction potential. In this case, we first approximate the initial operator H using the ansatz space of continuous functions in $H^1(\mathbb{R}^3)$, and then compute the discrete elliptic inverse $(-\Delta + zI)^{-1}$.

For example, the finite difference Helmholtz operator on the uniform $n \times n \times n$ tensor product grid (subject to homogeneous Dirichlet boundary conditions) is represented by a matrix $\mathbb{A} + zh^2\mathbf{I} \in \mathbb{R}^{n\times n\times n}$ with

$$\mathbb{A} := V^{(1)} \otimes I \otimes I + I \otimes V^{(2)} \otimes I + I \otimes I \otimes V^{(3)}\,,$$

and with $V^{(j)}, I \in \mathbb{R}^{n\times n}$, where I is the identity matrix and $V^{(j)} = \operatorname{tridiag}\{-1, 2, -1\}$, $j = 1, 2, 3$. Using sinc methods, we choose coefficients $t_k, c_k \in \mathbb{R}$ and then construct the rank-$(2M + 1)$ CP approximation in the form

$$\mathbb{B}_{(M)} = \sum_{k=-M}^{M} c_k \bigotimes_{\ell=1}^{3} \exp\left(-t_k\left(V^{(j)} + \frac{zh^2}{d} I\right)\right) \approx (\mathbb{A} + zh^2\mathbf{I})^{-1}\,,$$

providing exponential convergence in $r = 2M + 1$,

$$\left\|(\mathbb{A} + z\mathbf{I})^{-1} - \mathbb{B}_{(M)}\right\| \le Ce^{-sM/\log(M)}\,.$$

The choice of coefficients t_k, c_k is described in [141].

We assume that at each step m of the Newton iteration, we have rank-R_G and rank-R_F CP representations of tensors $\mathbf{G}_n \in \mathbb{R}^{n^3}$ and $\mathbf{F}_n \in \mathbb{R}^{n^3}$, respectively, at our disposal, as well as the good initial guess for the orthogonal components of the Tucker approximation to the next iterand $\psi_{m+1} = \mathbf{G}_m \star \mathbf{F}_m$. We apply the two-level approximation method [211]. Since the tensor ψ_{m+1} has maximal canonical rank $R_G R_F$, its rank-r Tucker approximation exhibits the cost

$$Q_{CT} = O(R_G R_F n r^2 + n r^2 \min\{r^2, n\})\,.$$

Computing the rank-R_F CP approximation to the corresponding core tensor of size $r \times r \times r$ (by ALS iteration) with the cost that does not depend on n,

$$Q_{FC} = O(N_{it} R_F r^2)\,,$$

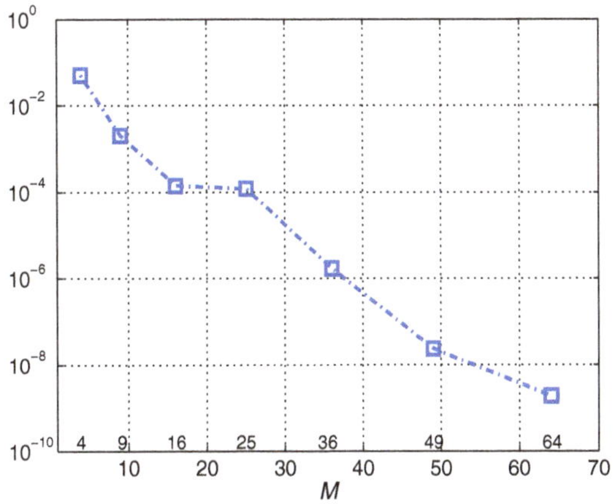

Fig. 5.13: CP approximation for the discrete elliptic inverse by $(2M + 1)$-term sinc quadrature.

where N_{it} is the number of ALS iterations and we return the current iterand ψ_{m+1} to the required format $\psi_{m+1} \in \mathcal{C}_{R_F}$. The overall numerical cost is estimated by $Q_{TC} + Q_{FC}$, which scales linearly in the univariate problem size n.

Similar complexity bounds can be derived for the FE/FD discretization.

Figure 5.13 illustrates the convergence of sinc approximation in M for the discrete elliptic inverse defined on $n \times n \times n$ grid with $n = 128$. Here M defines the number of terms in the sinc quadrature, so that the total rank is $r = 2M + 1$. Further numerical illustrations can be found in [204].

5.2.11 Linearized Hartree–Fock equation for finite lattice and quasiperiodic systems: tensor approach

In this section we briefly discuss the recent applications of tensor numerical methods for solving the linearized Hartree–Fock equation on finite lattices. The detailed presentation can be found in [187].

Here we discuss the grid based tensor approach to the solution of an elliptic eigenvalue problem for the 3D lattice structured systems in a bounding box. We follow [187] and focus on the application to the linearized Hartree–Fock equation for extended systems composed of atoms or molecules located at nodes of a $L_1 \times L_2 \times L_3$ finite lattice, for both open boundary conditions and a periodic supercell. The latter is useful because the structure of the respective Galerkin matrix (i.e., the Fock matrix) in the presence of defects can be treated as a small (local) perturbation to a periodic system. We consider the 3D model eigenvalue problem for the Fock operator confined to the core Hamiltonian part. We observed in numerical experiments that there is an irreducible difference between the spectral data for periodic and nonperiodic settings even for larger lattice size.

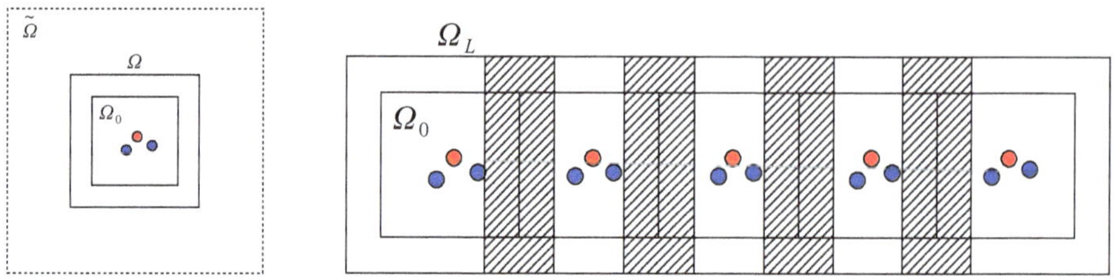

Fig. 5.14: 2D projection of the supercell for the $5 \times 1 \times 1$ chain in 3D.

We consider the matrix structure of the Galerkin discretization for the elliptic eigenvalue problem in the form

$$\mathcal{H}\varphi(x) \equiv [-\Delta + v(x)]\varphi(x) = \lambda\varphi(x)\,, \quad x \in \Omega \subset \mathbb{R}^d\,, \;\; d = 1, 2, 3\,, \tag{5.68}$$

where the potential $v(x)$ is constructed by replication of the reference kernel defined in the rectangular unit cell Ω_0 over a d dimensional rectangular $L_1 \times L_2 \times L_3$ lattice in a box, such that $\varphi \in H_0^1(\Omega)$, or in a rectangular supercell Ω with periodic boundary conditions. We focus on the important particular case of $v(x) = v_c(x)$ corresponding to the core Hamiltonian part in the Fock operator that constitutes the Hartree–Fock spectral problem arising in electronic structure calculations. In this case the electrostatic potential $v_c(x)$ is obtained as the large lattice sum of long range interactions defined by the Newton kernel.

The electrostatic potential generated by the core Hamiltonian is defined by a sum

$$v_c(x) = \sum_{\nu=1}^{M} \frac{Z_\nu}{\|x - a_\nu\|}\,, \quad Z_\nu > 0\,, \;\; x, a_\nu \in \mathbb{R}^3\,, \tag{5.69}$$

where M is the total number of nuclei in the system, and a_ν, Z_ν represent their Cartesian coordinates and the respective charge numbers.

Nuclear potential operator for a lattice system in a box. Given the potential sum $v_c(x)$ defined by (5.69) in the scaled unit cell $\Omega = [-\frac{b}{2}, \frac{b}{2}]^3$ of size $b \times b \times b$, seen in Figure 5.14 (left), we specify the smaller subdomain $\Omega_0 = [-\frac{b_0}{2}, \frac{b_0}{2}]^3 \subset \Omega$ (called the formation cell) whose interior contains all atomic centers in the unit cell included into the summation in (5.69).

Let us consider an interaction potential in a symmetric computational box (supercell)

$$\Omega_L = B_1 \times B_2 \times B_3\,, \quad \text{with} \quad B_\ell = \frac{1}{2}[-b_0 L_\ell - b, b_0 L_\ell + b]$$

consisting of a union of $L_1 \times L_2 \times L_3$ unit cells $\Omega_{\mathbf{k}}$, obtained by a shift of the reference domain Ω along the lattice vector $\mathbf{b_0}/2 + b_0\mathbf{k}$, where $\mathbf{k} = (k_1, k_2, k_3) \in \mathbb{Z}^3$, such that for $\ell = 1, 2, 3$,

$$k_\ell \in \mathcal{K}_\ell := \{0, 1, \dots, L_\ell - 1\}\,.$$

In this notation the choice $L_\ell = 1$ corresponds to the 3D one-layer system in the respective variable. Figure 5.14 (right) represents the 2D projection of the 3D computational domain Ω_L for the $L_1 \times 1 \times 1$ molecular chain with $L_1 = 5$. Dashed regions correspond to the overlapping parts between shifted unit cells.

For the discussion of complexity issues, we often consider a cubic lattice of equal sizes $L_1 = L_2 = L_3 = L$. By the construction, we set $b = nh$ and $b_0 = n_0 h$, where the mesh size $h > 0$ is chosen to be the same for all spatial variables.

In the most interesting case of extended system in a box, further called case (B), the potential $v_{c_L}(x)$, for $x \in \Omega_L$, is obtained by summation over all unit cells $\Omega_{\mathbf{k}}$ in Ω_L,

$$v_{c_L}(x) = \sum_{\nu=1}^{M_0} \sum_{\mathbf{k}\in\mathcal{K}^3} \frac{Z_\nu}{\|x - a_\nu - b\mathbf{k}\|}, \quad x \in \Omega_L . \tag{5.70}$$

Note that the direct calculation by (5.70) includes contributions of all summands at each of L^3 unit cells $\Omega_{\mathbf{k}} \subset \Omega_L$, $\mathbf{k} \in \mathcal{K}^3$, on a 3D lattice, which presupposes substantial numerical costs at least of the order of $O(L^3)$ per unit cell.

The fast calculation of (5.70) is implemented by using the tensor summation method introduced in [188, 189] (see also Section 5.5), which can be briefly described as follows. Let Ω_{N_L} be the $N_L \times N_L \times N_L$ uniform grid on Ω_L with the same mesh size h as above, we introduce the corresponding space of piecewise constant basis functions of the dimension N_L^3, where $N_L = n_0 L + n - n_0$. Given the reference canonical tensor $\mathbf{P}_R$, the resultant lattice sum is presented by the canonical tensor $\mathbf{P}_{c_L}$

$$\mathbf{P}_{c_L} = \sum_{\nu=1}^{M_0} Z_\nu \sum_{q=1}^{R} \left(\sum_{k_1\in\mathcal{K}_1} \mathcal{W}_{\nu(k_1)} \widetilde{\mathbf{p}}_q^{(1)} \right) \otimes \left(\sum_{k_2\in\mathcal{K}_2} \mathcal{W}_{\nu(k_2)} \widetilde{\mathbf{p}}_q^{(2)} \right) \otimes \left(\sum_{k_3\in\mathcal{K}_3} \mathcal{W}_{\nu(k_3)} \widetilde{\mathbf{p}}_q^{(3)} \right), \tag{5.71}$$

whose rank is uniformly bounded $R_c \le M_0 R$.

In the case of a lattice system in a box, we define the basis set on a supercell Ω_L (and on the extended domain $\widetilde{\Omega}_L$) by translation of the generating basis, defined in Ω_0 for the single molecule, by the lattice vector $b\mathbf{k}$, i.e., $\{g_\mu(x)\} \mapsto \{g_\mu(x + b\mathbf{k})\}$, $\mu = 1, \ldots, m_0$, where $\mathbf{k} = (k_1, k_2, k_3) \in \mathcal{K}^3$, assuming zero extension of $\{g_\mu(x + b\mathbf{k})\}$ beyond each local bounding box $\widetilde{\Omega}_{\mathbf{k}}$. The corresponding tensor representation of such functions is denoted by $\mathbf{G}_{\mathbf{k},\mu}$. The total number of basis functions for the lattice system is equal to $N_b = m_0 L^3$.

The matrix block entries of the $N_b \times N_b$ stiffness matrix V_{c_L}, corresponding to large basis set on a supercell Ω_L, will be numbered by a pair of multi-indices, $V_{c_L} = [V_{\mathbf{km}}]$, where each $m_0 \times m_0$ matrix block $V_{\mathbf{km}}$ is defined by

$$V_{\mathbf{km}}(\mu, \nu) = \langle \mathbf{G}_{\mathbf{k},\mu} \odot \mathbf{G}_{\mathbf{m},\nu}, \mathbf{P}_{c_L} \rangle , \quad \mathbf{k}, \mathbf{m} \in \mathcal{K}^3 . \tag{5.72}$$

This definition introduces the three-level block structure in the matrix V_{c_L}. We denote the number of overlapping neighboring cells by the overlap constant, L_0. For example, Figure 5.14 (right) corresponds to $L_0 = 2$.

Lemma 5.13 ([187]). *Assume that the overlap constant does not exceed L_0, then:*
(a) *The number of nonzero blocks in each block row (column) of the symmetric Galerkin matrix V_{c_L} does not exceed $(2L_0 + 1)^3$.*
(b) *The storage size is bounded by $m_0^2[(L_0 + 1)L]^3$.*
(c) *The cost for evaluation of each $m_0 \times m_0$ matrix block is bounded by $O(m_0^2 M_0 R N_L)$.*

The block L_0-diagonal structure of the matrix $V_{c_L} = [V_{\mathbf{km}}]$, $\mathbf{k}, \mathbf{m} \in \mathcal{K}^3$ described by Lemma 5.13 allows the essential saving in the storage costs.

However, the polynomial complexity scaling in L leads to severe limitations on the number of unit cells. These limitations can be relaxed if we look more precisely at the defect between matrix V_{c_L} and its block circulant version corresponding to the periodic boundary conditions. This defect can be split into two components corresponding to their local and nonlocal features:
(A1) The nonlocal effect indicates the asymmetry in the interaction potential sum on the lattice in a box.
(B1) The near boundary (local) defect affects only those blocks in $V_{c_L} = \{V_{\mathbf{km}}\}$ lying in the narrow L_0 width of $\partial\Omega_L$.

The defect in item (A1) can be diminished by a slight modification of the core potential to the shift invariant Toeplitz type form $V_{\mathbf{km}} = V_{|\mathbf{k}-\mathbf{m}|}$ by replication of the central unit cell to the whole lattice. In this way the L_0-overlap condition for the tensor $\mathbf{G_k}$ will impose the $(2L_0 + 1)$ block diagonal sparsity in the block Toeplitz matrix.

The boundary effect in item (B1) becomes relatively small for a large number of unit cells so that the block circulant part of the matrix V_{c_L} becomes dominating in relative norm as $L \to \infty$. However, the systematic perturbation in several eigenvalues for large values of L can be observed in the numerical tests.

The representation of the discrete Laplacian and the mass matrix, further called A_G and S_G respectively, is an easier task; see [187]. In particular, in the periodic case both matrices possess the three-level block circulant structure. In what follows, we consider the general multilevel structure of the 'potential' part V_{c_L} in the system matrix.

Definitions of multilevel block circulant and Toeplitz matrices. First, we recall the definition of multilevel block circulant and Toeplitz matrices. The extension of (one-level) block circulant matrices to multilevel structure is described as follows. We describe the main notions of multilevel block circulant (MBC) matrices with the particular focus on the three-level case.

Given a multi-index $\mathbf{L} = (L_1, L_2, L_3)$, we denote $|\mathbf{L}| = L_1 L_2 L_3$. A matrix class $\mathcal{BC}(d, \mathbf{L}, m_0)$ $(d = 1, 2, 3)$ of d-level block circulant matrices can be introduced by the following recursion:

Definition 5.14. For $d = 1$, define a class of one-level block circulant matrices by $\mathcal{BC}(1, \mathbf{L}, m) \equiv \mathcal{BC}(L_1, m)$, where $\mathbf{L} = (L_1, 1, 1)$. For $d = 2$, we say that a matrix

$A \in \mathbb{R}^{|\mathbf{L}|m_0 \times |\mathbf{L}|m_0}$ belongs to a class $\mathcal{BC}(d, \mathbf{L}, m_0)$ if

$$A = \mathrm{bcirc}(A_1, \ldots, A_{L_1}) \quad \text{with} \quad A_j \in \mathcal{BC}(d-1, \mathbf{L}_{[1]}, m_0), \ j = 1, \ldots, L_1 ,$$

where $\mathbf{L}_{[1]} = (L_2, L_3) \in \mathbb{N}^{d-1}$. Similar recursion applies to the case $d = 3$.

Similar to the case of one-level BC matrices, it can be seen that a matrix $A \in \mathcal{BC}(d, \mathbf{L}, m_0)$, $d = 1, 2, 3$, of size $|\mathbf{L}|m_0 \times |\mathbf{L}|m_0$ is completely defined (parametrized) by a dth order matrix valued tensor $\mathbf{A} = [A_{k_1 \ldots k_d}]$ of size $L_1 \times \cdots \times L_d$, ($k_\ell = 1, \ldots, L_\ell$, $\ell = 1, \ldots, d$), with $m_0 \times m_0$ matrix entries $A_{k_1 \ldots k_d}$, obtained by folding of the generating first column vector in A.

Recall that a symmetric block Toeplitz matrix $A \in \mathcal{BT}_s(L, m_0)$ is defined by [78],

$$A = \mathrm{BToepl}_s\{A_0, A_1, \ldots, A_{L-1}\} = \begin{bmatrix} A_0 & A_1^T & \cdots & A_{L-2}^T & A_{L-1}^T \\ A_1 & A_0 & \cdots & \vdots & A_{L-2}^T \\ \vdots & \vdots & \ddots & A_0 & \vdots \\ A_{L-1} & A_{L-2} & \cdots & A_1 & A_0 \end{bmatrix} \in \mathbb{R}^{Lm_0 \times Lm_0} , \tag{5.73}$$

where $A_k \in \mathbb{R}^{m_0 \times m_0}$ for $k = 0, 1, \ldots, L-1$, is a matrix of a general structure.

Similar to Definition 5.14, a matrix class $\mathcal{BT}_s(d, \mathbf{L}, m_0)$ of symmetric d-level block Toeplitz matrices can be introduced by the following recursion:

Definition 5.15. For $d = 1$, $\mathcal{BT}_s(1, \mathbf{L}, m_0) \equiv \mathcal{BT}_s(L_1, m_0)$ is the class of one-level symmetric block circulant matrices with $\mathbf{L} = (L_1, 1, 1)$. For $d = 2$ we say that a matrix $A \in \mathbb{R}^{|\mathbf{L}|m \times |\mathbf{L}|m_0}$ belongs to a class $\mathcal{BT}_s(d, \mathbf{L}, m_0)$ if

$$A = \mathrm{btoepl}_s(A_1, \ldots, A_{L_1}) \quad \text{with} \quad A_j \in \mathcal{BT}_s(d-1, \mathbf{L}_{[1]}, m_0), \ j = 1, \ldots, L_1 .$$

Similar recursion applies to the case $d = 3$.

It is known that similar to the case of circulant matrices, block circulant matrix in $\mathcal{BC}(L, m_0)$ is a unitary equivalent to the block diagonal one by means of Fourier transform via its equivalent Kronecker product representation defined by the associated matrix polynomial,

$$A = \sum_{k=0}^{L-1} \pi^k \otimes A_k =: p_A(\pi) , \tag{5.74}$$

where $\pi = \pi_L \in \mathbb{R}^{L \times L}$ is the periodic downward shift; see [78].

Now, we describe the block diagonal representation of a matrix $A \in \mathcal{BC}(L, m_0)$ in the form that is convenient for generalization to the multilevel block circulant matrices as well as for the description of FFT based implementation schemes. To that end, let us introduce the reshaping (folding) transform $\mathcal{T}_L$ that maps a $Lm_0 \times m_0$ matrix X (i.e., the first block column in A) to $L \times m_0 \times m_0$ tensor $B = \mathcal{T}_L X$ by plugging the ith $m_0 \times m_0$ block in X into a slice $B(i, :, :)$. The respective unfolding returns the initial matrix $X = \mathcal{T}_L' B$.

We denote by $\widehat{A} \in \mathbb{R}^{Lm_0 \times m_0}$ the first block column of a matrix $A \in \mathcal{BC}(L, m_0)$, with a shorthand notation

$$\widehat{A} = [A_0, A_1, \ldots, A_{L-1}]^{\mathrm{T}},$$

so that the $L \times m_0 \times m_0$ tensor $\mathcal{T}_L \widehat{A}$ represents slicewise all generating $m_0 \times m_0$ matrix blocks.

Proposition 5.16 ([187]). *For $A \in \mathcal{BC}(L, m_0)$ we have the block diagonal representation*

$$A = (F_L^* \otimes I_{m_0})\,\mathrm{bdiag}\{\bar{A}_0, \bar{A}_1, \ldots, \bar{A}_{L-1}\}(F_L \otimes I_{m_0}), \tag{5.75}$$

where

$$\bar{A}_j = \sum_{k=0}^{L-1} \omega_L^{jk} A_k \in \mathbb{C}^{m_0 \times m_0}$$

can be recognized as the jth $m_0 \times m_0$ matrix block in the block column $\mathcal{T}_L'(F_L(\mathcal{T}_L \widehat{A}))$, such that

$$\left[\bar{A}_0, \bar{A}_1, \ldots, \bar{A}_{L-1}\right]^{\mathrm{T}} = \mathcal{T}_L'(F_L(\mathcal{T}_L \widehat{A})).$$

A set of eigenvalues λ of A is then given by

$$\{\lambda | Ax = \lambda x,\ x \in \mathbb{C}^{Lm_0}\} = \bigcup_{j=0}^{L-1} \{\lambda | \bar{A}_j u = \lambda u,\ u \in \mathbb{C}^{m_0}\}. \tag{5.76}$$

The eigenvectors corresponding to the spectral sets

$$\Sigma_j = \{\lambda_{j,m} | \bar{A}_j u_{j,m} = \lambda_{j,m} u_{j,m},\ u_{j,m} \in \mathbb{C}^{m_0}\}, \quad j = 0, 1, \ldots, L-1, \quad m = 1, \ldots, m_0$$

can be represented in the form

$$U_{j,m} = (F_L^* \otimes I_m)\bar{U}_{j,m}, \quad \text{where} \quad \bar{U}_{j,m} = E_{[j]}\,\mathrm{vec}[u_{0,m}, u_{1,m}, \ldots, u_{L-1,m}], \tag{5.77}$$

with $E_{[j]} = \mathrm{diag}\{e_j\} \otimes I_{m_0} \in \mathbb{R}^{Lm_0 \times Lm_0}$, and $e_j \in \mathbb{R}^L$ being the jth Euclidean basis vector.

Block diagonal form of the system matrix. Following [187], we introduce the new data sparse block structure by imposing the low rank tensor factorizations within the diagonalized MBC matrix in the matrix class $\mathcal{BC}(d, \mathbf{L}, m_0)$, where $\mathbf{L} = (L_1, \ldots, L_d)$; see Definition 5.14.

The block diagonal form of an MBC matrix is well known in the literature; see e.g., [78]. A diagonalization of a d-level MBC matrix is based on representation via a sequence of cycling permutation matrices $\pi_{L_1}, \ldots, \pi_{L_d}$, $d = 1, 2, 3, \ldots$. Recall that the d dimensional Fourier transform (FT) can be defined via the Kronecker product of the univariate FT matrices (Kronecker rank-1 operator),

$$F_{\mathbf{L}} = F_{L_1} \otimes \cdots \otimes F_{L_d}.$$

Here we prove the diagonal representation in a form that is useful for the description of tensor based numerical algorithms. To that end we generalize the notations $\mathcal{T}_L$ and

$\widehat{A}$ to the class of multilevel matrices. We denote by $\widehat{A} \in \mathbb{R}^{|\mathbf{L}|m_0 \times m_0}$ the first block column of a matrix $A \in \mathcal{BC}(d, \mathbf{L}, m_0)$, with a shorthand notation

$$\widehat{A} = [A_0, A_1, \ldots, A_{L_1-1}]^{\mathrm{T}},$$

and define a $|\mathbf{L}| \times m_0 \times m_0$ tensor $\mathcal{T}_{\mathbf{L}}\widehat{A}$, which represents slicewise all generating $m_0 \times m_0$ matrix blocks in $\widehat{A}$ (reshaping of $\widehat{A}$). Note that in the case $m_0 = 1$, the matrix $\widehat{A} \in \mathbb{R}^{|\mathbf{L}| \times 1}$ represents the first column of A. Now the d dimensional Fourier transform $F_{\mathbf{L}}$ applies to $\mathcal{T}_{\mathbf{L}}\widehat{A}$ columnwise, while the backward reshaping of the resultant tensor, $\mathcal{T}'_{\mathbf{L}}$, returns an $|\mathbf{L}|m_0 \times m_0$ block matrix column. In the following we use the conventional matrix indexing and assume that the lattice $\mathbf{k}$-index runs as $k_\ell = 0, 1, \ldots, L_\ell - 1$.

Lemma 5.17 ([187]). *A matrix $A \in \mathcal{BC}(d, \mathbf{L}, m_0)$ can be converted to the block diagonal form by the Fourier transform $F_{\mathbf{L}}$,*

$$A = (F_{\mathbf{L}}^* \otimes I_{m_0})\, \mathrm{bdiag}_{m_0 \times m_0}\{\bar{A}_{\mathbf{0}}, \bar{A}_{\mathbf{1}}, \ldots, \bar{A}_{\mathbf{L-1}}\}(F_{\mathbf{L}} \otimes I_{m_0}), \tag{5.78}$$

where

$$\left[\bar{A}_{\mathbf{0}}, \bar{A}_{\mathbf{1}}, \ldots, \bar{A}_{\mathbf{L-1}}\right]^{\mathrm{T}} = \mathcal{T}'_{\mathbf{L}}(F_{\mathbf{L}}(\mathcal{T}_{\mathbf{L}}\widehat{A})).$$

Proof. First, we confine ourself to the case of three-level matrices, i.e., $d = 3$. We apply the Kronecker product decomposition (5.74) successively to each level of the block circulant A to obtain

$$\begin{aligned} A &= \sum_{k_1=0}^{L_1-1} \pi_{L_1}^{k_1} \otimes A_{k_1} \\ &= \sum_{k_1=0}^{L_1-1} \pi_{L_1}^{k_1} \otimes \left(\sum_{k_2=0}^{L_2-1} \pi_{L_2}^{k_2} \otimes A_{k_1 k_2} \right) = \sum_{k_1=0}^{L_1-1} \sum_{k_2=0}^{L_2-1} \pi_{L_1}^{k_1} \otimes \pi_{L_2}^{k_2} \otimes A_{k_1 k_2} \\ &= \sum_{k_1=0}^{L_1-1} \sum_{k_2=0}^{L_2-1} \sum_{k_3=0}^{L_3-1} \pi_{L_1}^{k_1} \otimes \pi_{L_2}^{k_2} \otimes \pi_{L_3}^{k_3} \otimes A_{k_1 k_2 k_3}, \end{aligned}$$

where $A_{k_1} \in \mathbb{R}^{L_2 L_3 m_0 \times L_2 L_3 m_0}$, $A_{k_1 k_2} \in \mathbb{R}^{L_3 m_0 \times L_3 m_0}$ and $A_{k_1 k_2 k_3} \in \mathbb{R}^{m_0 \times m_0}$.

Diagonalizing the periodic shift matrices $\pi_{L_1}^{k_1}$, $\pi_{L_2}^{k_2}$, and $\pi_{L_3}^{k_3}$ via the 1D Fourier transform, we arrive at the block diagonal representation

$$\begin{aligned} A &= (F_{\mathbf{L}}^* \otimes I_{m_0}) \left[\sum_{k_1=0}^{L_1-1} \sum_{k_2=0}^{L_2-1} \sum_{k_3=0}^{L_3-1} D_{L_1}^{k_1} \otimes D_{L_2}^{k_2} \otimes D_{L_3}^{k_3} \otimes A_{k_1 k_2 k_3} \right] (F_{\mathbf{L}} \otimes I_{m_0}) \\ &= (F_{\mathbf{L}}^* \otimes I_{m_0})\, \mathrm{bdiag}_{m_0 \times m_0}\{\mathcal{T}'_{\mathbf{L}}(F_{\mathbf{L}}(\mathcal{T}_{\mathbf{L}}\widehat{A}))\}(F_{\mathbf{L}} \otimes I_{m_0}), \end{aligned} \tag{5.79}$$

where the monomials of diagonal matrices $D_{L_\ell}^{k_\ell} \in \mathbb{R}^{L_\ell \times L_\ell}$, $\ell = 1, 2, 3$ are defined by (4.37). The generalization to the case $d > 3$ can be proven by a similar argument. □

Taking into account Corollary A.2 in [187] the multilevel symmetric block circulant matrix can be described in form (5.78), such that all real valued diagonal blocks remain symmetric.

Low rank tensor structure within diagonalized block matrix. In the particular case $d = 3$, the general block diagonal representation (5.79) allows the reduced storage cost for the coefficients tensor $[A_{k_1 k_2 k_3}]$ to the order of $O(|\mathbf{L}| m_0^2)$, where $|\mathbf{L}| = L_1 L_2 L_3$. Introduce the short notation $D_{\mathbf{L}}^{\mathbf{k}} = D_{L_1}^{k_1} \otimes D_{L_2}^{k_2} \otimes \cdots \otimes D_{L_d}^{k_d}$, then (5.79) takes a form

$$A = (F_{\mathbf{L}}^* \otimes I_{m_0}) \left(\sum_{\mathbf{k}=\mathbf{0}}^{\mathbf{L}-\mathbf{1}} D_{\mathbf{L}}^{\mathbf{k}} \otimes A_{\mathbf{k}} \right) (F_{\mathbf{L}} \otimes I_{m_0}) .$$

For large L the numerical cost becomes prohibitive. However, the above representation indicates that further storage and complexity reduction is possible ([187]), if the third order coefficients tensor $\mathbf{A} = [A_{k_1 k_2 k_3}]$, $k_\ell = 0, \ldots, L_\ell - 1$, with the matrix valued entries $A_{k_1 k_2 k_3} \in \mathbb{R}^{m_0 \times m_0}$, allows the low rank tensor factorization (approximation) in the multi-index $\mathbf{k} = (k_1, k_2, k_3)$, which can be described by a smaller than L^3 number of parameters.

To fix the idea, let us assume the existence of rank-1 separable tensor factorization,

$$A_{k_1 k_2 k_3} = A_{k_1}^{(1)} \odot A_{k_2}^{(2)} \odot A_{k_3}^{(3)}, \quad A_{k_1}^{(1)}, A_{k_2}^{(2)}, A_{k_3}^{(3)} \in \mathbb{R}^{m_0 \times m_0}, \quad \text{for} \quad k_\ell = 0, \ldots, L_\ell - 1 . \tag{5.80}$$

This assumption is motivated by existence of low rank canonical representations for the mass and Laplacian stiffness matrices for any d; see also Lemma 5.19 below.

Given $\ell \in \{1, \ldots, d\}$ and a matrix $G \in \mathbb{R}^{L_\ell \times L_\ell}$, define the *tensor prolongation* (lifting) mapping, $\mathcal{P}_\ell : \mathbb{R}^{L_\ell \times L_\ell} \to \mathbb{R}^{|\mathbf{L}| \times |\mathbf{L}|}$, by

$$\mathcal{P}_\ell(G) := \left(\bigotimes_{i=1}^{\ell-1} I_{L_i} \right) \otimes G \otimes \left(\bigotimes_{i=\ell+1}^{d} I_{L_i} \right) . \tag{5.81}$$

The following theorem introduces the new multilevel block circulant tensor structured matrix format, where the coefficient tensor $\mathbf{A}$ is represented via the low rank factorization.

Theorem 5.18 ([187]). *Assume the separability of a tensor $[A_{\mathbf{k}}]$ in the $\mathbf{k}$ space in the form (5.80), then the 3-level block circulant matrix A can be represented in the factorized block diagonal form as follows:*

$$A = (F_{\mathbf{L}}^* \otimes I_{m_0}) D_A (F_{\mathbf{L}} \otimes I_{m_0}) , \tag{5.82}$$

where the block diagonal matrix D_A with the block size $m_0 \times m_0$ is given by

$$D_A = \mathcal{P}_1(\operatorname{bdiag} F_{L_1} \mathbf{A}^{(1)}) \odot \mathcal{P}_2(\operatorname{bdiag} F_{L_2} \mathbf{A}^{(2)}) \odot \mathcal{P}_3(\operatorname{bdiag} F_{L_3} \mathbf{A}^{(3)}) ,$$

with tritensors $\mathbf{A}^{(\ell)} = [A_0^{(\ell)}, \ldots, A_{L_\ell - 1}^{(\ell)}]^{\mathrm{T}} \in \mathbb{R}^{L_\ell \times m_0 \times m_0}$, $\ell = 1, 2, 3$, defined by concatenation of ℓ-factors in (5.80).

The expansion (5.82) includes only 1D Fourier transforms thus reducing the representation cost to

$$O\left(m_0^2 \sum_{\ell=1}^{d} L_\ell \log L_\ell \right) .$$

Moreover, and even more important, the eigenvalue problem for the large matrix A now reduces to only $L_1 + L_2 + L_3 \ll L_1L_2L_3$ independent small $m_0 \times m_0$ matrix eigenvalue problems.

The block diagonal representation for $d = 3$ as above generalizes easily to the case of arbitrary dimension d. Furthermore, the rank-1 decomposition (5.80) was considered for the ease of exposition only. For instance, the separable representation (5.80) can be easily generalized to the case of canonical (CP) or Tucker formats in $\mathbf{k}$ space. In fact, both CP and Tucker formats provide the additive structure, which can be converted to the respective additive structure of the core coefficient in (5.82).

Note that in the practically interesting 3D case the use of MPS/TT type factorizations does not have the advantage over the Tucker format since the Tucker and MPS ranks in 3D usually are close to each other. Indeed, the HOSVD for a tensor of order 3 leads to the same sharp rank estimates for both the Tucker and TT tensor formats.

Block circulant structure in the periodic core Hamiltonian. Here we consider the periodic case, further called case (P), and describe the more refined sparsity pattern of the matrix V_{c_L} by using the d-level ($d = 1, 2, 3$) tensor structure in this matrix. The matrix block entries are numbered by a pair of multi-indices, $V_{c_L} = \{V_{\mathbf{km}}\}$, $\mathbf{k} = (k_1, k_2, k_3)$, where the $m_0 \times m_0$ matrix block $V_{\mathbf{km}}$ is defined by (5.72).

Following [189], we introduce the periodic cells $\mathcal{R} = \mathbb{Z}^d$, $d = 1, 2, 3$ for the $\mathbf{k}$ index, and consider a 3D B-periodic supercell $\Omega_L = B \times B \times B$, with $B = \frac{b}{2}[-L, L]$. The total electrostatic potential in the supercell Ω_L is obtained by, first, the lattice summation of the Coulomb potentials over Ω_L for (rather large) L, but restricted to the central unit cell Ω_0, and then by replication of the resultant function to the whole supercell Ω_L. Hence in this construction, the total potential sum $v_{c_L}(x)$ is designated at each elementary unit cell in Ω_L by the same value ($\mathbf{k}$-translation invariant). The electrostatic potential in each of the B periods can be obtained by copying the respective data from Ω_L.

Consider the case $d = 3$ in more detail. Recall that the reference value $v_{c_L}(x)$ will be computed at the central cell Ω_0, indexed by $(0, 0, 0)$, by summation over all contributions from L^3 elementary subcells in Ω_L. For technical reasons here and in the following we vary the summation index by $k_\ell = 0, \dots, L-1$, to obtain

$$v_0(x) = \sum_{k_1,k_2,k_3=0}^{L-1} \sum_{\nu=1}^{M_0} \frac{Z_\nu}{\|x - a_\nu(k_1, k_2, k_3)\|}, \quad x \in \Omega_0 . \tag{5.83}$$

The basis set in Ω_L is constructed by replication from the master unit cell Ω_0 to the whole periodic lattice. The tensor representation of the local lattice sum on the $n\times n\times n$ grid associated with Ω_0 takes a form

$$\mathbf{P}_{\Omega_0} = \sum_{\nu=1}^{M_0} Z_\nu \sum_{k_1,k_2,k_3=0}^{L-1} \sum_{r=1}^{R} \mathcal{W}_{\nu(\mathbf{k})} \widetilde{\mathbf{p}}_r^{(1)} \otimes \widetilde{\mathbf{p}}_r^{(2)} \otimes \widetilde{\mathbf{p}}_r^{(3)} \in \mathbb{R}^{n\times n\times n},$$

where the tensor $\mathbf{P}_{\Omega_0}$ of size $n \times n \times n$ allows the low rank expansion as in (5.71) with the reference tensor $\widetilde{\mathbf{P}}_R$ defined as $\mathbf{P}_R$ but in the extended domain $\widetilde{\Omega}$; see [187]. Here the Ω-windowing operator, $\mathcal{W}_{\nu(\mathbf{k})} = \mathcal{W}^{(1)}_{\nu(k_1)} \otimes \mathcal{W}^{(2)}_{\nu(k_2)} \otimes \mathcal{W}^{(3)}_{\nu(k_3)}$, restricts onto the $n \times n \times n$ unit cell by shifting the window via the lattice vector $\mathbf{k} = (k_1, k_2, k_3)$. This reduces both the computational and storage costs by a factor of L.

In the 3D case, we set $q = 3$ in the notation for a multilevel block circulant (BC) matrix. Similar to the case of one-level BC matrices, we note that a matrix $A \in \mathcal{BC}(3, \mathbf{L}, m)$ of size $|\mathbf{L}|m \times |\mathbf{L}|m$ is completely defined by a third order coefficients tensor $\mathbf{A} = [A_{k_1 k_2 k_3}]$ of size $L_1 \times L_2 \times L_3$, ($k_\ell = 0, \dots, L_\ell - 1$, $\ell = 1, 2, 3$) with $m \times m$ block matrix entries, obtained by folding of the generating first column vector in A.

Lemma 5.19 ([187]). *Assume that in case (P) the number of overlapping unit cells (in the sense of effective supports of basis functions) in each spatial direction does not exceed L_0. Then the Galerkin matrix $V_{c_L} = [V_{\mathbf{km}}]$ exhibits the symmetric, three-level block circulant Kronecker tensor product form, i.e., $V_{c_L} \in \mathcal{BC}(3, \mathbf{L}, m_0)$,*

$$V_{c_L} = \sum_{k_1=0}^{L_1-1} \sum_{k_2=0}^{L_2-1} \sum_{k_3=0}^{L_3-1} \pi_{L_1}^{k_1} \otimes \pi_{L_2}^{k_2} \otimes \pi_{L_3}^{k_3} \otimes A_{k_1 k_2 k_3}, \quad A_{k_1 k_2 k_3} \in \mathbb{R}^{m_0 \times m_0}, \tag{5.84}$$

where the number of nonzero matrix blocks $A_{k_1 k_2 k_3}$ does not exceed $(L_0 + 1)^3$. Similar properties hold for both the Laplacian and the mass matrix.

The required storage is bounded by $m_0^2(L_0+1)^3$ independent of L. The set of nonzero generating matrix blocks $\{A_{k_1 k_2 k_3}\}$ can be calculated in $O(m_0^2 L_0^3 n)$ operations.

The generating matrix blocks $\{S_{k_1 k_2 k_3}\}$ and $\{B_{k_1 k_2 k_3}\}$ for the mass and Laplacian stiffness matrices admit the rank-1 and rank-3 canonical separable representations respectively.

Furthermore, assume that the QTT ranks of the assembled canonical vectors do not exceed r_0. Then the numerical cost reduces to the logarithmic scale, $O(m_0^2 L_0^3 r_0^2 \log n)$.

In the Hartree–Fock calculations for lattice structured systems we deal with the multilevel, symmetric block circulant/Toeplitz matrices, where the first-level blocks, $A_0, \dots, A_{L_1-1}$, may have further block structures. In particular, Lemma 5.19 shows that the Galerkin approximation of the 3D Hartree–Fock core Hamiltonian H in a periodic setting leads to the symmetric, three-level block circulant matrix structures.

Figure 5.15 shows the difference between matrices V_{c_L} in periodic and nonperiodic cases.

In the next part, we discuss computational details of the FFT based eigenvalue solver in the example of a 3D linear chain of molecules.

Complexity analysis for spectral problems. Combining the block circulant representations (5.84) with the corresponding ones for the mass and Laplacian matrices, we are able to rewrite the eigenvalue problem for the Fock matrix in the Fourier space

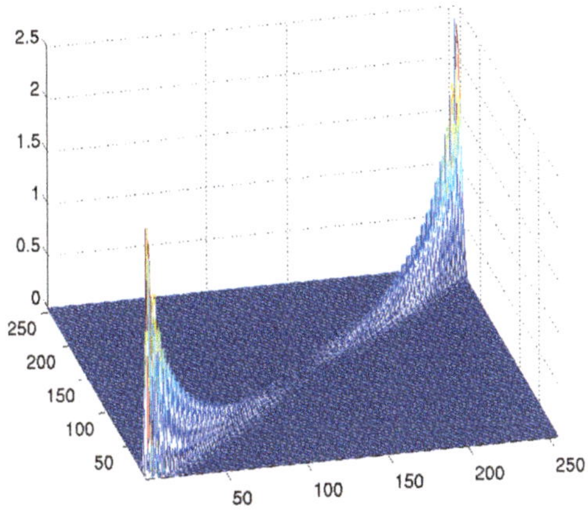

Fig. 5.15: Difference between matrices V_{c_L} in periodic and nonperiodic cases, $L = 64$.

as follows:

$$(\Delta_{c_L} + V_{c_L})U = \lambda S_{c_L} U\,, \tag{5.85}$$

where $U = (F_{\mathbf{L}} \otimes I_m)C$ and

$$\Delta_{c_L} + V_{c_L} - \lambda S_{c_L} = \sum_{\mathbf{k}=0}^{\mathbf{L}} D_{L_1}^{k_1} \otimes D_{L_2}^{k_2} \otimes D_{L_3}^{k_3} \otimes (B_{k_1k_2k_3} + A_{k_1k_2k_3} - \lambda S_{k_1k_2k_3})\,,$$

with the diagonal matrices $D_{L_\ell}^{k_\ell} \in \mathbb{R}^{L_\ell \times L_\ell}$, $\ell = 1, 2, 3$ and the compact notation

$$\sum_{\mathbf{k}=0}^{\mathbf{L}} = \sum_{k_1=0}^{L_1-1} \sum_{k_2=0}^{L_2-1} \sum_{k_3=0}^{L_3-1} .$$

The equivalent block diagonal form reads:

$$\operatorname{bdiag}_{m_0 \times m_0}\{\mathcal{T}'_{\mathbf{L}}[F_{\mathbf{L}}(\mathcal{T}_{\mathbf{L}}\widehat{B}) + F_{\mathbf{L}}(\mathcal{T}_{\mathbf{L}}\widehat{A})] - \lambda \mathcal{T}'_{\mathbf{L}}(F_{\mathbf{L}}[\mathcal{T}_{\mathbf{L}}\widehat{S})]\}U = 0\,. \tag{5.86}$$

The block structure specified by Lemma 5.19 allows us to apply the efficient eigenvalue solvers via FFT based diagonalization in the framework of Hartree–Fock calculations with the numerical cost $O(m_0^2 L^d \log L)$.

Remark 5.20. The low rank structure in the coefficients tensor defined by (5.80) allows us to reduce the factor $L^d \log L$ to $L \log L$ (for $d = 2, 3$) in the cost for assembling and storage of the Fock matrix. It was already observed in the proof of Lemma 5.19 that the respective coefficients in the overlap and Laplacian Galerkin matrices can be treated as the rank-1 and rank-3 tensors respectively. Clearly, the factorization rank for the nuclear part of the Hamiltonian does not exceed R. Hence, Theorem 5.18 can be applied in the generalized form as discussed after the formulation of that theorem.

Table 5.5 compares CPU times in seconds (MATLAB) for the full eigenvalue solver on a 3D $L \times 1 \times 1$ lattice in a box, and for the FTT based MBC diagonalization in the periodic supercell, all computed for $m_0 = 4$, $L = 2^p$ $(p = 7, 8, \ldots, 15)$. The number of basis functions (problem size) is given by $N_b = m_0 L$. We observe that the direct diagonalization is practically limited by the lattice size $L = 2^{10}$, while the eigenvalue problem for structured MBC matrices can be restricted by only the requirements on the d dimensional FFT in $\mathbf{k}$ space.

Tab. 5.5: CPU times (s): full eigenvalue solver versus FFT based MBC diagonalization for $L \times 1 \times 1$ lattice system, and with $m_0 = 4$, $L = 2^p$, $p = 7, 8, \ldots, 15$.

Matrix size $N_b = m_0 L$	512	1024	2048	4096	8192	16 384	32 768	65 536	131 072
Full eigenvalue solver	0.67	5.49	48.6	497.4	–	–	–	–	–
MBC diagonalization	0.10	0.09	0.08	0.14	0.44	1.5	5.6	22.9	89.4

Figure 5.16 represents the spectrum of the core Hamiltonian in a box (blue line) in comparison with that in a periodic supercell (black line) and with eigenvalues computed by our FFT based algorithm applied to the block circulant matrix (red line), labeled by ‘b.-c.’. We consider the $L \times 1 \times 1$ lattice structure with different number of cells $L = 128, 256$, where $m_0 = 4$. Note that in this figure the black and red lines coincide since our algorithm implementing the block circulant structure of the system matrix is numerically equivalent to the FFT based diagonalization.

The systematic difference between the eigenvalues in both box type and periodic structures can be observed even at the limit of large L. These interesting spectral pollution effects were discussed and theoretically analyzed in [56, 93, 285].

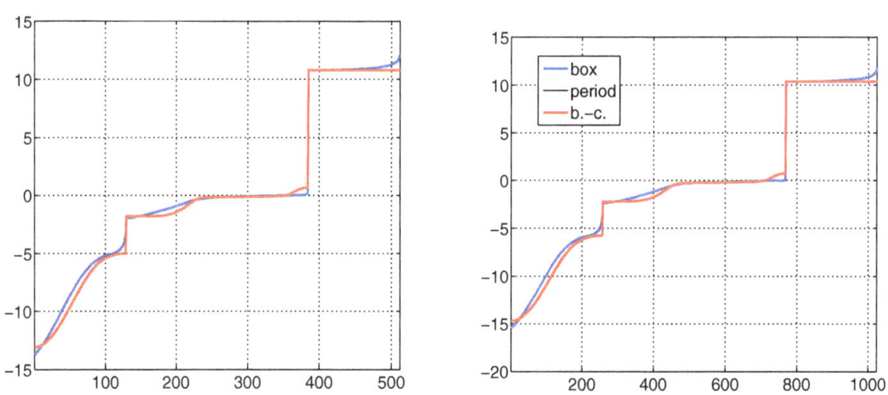

Fig. 5.16: Spectrum of the core Hamiltonian in a box and in a periodic supercell for $L = 128, 256$.

5.3 Real time dynamics by parabolic equations: tensor approach

5.3.1 General introduction

Let $\mathbb{W}$ be a complex Hilbert space and $\mathcal{H}$ be a self-adjoint positive definite operator with the domain $D(\mathcal{H})$ and the spectrum $\Sigma(\mathcal{H}) \in [\lambda_0, \infty)$, $\lambda_0 > 0$. Given $\sigma \in \{-1, i\}$, we consider the following initial value problem:

$$\frac{\partial \psi}{\partial t} = \sigma \mathcal{H} \psi(t) , \quad \psi(0) = \psi_0 \in D(\mathcal{H}) \subset \mathbb{W} . \tag{5.87}$$

The solution of (5.87) is represented by using operator exponential

$$\psi(t) = \mathrm{e}^{\sigma \mathcal{H} t} \psi_0 ,$$

however, in general the operator exponential that constitutes the solution operator of this parabolic problem, $S(t) = \mathrm{e}^{\sigma \mathcal{H} t}$, does not allow the accurate low rank tensor approximation.

In the case of nonhomogeneity, the right hand side solution of equation (5.87) is represented by

$$\psi(t) = \mathrm{e}^{\sigma \mathcal{H} t} \psi_0 + \int_0^t \mathrm{e}^{i \mathcal{H}(t-\tau)} f(\tau) d\tau .$$

In the following we focus on the special case $f = 0$. The general right hand side $f = f(x, t)$ can be included in our scheme.

In quantum mechanics, equations like (5.87) may represent the molecular or electronic Schrödinger equation in d dimensions that describes how the quantum state of a physical system evolves in time. In this case the many-particle Hamiltonian $\mathcal{H}$ is given by a sum of d dimensional Laplacian and a certain interaction potential, [258],

$$i \frac{\partial \psi}{\partial t} = \mathcal{H} \psi = \left(-\frac{1}{2} \Delta_d + V \right) \psi , \quad \psi(x, 0) = \psi_0(x) , \quad x \in \mathbb{R}^d ,$$

where $V \colon \mathbb{R}^d \to \mathbb{R}$ is (given) approximation to the potential energy surface (PES).

The practically important example of real time dynamics is given by the Fokker–Planck equation posed in high dimensional space for the probability density. This equation models the joint probability density distribution of noisy dynamical system configurations (e.g., positions of particles). The initial (stochastic) system of ODEs reads:

$$\frac{dx}{dt} = -V(x) + G\eta \in \mathbb{R}^d ,$$

where the noise satisfies $\langle \eta \rangle = 0$, and $\langle \eta_i \eta_j \rangle = \delta_{ij}$. The probability of finding configurations in some volume $x^* + dx$ is written as follows:

$$P\left(x \in \mathbb{B}_{|dx|}(x^*)\right) = \psi(x^*) dx ,$$

such that the *deterministic* real time parabolic PDE on the probability density, called the Fokker–Planck equation, reads as:

$$\psi(0) = \psi_0: \quad \frac{\partial \psi}{\partial t} = \frac{\partial}{\partial x} \cdot (V(x)\psi) + \frac{1}{2}\frac{\partial}{\partial x} \cdot \left(D \frac{\partial \psi}{\partial x} \right),$$

where $D = GG^{\mathrm{T}}$ and $\psi\colon \mathbb{R}^d \to \mathbb{R}$. It is known that for a linear system $V(x) = Vx$, the (unnormalized) steady probability density is given by the generalized Gaussian (see for example [257])

$$\psi|_{t\to\infty} = \exp(-x^{\mathrm{T}}Bx)\,, \tag{5.88}$$

where B is the solution of the following Lyapunov equation:

$$VB^{-1} + B^{-1}V^{\mathrm{T}} = -2D = -2GG^{\mathrm{T}}\,.$$

In Section 5.3.4, we consider the tensor based solution scheme for the simplified equation

$$\frac{\partial \psi}{\partial t} = -A\psi\,, \quad \psi(0) = \psi_0\,, \tag{5.89}$$

with the convection-diffusion elliptic operator given by

$$A\psi = -\varepsilon\Delta\psi + \mathrm{div}(\psi\mathbf{v})\,, \tag{5.90}$$

where $\mathbf{v}\colon \mathbb{R}^d \to \mathbb{R}^d$ is a given velocity field. Such equations arise in many stochastic models like polymeric liquids with Brownian motion [84, 86, 101, 174, 257], chemical master equations, etc.

In Section 5.3.7, we discuss in more detail the tensor numerical scheme for the solution of the so called chemical master equation (CME). This model describes the dynamics of probability density $\mathcal{P}(\mathbf{x}, t)$,

$$\mathcal{P}(\mathbf{x}, 0) = \mathcal{P}_0\,, \quad \frac{d\mathcal{P}(\mathbf{x}, t)}{dt} = \mathbf{A}\mathcal{P}(\mathbf{x}, t)\,, \quad \mathbf{x} \in \mathbb{R}^{n_1\times\cdots\times n_d}\,, \tag{5.91}$$

where $\mathcal{P}(\mathbf{x}, t)$ is the joint probability of the numbers of molecules of species $S_1, \ldots, S_d$, reacting in M channels, to take particular values $x_1, \ldots, x_d$ at time t. The semidiscrete equation is posed in d dimensional Euclidean space, where d is the number of reacting particles in the system. The finite state projection algorithm for the solution of CME was introduced in [281].

In the next section we present the general scheme for space-time separation of variables based on the Cayley transform techniques. This approach allows the TT/QTT rank estimates for the solutions of the d dimensional parabolic equations formulated as a global time-space scheme in dimension $d + 1$, where the time variable is considered as an additional dimension. This is the basic concept for the construction of the so called all-at-ones solvers in the rank structured tensor formats.

5.3.2 Basic approaches to time integration: approximating $e^{-t\mathcal{H}}\psi_0$

The traditional numerical schemes for solving the parabolic equations are based on the use of time stepping schemes. The main approaches to the construction of efficient time integrators in many dimensions concerned with the data sparse representations of the solution are as follows:

- Sparse grids in (x, t); see [128–130, 299, 316, 326].
- Discrete time propagation on 'Tucker/TT/QTT-manifold' $\mathcal{S}$ by projection onto the tangent space, using the Dirac–Frenkel variational principle:
  $$\left\langle \frac{dy}{dt} - Ay, \delta y \right\rangle = 0 , \quad \delta y \in T_y\mathcal{S} .$$
  Detailed discussion can be found in [230, 258, 259, 277] as well as in the recent papers [34, 154, 260, 260–262, 274, 294, 337].
- Greedy iterations (canonical format); see the recent papers on the topic [58, 59, 77, 101, 251, 342, 343].
- Time stepping by implicit scheme by using the TT/QTT tensor formats in combination with the ALS/DMRG/AMEn local solvers; see [84, 86, 181] for the detailed discussion.

An alternative approach to the time stepping algorithms is based on the global time-space discretization schemes in the dimension $\mathbf{d+1}$ where the so called all-at-ones solver can be applied. The methods that rely on the rank structured tensor representations have been developed recently:

- Time-space separation by QTT-Cayley transform and rank bounds estimate; see [113] and some theoretical papers [7, 105–107].
- QTT-Tucker tensor approximation combined with the ALS type solver on global (x, t) tensor manifold [84, 86].
- Numerics methods for the Fokker–Planck and chemical master equations [84, 86].

In the following discussion, we mostly focus on the tensor methods applied to the global (x, t) discretization scheme.

5.3.3 Rank bounds for space-time tensor approximation based on Cayley transform

In [113] the QTT-Cayley transform was proposed for computing dynamics and spectrums of high dimensional Hamiltonians with the focus on complex time dynamics. Here, we follow [113] and discuss the case of real time dynamics, i.e., $\sigma = -1$ in (5.87). For the ease of presentation, we further assume that the self-adjoint Hamiltonian operator $\mathcal{H}$ has the complete eigenbasis, $\mathbb{W} = \operatorname{span}\{\phi_n\}_{n=0}^{\infty}$, with the real eigenvalues $0 < \lambda_0 \le \lambda_1 \le \ldots$. An extension to the more general class of convection-diffusion operators is possible.

The idea of separation of the time and space variables via the Cayley transform is based on the series expansion for the solution operator

$$e^{-\mathcal{H}t} = (\mathcal{H} + I)^{-1} \sum_{p=0}^{\infty} L_p(t)\mathcal{C}^p(\mathcal{H}) , \tag{5.92}$$

where

$$\mathcal{C} = \mathcal{C}(\mathcal{H}) = \mathcal{H}(\mathcal{H} + I)^{-1}$$

is the Cayley transform of the operator $\mathcal{H}$, and $L_p(t) = L_p^{(0)}(t)$ is the Laguerre polynomial of degree p [18, 118, 331, 339].

This representation is based on the well known expansion [18] for the generating (reproducing) function of the Laguerre polynomial of degree p with a parameter α,

$$(1 - z)^{-\alpha-1} e^{\frac{tz}{z-1}} = \sum_{p=0}^{\infty} L_p^{(\alpha)}(t) z^p .$$

After the formal substitution $z \to \lambda(\lambda + 1)^{-1} := T(\lambda)$, $\lambda > 0$, and setting $\alpha = 0$, we obtain

$$e^{-\lambda t} = (\lambda + 1)^{-1} \sum_{p=0}^{\infty} L_p(t) T^p(\lambda) . \tag{5.93}$$

Hence, on every initial vector $\psi_0 \in D(\mathcal{H})$, we have

$$\psi_0 = \sum_{k=0}^{\infty} a_k \phi_k , \quad \text{with} \quad \sum_{k=0}^{\infty} |a_k|^2 \lambda_k^2 < \infty , \tag{5.94}$$

such that the solution operator can be represented by

$$e^{-\mathcal{H}t} = (\mathcal{H} + I)^{-1} \sum_{p=0}^{\infty} L_p(t) T^p(\mathcal{H}) , \tag{5.95}$$

where

$$T = T(\mathcal{H}) = \mathcal{H}(\mathcal{H} + I)^{-1}$$

is the Cayley transform of the operator $\mathcal{H}$. This global time-space representation can be used for separation of the time variable t from the spatial part of the solution.

This separation is based on the observation that the solution of our initial value problem subject (5.94) can be represented as

$$\psi(t) = \sum_{p=0}^{\infty} L_p(t) u_p \equiv (\mathcal{H} + I)^{-1} \sum_{p=0}^{\infty} L_p(t) \mathcal{C}^p \psi_0 , \tag{5.96}$$

where the elements u_p can be found from the recursion

$$u_0 = (\mathcal{H} + I)^{-1} \psi_0 : \quad u_{p+1} = \mathcal{H}(\mathcal{H} + I)^{-1} u_p , \quad p = 0, 1, \ldots$$

Now, as a computable approximation to the exact solution, we consider the m-term truncated series representation

$$\psi_m(t) = (\mathcal{H} + I)^{-1} \sum_{p=0}^{m} L_p(t)\mathcal{C}^p \psi_0 = \psi_0 + \mathcal{H} \sum_{p=0}^{m} (L_{p+1}(t) - L_p(t))u_p \,, \tag{5.97}$$

which effectively separates space and time variables.

We show that approximation (5.97) leads to an exponential convergence rate in m for the $\mathcal{H}$ analytical input data.

Definition 5.21. A vector $f = \sum_{k=0}^{\infty} a_k \phi_k \in D(\mathcal{H})$ is called analytical for $\mathcal{H}$ ($\mathcal{H}$-analytic) if there is a constant $C = C(f) > 0$ such that

$$\|\mathcal{H}^n f\| = \sqrt{\sum_{k=0}^{\infty} |a_k|^2 \lambda_k^{2n}} \le C^n n! \quad \text{for all} \quad n = 1, 2, 3, \ldots$$

The next theorem proves the exponential convergence of the m-term approximation (5.97); see [113], Remark 2.9.

Theorem 5.22. *Let ψ_0 be $\mathcal{H}$ analytic and let $r > 0$ be the convergence radius of the series $\sum_{k=0}^{\infty} \frac{s^k}{k!} \|\mathcal{H}^k \psi_0\|$. Then for every fixed $0 < s < r$, and $T > 0$, there exist $c, c_1 > 0$ independent of m, such that for all $m \in \mathbb{N}$,*

$$\|\psi(t) - \psi_m(t)\| \le c t^{\frac{1}{4}} \mathrm{e}^{\frac{t}{2}} m^{-1/4} \mathrm{e}^{-c_1\sqrt{m}} \|\psi_0\|_{s,\mathcal{H}} \,, \quad t \in [0, T] \,, \tag{5.98}$$

where $\|\psi_0\|_{s,\mathcal{H}} := \sum_{k=0}^{\infty} \frac{s^k}{k!} \|\mathcal{H}^k \psi_0\|$.

Proof. First, we note that the asymptotic properties of Laguerre polynomials yield

$$\|\psi(t) - \psi_m(t)\| \le c t^{-\frac{1}{4}} \mathrm{e}^{\frac{t}{2}} \sum_{p=m+1}^{\infty} p^{-3/4} \|u_p\| \,, \quad \text{for} \quad t \in [\varepsilon, T] \,, \tag{5.99}$$

where the iterand $u_{p+1} = \sum_{k=0}^{\infty} a_k (\frac{\lambda_k}{\lambda_k + 1})^p \phi_k$ admits the representation

$$\begin{aligned} u_{p+1} &= \sum_{k=0}^{\infty} a_k \mathrm{e}^{-\lambda_k s} \left(\frac{\lambda_k}{\lambda_k + 1}\right)^p \left(\sum_{n=0}^{\infty} \frac{\lambda_k^n s^n}{n!}\right) \phi_k \,, \\ &= \sum_{k=0}^{\infty} a_k \mathrm{e}^{-\lambda_k s} \left(\frac{\lambda_k}{\lambda_k + 1}\right)^p \sum_{n=0}^{\infty} \frac{s^n}{n!} \mathcal{H}^n \phi_k \,, \\ &= \sum_{n=0}^{\infty} \frac{s^n}{n!} \mathcal{H}^n \left(\sum_{k=0}^{\infty} a_k \Phi_s(\lambda_k) \phi_k \right) , \end{aligned} \tag{5.100}$$

with $\Phi_s(\lambda) := \mathrm{e}^{-\lambda s} (\frac{\lambda}{\lambda + 1})^p$. The simple variational analysis indicates that the function $\Phi_s(\lambda)$ takes its maximum at a point $\lambda_* \asymp \sqrt{p}$. Hence, taking into account that

$$\left\| \mathcal{H}^n \left(\sum_{k=0}^{\infty} a_k \Phi_s(\lambda_k) \phi_k \right) \right\| \le \max_{\lambda \in [\lambda_0, \infty)} \left| \Phi_s(\lambda) \right| \|\mathcal{H}^n \psi_0\| \,,$$

we arrive at the estimate $\|u_{p+1}\| \le c\mathrm{e}^{-c_1\sqrt{p}}\|\psi_0\|_{s,\mathcal{H}}$, implying

$$\|\psi(t) - \psi_m(t)\| \le ct^{\frac{1}{4}}\mathrm{e}^{\frac{t}{2}}\|\psi_0\|_{s,\mathcal{H}} \sum_{p=m+1}^{\infty} p^{-1/4}p^{-1/2}\mathrm{e}^{-c_1\sqrt{p}}$$
$$\le ct^{\frac{1}{4}}\mathrm{e}^{\frac{t}{2}}m^{-1/4}\mathrm{e}^{-c_1\sqrt{m}}\|\psi_0\|_{s,\mathcal{H}} ,$$

which completes our proof. □

Note that the constant $c_1 \approx s^{1/2}$ depends on s, while the constant c does not (see [113] for the discussion on the complex case $\sigma = i$).

In the following discussion we consider the semidiscrete scheme (i.e., already discretized in space), such that the operators $\mathcal{H}$ and $\mathcal{C}$ are substituted by a matrix $\mathbf{H}$ and $\mathbf{C}$ respectively. Assume that $\psi_m(t) \in \mathbb{W}_\mathbf{n}$ represents a dth order tensor obtained by the truncated series representation (5.97) composed of the discretized solutions u_p, $p = 0, 1, \dots, m$, and let $t_0, \dots, t_{N_t} \in [0, T]$ be the uniform discretization grid in time variable with a step size τ.

Given the rank truncation threshold $\varepsilon > 0$, then similar to Lemma 3.4 in [113], it can be derived that the choice $m = O(\log^2 \frac{1}{\varepsilon})$ implies that the ε QTT rank of a concatenated tensor

$$\mathbf{P}_m = [\psi_m(t_0), \dots, \psi_m(t_{N_t})]_{k=0}^{N_t} \in \mathbb{W}_\mathbf{n} \times \mathbb{R}^{N_t+1} , \quad t_k = k\tau ,$$

obtained by sampling of $\psi_m(t)$ on the time grid, is bounded by

$$\mathrm{rank}_{\mathrm{QTT}}(\mathbf{P}_m) \le \sum_{p=0}^{m}(p+1)\,\mathrm{rank}_{\mathrm{QTT}}(\mathbf{C}^p\psi_0) \le Cm^2\,\mathrm{rank}_{\mathrm{QTT}}(\mathbf{C}^m\psi_0) .$$

Suppose that $\mathrm{rank}_{\mathrm{QTT}}(\mathbf{C}^m\psi_0)$ is small, then the block two-diagonal system of equations defined by the implicit Euler scheme,

$$\psi_0 = \psi(0) , \quad (\mathbf{I} + \tau\mathbf{H})\psi_{k+1} - \psi_k = 0 , \quad k = 0, 1, \dots, N_t - 1 , \tag{5.101}$$

where $\psi_k \in \mathbb{W}_\mathbf{n}$ approximates the value of the true solution $\psi_m(t_k)$, can be assumed to have a low QTT rank solution represented by tensor $\mathbf{P}$, with $O(d \log N \log N_t)$ complexity scaling. In turn, the scheme (5.101) can be solved in the QTT format as the global system of equations with respect to the unknown space-time vector (tensor)

$$\mathbf{P} = [\psi_0, \psi_1, \dots, \psi_{N_t}] \in \mathbb{W}_\mathbf{n} \times \mathbb{R}^{N_t+1} \approx \mathbf{P}_m .$$

The solution of the global (x, t) system (5.101) living in large virtual dimensions $D = d \log N \log N_t$ can be approached by either tensor-truncated preconditioned or AMEn/DMRG type iteration with the asymptotic cost $O(d \log N \log N_t)$ (see [84, 86, 89] for more detail).

Finally we note that in the case of complex exponentials arising in molecular dynamics the Cayley transform based approximation to the solution operator $S(t) = \mathrm{e}^{it\mathcal{H}}$ remains stable only on the short time interval. The extension to the case of long time dynamics can be done by using the well known scaling and squaring approach based on the relation $[\mathrm{e}^{it\mathcal{H}}]^2 = \mathrm{e}^{i2t\mathcal{H}}$.

5.3.4 The TT/QTT based solver for the Fokker–Planck equation

In what follows, we discuss the rank structured tensor approach for the Fokker–Planck equation presented in [84].

We focus on the simplified case of scalar diffusion tensor $D_{ij} = \epsilon\Delta_{ij}$, which leads to the following form of the Fokker–Planck equation:

$$\frac{d\psi}{dt} = -A\psi\,, \quad \psi(0) = \psi_0\,; \quad A\psi = -\varepsilon\Delta\psi + \mathrm{div}(\psi\mathbf{v})\,, \tag{5.102}$$

where $\psi : \mathbb{R}^d \to \mathbb{R}$, and $\mathbf{v} = (v_1, \ldots, v_d)^{\mathrm{T}} : \mathbb{R}^d \to \mathbb{R}^d$ is a given velocity field.

The solution is sought in some box $[-a, a]^d$ and assumed to vanish at infinity. Thus, Dirichlet boundary conditions for sufficiently large a are employed. The equation (5.102) has a zero forcing term, thus the time evolution eventually converges to the nullspace of A, i.e., to the solution of the stationary equation

$$A\psi_* = 0\,. \tag{5.103}$$

In applications this stationary solution, normalized as

$$\int \psi_* dx_1 \ldots dx_d = 1\,, \tag{5.104}$$

has a meaning of the *probability density*, i.e., it has to be nonnegative.

The direct solution of the equation (5.102) by some approximate solver may not preserve nonnegativity. Time stepping schemes, however, after appropriate discretization of the equation (5.103) will preserve it. Another important feature of the nonstationary problem (5.89) is that the solution may have nontrivial dependence on t, for example the norm of the solution may grow in some time interval $[0, T]$ and only then start to decrease (so called 'shocks').

The analytical solution of equations (5.102), (5.103), and (5.104) is usually not known. However, if $\mathbf{v}$ is a potential field, i.e.,

$$\mathbf{v} = \mathrm{grad}(g)\,,$$

then the analytical solution of the stationary problem is given as

$$\psi = Ce^{-\frac{g}{\varepsilon}}\,, \tag{5.105}$$

where the constant C is defined to satisfy the normalization condition (5.104). From a practical point of view numerical solution of the Fokker–Planck equation with potential velocity field is not too interesting. However, it is very convenient to test the rank structured numerical algorithms for the solution of the Fokker–Planck equation on such kinds of input data with known solutions.

A particular model, also referred to as the Hookean spring model [101], is defined by

$$\phi = \frac{|q|^2}{2p^2} = \sum_{k=1}^{d} \frac{|q_k|^2}{2p^2}\,, \quad v_k = \frac{q_k}{p^2} \in \mathbb{R}^n\,. \tag{5.106}$$

In this case, the stationary solution is the Gaussian function

$$\psi = Ce^{-\frac{|q|^2}{2p^2\varepsilon}} = C\prod_{k=1}^{d} e^{-\frac{|q_k|^2}{2p^2\varepsilon}} .$$

Its TT and QTT rank bounds were obtained in Section 4.2.

To discretize (5.103) in spatial variables, we use a simple finite difference scheme. Take sufficiently large $a > 0$ and introduce a uniform tensor grid in $[-a, a]^d$ with $n = 2^L$ points in each variable, with a step size $h = \frac{2a}{n+1}$:

$$x(i) = -a + ih , \quad i = 1, \dots, n .$$

We consider the QTT tensor representation of the system matrix. The Laplace operator is discretized via a standard second order finite difference scheme such that the corresponding matrix takes the form

$$\Delta_d = \Delta_1 \otimes I \otimes \cdots \otimes I + \cdots + I \otimes \cdots \otimes \Delta_1 , \tag{5.107}$$

where

$$\Delta_1 = \frac{1}{h^2}\mathtt{tridiag}[1, -2, 1] ,$$

and I is a $n \times n$ identity matrix. The QTT representation of such a matrix with ranks bounded by 5 was obtained in [177]; see also Section 4.3. Thus, the storage of the discrete Laplace operator is negligible.

The second term is more involved. Since it is a divergence of a vector valued function it is convenient to use the central difference for the derivatives in each direction. The corresponding matrix T is then represented as

$$T = C_1\Lambda_1 + C_2\Lambda_2 + \cdots + C_d\Lambda_d , \tag{5.108}$$

where $C_k\Lambda_k$ is the discretization of the term $\frac{\partial}{\partial x_k}(v_k\psi_k)$, and v_k is the kth component of the vector field $\mathbf{v}$. The multiplication by v_k reduces to the multiplication by the diagonal matrix Λ_k obtained from v_k by taking values on a tensor grid. The diagonal elements of Λ_k are naturally indexed by a multi-index $(i_1, \dots, i_d)$, such that we have

$$\Lambda_k(i_1, \dots, i_d, i_1, \dots, i_d) = v_k(x(i_1), x(i_2), \dots, x_d(i_d)) . \tag{5.109}$$

The matrix C_k is a central-difference operator in the kth mode, i.e.,

$$C_k = I \otimes \cdots \otimes \underbrace{C}_{k} \otimes \cdots \otimes I ,$$

and C is a one dimensional central difference operator, i.e.,

$$C = \frac{1}{h}\mathtt{tridiag}[-1, 0, 1] . \tag{5.110}$$

The QTT ranks of the matrix (5.108) depend on the QTT ranks of the components of the vector field v_k taken on a considered grid. Then the following simple lemma holds:

Lemma 5.23. *([84]) Suppose that the QTT ranks of the functions v_k on a considered tensor grids are bounded by r. Then, the QTT ranks of the matrix T from (5.108) are bounded by $5rd$.*

Proof. Suppose that these ranks are bounded by r. Then the QTT ranks of the diagonal matrices Λ_k are also bounded by r. The QTT ranks of the matrices C_k are bounded by 5, which can be shown analogously to the Laplace operator. The statement of Lemma 5.23 follows from (5.108) using the multiplicativity and additivity properties of the QTT ranks. □

The proof of Lemma 5.23 is constructive and provides a way to compute the QTT representation of the matrix T using the QTT representation of the vector field $\mathbf{v}$. The QTT representation of the vector $\mathbf{v}$ can be obtained either by using known analytical representations [197, 288] (e.g., in the case where v_k is a sum of exponential, trigonometric, and/or polynomial functions), or by using adaptive cross approximation in the TT format [293], when these functions are given only pointwise.

The discretization in space on a uniform $n \times \cdots \times n$ tensor grid in $[-a, a]^d$, $n = 2^L$, leads to the system matrix

$$A \approx \Delta_d + T\,, \quad T = C_1\Lambda_1 + C_2\Lambda_2 + \cdots + C_d\Lambda_d\,,$$

with the multi-indexed matrices Δ_d and T defined above. This scheme ensures the approximation order $O(h^2 + \tau^2)$ in both time and space, provided that $h \approx \tau$.

A time propagation scheme with tensor truncation described in [84] can be applied to different time propagators in the form

$$\psi_{k+1} = T_\varepsilon(S\psi_k)\,, \quad \psi_k \approx \psi(t_k)\,, \quad t_k = \tau k,$$

where T_ε denotes the ε-truncation operator in the chosen tensor format. In our numerical examples we apply both the TT and QTT formats.

The most commonly used examples of the time propagator S are the following:
$S = I - \tau A$ (explicit Euler),
$S = (I + \tau A)^{-1}$ (implicit Euler),
$S = (I + \frac{\tau}{2}A)^{-1}(I - \frac{\tau}{2}A)$ (Crank–Nicolson).

In our numerical algorithms we use the Crank–Nicolson scheme. For discretization of the Fokker–Planck equation in time, we apply the Crank–Nicolson scheme with the step size τ and denote $t_k = \tau k$, $k = 0, 1, \ldots, N_t$, $y_k = \psi_k \approx \psi(t_k)$, and $f_k = f(t_k)$ in the case of nonzero right hand side. We simplify the notations by setting $N_i = N_x$ for $i = 1, \ldots, d$.

Given A, f_k, y_0 in the TT/QTT format, the system of discrete equations takes a form

$$\left(I + \frac{\tau}{2}A\right)y_{k+1} = \left(I - \frac{\tau}{2}A\right)y_k + \frac{\tau}{2}(f_k + f_{k+1}) =: F_{k+1}\,, \quad k = 0, 1, \ldots, N_t\,.$$

Two strategies suited for tensor methods can be applied:

(A) Time stepping by DMRG-TT iteration for

$$\left(I + \frac{\tau}{2}A\right) y_{k+1} = F_{k+1}, \quad k = 0, 1, \ldots, N_t .$$

(B) Global $O(\log N_x \log N_t)$ block solver in QTT format:

$$y_{k+1} - y_k + \frac{\tau}{2}Ay_{k+1} + \frac{\tau}{2}Ay_k = \frac{\tau}{2}(f_k + f_{k+1}),$$

which means solving the huge global $N_x^d \times N_t$ system in the QTT format,

$$\begin{bmatrix} I + \frac{\tau}{2}A & & & \\ -I + \frac{\tau}{2}A & I + \frac{\tau}{2}A & & \\ & \ddots & \ddots & \\ & & -I + \frac{\tau}{2}A & I + \frac{\tau}{2}A \end{bmatrix} \begin{bmatrix} y_1 \\ y_2 \\ \vdots \\ y_{N_t} \end{bmatrix} = \begin{bmatrix} (I - \frac{\tau}{2}A)\, y^0 \\ 0 \\ \vdots \\ 0 \end{bmatrix} + \frac{\tau}{2} \begin{bmatrix} f_0 + f_1 \\ f_1 + f_2 \\ \vdots \\ f_{N_t-1} + f_{N_t} \end{bmatrix}.$$

In this case the solution process can be based on either a preconditioned iteration applied to symmetrized system of equations or an ALS type iteration applied to the initial nonsymmetric system. Our numerical results are mainly based on the second approach in the form of the so called AMEn iteration [89], though in numerical examples below we compare both approaches.

In the following we briefly outline some numerical illustrations reported in [84]. In the first numerical experiment we consider the classical heat equation in different regimes determined by the input data.

5.3.5 Numerics for QTT based solver: heat equation

As a basic test we consider the heat equation

$$\begin{aligned} \frac{\partial u}{\partial t} - \Delta u &= f \quad \text{in } \Omega = [0, 1]^2 , \\ u|_{\partial\Omega} &= 0 , \\ u(0) &= g . \end{aligned}$$

The solution structure from the tensor point of view can differ significantly with the input data. We illustrate it on the following two choices of g and f.

For the case of a smooth analytic solution the convergence of the solution scheme with the refinement of the grids was investigated; see [84]. We choose $f = 0$, $g(x_1, x_2) = \sin(\pi x_1)\sin(\pi x_2)$. The analytical solution at the time t is $u^*(x_1, x_2) = g(x_1, x_2) \exp(-2\pi^2 t)$. The time interval is fixed to $[0, 1/2]$, and the residual tolerance for the TT solve algorithm is 10^{-6}. The numerical results in [84], Table 5.4, show that the convergence is a bit faster than the theoretical bound $\mathcal{O}(N_t^{-2} + N_x^{-2})$ of the Crank–Nicolson FD scheme. The solution time of the block QTT scheme (B) is almost independent on

Tab. 5.6: CPU times (s), relative residuals provided by the final solution, and the average QTT rank in the block and time stepping schemes.

	Block solution			Time stepping	
N_t	CPU time	$\frac{\|-\Delta u(1)-f\|}{\|f\|}$	rank	CPU time	$\frac{\|-\Delta u(1)-f\|}{\|f\|}$
2^{10}	83.23	7.34e-03	41.09	4547.1	7.38e-03
2^{12}	69.32	1.44e-04	39.40	3845.9	1.69e-03
2^{14}	60.40	3.67e-05	35.91	6232.4	1.58e-04
2^{16}	61.37	5.49e-05	32.98	12 707	7.24e-05

N_t, N_x (about 100–200 ms) for any test since the QTT ranks are uniformly bounded by a small constant.

Let us consider the practically interesting case of irregular data. Now, as the input, we take $f(x_1, x_2) = g = 1, x_1, x_2 \in (0, 1)$, i.e., the functions that have the discontinuity at the boundary. The time step should be small enough to resolve transitional processes near the boundary, otherwise, as the second order scheme is not monotonous, the oscillations occur, which usually have large tensor ranks.

The time interval is fixed to $[0, 1]$, the spatial grid size in each direction is $N_x = 256$, and the QTT approximation tolerance is 10^{-6}. We compare the time stepping solution scheme (A) with the block approach (B); see Table 5.6. For the number of time steps varying from 2^8 to 2^{16}, we present the computational times of both approaches, the measure of closeness of the final solution to the stationary one (the norm of the time derivative), and the QTT rank of the block solution.

We note that the solution time even decreases with the number of time steps in the block algorithm (and grows slower than linear in the time stepping case), since the smaller the time step, the smoother the solution is. Moreover, it leads to a smaller condition number of the matrices in the linear systems involved.

As for the rank, it is stable with respect to the number of time steps, which confirms Lemma 2.1 in [84] and manifests also a slight decrease due to the improving smoothness of the solution. This example demonstrates the advantages of the logarithmic scaling of our block QTT scheme, especially if we have to choose extremely small time steps.

5.3.6 Numerics for the Fokker–Planck problem: the dumbbell model discretized on large grids

We consider the *dumbbell model* discretized on large grids. It is a three dimensional Fokker–Planck model problem of form (5.89), (5.103) with

$$\mathbf{v} = Kq + \mathrm{grad}(\phi) \,,$$

$$K = \beta \begin{pmatrix} 0 & 1 & 0 \\ 0 & 0 & 0 \\ 0 & 0 & 0 \end{pmatrix}, \quad q = \begin{pmatrix} q_1 \\ q_2 \\ q_3 \end{pmatrix} \in [-a, a]^3 .$$

The potential energy ϕ is given as

$$\phi = \frac{1}{2}(q_1^2 + q_2^2 + q_3^2) + \frac{1}{2}\frac{z}{p^3} e^{-(q_1^2+q_2^2+q_3^2)/(2p^2)} ,$$

including the Hookean potential of one spring and the repulsion potential of the beads. In this case the velocity reads

$$\mathbf{v} = Kq - \frac{1}{2}q + \frac{\alpha}{2p^5} e^{-(x^2+y^2+z^2)/(2p^2)} q .$$

It follows that its TT ranks are bounded by 3. Indeed, the Gaussian

$$e^{-(x^2+y^2+z^2)/(2p^2)} = e^{-x^2/(2p^2)} e^{-y^2/(2p^2)} e^{-z^2/(2p^2)}$$

is a rank-1 function, so the rank of the total sum is not greater than 3. On the other side, the QTT ranks are greater. In fact, the QTT ranks of each coordinate vector are equal to 2 and the rank of the Gaussian function depends on the accuracy as $\mathcal{O}(|\log(\epsilon)|^{1/2})$, which is given in Section 4.2.

The following functional of the solution is interesting [84]:

$$\tau(t) = \int \psi(t) \, (q \otimes \operatorname{grad}(\phi)) \, dq .$$

In particular, we test

$$\eta(t) = -\frac{\tau_{12}}{\beta}, \quad \Psi(t) = -\frac{\tau_{11} - \tau_{22}}{\beta} .$$

The *dumbbell model* was discretized on large grids. The following parameters were fixed:

- $\beta = 1$, $\alpha = 0.1$, $p = 0.5$, computational domain $\Omega = [-10, 10]^3$,
- the problem was solved on the time interval $[0, 10]$, and the final solution was taken at the time $T = 10$. Relative tensor rounding (for approximations) and residual (for TT solve) accuracy $\varepsilon = 10^{-6}$.

In the results below, the quantics (Q) dimension is shown both for spatial d_x and time d_t discretizations. The grid size h and time step τ are computed as follows:

$$h = \frac{20}{2^{d_x} + 1}, \quad \tau = \frac{10}{2^{d_t}} .$$

In our dumbbell example, we split the global interval $[0; 10]$ on eight equal subintervals and solve the global system (B) on each of them. The time dimensions d_t presented below correspond to the discretization of each block system on each time

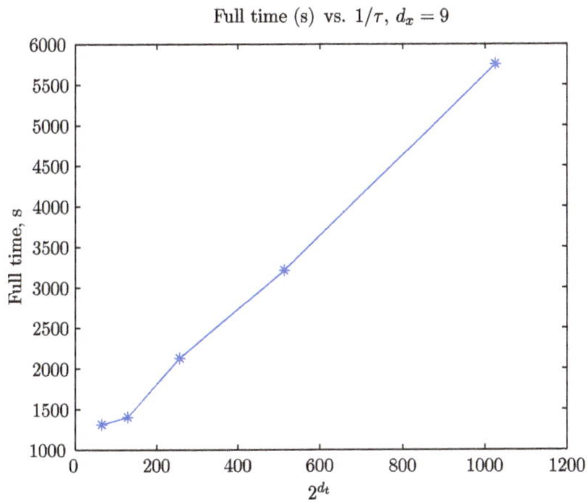

Fig. 5.17: TT solution time versus τ, $N_x = 2^{d_x}$, $N_t = 2^{d_t}$, Time $= O(r^2 N_t N_x)$.

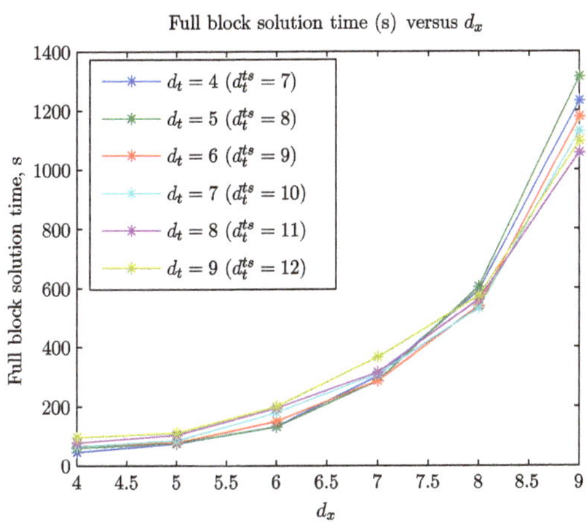

Fig. 5.18: Block QTT time versus d_t, d_x, $r = O(\log N_x)$, Time $= O(r^2 \log N_t \log N_x)$.

subinterval ($d_{t,m}$ in Algorithm 1 in [84]), thus the equivalent number of time steps in the time stepping procedure is eight times larger.

Figure 5.17 presents the computational times on each interval with respect to the Q time dimension d_t. The spacial dimension d_x is fixed to $d_x = 8$. In Figure 5.18 the total CPU times of the solutions (i.e., to compute the solution at the time $T = 10$) are given with respect to d_t and d_x. Note the difference between d_t shown here and in the previous experiment with the step-by-step procedure (the corresponding equivalent value in the time stepping scheme is presented in brackets as d_t^{ts}).

5.3.7 Chemical master equation in the QTT-Tucker format

In this section, we discuss the tensor numerical scheme via the QTT-Tucker format for the solution of chemical master equation (CME) as presented in [86].

Suppose that d species $S_1, \ldots, S_d$ react in M reaction channels. Denote the vector of their concentration $\mathbf{x} = (x_1, \ldots, x_d)$, $x_i \in \{0, \ldots, N_i - 1\}$. Each channel is specified by a stoichiometric vector $\mathbf{z}^m \in \mathbb{Z}^d$, where $\mathbf{z} = (z_1, \ldots, z_d)$, and a propensity function $w^m(\mathbf{x})$, $m = 1, \ldots, M$. Introduce the shift matrices

$$J^z = \begin{bmatrix} 0 & \cdots & 1 & & \\ & \ddots & & \ddots & \\ & & \ddots & & 1 \\ & & & \ddots & \vdots \\ & & & & 0 \end{bmatrix} \begin{matrix} \\ \\ \leftarrow \ \text{row} \quad N - z, \\ \\ \leftarrow \ \text{row} \quad N \end{matrix} \quad \text{if } z \geq 0; \qquad J^z = (J^{-z})^\top, \quad \text{if } z < 0.$$

Now the finite state approximation (FSP) of (5.91) can be written as a linear ODE, the so called CME, that is a deterministic difference equation on the joint probability density $P(x, t)$:

$$\frac{dP(t)}{dt} = AP(t)\,, \quad A = \sum_{m=1}^{M} (\mathbf{J}^{\mathbf{z}^m} - \mathbf{J}^0)\, \mathrm{diag}(w^m) P(t)\,, \qquad P(t) \in \mathbb{R}_+^{\prod_{i=1}^d N_i}\,, \tag{5.111}$$

$$\mathbf{J}^{\mathbf{z}} = J^{z_1} \otimes \cdots \otimes J^{z_d}\,,$$

where $w^m = \{w^m(\mathbf{x})\}$ and $P(t) = \{P(\mathbf{x}, t)\}$, $\mathbf{x} \in \bigotimes_{i=1}^d \{0, \ldots, N_i - 1\}$, are the corresponding values of w^m and P stacked into vectors, $\mathrm{diag}(w^m)$ is a diagonal matrix with the values of w^m stretched along the diagonal, and $\otimes$ means the rank-1 matrix format.

To employ tensor decompositions, we need to present all initial data in our favorable format. Assuming the tensor separability of each propensity function w^m, we obtain from (5.111) the tensor rank estimate of the CME operator.

Lemma 5.24. *([86]) There holds*

$$\mathrm{rank}(A) \leq \sum_{m=1}^{M} 2 \cdot \mathrm{rank}(w^m)\,. \tag{5.112}$$

Proof. Both $\mathbf{J}^{\mathbf{z}^m}$ and $\mathbf{J}^0$ are (rank-1) Kronecker products. The difference $\mathbf{J}^{\mathbf{z}^m} - \mathbf{J}^0$ is therefore of rank 2. Given a separable form $w^m = \sum_{\alpha} W^{(1)}_{\alpha_1} \otimes W^{(2)}_{\alpha_1,\alpha_2} \otimes \cdots \otimes W^{(d)}_{\alpha_{d-1}}$, the matrix $\mathrm{diag}(w^m)$ is constructed without changing the rank,

$$\mathrm{diag}(w^m) = \sum_{\alpha} \mathrm{diag}(W^{(1)}_{\alpha_1}) \otimes \mathrm{diag}(W^{(2)}_{\alpha_1,\alpha_2}) \otimes \cdots \otimes \mathrm{diag}(W^{(d)}_{\alpha_{d-1}})\,.$$

The product of $\mathbf{J}^{\mathbf{z}^m} - \mathbf{J}^0$ and $\mathrm{diag}(w^m)$ multiplies the ranks, and finally we sum the terms corresponding to all reactions. □

Now, the CME (5.111) reads:

$$\frac{dP(t)}{dt} = AP = (A^+ + A^-)P\,,$$
$$A^+ = -\sum_{m=1}^{d} \nabla_m^- \operatorname{diag}(w^{m+})\,, \quad A^- = \sum_{m=1}^{d} \nabla_m^+ \operatorname{diag}(w^{m-})\,, \tag{5.113}$$

which has a very close form to the diffusion equation discretized using the finite difference scheme.

The general Lemma 5.24 can be applied, giving the minimal tensor rank $4d$. However, we would like to consider typical gene networks in more detail and prove refined results.

In such cases, when the mth destruction propensity depends only on x_m, its rank-1 decomposition reads:

$$w^{m-}(\mathbf{x}) = e_1(x_1)\dots\hat{w}^{m-}(x_m)\dots e_d(x_d)\,,$$

where $e_i(x_i) = 1\ \forall\ x_i = 0, \dots, N_i - 1$. Now, the destruction part of the operator, A^- in (5.113), has the Laplace like form

$$A^- = D_1 \otimes J^0 \cdots \otimes J^0 + \cdots + J^0 \otimes \cdots \otimes D_d\,, \quad D_m = (J^1 - J^0)\operatorname{diag}(\hat{w}^{m-})\,,$$

which is proven to have the TT rank 2 [177]; see Section 4.3.

The creation part is usually more complicated (it contains feedbacks between species), and depends on several variables. In the cascade networks (including the toggle switch), the mth creation propensity depends on x_{m-1} (or x_{m+1}), and probably on x_m. Thus, the corresponding operator part sums the two-variate terms,

$$A^+ = D_1^1 \otimes J^0 \cdots \otimes J^0 + D_1^2 \otimes D_2^2 \otimes J^0 \cdots \otimes J^0 + \cdots + J^0 \cdots \otimes D_{d-1}^d \otimes D_d^d\,,$$

where $D_{m-1}^m = \operatorname{diag}(\hat{w}^{m+})$, $D_m^m = -(J^0 - J^{-1})$. Up to now, we assume that w^{m+} does not depend on x_m. For example, the Michaelis–Menten rate in a cascade reads $w^{m+}(\mathbf{x}) = e_1(x_1)\dots\hat{w}^{m+}(x_{m-1})\cdot e_m(x_m)\dots e_d(x_d)$, $\hat{w}^{m+}(x_{m-1}) = \frac{\alpha}{\beta + x_{m-1}}$. The generalization will be given in Remark 5.26.

In the rest of this section, we are going to discuss the structure of TT factors. Contrary to the $\mathcal{O}(d)$ canonical rank, provided by Lemma 5.24, the following Lemma 5.25 shows that a sum like A^+ can be represented as a rank-3 TT decomposition, thus with the linear in d memory cost. The matrix product form is especially convenient for such analysis, but the variables x_k, x_k' would take too much space, and we omit them for brevity. That is, by E_k, F_k^k, F_k^{k+1} we will mean not the whole matrices, but the elements $E_k(x_k, x_k')$, $F_k^k(x_k, x_k')$, $F_k^{k+1}(x_k, x_k')$.

Lemma 5.25 ([86]). *Given the matrices E_k, F_k^k, F_k^{k+1}. The cascadic sum*

$$H(\mathbf{x}, \mathbf{x}') = F_1^1 \left(\prod_{k=2}^{d} E_k\right) + \sum_{i=2}^{d} \left(\prod_{k=1}^{i-2} E_k\right) \cdot F_{i-1}^i \cdot F_i^i \cdot \left(\prod_{k=i+1}^{d} E_k\right) \tag{5.114}$$

possesses an explicit exact rank-3 TT decomposition $H(\mathbf{x}, \mathbf{x}') = H^{(1)} \dots H^{(d)}$, where

$$H^{(1)} = \begin{bmatrix} E_1 \\ F_1^2 \\ F_1^1 \end{bmatrix}^\top, \quad H^{(k)} = \begin{bmatrix} E_k & F_k^{k+1} & 0 \\ 0 & 0 & F_k^k \\ 0 & 0 & E_k \end{bmatrix}, \quad H^{(d-1)} = \begin{bmatrix} F_{d-1}^d & 0 \\ 0 & F_{d-1}^{d-1} \\ 0 & E_{d-1} \end{bmatrix}, \quad H^{(d)} = \begin{bmatrix} F_d^d \\ E_d \end{bmatrix}, \tag{5.115}$$

$k = 2, \dots, d-2$. For the Tucker decomposition the same rank-3 bound holds.

Proof. We begin to split the dimensions recurrently, extracting the linearly independent elements, in the same way as in the TT-SVD algorithm [289]. So, the first step reads:

$$H(\mathbf{x}, \mathbf{x}') = \begin{bmatrix} E_1 & F_1^2 & F_1^1 \end{bmatrix} \begin{bmatrix} F_2^3 F_3^3 \dots E_d + \dots + E_2 \dots F_{d-1}^d F_d^d \\ F_2^2 E_3 \dots E_d \\ E_2 \dots E_d \end{bmatrix}.$$

The first term here is exactly the first TT factor of the decomposition. Now suppose we have the following form:

$$H_k(x_k, \dots, x_d, x_k', \dots, x_d') = \begin{bmatrix} F_k^{k+1} F_{k+1}^{k+1} \dots E_d + \dots + E_2 \dots F_{d-1}^d F_d^d \\ F_k^k E_{k+1} \dots E_d \\ E_k \dots E_d \end{bmatrix}. \tag{5.116}$$

We split the kth dimension from each row in the same manner,

$$H_k(x_k, \dots, x_d, x_k', \dots, x_d') = \begin{bmatrix} E_k & F_k^{k+1} & 0 \\ 0 & 0 & F_k^k \\ 0 & 0 & E_k \end{bmatrix} \begin{bmatrix} F_{k+1}^{k+2} F_{k+2}^{k+2} \dots E_d + \dots + E_{k+1} \dots F_{d-1}^d F_d^d \\ F_{k+1}^{k+1} E_{k+2} \dots E_d \\ E_{k+1} \dots E_d \end{bmatrix},$$

and derive the kth TT factor (the first term). The remaining dimensions are presented in the same form as (5.116), so we can continue the splitting. The last two factors are separated as follows:

$$\begin{bmatrix} F_{d-1}^d F_d^d \\ F_{d-1}^{d-1} E_d \\ E_{d-1} E_d \end{bmatrix} = \begin{bmatrix} F_{d-1}^d & 0 \\ 0 & F_{d-1}^{d-1} \\ 0 & E_{d-1} \end{bmatrix} \begin{bmatrix} F_d^d \\ E_d \end{bmatrix},$$

giving the $(d-1)$th and dth TT factors, respectively. We see that all the TT ranks are equal to 3, except that the $(d-1)$th is equal to 2, which confirms the claim of the lemma. To obtain the Tucker rank estimate, it is sufficient to note that each TT factor contains only three independent elements and follow to the TT-to-Tucker procedure described in [87], see also Section 4.3.11. □

Remark 5.26. In (5.114), each summand is a rank-1 tensor. However, we can generalize it to the case when the neighboring terms are summed from several components,

$$\sum_{\alpha_k=1}^{r_k} F^k_{k-1,\alpha_k}(x_{k-1}, x'_{k-1}) \cdot F^k_{k,\alpha_k}(x_k, x'_k) \quad \text{instead of} \quad F^k_{k-1}(x_{k-1}, x'_{k-1}) \cdot F^k_k(x_k, x'_k)$$

(i.e., the TT rank of each propensity is not equal to 1). In this case, we can collect respectively the row and column vectors

$$F^k_{k-1}(x_{k-1}, x'_{k-1}) = \begin{bmatrix} F^k_{k-1,1}(x_{k-1}, x'_{k-1}) \\ \vdots \\ F^k_{k-1,r_k}(x_{k-1}, x'_{k-1}) \end{bmatrix}^\top, \quad F^k_k(x_k, x'_k) = \begin{bmatrix} F^k_{k,1}(x_k, x'_k) \\ \vdots \\ F^k_{k,r_k}(x_k, x'_k) \end{bmatrix},$$

and the constructions (5.115) will be considered as block matrices, with the sizes (i.e., the TT ranks) $(2 + r_k) \times (2 + r_{k+1})$. Counting the linearly independent elements in each TT factor, we conclude that the kth Tucker rank is bounded by $1 + r_k + r_{k+1}$.

We refer to [178, 181, 182] for other rank estimates in CME problem.

Corollary 5.27 ([86]). *The CME operator in the cascade model admits an explicit exact TT decomposition of rank 4, $A(\mathbf{x}, \mathbf{x}') = A^{(1)} \dots A^{(d)}$, where*

$$A^{(1)} = \begin{bmatrix} J^0 \\ D_1^2 \\ D_1^1 \\ D_1 \end{bmatrix}^\top, \quad A^{(k)} = \begin{bmatrix} J^0 & D_k^{k+1} & 0 & D_k \\ 0 & 0 & D_k^k & 0 \\ 0 & 0 & J^0 & 0 \\ 0 & 0 & 0 & J^0 \end{bmatrix},$$

$$A^{(d-1)} = \begin{bmatrix} D_{d-1}^d & J^0 & D_{d-1} \\ 0 & 0 & D_{d-1}^{d-1} \\ 0 & J^0 & 0 \\ 0 & 0 & J^0 \end{bmatrix}, \quad A^{(d)} = \begin{bmatrix} D_d^d \\ D_d \\ J^0 \end{bmatrix},$$

$k = 2, \dots, d-2$. The same rank-4 bound holds for the Tucker decomposition.

Proof. The rank-2 decomposition of the destruction operator was already discussed. For the creation part we use Lemma 5.25, by setting $E_k = J^0$, $F^k_{k-1} = \operatorname{diag}(\hat{w}^{m+})$, and $F^k_k = -(J^0 - J^{-1})$.

The straightforward TT addition gives the rank-5 structure, but due to the fact that J^0 encounters both A^+ and A^-, the rank can be reduced as follows. The first factor reads

$$\begin{bmatrix} J^0 & D_1^2 & D_1^1 & J^0 & D_1 \end{bmatrix} = \begin{bmatrix} J^0 & D_1^2 & D_1^1 & D_1 \end{bmatrix} \begin{bmatrix} 1 & 0 & 0 & 1 & 0 \\ 0 & 1 & 0 & 0 & 0 \\ 0 & 0 & 1 & 0 & 0 \\ 0 & 0 & 0 & 0 & 1 \end{bmatrix}.$$

Multiplying the latter mode independent term with the second rank-5 factor, we obtain

$$\begin{bmatrix} J^0 & D_2^3 & 0 & J^0 & D_2 \\ 0 & 0 & D_2^2 & 0 & 0 \\ 0 & 0 & J^0 & 0 & 0 \\ 0 & 0 & 0 & 0 & J^0 \end{bmatrix} = \begin{bmatrix} J^0 & D_2^3 & 0 & D_2 \\ 0 & 0 & D_2^2 & 0 \\ 0 & 0 & J^0 & 0 \\ 0 & 0 & 0 & J^0 \end{bmatrix} \begin{bmatrix} 1 & 0 & 0 & 1 & 0 \\ 0 & 1 & 0 & 0 & 0 \\ 0 & 0 & 1 & 0 & 0 \\ 0 & 0 & 0 & 0 & 1 \end{bmatrix}.$$

We see that after the linear dependency reduction, the same scalar factor arose as in the first step. So we can continue the process and come to the form claimed by the corollary. □

5.3.8 Discretization of CME. Analysis of the rank structure in Hamiltonian

For discretization of the CME equation in time, we use the Crank–Nicolson scheme with the step size τ and denote $t_k = \tau k$, $k = 0, 1, \ldots, N_t$, $y_k = P_k \approx P(t_k)$, and $f_k = f(t_k)$ in the case of nonzero right hand side. We simplify the notations by setting $N_i = N_x$ for $i = 1, \ldots, d$. Given A, f_k, y_0 in the TT/QTT format, the system of discrete equations takes a form

$$\left(I + \frac{\tau}{2}A\right) y_{k+1} = \left(I - \frac{\tau}{2}A\right) y_k + \frac{\tau}{2}(f_k + f_{k+1}) =: F_{k+1}\,, \quad k = 0, 1, \ldots, N_t\,. \tag{5.117}$$

We apply the approach based on the global $O(\log N_x \log N_t)$ block solver in QTT format applied to the equation

$$y_{k+1} - y_k + \frac{\tau}{2}Ay_{k+1} + \frac{\tau}{2}Ay_k = \frac{\tau}{2}(f_k + f_{k+1})\,,$$

which means solving the huge global $N_x^d \times N_t$ linear system of equations in the QTT-Tucker format,

$$\begin{bmatrix} I + \frac{\tau}{2}A & & & \\ -I + \frac{\tau}{2}A & I + \frac{\tau}{2}A & & \\ & \ddots & \ddots & \\ & & -I + \frac{\tau}{2}A & I + \frac{\tau}{2}A \end{bmatrix} \begin{bmatrix} y_1 \\ y_2 \\ \vdots \\ y_{N_t} \end{bmatrix} = \begin{bmatrix} (I - \frac{\tau}{2}A)\,y^0 \\ 0 \\ \vdots \\ 0 \end{bmatrix} + \frac{\tau}{2} \begin{bmatrix} f_0 + f_1 \\ f_1 + f_2 \\ \vdots \\ f_{N_t-1} + f_{N_t} \end{bmatrix}.$$

Our numerical tests are based on the second approach in the form of the so called AMEn iteration [89].

We present numerical tests (reported in [86]) for a high dimensional cascade problem from [4, 162], which occurs when adjacent genes produce proteins that influence the expression of a succeeding gene; see Figure 5.19. Tensor properties of this model,

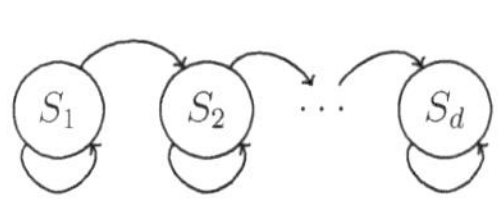

Fig. 5.19: Cascade signaling network.

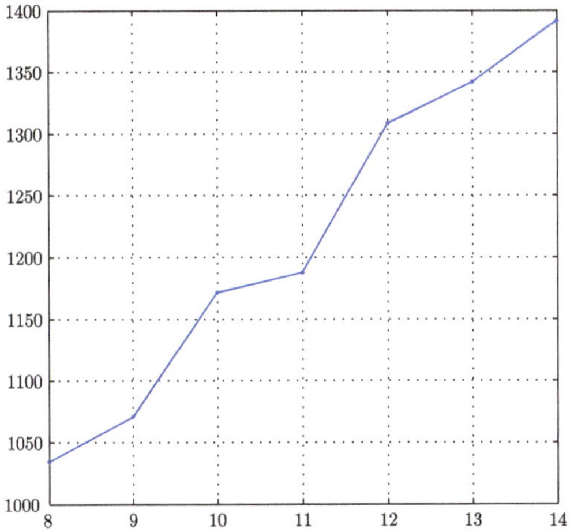

Fig. 5.20: CPU time (s) versus $\log N_t$.

including tensor ranks of CME matrices, have been analyzed theoretically in the previous section; see also [86] for the numerical estimates.

The model parameters are specified as follows:
- $d = 20, M = 40$;
- for $m = 1$: $w^m(\mathbf{x}) = 0.7$, $\mathbf{z}^m = -\delta_m$: generation of the first protein;
- for $m = 2, \dots, 20$: $w^m(\mathbf{x}) = \frac{x_{m-1}}{5+x_{m-1}}$, $\mathbf{z}^m = -\delta_m$: succeeding creation reactions;
- for $m = 21, \dots, 40$: $w^m(\mathbf{x}) = 0.07 \cdot x_{m-20}$, $\mathbf{z}^m = \delta_{m-20}$: destruction reactions.

Here Δ_m is the mth identity vector, $N_i = 63$, hence the full grid problem size is about 64^{20}. The linear QTT format was used for state and time. The dynamical problem was solved until $T = 400$ via the restarted global state-time solver after each $T_0 = 15$ time subintervals. This means that the global state-time solver applies to (x, t) discretization successively on each coarse time interval of size $T_0 = 15$, so that number of intervals is T/T_0. The initial state was chosen by $P(0) = \delta_1 \otimes \cdots \otimes \delta_1$, i.e., all copy numbers are zeros. The solution threshold in the discrete L_2 norm for the QTT rank truncation was chosen as $\varepsilon = 10^{-5}$.

Figure 5.20 illustrates the $\log N_t$ scaling of the solution time for the varying number of time steps, $N_t = 2^8, 2^9, \dots, 2^{14}$. Figure 5.21 shows the convergence of the transient solution to the stationary one. This confirms that the chosen time interval T is large enough (long time dynamics) to catch the steady state with the required precision.

We conclude that presented results demonstrate the high performance of QTT and QTT-Tucker formats in tensor computations for multidimensional CMEs. For many interesting cases the theoretical rank bounds for the CME matrices have been proven [86]. Numerical tests confirm the logarithmic complexity scaling of the global

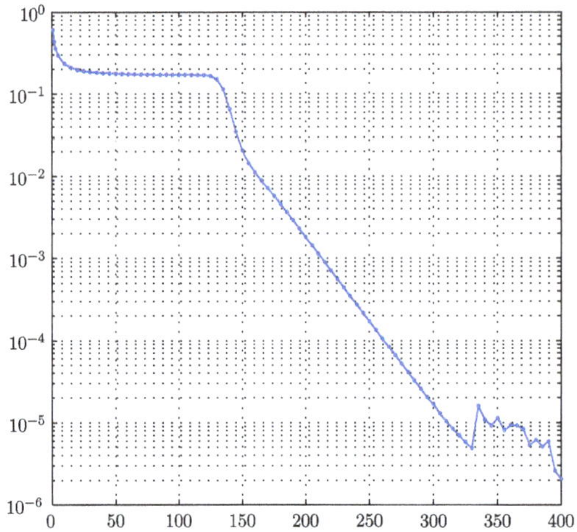

Fig. 5.21: Closeness to the kernel $\frac{\|AP\|}{\|P\|}(t)$.

space-time tensor schemes in both time and space discretization parameters, indicating the potential advantages of this approach for high dimensional dynamical simulations.

5.3.9 Towards application to spin models

Similar operators arise also in the one dimensional spin systems modeling with nearest neighbor interactions. For example, the Heisenberg (XYZ) model with open boundary conditions [170], acting on $\bigotimes_{i=1}^{d} \mathbb{C}^2$, reads:

$$H = j_x H_{xx} + j_y H_{yy} + j_z H_{zz} + \lambda H_x \,,$$

$$H_{\mu\nu}(\mathbf{x}, \mathbf{x}') = \sum_{i=2}^{d} \left(\prod_{k=1}^{i-2} E_k \right) \cdot P_\mu \cdot P_\nu \cdot \left(\prod_{k=i+1}^{d} E_k \right),$$

$$H_\mu(\mathbf{x}, \mathbf{x}') = \sum_{i=1}^{d} \left(\prod_{k=1}^{i-1} E_k \right) \cdot P_\mu \cdot \left(\prod_{k=i+1}^{d} E_k \right),$$

where E_k are the identity matrices, P_μ are the Pauli matrices ($\mu = x, y, z$),

$$P_x = \begin{pmatrix} 0 & 1 \\ 1 & 0 \end{pmatrix}, \quad P_y = \begin{pmatrix} 0 & -i \\ i & 0 \end{pmatrix}, \quad P_z = \begin{pmatrix} 1 & 0 \\ 0 & -1 \end{pmatrix},$$

and j_μ, λ are scalars. Lemma 5.25 and Remark 5.26 can be applied straightforwardly, giving the following rank estimate:

Lemma 5.28. *([86]) The Heisenberg (XYZ) Hamiltonian admits an explicit rank-7 TT (or Tucker) representation.*

Proof. Assembling the factors

$$F_{k-1}^k = \begin{bmatrix} j_x P_x & j_y P_y & j_z P_z \end{bmatrix}, \quad F_k^k = \begin{bmatrix} P_x & P_y & P_z \end{bmatrix}^\top,$$

we reduce the problem to that described by Lemma 5.25 and Remark 5.26, with $r_k = 3$. Taking into account the Laplace like rank-2 structure in λH_x, we obtain the rank estimate by $(2+3)+2=7$. □

If some of j_x, j_y, j_z are equal to zero, the reduced models appear, for example the Heisenberg (XY) Hamiltonian with $j_z = 0$ and TT rank 6, or the Ising (ZZ) model with $j_x = j_y = 0$ and TT rank 5.

5.4 Rank structured approximation to stochastic and parametric PDEs

5.4.1 Problem setting

Following [227], we consider parametric, elliptic problems that are posed in the physical domain $D := (0,1)^{d_0}$ of dimension $d_0 = 1,2,3$, and which depend on a vector of M parameters that take values in the hypercube in the M dimensional parametric space $\Gamma := (-1,1)^M \equiv I^M$, $M \in \mathbb{N}_+$. To formulate the problems, we introduce the tensor product Hilbert space ([308])

$$V := V_y \otimes V_x \quad \text{with} \quad V_y := L^2(\Gamma) = \bigotimes_{m=1}^{M} L^2(I), \quad V_x := H_0^1(D).$$

We are given a parametric elliptic operator

$$\mathcal{A}(y) := -\mathrm{div}_x\,(a(y,x)\mathrm{grad}_x) \quad \text{and} \quad f \in L^2(D), \quad y \in \Gamma,$$

where the coefficient $a(y,x)$ is a smooth function of $x \in D$ and the parameter vector $y = (y_1, \ldots, y_M) \in \Gamma$ with a possibly large number M of parameters.

We are interested in the efficient numerical solution of the parametric elliptic problem: for every $y \in \Gamma$, find $u_M \in V$, such that

$$\mathcal{A} u_M(y,x) = f(x) \ \text{in} \ D, \qquad u_M(y,x) = 0 \ \text{on} \ \partial D. \tag{5.118}$$

In this problem setting the dimension M of the parametric space corresponds to the truncation parameter in the Karhunen–Loève expansion (e.g., [215, 216, 324]). In discretizations of diffusion problems with random inputs, the dimension M of the parameter space could become very large. We consider the two classes of problems that are traditionally analyzed in the literature:

(a) For random field, that is linear in the stochastic variable, we have the truncated Karhunen–Loéve expansion (the so called additive case)

$$a_M(y,x) := a_0(x) + a_y(y,x), \quad \text{where } a_y(y,x) = \sum_{m=1}^{M} a_m(x) y_m, \tag{5.119}$$

with $a_m \in L^\infty(D)$, $m = 1, \ldots, M$.

(b) In the log additive case the coefficient function is defined by (see also Section 5.4.4)

$$a_M(y,x) := \exp\left(a_0(x) + \sum_{m=1}^{M} a_m(x) y_m \right) > 0 .$$

Concerning the coefficient function $a_M(y,x)$, we assume that for all $x \in D$, there exists $a_{\min} > 0$, such that

1. $a_{\min} \leq a_0(x) < \infty$,
2. $|\sum_{m=1}^{M} a_m(x) y_m| < \gamma a_{\min}$ with $\gamma < 1$, and for $|y_m| < 1$ $(m = 1, \ldots, M)$.

Conditions (1)–(2) imply the strong ellipticity of the variational problem (5.118) uniformly in y, that is

$$a_M(y,x) \geq (1-\gamma) a_{\min} > 0 . \tag{5.120}$$

Hence, under Assumptions (1)–(2), we have the unique solvability for the corresponding weak formulation (see [227] for the details): for any $f \in H^{-1}(D)$ and for any $y \in \Gamma$, there exists a unique solution $u_M(y,\cdot) \in H_0^1(D)$ of the problem:

$$\text{Find } u_M(y,\cdot) \in V_x , \text{ such that } A(u_M(y,\cdot), w) = \int_D f(x) w(x) dx \quad \forall\, w \in V_x . \tag{5.121}$$

Here for $y \in \Gamma$ we introduce the associated parametric bilinear form in the physical space V_x,

$$A(z,w) := \langle \mathcal{A}(y) z, w \rangle_{L^2(D)} = \int_D a_M(y,x) \nabla_x z \cdot \nabla_x w dx \quad \forall\, z, w \in V_x .$$

The idea of separable representation in the parametric space for PDEs with random input were already elaborated in [43, 44, 324, 325, 341]. Exploiting sparsity in the solutions, polynomial chaos expansion leads to superalgebraic (subexponential) convergence rates of both Galerkin and collocation approximations in terms of the number N of deterministic diffusion problems to be solved ([44, 341]), in the case that the Karhunen–Loéve expansion of the input random field converges exponentially. Furthermore, the algebraic convergence rates of best N-term polynomial approximations of the parametric solution has been observed [70]. It is worth noting that N-term truncated polynomial chaos expansions are separable expansions.

In the paper [227] the concept of rank structured approximation was first elaborated for tensor product approximation to solution of PDEs living in the physical space coupled with multidimensional parametric space. It was shown that the FEM-Galerkin approximation in a tensor product basis allows for approximate low tensor rank representations of arising stiffness matrices and right hand sides. The principal idea of the approach is the iterative solution of a single coupled system of discrete, multiparametric elliptic equations projected onto the nonlinear manifold of low rank tensor structured vectors. In particular, the canonical tensor format was applied. The numerical cost of the matrix vector multiplication in our setting scales linearly in M, and at most quadratically in the univariate discrete problem size. In what follows, we sketch the main results in [227] and in [223].

Note that the idea of rank structured tensor approximation for parametric PDEs was further developed in recent years; see for example [67, 215, 216, 272, 273] among others.

5.4.2 Stochastic collocation: canonical tensor discretization in the additive case

Following [227], we consider the case of tensor product basis functions $\{\phi_{\mathbf{j}}\}$,

$$\phi_{\mathbf{j}}(y, x) = \phi_{j_0}(x) \prod_{m=1}^{M} \phi_{j_m}(y_m)\,, \quad \mathbf{j} \in \mathcal{J} = J_0 \times J^M\,, \quad J := \{1, \dots, n\}\,,$$

where $\{\phi_{j_0}(x)\}$ is chosen as the basis set in the Galerkin subspace $X_{N_0} = (X_{n_0})^{d_0} \in H_0^1(D) = (H_0^1(0,1))^{d_0}$ of tensor product piecewise linear functions in variable x. In turn, ϕ_{j_m} are piecewise polynomials in variable $y_m \in I$ $(m = 1, \dots, M)$ that span the tensor product space $\mathbb{Y}_M = (Y_n)^M$, where Y_n is either (A) the set of the univariate Legendre polynomials of degree $n-1$ in y_m $(m = 1, \dots, M)$, or (B) the space of piecewise constant basis functions in variable y_m, corresponding to the equidistant grid of size $h = 2/n$.

To build the Galerkin approximation of the initial boundary value problem (5.121) on the tensor product Hilbert space

$$V_n := \mathbb{Y}_M \otimes X_{N_0} \subset V = \bigotimes_{m=1}^{M} L^2(I) \otimes H_0^1(D)\,,$$

we search for the solution in the form $u_M^h = \sum_{\mathbf{j}\in\mathcal{J}} u_{\mathbf{j}} \phi_{\mathbf{j}} \in V_n$ that satisfies

$$A_M(u_M^h, v) = f(v) \quad \forall\, v \in V_n\,. \tag{5.122}$$

Conditions (1) and (2) above imply the following simple result:

Lemma 5.29. *([227]) The Galerkin equation* (5.122) *has a unique solution, which is quasioptimal,*

$$\|u_M - u_M^h\|_V \le C \inf_{v\in V_n} \|u_M - v\|_V\,,$$

where the constant $C > 0$ does not depend on M, n, and N_0.

Proof. The bilinear form A_M is coercive and continuous uniformly in M. Then the result follows by the Lax–Milgram lemma. □

To derive the Galerkin matrix equation, let us choose some rank-1 test and trial functions in the set $\{\phi_{\mathbf{j}}\}$, $\mathbf{j} \in \mathcal{J}$,

$$u(y, x) = u^{(0)}(x)\Pi_{m=1}^{M} u^{(m)}(y_m) , \quad v(y, x) = v^{(0)}(x)\Pi_{m=1}^{M} v^{(m)}(y_m) ,$$

then the associated bilinear form can be written as follows:

$$A_M(u, v) = A_0(u, v) + A_y(u, v)$$

with the separable representations

$$\begin{aligned}
A_0(u, v) &= \left[\sum_{i=1}^{d_0} \left\langle a_0(x) \frac{\partial}{\partial x_i} u^{(0)}(x), \frac{\partial}{\partial x_i} v^{(0)}(x) \right\rangle_{L^2(D)} \right] \\
&\quad \cdot \prod_{\ell=1}^{M} \left\langle u^{(\ell)}(y_\ell), v^{(\ell)}(y_\ell) \right\rangle_{L^2(I)} , \\
A_y(u, v) &= \sum_{m=1}^{M} \left[\sum_{i=1}^{d_0} \left\langle a_m(x) \frac{\partial}{\partial x_i} u^{(0)}(x), \frac{\partial}{\partial x_i} v^{(0)}(x) \right\rangle_{L^2(D)} \right] \\
&\quad \cdot \prod_{\ell=1}^{M} \left\langle y_\ell^{\Delta_{m\ell}} u^{(\ell)}(y_\ell), v^{(\ell)}(y_\ell) \right\rangle_{L^2(I)} ,
\end{aligned}$$

where $\Delta_{m\ell}$ is the Kronecker delta of the discrete argument.

With the notation $U = \{u_{\mathbf{j}}\}_{\mathbf{j}\in\mathcal{J}} \in \mathbb{R}^{\mathcal{J}}$ for the coefficient tensor, the Galerkin system of linear algebraic equations now reads:

$$\mathbf{A}U \equiv \left(\mathbf{A}_0 + \sum_{m=1}^{M} \mathbf{A}_m \right) U = F , \quad U, F \in \mathbb{R}^{\mathcal{J}} , \tag{5.123}$$

with the following tensor product representation of the stiffness matrix and of the respective right hand side:

$$\mathbf{A}_m = \bigotimes_{\ell=0}^{M} A_m^{(\ell)} , \qquad F = \bigotimes_{\ell=0}^{M} F^{(\ell)} , \tag{5.124}$$

where

$$A_m^{(\ell)} \in \mathbb{R}^{n\times n}, \ F^{(\ell)} \in \mathbb{R}^{n} \quad \ell = 1, \dots, M , \quad A_m^{(0)} \in \mathbb{R}^{N_0\times N_0}, \ F^{(0)} \in \mathbb{R}^{N_0} ,$$

such that for $m = 0, \dots, M$, we have

$$A_m^{(0)} = \left\{ \sum_{i=1}^{d_0} \left\langle a_m(x) \frac{\partial}{\partial x_i} \phi_{p_0}(x), \frac{\partial}{\partial x_i} \phi_{q_0}(x) \right\rangle_{L^2(D)} \right\}_{p_0,q_0=1}^{N_0} , \quad F^{(0)} = \left\{ \langle f, \phi_{q_0} \rangle_{L^2(D)} \right\}_{q_0=1}^{N_0} ,$$

and

$$A_m^{(\ell)} = \left\{ \left\langle y_\ell^{\Delta_{m\ell}} \phi_p(y_\ell), \phi_q(y_\ell) \right\rangle_{L^2(I)} \right\}_{p,q=1}^{n}, \quad F^{(\ell)} = \left\{ \langle 1, \phi_q(y_\ell) \rangle_{L^2(I)} \right\}_{q=1}^{n},$$
$$\ell = 1, \ldots, M.$$

The next lemma characterizes the rank structured representation of the matrix **A** and loading vector F in (5.123).

Lemma 5.30 ([227]). *The matrix* **A** *belongs to the class of rank-*$(M+1)$ *canonical format, while and* $F \in \mathcal{C}_{1,\mathbf{n}}$ *with* $\mathbf{n} = (N_0, n, \ldots, n)$*. The storage requirements to represent the matrix* **A** *and the rank-*1 *vector* F *are estimated by*

$$Q(\mathbf{A}) = O(N_0 M + n^\alpha M), \quad \alpha = 1, 2, \quad Q(F) = O(N_0 + nM)$$

respectively. Here $\alpha = 1$ *corresponds to the piecewise constant elements, and* $\alpha = 2$ *appears in the case of Legendre polynomials. The matrix-times-vector multiplication of* **A** *with a rank-*1 *tensor in* $\mathcal{C}_{1,\mathbf{n}}$ *scales linearly in* M*,* $O(N_0 M + nM)$*.*

We summarize that a parametric linear system in $\mathbb{R}^N$, where N is the grid size in x, obtained by FEM or FD approximation in x reads:

$$A(y)u(y) = f, \quad f \in \mathbb{R}^N, \; u(y) \in \mathbb{R}^N, \quad y \in \Gamma, \tag{5.125}$$

with the parameter dependent matrix

$$A(y) = A_0 + \sum_{m=1}^{M} A_m y_m, \quad A_m \in \mathbb{R}^{N \times N},$$

where A_m are $N \times N$ Galerkin stiffness matrices. Using the simplest collocation on a 1D grid, where n is the grid size in the parameter y_m, $m = 1, \ldots, M$,

$$\{y_m^{(k)}\} =: \Gamma_n \in [-1, 1], \quad k = 1, \ldots, n,$$

the problem is reduced to n^M linear systems

$$A(i_1, \ldots, i_M)u(i_1, \ldots, u_M) = f, \quad 1 \le i_k \le n,$$

which can be written as one assembled large linear system in the form (5.123),

$$\mathbb{A}\mathbf{u} = \mathbf{f}, \quad \mathbf{u}, \mathbf{f} \in \mathbb{R}^{Nn^M}, \quad \mathbb{A} \in \mathbb{R}^{Nn^M \times Nn^M}.$$

Here $\mathbb{A}$ is a $Nn^M \times Nn^M$ matrix,

$$\mathbb{A} = A_0 \times I \times \cdots \times I + A_1 \times D_1 \times I \times \cdots \times I + \cdots + A_M \times I \times \cdots \times D_M,$$

and D_m, $m = 1, \ldots, M$, is a $n \times n$ diagonal matrix with positions of collocation points $y_m^{(k)} \in \Gamma_n$ on the diagonal. Here $\mathbf{u}, \mathbf{f} \in \mathbb{R}^{Nn^M}$ may have the rank structured representations, in particular,

$$\mathbf{f} = f \times e \times \cdots \times e, \quad e = (1, \ldots, 1)^{\mathrm{T}} \in \mathbb{R}^n.$$

It can be easily seen that

$$\mathrm{rank}_C(\mathbb{A}) \leq M\,,$$

and the vector of right hand side has tensor rank 1.

In what follows, we first consider the solution of this system in the low rank canonical format.

5.4.3 Preconditioned rank truncated iteration

Given $\varepsilon > 0$, we apply the truncated preconditioned iteration in the canonical format

$$\tilde{\mathbf{u}}^{(k+1)} := \mathbf{u}^{(k)} - \omega \mathbb{B}_k^{-1}(\mathbb{A}\mathbf{u}^{(k)} - \mathbf{f})\,, \quad \mathbf{u}^{(k+1)} = T_\varepsilon(\tilde{\mathbf{u}}^{(k+1)}) \to \mathbf{u}\,,$$

where T_ε is the rank truncation operator preserving accuracy ε.

Recall that in the additive case we have $\mathrm{rank}_C(\mathbb{A}) \leq M$. A good choice of a (rank-1) preconditioner is given by ([227])

$$\mathbb{B}_0^{-1} = A_0^{-1} \times I \times \cdots \times I\,,$$

where A_0 is the parameter independent part in the system matrix or its spectrally equivalent approximand.

In a log additive case and QTT tensor representation, an adaptive preconditioner iteration step k can be applied ([223] and Section 5.4.4),

$$\mathbb{B}_k^{-1} = A(y_k^*)^{-1} \times I \times \cdots \times I\,, \quad y_k^* = \mathrm{argmin}_{\mathrm{QTT}}(\|\mathbf{f} - \mathbb{A}\mathbf{u}^{(k)}\|)\,,$$

so that $\mathbb{B}_0$ corresponds to the particular choice $y^* = 0$.

The spectral equivalence, $\mathbb{B}_0 \sim \mathbb{A}$, was proven in both additive and log additive cases; see [223, 227].

Here we discuss the additive case where the convenient choice of preconditioner is given by $A_0 = \Delta_{(d)}$. First, we recall the results on the low tensor rank Laplacian inverse. Let $\Delta_{(d)}$ be the FD negative Laplacian on $H_0^1([0,1]^d)$, $d = 2, 3, \ldots$, then the sinc quadrature rank-R, ($R = 2M + 1$) approximation of $\Delta_{(d)}^{-1}$ reads:

$$\mathbb{B}_M := \sum_{k=-M}^{M} c_k \bigotimes_{\ell=1}^{d} \exp(-t_k \Delta^{(\ell)}) \approx (\Delta_{(d)})^{-1}\,, \quad \Delta^{(\ell)} = \Delta \in \mathbb{R}^{n\times n}\,. \tag{5.126}$$

We have the exponential convergence in R ([110]):

$$\left\|(\Delta_{(d)})^{-1} - \mathbb{B}_M\right\| \leq C_0 e^{-\pi\sqrt{M}}\,, \quad \text{for } t_k = e^{k\mathfrak{h}}\,, \ c_k = \mathfrak{h} t_k\,, \ \mathfrak{h} = \frac{\pi}{\sqrt{M}}\,. \tag{5.127}$$

The ε rank of $\Delta_{(d)}^{-1}$ is $O(|\log \varepsilon|^2)$, uniformly in d.

The matrix vector multiplication of $\mathbb{B}_M$ with rank-1 vector in $\mathbb{R}^{n^{\otimes d}}$ takes $\mathcal{O}(dRn \log n)$ operations by the diagonalization

$$\exp(-t_k \Delta^{(\ell)}) = F_\ell' \cdot D_\ell \cdot F_\ell\,, \quad D_\ell = \mathrm{diag}\{e^{-t_k \lambda_1^{(\ell)}}, \ldots, e^{-t_k \lambda_n^{(\ell)}}\}\,,$$

where F_ℓ is the $n \times n$ matrix of ℓ-mode sin-transform, and $\lambda_i^{(\ell)}$ $(i = 1, \ldots, n)$ are the eigenvalues of the 1D Laplacian $\Delta^{(\ell)}$. Furthermore, the equivalent representation by using tensor product FFT holds:

$$\Delta_{(d)}^{-1} \approx \bigotimes_{\ell=1}^{d} F_\ell^{\mathrm{T}} \sum_{k=-M}^{M} c_k \bigotimes_{\ell=1}^{d} \operatorname{diag}\{\mathrm{e}^{-t_k\lambda_1^{(\ell)}}, \ldots, \mathrm{e}^{-t_k\lambda_n^{(\ell)}}\} \bigotimes_{\ell=1}^{d} F_\ell =: \mathbb{L}_M .$$

Note that the above decomposition is similar to the sinc approximation of the Hilbert tensor, now applied to the $n^{\otimes d}$ tensor $\Lambda = [1/(\lambda_{i_1}^{(1)} + \cdots + \lambda_{i_d}^{(d)})]$.

The QTT approach for the collective treatment of a family of matrix exponentials was described in Section 4.3.9.

Now we discuss the stability of spectral equivalence of the preconditioner with respect to the sinc quadrature based low rank approximation.

Introduce the class of rank-R Kronecker product preconditioners defined by approximate inverse $\mathbb{B}_R$ as above. Here the coefficients a_ℓ can be chosen from the optimization of the condition number in

$$C_1 \langle \mathbb{A} U, U\rangle \le \langle \mathcal{L}_0 U, U\rangle \le C_2 \langle \mathbb{A} U, U\rangle \quad \forall\, U \in \mathbb{V}_n , \tag{5.128}$$

so that the matrix $\mathcal{L}_0$ is supposed to be spectrally close/equivalent to the initial stiffness matrix $\mathbb{A}$ (here we use $\mathcal{L}_0 = \Delta_{(d)}$).

The following lemma proves the spectral equivalence estimates; see also [208, 227]:

Lemma 5.31 (Stability of spectral equivalence). *Suppose that the constants C_0, C_1, C_2 are determined by* (5.127) *and* (5.128) *respectively. Choose M such that the inequality $C_0\|\mathcal{L}_0\|\|\mathbb{B}_R^{-1}\|\mathrm{e}^{-\pi\sqrt{M}} < q(M)C_1$ holds with $q < 1$, then*

$$\widetilde{C}_1 \langle \mathbb{A} U, U\rangle \le \left\langle \mathbb{B}_R^{-1} U, U\right\rangle \le \widetilde{C}_2 \langle \mathbb{A} U, U\rangle \quad \forall\, U \in \mathbb{V}_n , \tag{5.129}$$

with some spectral equivalence constants $\widetilde{C}_1, \widetilde{C}_2 > 0$, which allow the following bound on the condition number:

$$\frac{\widetilde{C}_2}{\widetilde{C}_1} \le \frac{1}{1-q(M)}\frac{C_2}{C_1} + \frac{q(M)}{1-q(M)} .$$

Proof. The matrices A_ℓ, are symmetric and positive definite, hence the Laplace transform and related exponential factors in (5.126) are correctly defined. Moreover, all terms in the sum $\mathcal{L}_0 = A_1 \otimes \cdots \otimes I + \cdots + I \otimes \cdots \otimes A_d$ mutually commute providing the corresponding factorization of the matrix exponential

$$\mathrm{e}^{-t_k\mathcal{L}_0} = \prod_{\ell=1}^{d} \mathrm{e}^{-t_k I\otimes\cdots\otimes A_\ell\otimes\cdots\otimes I} , \quad (k = -M, \ldots, M) .$$

Then using the property of matrix exponential we obtain

$$\prod_{\ell=1}^{d} \mathrm{e}^{-t_k I\otimes\cdots\otimes A_\ell\otimes\cdots\otimes I} = \bigotimes_{\ell=1}^{d} \mathrm{e}^{-t_k A_\ell} ;$$

see [198, Theorem 5.3]. We have

$$\left\|\mathcal{L}_0 - \mathbb{B}_R^{-1}\right\| = \left\|-\mathcal{L}_0(\mathcal{L}_0^{-1} - \mathbb{B}_R)\mathbb{B}_R^{-1}\right\| \le \|\mathcal{L}_0\| \left\|\mathcal{L}_0^{-1} - \mathbb{B}_R\right\| \left\|\mathbb{B}_R^{-1}\right\| .$$

Using these inequalities the constants $\widetilde{C}_1, \widetilde{C}_2 > 0$ can be estimated by

$$C_1 - C_0\|\mathcal{L}_0\|\|\mathbb{B}_R^{-1}\|\mathrm{e}^{-\pi\sqrt{M}} \le \widetilde{C}_1 , \quad \widetilde{C}_2 \le C_2 + C_0\|\mathcal{L}_0\|\|\mathbb{B}_R^{-1}\|\mathrm{e}^{-\pi\sqrt{M}} ,$$

with $C_0 > 0$ defined in (5.127). Combining the above inequality with the error estimate (5.127) and with (5.128) leads to the desired bound. □

Lemma 5.31 indicates that the rank-R preconditioner $\mathbb{B}_R^{-1}$ has linear (or quadratic) complexity scaling in the univariate problem size n, providing at the same time the condition number of order C_2/C_1 as in (5.128), as soon as the estimate

$$C_0\|\mathcal{L}_0\| \left\|\mathbb{B}_R^{-1}\right\| \mathrm{e}^{-\pi\sqrt{M}} < C_1$$

holds. The latter is valid for $R = O(|\log(q(M)/C_1)|^2) = O((\log n)^2)$. Note that the modified sinc quadrature leads to the improved convergence rate in (5.127), $C_0\mathrm{e}^{-\alpha M/\log(M)}$, with $\alpha = \log(\mathrm{cond}(\mathcal{L}_0))$, providing the improved rank estimate $R = O(\log(|\log(q(M))| /C_1)) = O((\log n)^2)$.

By the way, we comment that in the case of highly variable coefficients $a(y, x)$ the so called 'reciprocal' preconditioner

$$\left(\nabla^{\mathrm{T}} a \nabla\right)^{-1} \approx P := \Delta^{-1}\left(\nabla^{\mathrm{T}} \frac{1}{a} \nabla\right)\Delta^{-1} \tag{5.130}$$

may do the job. Indeed it can be proven that $\Delta^{-1}(\nabla^{\mathrm{T}}\frac{1}{a}\nabla)\Delta^{-1}(\nabla^{\mathrm{T}} a\nabla) = I + R$, where for $d = 1$ $\mathrm{rank}(R) = 1$, and for $d \ge 2$, we have $\mathrm{rank}(R) = const$ in the case of piecewise constant coefficient with one interface. Numerically, it demonstrates surprisingly good clustering properties; see [82, 83].

We note that the rank structured iteration methods for linear systems in rather general settings were discussed in [88, 152, 208, 240]. The low rank approximations (by using sinc quadrature) to the Laplacian inverse and to the more general class of operator valued functions have been considered in [112, 122] and in [110, 112] respectively.

5.4.4 Stochastic collocation in log additive case: using TT tensor format

In the log additive case the dependence on y is no longer affine. Here we follow [223]. For the ease of exposition, we let $d = 1$. Applying collocation to (5.125) we obtain n^M linear systems (using piecewise linear FEM on the uniform grid of size N),

$$A(j_1, \dots, j_M)u(j_1, \dots, j_M) = f , \quad 1 \le j_m \le n ,$$

where

$$A(i,j,y) = \int_D b(x,y)\frac{\partial\phi_i}{\partial x}\frac{\partial\phi_j}{\partial x}dx\,, \quad y \in \Gamma_n\,, \quad D = [0,1]\,, \quad i,j = 1,\ldots,N\,.$$

We use the simple quadrature to obtain

$$A(i,i,y) = \frac{1}{4}(b(x_{i-1},y) + 2b(x_i,y) + b(x_{i+1},y))\,,$$

$$A(i,i-1,y) = \frac{1}{2}(b(x_{i-1},y) + b(x_i,y))\,, \quad A(i-1,i,y) = A(i,i-1,y)\,,$$

for $i = 1,\ldots,N$, with

$$b(x,y) = \mathrm{e}^{a(x,y)} = \mathrm{e}^{a_0(x)}\prod_{m=1}^{M}\mathrm{e}^{a_m(x)y_m}\,, \quad y \in \Gamma_n\,.$$

This construction leads to the linear system of equations

$$\mathbb{A}\mathbf{u} = \mathbf{f}\,,$$

where there is still good low rank approximations to the stiffness matrix of the form

$$\mathbb{A} \approx \sum_{k=1}^{R}\bigotimes_{m=0}^{M}A_{mk}\,, \quad A_{mk} \in \mathbb{R}^{(M+1)\times n}\,.$$

Matrices $A_{mk} \in \mathbb{R}^{(M+1)\times n}$ can be precomputed for fast calculations.

In the log additive case, the rank estimate for $\mathbb{A}$ is a nontrivial task. We have only local pointwise bounds proven in [223].

Lemma 5.32 ([223]). *For quadrature FEM discretization of 1D PDE by piecewise linear elements in the log additive case, we have:*
(a) $\mathrm{rank}_C(A(i,j,y)) \leq 3$ $(i,j \leq N,\ y \in \Gamma_n) \Rightarrow \mathbb{A} \in C_{\mathrm{loc}}[3] \subset \mathrm{QTT}_{loc}[3]$.
(b) $\mathrm{rank}_{\mathrm{QTT}}(A(i,j,y)) \leq 3$ $i,j \leq N,\ y \in \Gamma_n, \Rightarrow \mathbb{A} \in \mathrm{QTT}_{\mathrm{loc}}[3]$.
(c) *Moreover,*

$$\mathrm{rank}_C(\mathbb{A}) \leq 7N\,, \quad \textit{uniform in } n,\ M \tag{5.131}$$

independently of n and M.

Proof. We split the local matrix into three terms

$$A(y) = D(y) + Z(y) + Z^{\top}(y)\,, \quad y \in \Gamma_n\,,$$

where $D(y)$ is a diagonal of A and Z is the first subdiagonal. $D(y)$ is represented as

$$D(y) = \sum_{i=1}^{N}A(i,i,y)e_ie_i^{\top} = \frac{1}{4}(C_1(y) + 2C_2(y) + C_3(y))\,,$$

where $C_2(y)$ takes the form

$$C_2(y) = \sum_{i=1}^{N}e_ie_i^{\top}\mathrm{e}^{a_0(x_i)}\prod_{m=1}^{M}\mathrm{e}^{a_m(x_i)y_m}\,. \tag{5.132}$$

$C_2(y)$, $y \in \Gamma_n$, is a $Nn^m \times Nn^m$ diagonal matrix, and each summand in (5.132) has tensor rank 1. For QTT format in variable y_m, TT ranks will be equal to 1, since it is an exponential function; see Section 4.2. Then simple calculations complete the proof. □

A theoretical estimate of the separation rank in the discretized solution is a challenging problem in both additive and log additive cases. In the following we discuss several examples when the explicit rank estimate is possible.

First, we consider the rank bound for the solution in the case $d = 1$. Define

$$v = -\Delta_x^{-1} f\,, \quad \sigma_m = \frac{\|a_m\|}{\sum_{m=1}^{M} \|a_m\|} > 0\,,$$

and introduce the reassembled coefficients

$$b_m(y_m, x) = \sigma_m a_0(x) + a_m(x) y_m\,, \quad (m = 1, \ldots, M)\,.$$

The following statement collects various rank bounds for the solution and its gradient; see [209]:

Proposition 5.33. *Let $d = 1$, assume $\nabla_x u_M(y, x) \in C(D)$ for all $y \in \Gamma$, $\nabla_x v(x) \in C(D)$, and that there exists $a_{\min} > 0$, such that*
(A) $a_{\min} \leq a_0(x) < \infty$,
(B) $|\sum_{m=1}^{M} a_m(x) y_m| \leq \gamma a_{\min}$ *with $\gamma < 1$, and for $|y_m| < 1$ $(m = 1, \ldots, M)$.*

Then for ε rank we have

$$\operatorname{rank}(\nabla_x u_M) \leq C|\log \varepsilon| \quad \textit{(additive)}\,; \quad \operatorname{rank}_{C_{\mathrm{loc}}}(\nabla_x u_M) = 1 \quad \textit{(log additive)}\,.$$

Proof. We have $\nabla_x u_M(y, x) = \frac{1}{a_M(y,x)}(C_0 + \nabla_x v(x))$. Then, in the additive case, there exist $c_k, t_k \in \mathbb{R}_{>0}$, such that

$$\left\| \nabla_x u_M(y, x) - \sum_{k=-K}^{K} c_k \prod_{m=1}^{M} \mathrm{e}^{-t_k b_m(y_m, x)} (C_0 + \nabla_x v(x)) \right\|_{L^\infty} \leq C \mathrm{e}^{-\beta K / \log K}\,,$$

where $\beta > 0$ and C do not depend on M and K. In the log additive case we have $\operatorname{rank}(a_M(y, x)) = 1$, which completes the proof. □

Note that the discrete analogy of Proposition 5.33 leads to similar rank bounds.

The following result indicates that in the case of constant coefficients $a_m(x) = \overline{a}_m$, $(m = 0, 1, \ldots, M)$ we can achieve the exponential convergence rate in the number of terms based on the explicit rank estimate of the formatted low rank approximation to the solution [227]. In the case of spatially homogeneous coefficients, we denote the associated coefficient function and bilinear form by $\overline{a}_M$ and $\overline{A}_M(\cdot, \cdot)$ respectively.

Proposition 5.34 ([227]). *Assume that the stochastic coefficients are constant, $a_m(x) = \overline{a}_m$ $(m = 0, 1, \ldots, M)$. Then the solution of equation (5.118) can be presented by the explicit formula*

$$u_M(y, x) = \frac{1}{\overline{a}_0 + \sum_{m=1}^{M} \overline{a}_m y_m} (-\Delta_x)^{-1} f(x)\,, \tag{5.133}$$

where Δ_x is the Laplace operator on $H_0^1(D)$ in variable $x \in \mathbb{R}^{d_0}$.

Under initial Assumptions (1)–(2) on $a_M(x,y)$ as in Section 5.4.1, the multivariate coefficient function in (5.133) *allows the following R-term separable approximation:*

$$G(y) := \frac{1}{\overline{a}_0 + \sum_{m=1}^{M} \overline{a}_m y_m} \approx \sum_{k=1}^{R} g_1^k(y_1)\dots g_M^k(y_M)\,, \quad \forall\, y \in I^M\,,$$

which converges exponentially in R,

$$\left\| G(y) - \sum_{k=1}^{R} g_1^k(y_1)\dots g_M^k(y_M) \right\|_{L^\infty(\Gamma)} \le C e^{-\beta R/\log R}\,, \tag{5.134}$$

where $\beta > 0$ does not depend on M and R.

Proposition 5.34 implies that the number of terms R in the above constructed decomposition, which allows us to achieve an accuracy $\varepsilon = Ce^{-C_1 M^\kappa}$, $\kappa = 1/d_0$, is equal to $R = O(M^\kappa \log M)$, which follows from the analysis of the equation

$$\frac{R}{\log R} = M^\kappa\,, \quad M \gg 1\,.$$

Hence, in this case we arrive at even better than linear logarithmic complexity in M (sublinear complexity for $d_0 \ge 2$). Note that an approximation of $1/x$ by the $(2L+1)$-term exponential sums on a semiaxis with the convergence rate $\exp(-\beta\sqrt{L})$ was first proven in [50].

One can expect that in the case of 'smooth' coefficients $a_m(x)$ the situation might be very similar to that described in Proposition 5.34. Moreover, the operator with constant coefficients, $a_m(x) = \overline{a}_m$, is a good candidate for the spectrally close preconditioner.

5.4.5 Numerics to rank structured solution of sPDEs: additive and log additive cases

In the first part of the presented numerical illustrations, we test a preconditioned tensor truncated iteration in $(d+M)$ dimensional parametric space implemented in low rank canonical format in the range of stochastic dimension $M \le 100$. The following numerics have been reported in [227].

We use the $\mathcal{S}$ truncated preconditioned iteration for solving $1D$ sPDE in physical space. The numerical complexity scales linearly in M. We set $\mathcal{S} = \mathcal{C}_R$ and $\mathbb{B}^{-1} := \mathbb{A}(0)^{-1}$, and then fix $M = 20$ and the total grid size $N^{\otimes(M+d)}$, $N = 63$, $R \le 5$. We solve the full system of equations

$$\mathcal{A}(y)U(y) = F$$

in the case of variable coefficients with exponential decay

$$a_m(x) = 0.5\, e^{-m} \sin(mx)\,, \quad m = 1, 2, \dots, M\,, \quad x \in (0, \pi)\,.$$

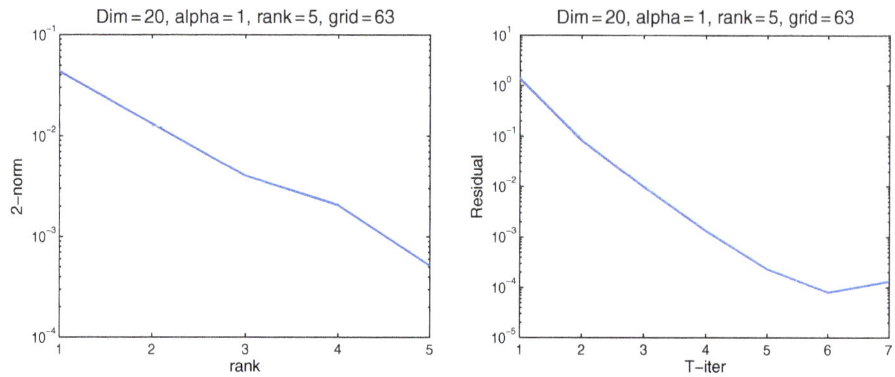

Fig. 5.22: Approximation error vs. rank parameter.

In Figure 5.22 (left) we observe the fast (almost exponential in R) decay of the approximation error in the rank parameter. The preconditioned CG iteration demonstrates linear convergence rate until the residual achieves the level of rank truncation error as indicated in Figure 5.22 (right).

As the second example, we consider preconditioned truncated iteration in the case of highly oscillating coefficients. This example serves to demonstrate that in practice the rank approximation to the parametric solution does not depend on the smoothness in the coefficients in the Karhunen–Loéve expansion.

We compare smooth and random coefficient in y. In the smooth case we set

$$a(y,x) = a(y) := 1 + \sum_{m=1}^{M} a_m y_m \quad \text{with} \quad y = \|\mathbf{a}\|_{\ell_1} := \sum_{m=1}^{M} |a_m| < 1\,,$$

for the truncated sequence of (spatially homogeneous) coefficients $a_m = (1+m)^{-\alpha}$, $(m = 1, \ldots, M)$ with algebraic decay rates $\alpha = 2, 3, 5$. In this case the zero order sPDE of the form

$$a(y)u(y) = f \tag{5.135}$$

can be solved directly.

The highly oscillating random coefficient is given by

$$a(y) = 1 + \sum_{m=1}^{M} a_m y_m H(y_m - c_m(y_m))\,,$$

where the piecewise constant function $c_m(y_m)$ is defined by a random n vector at $[-1,1]$. Here the function $H\colon \mathbb{R} \to \{-1,1\}$ is specified by $H(x) = -1$ for $x < 0$ and $H(x) = 1$ for $x \geq 0$.

Figure 5.23 illustrates the approximation error versus rank R (left) and the five canonical vectors in variable y_1 (right) for the solution of (5.135) with $M = 20$.

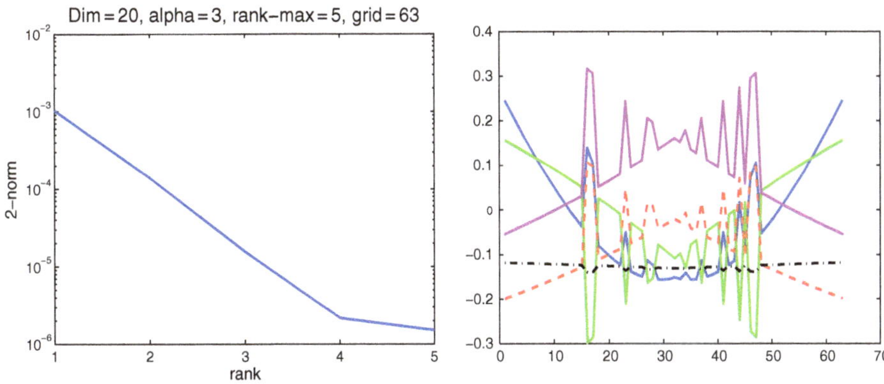

Fig. 5.23: Approximation error versus rank R (left) and the five canonical vectors in variable y_1 (right) for the solution of (5.135), $M = 20$.

We observe the exponential convergence in the rank parameter r, same as in the smooth case. The canonical vectors are highly oscillating, however this does not affect the separability properties of the parametric solution.

In the next numerical examples we study the rank truncated preconditioned iteration in the QTT format. The matrix QTT ranks are tested in both additive and log additive cases; see [223] for the detailed exposition.

Consider a stratified 2D dimensional sPDE in the two cases:
1. Polynomial decay: $a_m(x) = \frac{0.5}{(m+1)^2} \sin mx$, $x \in [-\pi, \pi]$, $m = 1, \ldots, M$.
2. Exponential decay: $a_m(x) = e^{-0.7m} \sin mx$, $x \in [-\pi, \pi]$, $m = 1, \ldots, M$.

The parametric space is discretized on a uniform mesh in $[-1, 1]$ with 2^p points in each spatial direction, where $p = 8$.

Table 5.7 demonstrates QTT rank of the matrix for 2D SPDE in the log additive case and with the exponential decay in coefficients. The univariate grid size is fixed to $N = 128$.

Figure 5.24 illustrates the results for the 2D example with stratified coefficient. We tested two different truncation parameters, where we use the one-point preconditioner. Residual versus iteration number, ranks versus iteration number, and CPU time scaling in M are presented. We observe almost linear dependence of the QTT rank on the parametric dimension M.

Tab. 5.7: QTT rank of the matrix: 2D SPDE, log additive case, exponential decay $N = 128$.

M	QTT rank(10^{-7})	QTT rank(10^{-3})
5	33	11
10	43	21
20	51	23
40	50	25

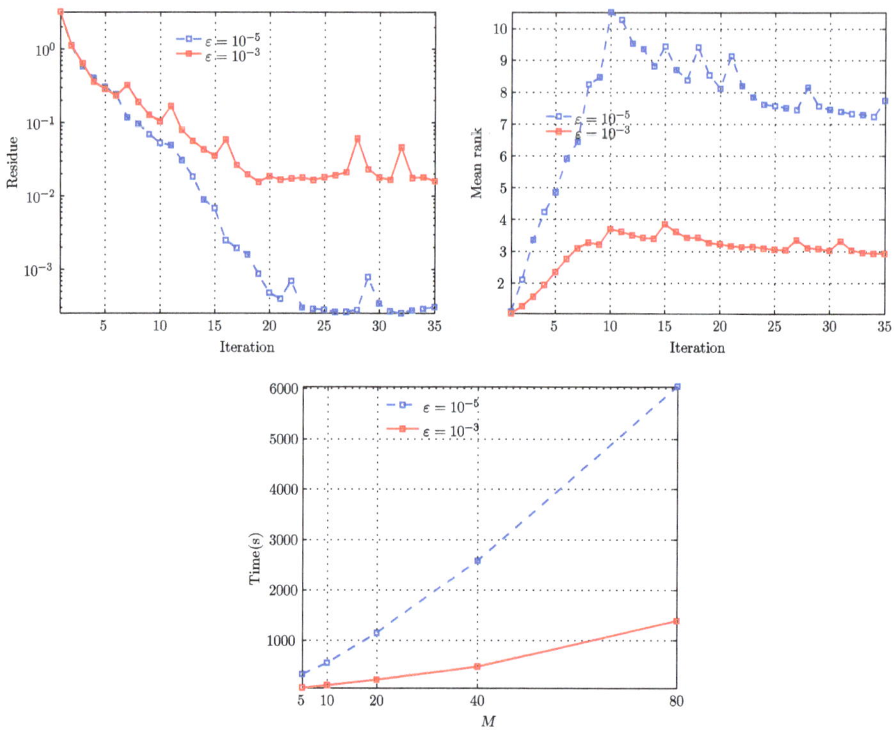

Fig. 5.24: Upper left: Residue versus iteration. Upper right: QTT ranks versus iteration. Bottom: CPU time versus M, for 35 iterations, log additive case.

5.5 Range separated tensor format: breaking through the complexity of many-particle modeling

5.5.1 Main motivations

The new class of range separated (RS) tensor formats was recently introduced in [30], see also the recent paper [31]. This development was motivated by a range of demanding applications related to many-particle modeling. In such applications the nonstructured location of interacting particles is described by highly nonregular functions with many local cusps. Clearly, the traditional tensor formats are not useful for representation of such unstructured data. However, the idea of range separated decomposition of the target interaction potential in $\mathbb{R}^d$ allows us to split the difficulties and approximate the long range part in the potential by using the low rank canonical/Tucker tensors. This has led to the new class of grid based tensor numerical methods, which allow efficient computations with many-particle systems at quasioptimal costs. It is worth noting that the theoretical justification of the approach is based on the construction of range separated splitting to the potential of interest in the frequency domain, which indicates the basic duality between the well structured nature of the long range part in the multiparticle interaction and the narrow band structure of that in the Fourier space.

There is an interesting prehistory to the invention of RS tensor formats. The canonical (CP) tensor approximation of the single Newton kernel $\frac{1}{\|\mathbf{x}\|}$ discretized on large $n \times n \times n$ tensor grids is the important ingredient in the Hartree–Fock calculations, and in particular in the fast rank structured tensor computation of the electrostatic potentials in large lattice type atomic systems; see Sections 5.2 and 2.4.4. Recall that the full third order tensor $\mathbf{P}$ representing the Newton kernel

$$\mathbf{P} := [p_{\mathbf{i}}] \in \mathbb{R}^{n\times n\times n}, \quad p_{\mathbf{i}} = \int_{\mathbb{R}^3} \frac{\psi_{\mathbf{i}}(\mathbf{x})}{\|\mathbf{x}\|} d\mathbf{x}, \quad \psi_{\mathbf{i}}(\mathbf{x}) = \prod_{\ell=1}^{d} \psi_{i_\ell}^{(\ell)}(x_\ell), \ \{i_\ell\}_1^{n_\ell},$$

$\ell = 1, 2, 3$, can be approximated by the R-term canonical representation

$$\mathbf{P} \approx \mathbf{P}_R = \sum_{k=-M}^{M} a_k \bigotimes_{\ell=1}^{3} \mathbf{b}^{(\ell)}(t_k) = \sum_{q=1}^{R} \mathbf{p}_q^{(1)} \otimes \mathbf{p}_q^{(2)} \otimes \mathbf{p}_q^{(3)} \in \mathbb{R}^{n\times n\times n}, \quad a_k, t_k \in \mathbb{R}, \tag{5.136}$$

where $R = 2M + 1 \approx C|\log \varepsilon|$. Here the basic parameters are determined by using the sinc quadrature approximation to the Laplace–Gauss integral transform of the analytic function $p(z) = \frac{1}{z}$ by substitution $z = \|x\| = \sqrt{x_1^2 + x_2^2 + x_3^2}$,

$$\frac{1}{z} = \frac{2}{\sqrt{\pi}} \int_{\mathbb{R}_+} e^{-t^2 z^2} dt \approx \sum_{k=-M}^{M} a_k e^{-t_k^2 \|x\|^2} = \sum_{k=-M}^{M} a_k \prod_{\ell=1}^{3} e^{-t_k^2 x_\ell^2}.$$

It was shown [141], see also Section 2.4.4, that in the range $0 < h \le \|x\| \le A < \infty$ the sinc quadrature approximation converges exponentially fast in M:

$$\left| \frac{1}{\|x\|} - \sum_{k=-M}^{M} a_k e^{-t_k^2 \|x\|^2} \right| \le \frac{C}{a} e^{-\beta\sqrt{M}}, \quad \text{with some } C, \beta > 0,$$

where the quadrature points and weights are given for example by $(a(t_k) = \frac{2}{\sqrt{\pi}})$

$$t_k = k\mathfrak{h}_M, \quad a_k = a(t_k)\mathfrak{h}_M, \quad \mathfrak{h}_M = C_0 \log(M)/M, \quad C_0 > 0. \tag{5.137}$$

The numerical calculation of the tensor $\mathbf{P}_R$ can be implemented efficiently [38]. For example, Table 5.8 presents the CPU times (MATLAB) to compute canonical tensor $\mathbf{P}_R$ with tolerance $\varepsilon = 10^{-7}$ versus the different grid size.

Tab. 5.8: CPU times to compute canonical tensor $\mathbf{P}_R$, tol. $\varepsilon = 10^{-7}$.

Grid size n^3	8192^3	$16\,384^3$	$32\,768^3$	$65\,536^3$	$131\,072^3$
Time (s)	6	16	61	241	1000
CP rank R	34	37	39	41	43
Compression rate	$2 \cdot 10^6$	$7 \cdot 10^6$	$2 \cdot 10^7$	$1 \cdot 10^8$	$4 \cdot 10^8$

This kind of sinc based CP tensor approximation applies to a wide class of long range radial basis functions $p(\|x\|)$ in $\mathbb{R}^3$, for example to the Slater, Yukawa, Lennard-Jones, or Van der Waals interaction potentials:

$$\text{Slater function:} \quad p(\|x\|) = \exp(-\lambda\|x\|)\,, \quad \lambda > 0\,,$$

$$\text{Yukawa kernel:} \quad p(\|x\|) = \frac{\exp(-\lambda\|x\|)}{\|x\|}\,, \quad \lambda > 0\,,$$

$$\text{Lennard-Jones potential:} \quad p(\|x\|) = 4\epsilon\left[\left(\frac{\sigma}{\|x\|}\right)^{12} - \left(\frac{\sigma}{\|x\|}\right)^{6}\right].$$

The simplified version of the Lennard-Jones potential, the Buckingham function reads:

$$p(\|x\|) = 4\epsilon\left[e^{\|x\|/r_0} - \left(\frac{\sigma}{\|x\|}\right)^{6}\right].$$

The electrostatic dipole-dipole potential energy (Van der Waals forces) is given by

$$p(\|x\|) = \frac{C_0}{\|x\|^3}\,.$$

The sinc based tensor approximation also applies to the so called Stokeslet. We summarize that the low rank CP approximations to the above radial basis functions by Gaussian sums can be derived by using sinc quadrature techniques.

5.5.2 Rank structured lattice sum of interaction potentials

Calculation of large $L \times L \times L$ lattice sums of long range electrostatic potentials, $\frac{1}{\|x\|}$, (Newton kernels),

$$V_c = \sum_{a=1}^{A} \frac{Z_a}{\|x - x_a\|}\,, \quad A = L^3\,,$$

is a challenging computational problem.

Consider a sum of potentials on a $L \times L \times L$ lattice,

$$v_{c_L}(x) = \sum_{k_1,k_2,k_3=1}^{L} \frac{Z}{\|x - a_{k_1,k_2,k_3}\|}\,, \quad x \in \Omega_L = \bigcup_{k_1,k_2,k_3=1}^{L} \Omega_{\mathbf{k}} \in \mathbb{R}^3\,,$$

where $\Omega_{\mathbf{k}}$ represent the unit cell in the lattice. Summation of long range potentials in $\mathbb{R}^3$ is a classical problem in many-particle modeling for crystalline type systems and for biomolecules, in molecular and many particle dynamics, etc. The classical methods are based on the Ewald summation techniques [96] combined with the FFT and fast multipole methods (FMM), [126]. Classical methods for calculation of long range potential sums on 3D lattices scale linearly logarithmically in the volume size i.e., $O(L^3 \log L)$. A special method for periodic systems is discussed in [256].

The Ewald summation approach is based on a specific local-global decomposition of the Newton kernel

$$\frac{1}{r} = \frac{\tau(r)}{r} + \frac{1-\tau(r)}{r}, \qquad r = \|x\|,$$

where the traditional choice of the cutoff function τ is the complementary error function

$$\tau(r) = \operatorname{erfc}(r) := \frac{2}{\sqrt{\pi}} \int_r^{\infty} \exp(-t^2)\,dt .$$

The Ewald summation techniques allow to achieve complexity $O(L^3 \log L)$ for calculation of the interaction energy for the $L \times L \times L$ lattice structured systems,

$$E_{\text{nuc}} = \sum_{i=1}^{L^3} \sum_{j<i}^{L^3} \frac{Z_i Z_j}{\|x_i - x_j\|},$$

instead of $O(L^6)$ scaling as for the direct summation. The grid based lattice summation approach by assembled canonical/Tucker tensors allows us to reduce the numerical cost down to $O(L \log L)$ vs. $O(L^3 \log L)$; see [188, 189].

Let $\mathcal{W}_{(\mathbf{k})} = \mathcal{W}_{(k_1)} \otimes \mathcal{W}_{(k_2)} \otimes \mathcal{W}_{(k_3)}$ be the windowing (restriction) operation on the shifted reference potential $\mathbf{P}_R$ defined in Ω_L; see [189].

Theorem 5.35 ([188]). *The projected tensor of the interaction potential $v_{c_L}(x)$, $x \in \Omega_L$, representing the full lattice sum over L^3 charges, can be represented by the canonical tensor $\mathbf{P}_{c_L}$ with the rank $R = \operatorname{rank}(\mathbf{P}_R)$,*

$$\mathbf{P}_{c_L} = \sum_{q=1}^{R} \left(\sum_{k_1=1}^{L} \mathcal{W}_{(k_1)} \mathbf{p}_q^{(1)} \right) \otimes \left(\sum_{k_2=1}^{L} \mathcal{W}_{(k_2)} \mathbf{p}_q^{(2)} \right) \otimes \left(\sum_{k_3=1}^{L} \mathcal{W}_{(k_3)} \mathbf{p}_q^{(3)} \right) \in \mathbb{R}^{n_L \times n_L \times n_L}. \tag{5.138}$$

The numerical cost and storage size are estimated by $O(RLn_L)$ and $O(Rn_L)$ respectively, where $n_L = O(L)$ is the univariate grid size.

Proof. The sum of potentials is defined on the domain Ω_L,

$$v_{c_L}(x) = \sum_{k_1,k_2,k_3=1}^{L} \frac{Z}{\|x - b\mathbf{k}\|}, \qquad x \in \Omega_L . \tag{5.139}$$

Then the projected tensor representation of $v_{c_L}(x)$ takes the form (omitting factor Z)

$$\mathbf{P}_{c_L} = \sum_{k_1,k_2,k_3=1}^{L} \mathcal{W}_{(\mathbf{k})} \mathbf{P}_R = \sum_{q=1}^{R} \sum_{k_1,k_2,k_3=1}^{L} \mathcal{W}_{(\mathbf{k})} \left(\mathbf{p}_q^{(1)} \otimes \mathbf{p}_q^{(2)} \otimes \mathbf{p}_q^{(3)} \right) \in \mathbb{R}^{n_L \times n_L \times n_L},$$

where the 3D shift vector is defined by $\mathbf{k} \in \mathbb{Z}^{L \times L \times L}$. □

The representation in Theorem 5.35 indicates that the CP tensor rank of the lattice sum remains the same as that for a single reference kernel $\mathbf{P}_R$ representing $p(\|x\|)$. This provides the constructive algorithm for calculating large lattice sums of electrostatic

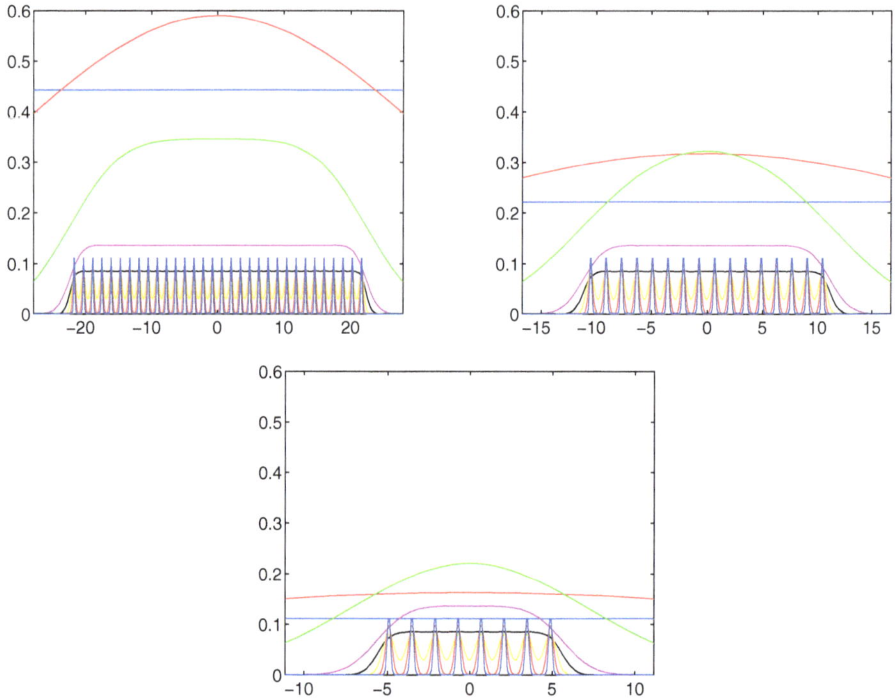

Fig. 5.25: Assembled canonical vectors for the sum of electrostatic potentials over a $32 \times 16 \times 8$ lattice.

potentials with $O(L \log L)$ cost. It can be shown that the canonical skeleton vectors $\mathcal{W}_{(k_\ell)}\mathbf{p}_q^{(\ell)}$ allow the low rank QTT representation that leads to the desired cost estimate; see Remark 5.36.

Figure 5.25 represents assembled x, y, and z axis canonical skeleton vectors for a cluster of $32 \times 16 \times 8$ hydrogen atoms. This corresponds to lattice sums of 4096 electrostatic potentials with the absolute accuracy 10^{-7}.

In some cases tensor sums on composite geometries can be reduced to the application of tensor summation method to a small number of regular lattice sums. Indeed, assume that the target lattice $\mathcal{L}$ can be split into the union of several sublattices, $\mathcal{L}_q$, all represented on the same fine 3D rectangular grid, $\mathcal{L} = \bigcup \mathcal{L}_q$. The resulting potential agglomerates all sums over q lattices, $A_{\text{sum}} = A_1 + A_2 + \cdots + A_q$. Figure 5.26 ([188]) illustrates the electrostatic potential for the system of particles located in the 'L' shaped and 'O' shaped lattices.

The summation scheme also applies to a hexagonal lattice that can be split into a union of two rectangular lattices; see [188].

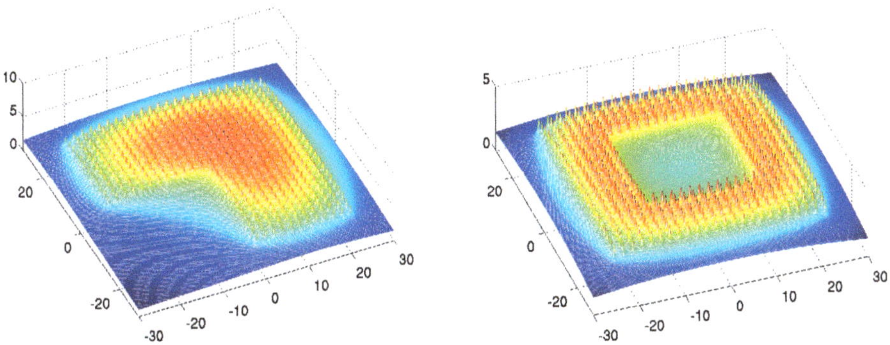

Fig. 5.26: Electrostatic potentials for 'L' and 'O' shaped lattice structures.

In the presence of multiple defects the initial canonical rank may increase considerably. In this case the low rank Tucker tensor approximation of the resultant tensor sum via the canonical-to-Tucker algorithm can be applied.

Note that the tensor summation techniques can also be implemented directly in the Tucker tensor format, thus preserving the Tucker rank of the generating Newton kernel as shown in the following representation:

$$\mathbf{T}_{C_L} = \sum_{m_1=1}^{r_1} \sum_{m_2=1}^{r_2} \sum_{m_2=1}^{r_2} b_{m_1,m_2,m_3} \left(\sum_{k_1=1}^{L} \mathcal{W}_{(k_1)} \tilde{\mathbf{t}}_{m_1}^{(1)} \right) \otimes \left(\sum_{k_2=1}^{L} \mathcal{W}_{(k_2)} \tilde{\mathbf{t}}_{m_2}^{(2)} \right) \otimes \left(\sum_{k_3=1}^{L} \mathcal{W}_{(k_3)} \tilde{\mathbf{t}}_{m_3}^{(3)} \right).$$

Remark 5.36. The next step for the enhancement of the tensor summation techniques may be the application of the QTT based approximations within the lattice summation, which allows the storage in log complexity. Given the rank-R tensor representing $\frac{1}{\|x\|}$, a lattice sum of single charges can be presented by rank-R QTT-canonical tensor

$$\mathbf{P}_{C_L} = \sum_{q=1}^{R} \left(\mathcal{Q} \sum_{k_1=0}^{L-1} \mathcal{W}_{(k_1)} \mathbf{p}_q^{(1)} \right) \otimes \left(\mathcal{Q} \sum_{k_2=0}^{L-1} \mathcal{W}_{(k_2)} \mathbf{p}_q^{(2)} \right) \otimes \left(\mathcal{Q} \sum_{k_3=0}^{L-1} \mathcal{W}_{(k_3)} \mathbf{p}_q^{(3)} \right),$$

where the QTT rank of each skeleton vector is bounded by $r_{\mathrm{QTT}} \leq C \log(L/\varepsilon)$; see [189]. The computational cost is then bounded by $O(Rr_{\mathrm{QTT}}^3 L)$, while the storage size scales as $O(R \log^2(L/\varepsilon))$.

5.5.3 Basic idea and general definition of range separated formats

The new range separated (RS) tensor format was recently proposed and analyzed in [30] in relation to the rank structured tensor approximation of highly nonregular functions with multiple singularities in $\mathbb{R}^3$, sampled on the fine $n \times n \times n$ grid. These may be the electrostatic potentials of a large atomic system like a biomolecule or in many-particles dynamics, as well as the radial basis (RB) functions interpolant in the

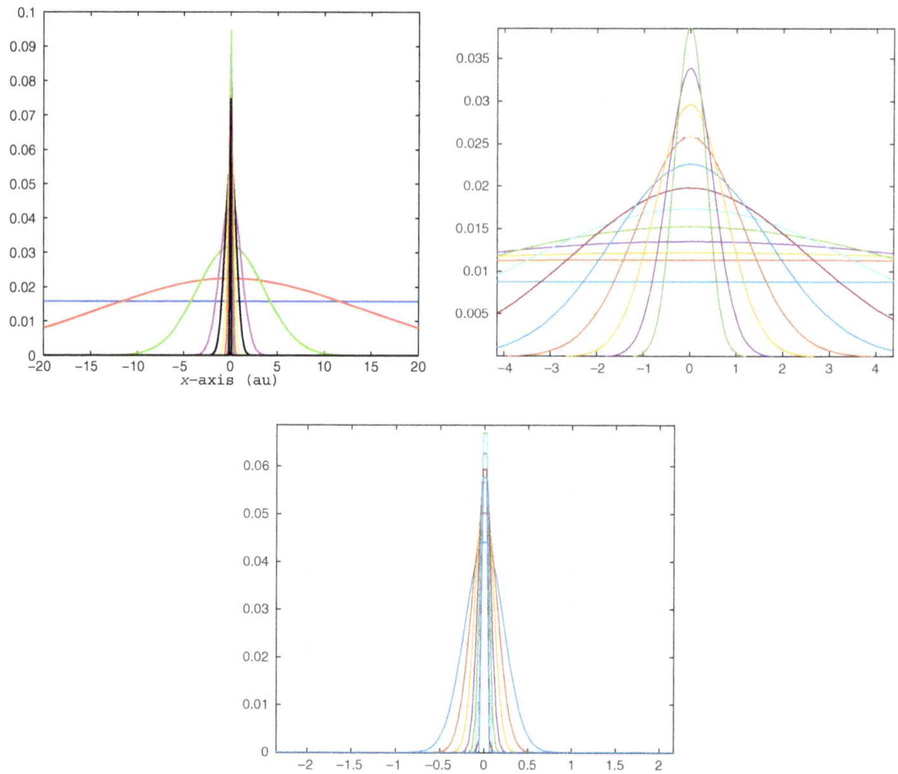

Fig. 5.27: Short and long range skeleton canonical vectors of the reference Newton kernel $1/\|x\|$.

modeling of multidimensional scattered data. In such cases the corresponding RB functions can be centered at the rather unstructured set of points in $\mathbb{R}^d$.

As described in [30], the mathematical construction is motivated by the observation that the set of skeleton vectors $\mathbf{p}_q^{(\ell)}$, generated by sampling Gaussians with different exponential t_k^2 onto the tensor grid, can be split into the two groups of the short and long range functions.

Figure 5.27 shows the canonical vectors $\mathbf{p}_q^{(1)}$ of the symmetric tensor $\mathbf{P}_R$ (upper left) and the corresponding long (upper right) and short (bottom) range components.

Hence, the construction of RS tensor formats for approximating functions with multiple cusps is based on the *rank structured splitting* of the generating kernel into the short and long range parts. The example of such splitting for the reference Newton kernel $1/\|x\|$ is presented in Figure 5.28. This corresponds to separation of the canonical skeleton vectors depicted in Figure 5.27.

We confine ourselves to the case of the Newton kernel, so that the sum in (5.136) reduces to $k = 0, 1, \ldots, M$ (due to symmetry argument). We observe that the sequence of quadrature points $\{t_k\}$ can be split into two subsequences, $\mathcal{T} := \{t_k | k = 0, 1, \ldots, M\} =$

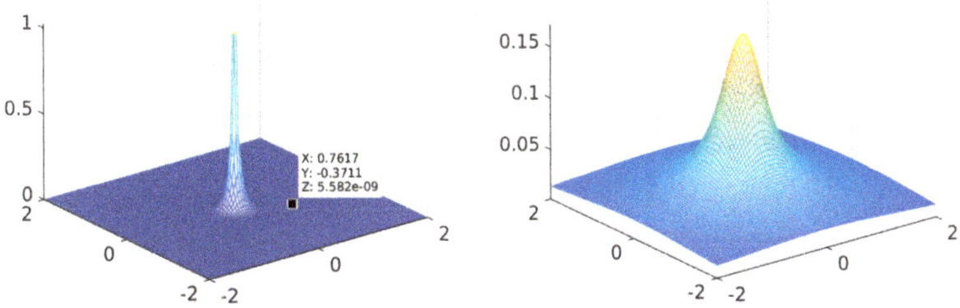

Fig. 5.28: Short and long range parts of the reference Newton kernel $1/\|x\|$.

$\mathcal{T}_l \cup \mathcal{T}_s$, with

$$\mathcal{T}_l := \{t_k \,|\, k = 0, 1, \ldots, R_l\}, \quad \text{and} \quad \mathcal{T}_s := \{t_k \,|\, k = R_l + 1, \ldots, M\}. \tag{5.140}$$

Here $\mathcal{T}_l$ includes quadrature points t_k condensed 'near' zero, hence generating the long range Gaussians (low pass filters), and $\mathcal{T}_s$ accumulates the increasing sequence in the $(M \to \infty)$ sequence of 'large' sampling points t_k with the upper bound $C_0^2 \log^2(M)$, corresponding to the short range Gaussians (high pass filters). Note that the quasioptimal choice of the constant $C_0 \approx 3$ was determined in [38]. We further denote $\mathcal{K}_l := \{k \,|\, k = 0, 1, \ldots, R_l\}$ and $\mathcal{K}_s := \{k \,|\, k = R_l + 1, \ldots, M\}$.

Splitting (5.140) generates the additive decomposition of the canonical tensor $\mathbf{P}_R$ into the short and long range parts,

$$\mathbf{P}_R = \mathbf{P}_{R_s} + \mathbf{P}_{R_l},$$

where

$$\mathbf{P}_{R_s} = \sum_{t_k \in \mathcal{T}_s} \mathbf{p}_k^{(1)} \otimes \mathbf{p}_k^{(2)} \otimes \mathbf{p}_k^{(3)}, \quad \mathbf{P}_{R_l} = \sum_{t_k \in \mathcal{T}_l} \mathbf{p}_k^{(1)} \otimes \mathbf{p}_k^{(2)} \otimes \mathbf{p}_k^{(3)}. \tag{5.141}$$

The choice of the critical number $R_l = \#\mathcal{T}_l - 1$ (or equivalently, $R_s = \#\mathcal{T}_s = M - R_l$), which specifies the splitting $\mathcal{T} = \mathcal{T}_l \cup \mathcal{T}_s$, is determined by the active support of the short range components such that one can cut off the functions $\mathbf{p}_k(x)$, $t_k \in \mathcal{T}_s$, outside of the sphere B_σ of radius $\sigma = \sigma_\delta > 0$, subject to a certain threshold $\delta > 0$. We denote the number of grid points in the interval $[0, \sigma]$ by $\gamma = \gamma_\delta$.

The development of the RS tensor format was motivated by the attempt to simplify the grid based calculation of a large sum of the long range electrostatic potentials

$$V = \sum_{\nu=1}^{N_0} \frac{Z_\nu}{\|x - x_\nu\|},$$

in a form of a low rank canonical tensor plus a sum of CP tensors localized around the particle centers x_ν, $\nu = 1, \ldots, N_0$. Indeed, by using the splitting (5.141), we arrive at the decomposition to the potential V, meshed up on a fine $n \times n \times n$ grid,

$$V \rightsquigarrow \mathbf{V} := \sum_{\nu=1}^{N_0} \mathbf{P}_{R,\nu} = \sum_{\nu=1}^{N_0} \mathbf{P}_{R_s,\nu} + \sum_{\nu=1}^{N_0} \mathbf{P}_{R_l,\nu}. \tag{5.142}$$

Our analysis in [30] indicated that the last sum in (5.142) can be converted to the low rank CP/Tucker tensor that motivated us to introduce the RS tensor representation.

The following definition of the RS canonical tensor format was described in [30]:

Definition 5.37 (RS-canonical tensors). The RS-canonical tensor format defines the class of d-tensors $\mathbf{A} \in \mathbb{R}^{n_1\times\cdots\times n_d}$, represented as a sum of a rank-R CP tensor $\mathbf{U}_{\text{long}} = \sum_{k=1}^{R} \xi_k \mathbf{u}_k^{(1)} \otimes \cdots \otimes \mathbf{u}_k^{(d)}$, and a cumulated CP tensor $\mathbf{U}_{\text{short}} = \sum_{\nu=1}^{N_0} c_\nu \mathbf{U}_\nu$, such that

$$\mathbf{A} = \sum_{k=1}^{R} \xi_k \mathbf{u}_k^{(1)} \otimes \cdots \otimes \mathbf{u}_k^{(d)} + \sum_{\nu=1}^{N_0} c_\nu \mathbf{U}_\nu\,, \tag{5.143}$$

where $\mathbf{U}_{\text{short}}$ is generated by the localized reference CP tensor $\mathbf{U}_0$, i.e., $\mathbf{U}_\nu = \text{Replica}(\mathbf{U}_0)$, with $\text{rank}(\mathbf{U}_\nu) = \text{rank}(\mathbf{U}_0) \leq R_0$, where, given the threshold $\delta > 0$, the effective support of $\mathbf{U}_\nu$ is bounded by $\text{diam}(\text{supp}\,\mathbf{U}_\nu) \leq 2\gamma_\delta$ in the index size.

Each RS-canonical tensor is therefore uniquely defined by the following parametrization: rank-R canonical tensor $\mathbf{U}_{\text{long}}$, the rank-R_0 reference canonical tensor $\mathbf{U}_0$ with the small mode size bounded by 2γ, list $\mathcal{J}$ of the coordinates and weights of N_0 particles in $\mathbb{R}^d$.

The analysis in the next section provides bounds on the storage complexity for the RS tensors.

5.5.4 Rank and complexity estimates for long range part: sketching the proof

The following theorem provides the bound on the storage of RS tensors and estimates the Tucker rank of the long range part for the class of RS tensors associated with decomposition (5.142) for the electrostatic potential of N_0 charged particles.

Theorem 5.38 ([30]). *The storage size for RS-canonical tensors is estimated by*

$$stor(\mathbf{A}) \leq dRn + (d+1)N_0 + dR_0\gamma\,.$$

Given $\mathbf{i} \in \mathcal{I}$, denote by $\overline{\mathbf{u}}_{i_\ell}^{(\ell)}$ the row vector with index i_ℓ in the side matrix $U^{(\ell)} \in \mathbb{R}^{n_\ell\times R}$, and let $\xi = (\xi_1,\ldots,\xi_d)$. Each entry of an RS-CP tensor can be calculated at $O(dR + 2d\gamma R_0)$ cost as a sum of long and short range contributions by

$$a_{\mathbf{i}} = \left(\odot_{\ell=1}^{d} \overline{\mathbf{u}}_{i_\ell}^{(\ell)}\right)\xi^{\mathrm{T}} + \sum_{\nu\in\mathcal{L}(\mathbf{i})} c_\nu \mathbf{U}_\nu(\mathbf{i})\,, \quad \textit{at the expense} \quad O(dR + 2d\gamma R_0)\,.$$

Moreover, the ε rank $\mathbf{r}_0$ of the Tucker approximation to the long range CP tensor $\mathbf{U}$ is bounded by

$$|\mathbf{r}_0| := \text{rank}_{\text{Tuck}}(\mathbf{U}) \leq C\,b\,\log^{3/2}(|\log(\varepsilon/N_0)|)\,.$$

Proof. We present the proof for the completeness. The proof is based on the Fourier analysis of the Gaussians in the canonical sinc based approximation; see [29]. We consider the Gaussian in normalized form $G_\sigma(x) = e^{-\frac{x^2}{2\sigma^2}}$ so that the relation $e^{-t_k^2 x^2} = e^{-\frac{x^2}{2\sigma^2}}$

holds, i.e., if we set the quadrature parameters in (5.137) by

$$t_k = \frac{1}{\sqrt{2}\sigma_k}, \quad \text{with } t_k = k\mathfrak{h}_M, \quad k = 0, 1, \ldots, M,$$

where $\mathfrak{h}_M = C_0 \log M/M$. The choice of R_s is based on the bound of the L^1 norm

$$a_k \int_a^\infty e^{-\frac{x^2}{2\sigma_k^2}} \le \frac{\varepsilon}{2} < 1, \quad a_k = \mathfrak{h}_M.$$

This allows us to select all σ_k that satisfy this criteria.

We sketch the proof to the following steps: (A) We represent all shifted Gaussian functions, contributing to the total sum, in the fixed set of basis functions by using truncated Fourier series. (B) We prove that, on the 'long range' index set $k \in \mathcal{T}_l$, the parameter σ_k remains uniformly bounded in N from below, implying the uniform bound on the number of terms in the ε-truncated Fourier series. (C) The summation of functions presented in the fixed Fourier basis set does not enlarge the Tucker rank, but only affects the Tucker core. The dependence on size of computational domain b appears in the explicit form.

Specifically, let us consider the rank-1 term in the splitting (5.141) with maximal index $k \in \mathcal{T}_l$. Taking into account the asymptotic choice $M = \log^2 \varepsilon$, where $\varepsilon > 0$ is the accuracy of the sinc quadrature, we obtain the relation

$$\max_{k \in \mathcal{T}_l} t_k = R_l \mathfrak{h}_M = \frac{M}{2} C_0 \log(M)/M \approx \log(M) = 2\log(|\log(\varepsilon)|). \tag{5.144}$$

Now we consider the Fourier transform of the univariate Gaussian on $[-b, b]$,

$$G_\sigma(x) = e^{-\frac{x^2}{2\sigma^2}} = \sum_{m=0}^{M} \alpha_m \cos\left(\frac{\pi m x}{b}\right) + \eta, \quad \text{with} \quad |\eta| = \left| \sum_{m=M+1}^{\infty} \alpha_m \cos\left(\frac{\pi m x}{b}\right) \right| < \varepsilon,$$

where

$$\alpha_m = \frac{\int_{-b}^{b} e^{-\frac{x^2}{2\sigma^2}} \cos\left(\frac{\pi m x}{b}\right) dx}{|C_m|^2},$$

with

$$|C_m|^2 = \int_{-b}^{b} \cos^2\left(\frac{\pi m x}{b}\right) dx = \begin{cases} 2b, & \text{if } m = 0, \\ b, & \text{otherwise}. \end{cases}$$

Following arguments in [84] one obtains

$$\alpha_m = \left(\sigma e^{-\frac{\pi^2 m^2 \sigma^2}{2a^2}} - \xi_m\right) / |C_m|^2, \quad \text{where} \quad 0 < \xi_m < \varepsilon.$$

Truncation of the coefficients α_m at $m = m_0$ such that $\alpha_{m_0} \le \varepsilon$, leads to the bound

$$m_0 \ge \frac{\sqrt{2}}{\pi} \frac{b}{\sigma} \log^{0.5}\left(\frac{\sigma}{(1 + |C_M|^2)\varepsilon}\right) = \frac{\sqrt{2}}{\pi} \frac{b}{\sigma} \log^{0.5}\left(\frac{\sigma}{1 + b} \frac{1}{\varepsilon}\right)$$

for all admissible $\sigma = \sigma_k$. On the other hand, (5.144) implies

$$1/\sigma_k \leq c \log(|\log \varepsilon|), \quad k \in \mathcal{T}_l, \quad \text{i.e.,} \quad 1/\sigma_{R_l} \approx \log(|\log \varepsilon|),$$

which ensures the following estimate on m_0:

$$m_0 = O(b \log^{3/2}(|\log \varepsilon|)). \tag{5.145}$$

Now following [189], we represent the Fourier transform of the shifted Gaussians by

$$G_\sigma(x - x_\nu) = \sum_{m=0}^{M} \alpha_m \cos\left(\frac{\pi m(x - x_\nu)}{b}\right) + \eta_\nu, \quad |\eta_\nu| < \varepsilon,$$

which requires only the double number of trigonometric terms compared with the single Gaussian analyzed above. To compensate the possible increase in $|\sum_\nu \eta_\nu|$, we refine $\varepsilon \mapsto \varepsilon/N$. These estimates also apply to all Gaussian functions presented in the long range sum since for $k \in \mathcal{T}_l$ they have larger values of σ_k than σ_{R_l}. Indeed, the number of summands in the long range part is of the order $R_l = M/2 = O(\log^2 \varepsilon)$. Combining these arguments with (5.145) proves the resulting estimate. □

Figure 5.29 illustrates the localization properties in the frequency domain of the short and long range components discretized on a $n \times n \times n$ grid.

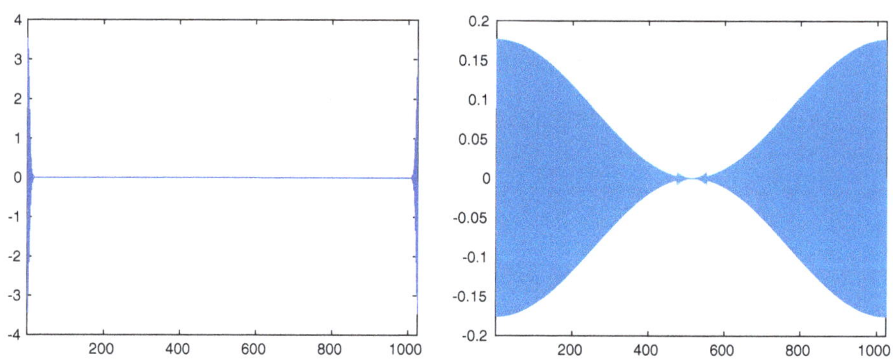

Fig. 5.29: Fourier coefficients of the long range (left) and short range (right) parts in the discrete Gaussians, $n = 1024$.

Figure 5.30 (left) demonstrates the fast exponential decay in the rank parameter for the long-range part in the tensor representing the electrostatic potential of N-particle systems. Figure 5.30 (right) shows the full rank representation to the complete electrostatic potential (including the short-range part).

The main beneficial properties of the RS tensors are described in the following:

Proposition 5.39 ([30], properties of RS tensors). *The following operations on RS canonical/Tucker tensors can be realized efficiently:*

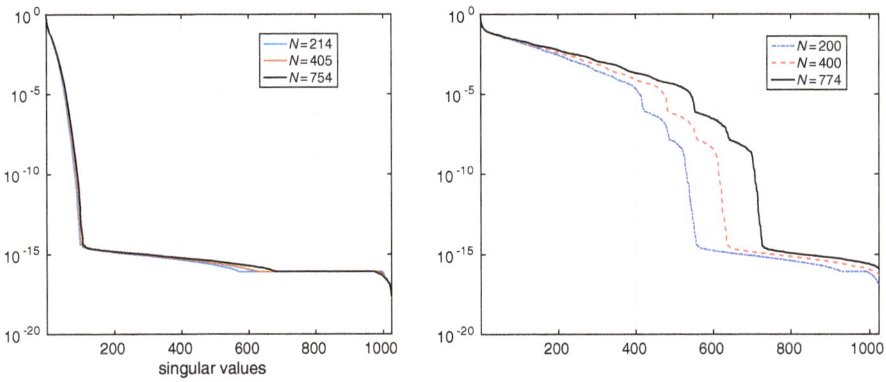

Fig. 5.30: Mode-1 singular values of side matrices for the full sum and long range part $\mathbf{P}_l$ versus N_0, $R_l = 12$.

(a) *Storage and real space representation on a fine rectangular (sub)grid.*
(b) *Construction of functional interpolants in $\mathbb{R}^d$ via radial basis functions.*
(c) *Computation of scalar products.*
(d) *Summation of many-particle interaction potentials meshed up on a fine grid in $\mathbb{R}^d$.*
(e) *Computation of interaction energy, gradients, and forces for many-particle system.*

The principal practical question in the construction of the RS formatted tensors is how to split the short and long range parts when trading off between the overlap in the localized components and the tensor rank of the globally supported long range component. Figure 5.31 illustrates the effective supports of the cumulated canonical tensors (left), embedded into the computational box, and zooms into the short range canonical vectors for $k = 1, \ldots, 11$, on a log scale (right). This example corresponds to the weighted sum of the Newton kernels.

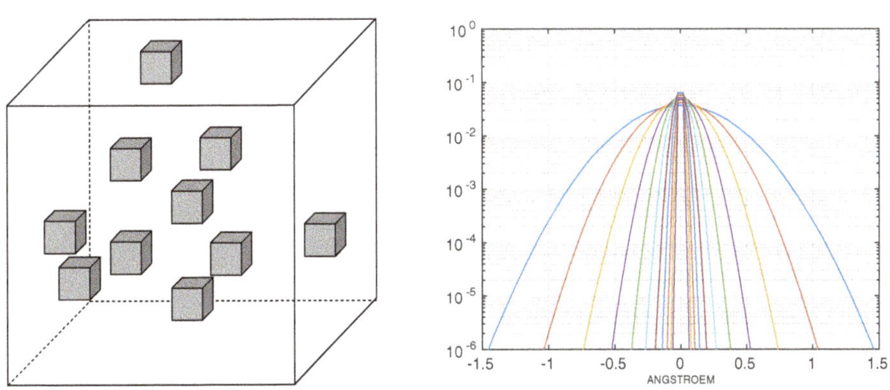

Fig. 5.31: Effective supports of the CCT (left); short range canonical vectors for $k = 1, \ldots, 11$ on log scale (right).

The classes of RS-Tucker and RS-TT tensors can be defined completely similarly.

Table 5.9 represents the Tucker ranks $\mathbf{r} = (r_1, r_2, r_3)$ for the long range part of N_0-nuclei electrostatic potential in a protein, with $\epsilon = 10^{-6}$ and for $N_0 = 200, 400, 782$. We observe the stability of the Tucker rank in the number of particles as predicted by theory.

Tab. 5.9: Tucker ranks $\mathbf{r} = (r_1, r_2, r_3)$ for the long range part of N_0-nuclei electrostatic potential in a protein, $\epsilon = 10^{-6}$.

N_0/R_l	8	9	10	11	12	13
200	10, 10, 11	13, 12, 12	18, 15, 16	23, 19, 21	32, 24, 27	42, 30, 34
400	11, 10, 11	14, 13, 14	19, 16, 20	26, 21, 26	35, 27, 36	47, 34, 47
782	11, 11, 12	15, 14, 15	20, 18, 20	28, 26, 27	39, 35, 37	52, 46, 50

In what follows, we sketch some initial applications of the RS format. The extended discussion of possible applications can be found in [30].

5.5.5 Sketch of initial applications: electrostatic potential of large biomolecules

The basic Poisson–Boltzmann (PB) equation describes the electrostatic potential of a biomolecule. The linearized PB equation reads as ([166])

$$-\nabla \cdot (\epsilon \nabla u) + \kappa^2 u = \rho_f, \quad \text{in} \quad \Omega, \tag{5.146}$$

where u is the target electrostatic potential of a N-atomic molecule, and

$$\rho_f = \sum_{k=1}^{N} z_k \delta(\|x - x_k\|)$$

is the scaled singular charge distribution supported at the atomic positions $x_k \in \Omega_m$, where δ is the Dirac delta and $\kappa = 0$, $\epsilon = 1$ in the molecular region Ω_m (see Figure 5.32). The interface conditions on the interior boundary, $\Gamma = \partial\Omega_m$, arise from the dielectric theory:

$$[u] = 0, \quad \left[\epsilon \frac{\partial u}{\partial n}\right] \quad \text{on} \quad \Gamma.$$

New regularization schemes for the Poisson–Boltzmann equation are based on the RS tensor approach; see [30].

We introduce the additive splitting in Ω:

$$u = u^r + u^s, \quad \text{where} \quad u^s = 0 \quad \text{in} \quad \Omega_s.$$

Then the singular component satisfies the equation

$$-\epsilon_m \Delta u^\rho = \rho_f \quad \text{in} \quad \Omega_m; \quad u^\rho = 0 \quad \text{on} \quad \Gamma. \tag{5.147}$$

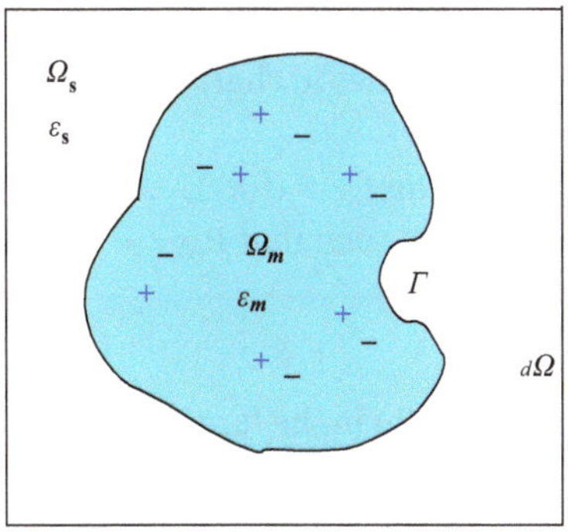

Fig. 5.32: Computational domain for the PB equation: solute (molecule) region Ω_m, solvent Ω_s.

To facilitate solving of (5.147) we use the free space singular potential

$$P_0(x) = \sum_{k=1}^{N} z_k/\|x - x_k\|: \quad \epsilon_m \Delta P_0 = \rho_f \quad \text{in} \quad \mathbb{R}^3 ,$$

in the form of RS representation (we mean the functions are discretized on a $n \times n \times n$ grid) and let P^m be the restriction of $P_0(x)$ onto Ω_m, that is

$$P^m = P_{0|\overline{\Omega}_m} \quad \text{in} \quad \overline{\Omega}_m ; \quad P^m = 0 \quad \text{in} \quad \Omega_s .$$

Now we set $u^\rho = P^m + u^h$. A harmonic function u^h compensates the discontinuity of P^m on Γ,

$$\Delta u^h = 0 \quad \text{in} \quad \Omega_m ; \quad u^h = -P^m \quad \text{on} \quad \Gamma .$$

Finally, equation (5.146) is transformed to that for the regular potential u^r ([30]):

$$-\nabla \cdot (\epsilon \nabla u^r) + \kappa^2 u^r = 0 \quad \text{in} \quad \Omega ,$$

$$[u^r] = 0 , \quad \left[\epsilon \frac{\partial u^r}{\partial n}\right] = -\epsilon_m \frac{\partial u^\rho}{\partial n}, \quad \text{on} \quad \Gamma .$$

The benefit of this splitting scheme comes from the fact that the interface data $\partial_n u^\rho{}_{|\Gamma}$ depends only on the long range part in $P_0(x)$, which is smooth and allows the low rank tensor representation. Moreover, the discretization and solving of the equation with highly singular right hand side ρ_f can be completely avoided.

The second approach to solving the PB equation is based on the range separated splitting to the discretized Dirac delta; [205], see Section 5.8.1 for more details. Numerical examples on the application of this approach to the solution of the Poisson–Boltzmann equation are reported in [32].

5.5.6 Scattered data modeling and tensor approximation of large covariance matrices

In scattered data modeling the problem is in a low parametric approximation of multivariate functions $f\colon \mathbb{R}^d \to \mathbb{R}$ by sampling at a finite set $\mathcal{X} = \{x_1, \ldots, x_N\} \subset \mathbb{R}^d$ of piecewise distinct points. Here the function f might be the surface of a solid body, the solution of a PDE, many-body potential field, multiparametric characteristics of physical systems, or some other multidimensional data, etc.

The traditional way of recovering f from a sampling vector $f_{|\mathcal{X}} = (f(x_1), \dots, f(x_N))$ is the construction of a functional interpolant $P_N : \mathbb{R}^d \to \mathbb{R}$ such that $P_{N|\mathcal{X}} = f_{|\mathcal{X}} =: \mathbf{f} \in \mathbb{R}^N$, i.e.,

$$P_N(x_j) = f(x_j), \quad \forall\, 1 \le j \le N. \tag{5.148}$$

Using radial basis (RB) functions one can find interpolants P_N in the form

$$P_N(x) = \sum_{j=1}^{N} c_j p(\|x - x_j\|) + Q(x), \quad Q \text{ is some smooth function, say } Q = 0, \tag{5.149}$$

where $p = p(r) : [0, \infty) \to \mathbb{R}$ is a fixed RB function and $r = \|\cdot\|$ is the Euclidean norm on $\mathbb{R}^d$. For example, the following RB functions are commonly used:

$$p = r^\nu, \quad (1 + r^2)^\nu, \ (\nu \in \mathbb{R}), \quad \exp(-r^2), \quad r^2 \log(r).$$

The other examples of RB functions are defined by Green's kernels or by the class of Matérn functions.

We discuss the following computational tasks (A) and (B), see [30]:

(A) Fixed coefficient vector $\mathbf{c} = (c_1, \dots, c_N)^\mathrm{T} \in \mathbb{R}^N$, efficiently representing the interpolant $P_N(x)$ on the fine tensor grid in $\mathbb{R}^d$ providing:
   (a) $O(1)$-fast point evaluation of P_N in the computational volume Ω, and
   (b) computation of various integrodifferential operations on that interpolant (say gradients, forces, scalar products, convolution integrals, etc.).

(B) Finding the coefficient vector $\mathbf{c}$ that solves the interpolation problem (5.148) in the case of large number N.

Problem (A) exactly fits our RS tensor framework so that the RS tensor approximation solves the problem with low computational costs.

Problem (B): Suppose that we use some favorable preconditioned iteration for solving coefficient vector $\mathbf{c} = (c_1, \dots, c_N)^\mathrm{T}$,

$$A_{p,\mathcal{X}}\mathbf{c} = \mathbf{f}, \quad \text{with} \quad A_{p,\mathcal{X}} = A_{p,\mathcal{X}}^\mathrm{T} = [p(\|x_i - x_j\|)]_{1 \le i,j \le N} \in \mathbb{R}^{N \times N}, \tag{5.150}$$

with the distance dependent symmetric system matrix $A_{p,\mathcal{X}}$. We assume $\mathcal{X} = \Omega_h$ is the $n^{\otimes d}$ set of grid points located on tensor grid, i.e., $N = n^d$. Introduce the d-tuple multi-index $i \mapsto \mathbf{i} = (i_1, \dots, i_d)$ and $j \mapsto \mathbf{j} = (j_1, \dots, j_d)$, and reshape $A_{p,\mathcal{X}}$ into the tensor form

$$A_{p,\mathcal{X}} \mapsto \mathbf{A} = [a(i_1, j_1, \dots, i_d, j_d)] \in \bigotimes_{\ell=1}^{d} \mathbb{R}^{n \times n},$$

which can be decomposed by using the RS based splitting

$$\mathbf{A} = \mathbf{A}_{R_s} + \mathbf{A}_{R_l},$$

generated by the RS representation of the weighted potential sum in (5.149). Here $\mathbf{A}_{R_s}$ is (almost) diagonal matrix, while $\mathbf{A}_{R_l} = \sum_{k=1}^{R_l} A_k^{(1)} \otimes \cdots \otimes A_k^{(d)}$ is the low Kronecker rank matrix. This implies a bound on the storage, $O(N + dR_l n)$, and ensures a fast matrix vector multiplication. Introducing the additional rank structured representation in $\mathbf{c}$, the solution of (5.150) can be further simplified.

The above approach can be applied to the data sparse representation for the class of large covariance matrices in the spacial statistics; see for example [254, 255, 271].

5.6 Tensor methods for quasiperiodic systems versus geometric homogenization

Tensor approaches to the numerical solution elliptic equations with highly oscillating quasiperiodic coefficients are essentially based on the use of separation in physical variables and the QTT tensor approximation of functions with oscillating features. Our approach provides the alternative to the classical methods of geometric homogenization, which until now have been the main choice for numerical approximation of quasiperiodic systems [14, 35, 60, 175].

5.6.1 Fast integration of highly oscillating functions

We begin with the discussion of the QTT based superfast quadratures for integration of highly oscillating functions.

The QTT based quadratures of $O(\log n) = O(|\log \varepsilon|)$ complexity are based on the QTT representation of integrand $f(x)$ and weight function $w(x)$ with moderate QTT ranks. Using the simple equidistant trapezoidal rule, we arrive at the fast integration scheme, which can be implemented in the form of a scalar product in the QTT tensor format

$$I = \int_{-1}^{1} w(x) f(x) dx \approx I_n(f) := h \sum_{i=1}^{n} w(x_i) f(x_i) : = \langle \mathbf{W}, \mathbf{F} \rangle_{\mathrm{QTT}}, \quad \mathbf{W}, \mathbf{F} \in \otimes_{\ell=1}^{L} \mathbb{R}^2 .$$

Indeed, assume that both integrands, discretized on large uniform grid of size n, can be approximated in the QTT format by small average rank-r tensors $\mathbf{W}, \mathbf{F}$, then for the rectangular n-point quadrature with $n = 2^L$ we have $|I - I_n| = O(n^{-\alpha})$, at the storage expense $O(r^2 \log n)$ and computational time $O(r^2 \log n)$. Here $\mathbf{W}, \mathbf{F}$ are the L dimensional tensors.

Example 5.40 illustrates the uniform bound on the QTT rank for nontrivial highly oscillating functions; see Figure 5.33. Here and in the following the threshold error like ϵ_{QTT} corresponds to the Euclidean norm.

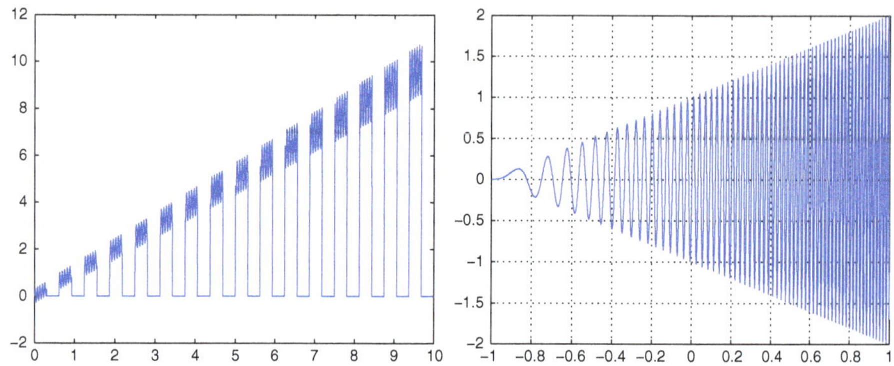

Fig. 5.33: Visualizing functions f_3 (left) and f_4 (right).

Example 5.40. Highly oscillated and singular functions on $[0, A]$, $\omega = 100$, $\epsilon_{\mathrm{QTT}} = 10^{-6}$,

$$f_3(x) = \begin{cases} x + a_k \sin(\omega x)\,, & x \in 10\left(\frac{k-1}{p}; \frac{k-0.5}{p}\right] \\ 0\,, & x \in 10\left(\frac{k-0.5}{p}; \frac{k}{p}\right] \end{cases}$$

$$f_4(x) = (x+1)\sin(\omega(x+1)^2)\,, \quad x \in [0,1]\,, \quad \text{(Fresnel integral)}\,,$$

where function $f_3(x)$, $x \in [0, 10]$, $k = 1, \dots, p$, $p = 16$, $a_k = 0.3 + 0.05(k-1)$, is recognized on three different scales. The average QTT rank over all mode ranks for the corresponding functional vectors are given in Table 5.10. The maximum rank over all the fibers is nearly the same as the average one.

Note that 1D and 2D numerical quadratures based on interpolation by Chebyshev polynomials have been developed [155, 344, 345] in the framework of Chebfun techniques. Taking into account that a Chebyshev polynomial sampled on a Chebyshev grid has exact rank-2 QTT representation [197], suggests the efficient numerical integration by Chebyshev interpolation using the QTT approximation.

The next examples are related to the fast QTT based computation of the oscillatory integrals arising in the representation of the so called retarded potentials; see [226, 228].

Tab. 5.10: Average QTT ranks of N vectors generated by f_3 and f_4.

$N \setminus \bar{r}$	$r_{\mathrm{QTT}}(f_3)$	$r_{\mathrm{QTT}}(f_4)$
2^{14}	3.5	6.5
2^{15}	3.6	7.0
2^{16}	3.6	7.5
2^{17}	3.6	7.9

We apply the twofold QTT: fast quadratures for highly oscillating integrals plus QTT approximation in ω.

We compute function of $\omega \in [\omega_{\min}, \omega_{\max}]$ for various $f(x)$, $g(x)$ ([228]),

$$I(\omega, f) := \int_{\Omega} f(x) e^{i\,\omega g(x)} dx\,,$$

by reducing this task to the real valued integrals

$$I_{\Re e}(\omega, f) := \int_{\Omega} f(x) \cos(\omega g(x)) dx\,, \quad I_{\Im m}(\omega, f) := \int_{\Omega} f(x) \sin(\omega g(x)) dx\,.$$

There are many practically interesting examples of the exotic oscillators like $h_\omega(x) = e^{i\,\omega g(x)}$ and beyond; see [171, 228].

Example 5.41. Examples of methods for exotic oscillators, which are not of the form $h_\omega(x) = e^{i\,\omega g(x)}$, include:
(A) Levin type methods for the case of the Bessel oscillator $h_\omega(x) = J_\nu(\omega x)$.
(B) Filon type methods for oscillators of the form $h_\omega(x) = \nu(\sin(\omega\theta(x)))$; see [171].

For most of other types of exotic oscillators such methods are not available so far. In Table 5.11 we show that our approach can also be applied in these (and other) cases in the same way as before. Here $N = 2^L$ denotes the grid size for the accurate QTT interpolation in ω.

Tab. 5.11: QTT ranks of M vectors related to the function $I_{h_\omega,5}(\omega)$ sampled on a uniform grid in $[\omega_{\min}, \omega_{\max}]$.

N	$[\omega_{\min}, \omega_{\max}]/h_\omega(x)$	$J_{11}(\omega x)$	$J_8^2(\omega x^2)$	$\cos(\sin(\omega x)+1)$	$\Gamma(0.5 \cdot \sin(\omega x) + 2)$
2^{40}	[0, 500]	5.4	5.7	7.2	7.3
2^{50}	[0, 500]	4.9	5.2	6.5	6.5
2^{60}	[0, 500]	4.5	4.7	6.0	5.9

The low rank QTT (cross) approximation [290] of huge N vectors for a function of ω given by

$$I_{h_\omega,k}(\omega) := \int_{-1}^{1} T_k(x) h_\omega(x) dx\,, \quad \omega \in [\omega_{\min}, \omega_{\max}]\,, \quad 1 \le k \le M$$

was observed in all tested cases, where T_k are Chebyshev polynomials up to the order M.

The QTT quadratures can also be applied to highly oscillating d dimensional integrals:

$$I_{\Re e}(\omega, L_{j_1,\ldots,j_d}) = \int_{[-1,1]^d} L_{j_1}(y_1) \ldots L_{j_d}(y_d) \cos(\omega g(y)) dy\,.$$

Tab. 5.12: Effective QTT ranks of N vectors related to the function $I_{\mathcal{R}e}(\omega, L_{2,5})$ for $d = 2$ sampled on a uniform grid.

N	$[\omega_{\min}, \omega_{\max}]$	$g(x) = x_1 + x_2$	$g(x) = \sin(x_1)/\sqrt{x_1 x_2 + 3}$
2^{30}	[0, 100]	5.2	4.4
2^{40}	[0, 100]	4.6	3.9
2^{50}	[0, 100]	4.2	3.5

Tab. 5.13: QTT ranks of N vectors related to the function $I_{\mathcal{R}e}(\omega, L_{2,5,3})$ for $d = 3$ sampled on a uniform grid in $[\omega_{\min}, \omega_{\max}]$.

N	$[\omega_{\min}, \omega_{\max}]$	$g(x) = x_1 + x_2 + x_3$	$g(x) = \sin(x_1 x_3)/\sqrt{x_1 x_2 + 3}$
2^{30}	[0, 50]	6.1	4.3
2^{40}	[0, 50]	5.5	3.9
2^{50}	[0, 50]	5.3	3.7

We used a tensorized version of the Gauss–Legendre quadrature. The QTT ranks of different multivariate functions are presented in Tables 5.12 and 5.13; see [228] for the detailed discussion.

5.6.2 Elliptic PDEs with oscillating features: analysis in 1D

In this section we describe the alternative to the classical periodic (geometric) homogenization methods for solving PDEs with oscillating features. The main idea of the approach, proposed in [224, 225], is the low rank QTT tensor representation of the coefficients and the solution discretized on a large tensor grid in $\mathbb{R}^d$, and then solving the resultant large linear system of equations in the tensor format. The numerical cost of such an approach scales logarithmically in the grid size.

We consider the traditional FEM-Galerkin piecewise linear approximation in $\Omega = (0, 1)^d$ to the problem:

$$\text{Find} \quad u_\epsilon \in H_0^1(\Omega): \qquad -\operatorname{div}(a_\epsilon \nabla u_\epsilon) = f \quad \text{in } \Omega, \qquad f \in L^2(\Omega), \tag{5.151}$$

with oscillating or lattice structured/replicated coefficients in $\Omega = \cup_{\mathbf{i}} \Pi_{\mathbf{i}}^\epsilon$,

$$a_\epsilon(x) := \widehat{A}\left(\frac{x - x_i}{\epsilon}\right) \quad \text{in the scaled unit cell } \Pi_{\mathbf{i}}^\epsilon .$$

The main difficulties in numerical FEM simulations are caused by huge grid size of the order of $(n_0/\epsilon)^{-d}$ required to resolve the oscillating coefficients, solution, and the right hand side. Another problem is related to the adaptive error control.

For example, let $d = 3$, $n_0 = 100$, $\epsilon = 0.01$, then for a $N \times N$ stiffness matrix we have $N = 10^{12}$.

Homogenization methods for systems with periodic coefficients (see for example [14]) suggest asymptotically, as $\epsilon \to 0$, the error bound

$$\|u_\epsilon - u_{\text{homo}}\| \leq C\sqrt{\epsilon}\,,$$

which has practical significance only for a small parameter ϵ. Homogenization methods propose the rough, strictly ϵ-dependent approximation in the case of perfect periodic coefficients, but possibly 'defected' boundary conditions (say Dirichlet or Neumann instead of periodic).

However, in many applications the 'defected' periodic coefficients arise, for example in material science, in numerical modeling of lattice structured atomic systems, in uncertainty quantification, in multidimensional data analysis, etc. Several principal questions arise:

- How do we treat the defected periodic coefficients?
- What if ϵ is not too small?
- How do we obtain the high precision approximate solution uniformly in ϵ?
- How do we control the approximation error efficiently (adaptively)?

Some examples of 'defected' periodic coefficients in the form of modulated oscillating functions are presented in Figure 5.34.

It was shown in [224, 225] that the QTT approximation for a class of 'modulated periodic' coefficients (see examples above) leads to algorithms with logarithmic complexity scaling in ϵ, $O(|\log \epsilon|)$. In particular, for all examples above the averaged QTT ε rank was of the order of $O(1)$ uniformly in the grid size. The a posteriori error estimates for the problems with highly oscillating coefficients were also analyzed in [224, 225]. The a posteriori error estimates for the general classes of elliptic problems were discussed in detail in [264, 282, 310, 311].

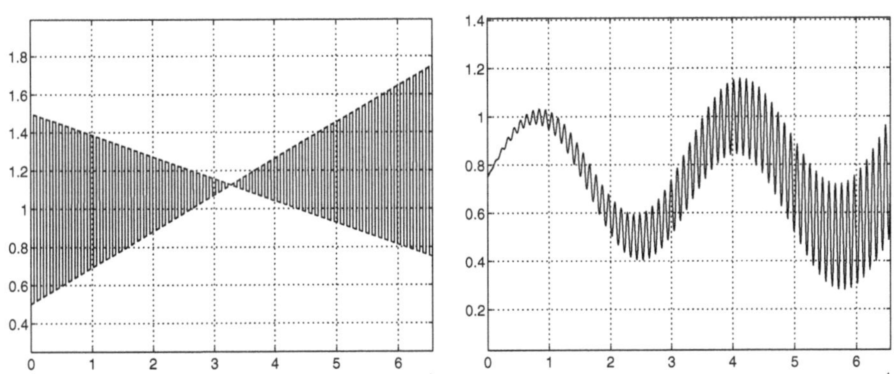

Fig. 5.34: Examples of modulated periodic and piecewise periodic coefficients in 1D.

To apply the QTT approximation to PDEs in periodic homogenization and beyond, we assume that:

(A) the solution of *homogenized approximation*,

$$\mathbb{A}_0 \mathbf{u}_0 = \mathbf{f}\,, \tag{5.152}$$

corresponding to the simplified piecewise constant coefficient of the form

$$a_0(x) = \mathrm{mean}\{a_\epsilon(x)\} + C_0\,,$$

is cheap;

(B) for the stiffness matrix of the FEM-Galerkin approximation to (5.151), $\mathbb{A}_\epsilon$, the spectral equivalence holds

$$\lambda_0 \mathbb{A}_0 \le \mathbb{A}_\epsilon \le \lambda_1 \mathbb{A}_0\,, \quad \lambda_0, \lambda_1 > 0\,,$$

with λ_1/λ_0 uniformly bounded in ϵ; and

(C) the system matrix $\mathbb{A}_\epsilon$, the vector of right hand side $\mathbf{f}$, and matrices $\mathbb{A}_0$, $\mathbb{A}_0^{-1}$ allow the low rank QTT approximation.

The basic idea of the new rank structured tensor approach can be formulated as the following computational scheme ([224]):
Solve the FEM-Galerkin approximation of (5.151)

$$\mathbb{A}_\epsilon \mathbf{u}_\epsilon = \mathbf{f}\,, \quad \mathbb{A}_\epsilon \in \mathbb{R}^{N^d \times N^d}\,, \quad \mathbf{u}_\epsilon, \mathbf{f} \in \mathbb{R}^{N^d}\,, \qquad N = 2^L \tag{5.153}$$

by rank truncated preconditioned iteration in quantized tensor space $\mathbb{Q}_L$:

$$(\mathbb{E} + \beta \mathbb{A}_0^{-1} \mathbb{A}_\epsilon - \mathbb{E})\mathbf{u} \equiv (\mathbb{E} + \mathbb{B})\mathbf{u} = \beta \mathbf{u}_0\,, \quad \beta > 0\,,$$

assuming that for $\mathbb{B} = \beta \mathbb{A}_0^{-1} \mathbb{A}_\epsilon - \mathbb{E}$, there holds $\|\mathbb{B}\| = q < 1$, and $\mathrm{rank}_{QTT}(\mathbb{B})$ is small.

To that end perform the tensor truncated preconditioned iteration

$$\text{Given } \mathbf{u}_0 \in \mathbb{Q}_L^{(r)}: \quad \mathbf{u}_{k+1} = \mathcal{T}_\varepsilon\left(\beta \mathbf{u}_0 - \mathbb{B}\mathbf{u}_k\right)\,, \quad k = 0, 1, 2, \ldots$$

that provides geometric convergence rate

$$\|\mathbf{u}_{k+1} - \mathbf{u}_k\| \le q^k \|\mathbf{u}_0\|\,.$$

The uniform condition number estimate is proven in [224].

Lemma 5.42. *The condition number of the preconditioner $\mathbb{A}_0$ defined by $a_0(x)$ is bounded by*

$$\mathrm{cond}\{\mathbb{A}_0^{-1}\mathbb{A}_\epsilon\} \le C \max \frac{1 + q(x)}{1 - q(x)}\,, \quad \textit{with} \quad q(x) := \frac{a^+(x) - a^-(x)}{a^+(x) + a^-(x)}\,.$$

5.6.3 QTT matrix representation and numerics for the QTT tensor solver

It is worth noting that the QTT rank for the system matrix $\mathbb{A}_\epsilon$ can be estimated a priori in terms of the QTT ranks for the corresponding equation coefficient discretized on a fine spatial grid. For example the 1D stiffness matrix $A[a_\epsilon]$ given by

$$(A[a_\epsilon])_{i,i'} = \left(a_\epsilon(x)\nabla\phi_i(x), \nabla\phi_{i'}(x)\right)_{L_2(\Omega)}, \quad i, i' = 1, \dots, N,$$

$$A[a_\epsilon] = \frac{1}{h}\begin{bmatrix} a_1 + a_2 & -a_2 & & & \\ -a_2 & a_2 + a_3 & -a_3 & & \\ & \ddots & \ddots & \ddots & \\ & & -a_{N-1} & a_{N-1} + a_N & -a_N \\ & & & -a_N & 2a_N \end{bmatrix},$$

specified by the coefficient vector $\mathbf{a} = [a_i] \in \mathbb{R}^N$, obtained by the cell centered collocation, i.e., $a_i = a_\epsilon(x_{i-1/2})$, $i = 1, \dots, N$, the explicit rank estimate was presented in Section 4.3.10.

Further results on the QTT rank analysis for the class of stiffness matrices for elliptic operators can be found in [79, 82, 177, 179–181].

In what follows, we present some numerical illustrations that confirm the logarithmic scaling of the computational cost in the grid size, that is inverse proportional to ϵ. These examples correspond to periodic and 'defected' periodic coefficients in the form of smoothly modulated oscillating functions in the case of (nonperiodic) Dirichlet boundary conditions. As mentioned above, the equation coefficients are given by

$$a_\epsilon(x) = C + g(x)\sin(\omega x^m), \quad m = 1, 2, 3, \dots, \tag{5.154}$$

corresponding to Figure 5.35: (upper left): $C = 1, g(x) = 1, m = 1$; (upper right): $C = 1$, $g(x) =$ step function, $m = 1$; (bottom): $C = 1, g(x) = 1, m = 3$. $\omega = 2\pi\,32$, $\epsilon = \omega^{-1}$.

Table 5.14 presents CPU (s), iteration history, and QTT ranks with $\omega = 1/\epsilon = 2\pi 64$, $\varepsilon_{QTT} = 10^{-7}$. We observe the numerical error $\|u_\epsilon - \mathbf{u}_\epsilon\|_1 \doteq 10^{-5} - 10^{-6}$.

The QTT rank truncated iteration ensures the logarithmic cost, $O(\log N)$, where $N = 1/\epsilon$, providing the guaranteed approximation error proportional to the power of $1/N$, which essentially outperforms the capability of classical homogenization methods. We observe:

Tab. 5.14: CPU (s), iteration history, and QTT ranks for the three types of oscillating coefficients, and QTT ranks of the solution.

Grid size 2^L	2^{13}, (it)	2^{14}, (it)	2^{15}, (it)	2^{16}, (it)	2^{17}, (it)	$r(\mathbf{a}_\epsilon)$	$r(\mathbf{u}_\epsilon)$
$C + \sin(\omega x)$	0.97, (5)	1.2, (5)	1.3, (5)	2.0, (6)	2.1, (6)	2.67	3.7
4-steps coef.	3.4, (9)	4.3, (9)	4.5, (9)	6.7, (9)	14.3, (14)	2.9	4.96
$C + \sin(\omega x^3)$	5.3, (5)	10.0, (6)	9.95, (6)	11.98, (6)	16.2, (5)	7.53	8.24

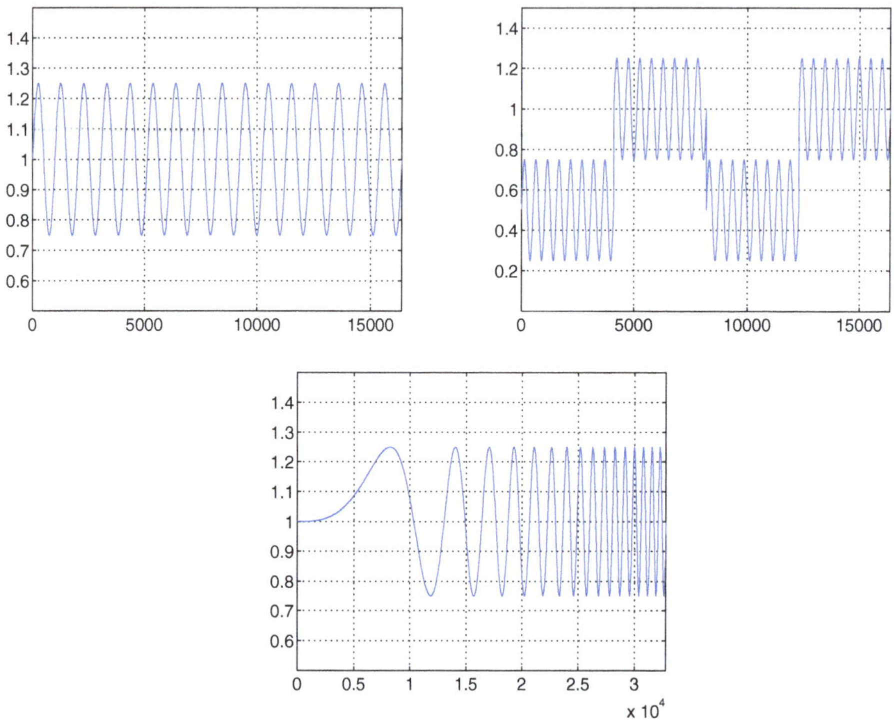

Fig. 5.35: Examples of periodic and nonperiodic oscillating coefficients.

- uniform convergence rate of preconditioned iteration,
- approximation of order $O(h^2)$ uniformly in ϵ, and
- logarithmic complexity of order $O(\log n) = O(\log \frac{1}{\epsilon})$.

Figure 5.36 demonstrates the performance of preconditioned steepest descent (SD) iteration in QTT truncated tensor format for the periodic, step, and cubically oscillating coefficients.

Figure 5.37 shows the difference between exact and homogenized solutions (no convergence in gradients observed).

5.6.4 Multidimensional problems

Low rank reduction in geometric homogenization reduces the complexity in the $d \geq 2$ dimensional case in two steps: first by separation of spacial variables and then by QTT approximation to the univariate functions discretized on the fine spacial grid,

$$O(1/\epsilon^d) \to \quad O(d/\epsilon) \to O(d|\log \epsilon|)\,.$$

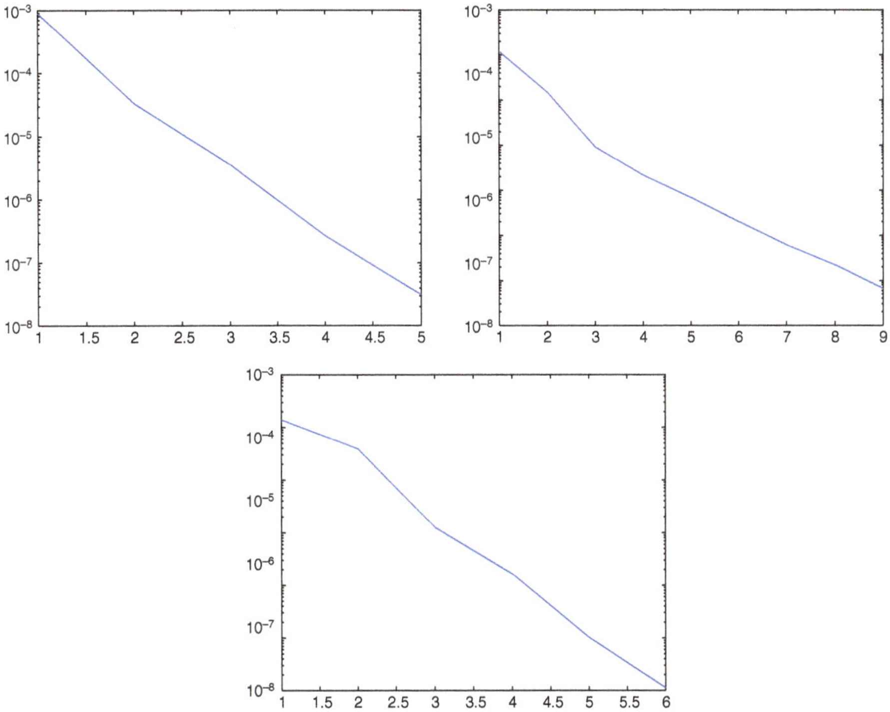

Fig. 5.36: Preconditioned SD iteration history for the periodic, jumping oscillating, and cubically oscillating coefficients.

Given quasiperiodic coefficients $a_\epsilon(x) > 0$, $d \geq 2$, $x \in [0, 1]^d$, consider the equation

$$\mathcal{A} u_\epsilon := -\operatorname{div}(a_\epsilon(x) \operatorname{grad} u_\epsilon) = f(x)\,, \quad x \in \Omega = (0,1)^d\,, \quad u_{\epsilon|\partial\Omega} = 0\,, \tag{5.155}$$

assuming that the multivariate functions $f(x_1, \ldots, x_d)$ and $a_\epsilon(x)$ can be approximated with the low separation rank.

We apply the Galerkin discretization of equation (5.155) by means of tensor product piecewise linear finite elements

$$\{\phi_i(x) := \phi_{i_1}(x_1) \ldots \phi_{i_d}(x_d)\}, \quad i = (i_1, \ldots, i_d), \quad i_\ell \in \mathcal{I}_\ell = \{1, \ldots, n_\ell\},\ \ell = 1, \ldots, d.$$

The univariate grid size is chosen proportional to the frequency parameter $n_\ell = O(1/\epsilon)$ such that the total problem size is estimated by $N = O(1/\epsilon^d)$.

In the case $d = 2$, we assume that the scalar diffusion coefficient $a(x_1, x_2)$ can be represented in the separate form

$$a(x_1, x_2) = \sum_{k=1}^{R} a_k^{(1)}(x_1) a_k^{(2)}(x_2) > 0\,,$$

with a small rank parameter R. In particular, for $R = 1$, we have

$$a(x_1, x_2) = a^{(1)}(x_1) a^{(2)}(x_2)\,.$$

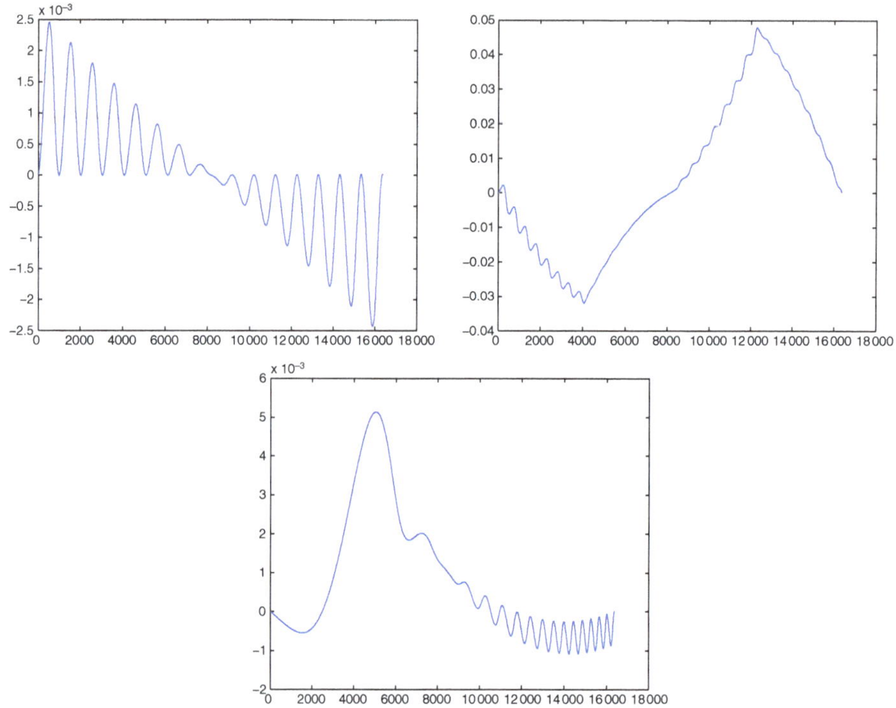

Fig. 5.37: Difference between exact and homogenized solutions.

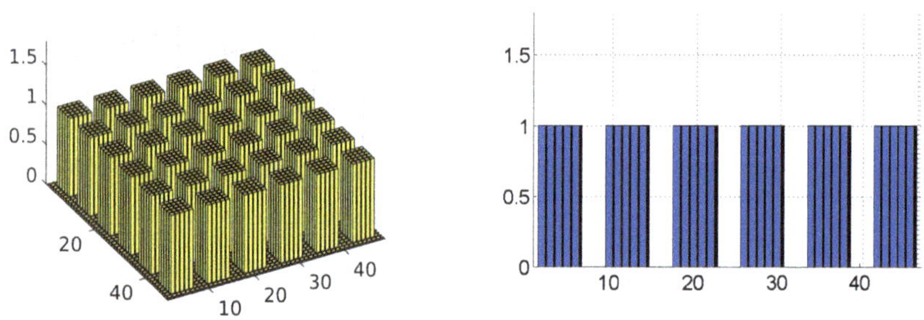

Fig. 5.38: Example of a periodic oscillating coefficient (left) and the respective 1D factor, $a(\cdot)$ (right).

Examples of a periodic oscillating coefficient with $L = 6$, (left) and the respective 1D factor, $a^{(1)}(\cdot)$ (right) are presented in Figure 5.38.

In the case $R = 1$ the Galerkin matrix can be represented in the Kronecker product form. Indeed, the Galerkin stiffness matrix $A = [a_{ij}] \in \mathbb{R}^{N\times N}$ takes a form of the rank-2 Kronecker product representation (see Subsection 4.3.10)

$$A = [a_{ij}] = A_1 \otimes M_2 + M_1 \otimes A_2 \ .$$

Here $A_1 = [a_{i_1 j_1}] \in \mathbb{R}^{n_1 \times n_1}$ and $A_2 = [a_{i_2 j_2}] \in \mathbb{R}^{n_2 \times n_2}$ are the univariate stiffness matrices, and $M_1 = [m_{i_1 j_1}] \in \mathbb{R}^{n_1 \times n_1}$ and $M_2 = [m_{i_2 j_2}] \in \mathbb{R}^{n_2 \times n_2}$ represent the weighted mass matrices. By lumping of the mass matrices the matrix A can be reduced to the simple form

$$A \mapsto A = A_1 \otimes D_2 + D_1 \otimes A_2 ,$$

where the factors D_1, D_2 are the diagonal matrices.

In the case of discrete Laplacian, the corresponding Galerkin matrix simplifies to

$$A \mapsto A = A_1 \otimes I_2 + I_1 \otimes A_2 ,$$

which provides a prototype preconditioner for solving the linear system for $R > 1$

$$A\mathbf{u} = \mathbf{f} \quad \text{with} \quad \mathbf{f} = \sum_{k=1}^{R_f} \mathbf{f}_k^{(1)} \otimes \mathbf{f}_k^{(2)} , \quad \mathbf{f}_k^{(\ell)} \in \mathbb{R}^{n_\ell} . \tag{5.156}$$

In the general case $d \geq 2$ and $R \geq 1$ we have the following bound for the Kronecker rank of A:

$$\operatorname{rank}_{\text{Kron}}(A) = d\,R .$$

Recall that the existence of the low rank solution to equation (5.156) in the Laplacian case can be proven by the sinc quadrature approximation to the Laplace transform

$$A^{-1} = \int_{\mathbb{R}_+} \mathrm{e}^{-tA} dt \approx B_M := \sum_{k=-M}^{M} c_k \mathrm{e}^{-t_k A} = \sum_{k=-M}^{M} c_k \mathrm{e}^{-t_k A_1} \otimes \mathrm{e}^{-t_k A_2} ,$$

with the proper choice of quadrature parameters c_k, t_k, M.

Now the Kronecker product approximation to A^{-1} can be applied directly to the right hand side

$$\mathbf{u} = A^{-1}\mathbf{f} \approx \sum_{k=-M}^{M} c_k \sum_{m=1}^{R_f} \mathrm{e}^{-t_k A_1} \mathbf{f}_m^{(1)} \otimes \mathrm{e}^{-t_k A_2} \mathbf{f}_m^{(2)} .$$

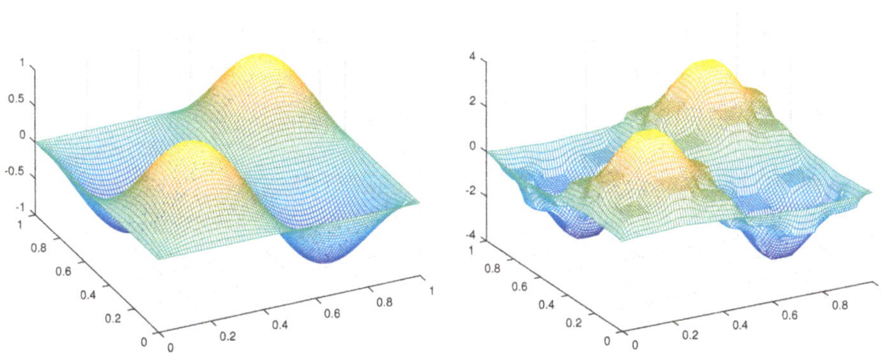

Fig. 5.39: Examples of the right hand side and the respective solutions for coefficients in Figure 5.38.

In what follows, we present some numerical illustrations on rank analysis of the solution in the 2D case. Figure 5.39 demonstrates examples of the solution (right) in the case of periodic oscillating coefficients (as in Figure 5.38) and the corresponding right hand side (left).

The behavior of singular values in the SVD decomposition of the solution in the case of a 8×8 periodic coefficient is shown in Figure 5.40. This indicates that the ϵ

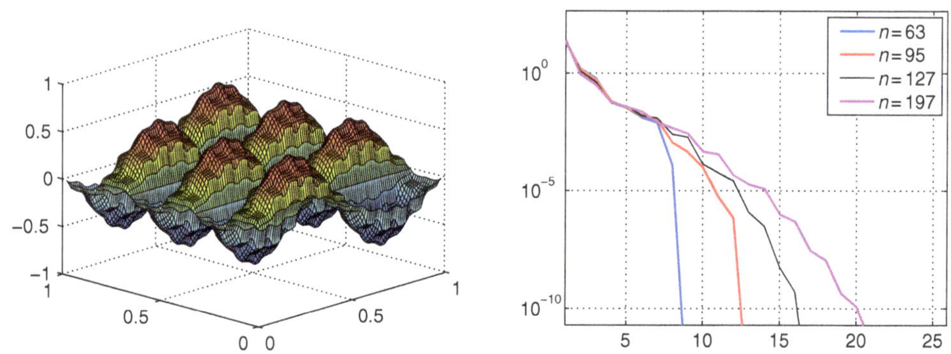

Fig. 5.40: Rank decomposition of the solution for a 8×8 periodic coefficient.

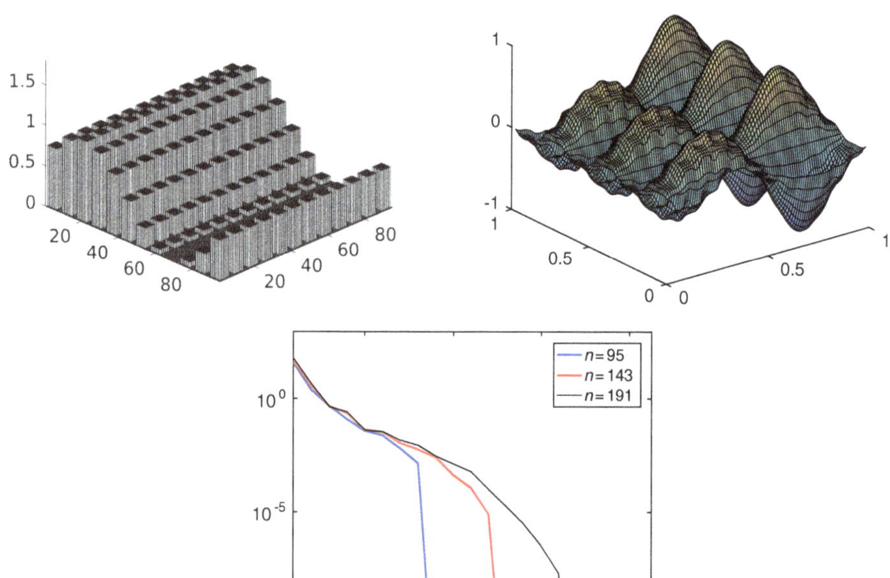

Fig. 5.41: Rank decomposition of the solution in the case of a 12×12 modulated periodic coefficient.

separation rank of the solution, considered as the two dimensional $n \times n$ data array, scales logarithmically in ϵ.

We observe from numerical tests that the rate of exponential decay in the approximation error with respect to the rank parameter does not depend on the $L \times L$ lattice size in the coefficient, i.e., on $\epsilon = 1/L$.

We also demonstrate the numerical results for our tensor approach applied to modulated periodic coefficients. We consider the solution discretized on a 400×400 grid, with the right hand side $f_1(x_1, x_2) = \sin(2\pi x_1)\sin(3\pi x_2)$.

The PCG solver applied to equation (5.156) with the discrete Laplacian as the preconditioner demonstrates robust convergence rate $q \ll 1$ uniformly in ϵ: it requires about six iterations to achieve the residual 10^{-6}. We observe that the bound $\operatorname{rank}(\mathbf{u}_\epsilon) \leq 10$ holds uniformly in ϵ. The corresponding solution and the behavior of singular values for the respective matrix representing the solution as the 2D array are presented in Figure 5.41.

5.7 A numerical scheme for stochastic homogenization problems

5.7.1 Elliptic problem in periodic supercells

The transition from geometric to stochastic homogenization addresses several challenging computational problems. The first attempt at FEM simulations in the stochastic homogenization theory was recently reported in [193]. In what follows, we discuss the main aspects of the economical numerical scheme presented in [193].

For given $f \in L^2(\Omega)$ such that $\int_\Omega f(x)dx = 0$, $x = (x_1, x_2)$, we consider the 2D elliptic boundary value problem on $\Omega := [0, 1)^2$, subject to periodic boundary conditions on $\Gamma = \partial\Omega$,

$$\mathcal{A}\varphi := -\nabla \cdot \mathbb{A}(x)\nabla\varphi = f(x), \quad \mathbb{A}(x) = \begin{pmatrix} a(x) & 0 \\ 0 & a(x) \end{pmatrix}. \tag{5.157}$$

In asymptotic analysis of the problems of stochastic homogenization (SH) the coefficient and the right hand side are chosen in a specific way, which will be specified later on; see [114–116, 252] for the particular problem setting.

Given $L = 1, 2, 3, \ldots$, we generate decomposition of the domain Ω into L^2 equal unit cells G_s, $s = 1, \ldots, L^2$ (possibly overlapping), each of size $\frac{2\alpha}{L} \times \frac{2\alpha}{L}$, whose centers are distributed randomly within the supercell Ω (by Poisson distribution), taking into account periodicity in both spacial variables x_1 and x_2 (stochastic realizations). Here the overlap factor satisfies $\alpha \leq 1/2$. Stochastic characteristics of the system can be estimated in the limit of large number L and large number of stochastic realizations.

We consider a sequence of random coefficient distributions $\{G_s\}_m$, numbered by $m = 1, \ldots, M$, where the particular set $\{G_s\} = \{G_s\}_m$ for fixed m will be called realization. For any fixed realization define the covered domain

$$\widehat{G} = \widehat{G}_m := \bigcup_{s=1}^{L^2} G_s , \tag{5.158}$$

and the respective coefficient

$$\widehat{a}(x) = \widehat{a}_m(x) = \begin{cases} 1 & \text{if} x \in \widehat{G}_m , \\ 0 & \text{otherwise} . \end{cases} \tag{5.159}$$

The stochastic model is specified by the choice of the overlap constant $\alpha \in (0, 1/2]$ and the scaling factor $\lambda \in (0, 1]$, such that the differential equation (5.157) takes a form

$$\mathcal{A}_{\lambda,m}\varphi = -\lambda\Delta\varphi - (1-\lambda)\nabla \cdot \widehat{a}_m(\cdot)\nabla\varphi = f , \tag{5.160}$$

where $\widehat{a}_m(x)$ is defined by (5.159). The constant λ will be fixed in the interval $0.1 \le \lambda \le 0.8$, hence we use the simplified notation $\mathcal{A}_m = \mathcal{A}_{\lambda,m}$ or just $\mathcal{A}$ (if m is fixed) so that

$$\mathcal{A}_m = -\nabla \cdot \mathbb{A}_m(x)\nabla ,$$

where the corresponding 2×2 coefficient matrix $\mathbb{A}_m(x)$ takes a form

$$\mathbb{A}_m(x) = \lambda \begin{pmatrix} 1 & 0 \\ 0 & 1 \end{pmatrix} + (1-\lambda) \begin{pmatrix} \widehat{a}_m(x) & 0 \\ 0 & \widehat{a}_m(x) \end{pmatrix} := \begin{pmatrix} a_m(x) & 0 \\ 0 & a_m(x) \end{pmatrix} , \quad x \in \Omega . \tag{5.161}$$

In what follows, we also use the notation $\widehat{\mathbb{A}}_m(x)$ for the 'stochastic part' of a matrix associated with the diagonal coefficient $\widehat{a}_m(x)$, i.e.,

$$\widehat{\mathbb{A}}_m(x) = \begin{pmatrix} \widehat{a}_m(x) & 0 \\ 0 & \widehat{a}_m(x) \end{pmatrix} .$$

The choice of the right hand side in (5.160) will be fixed later on.

Figure 5.42 illustrates an example of the stochastic coefficient $a_m(x)$ in the case $L = 8$, $\lambda = 0.1$ and $\alpha = 1/4$, visualized on a $n_1 \times n_1$ grid with $n_1 = 97$ (top). In Figure 5.42, bottom, we show the covered domain defining the jumping coefficient for the problem with $L = 64$, $n_1 = 513$.

The problem setting remains verbatim also in the 3D case and for higher dimensions.

5.7.2 Generation of stiffness matrix by using Kronecker products of univariate operators

Here we describe the fast matrix generation scheme presented in [193]. This scheme is essentially based on the use of Kronecker product constructions.

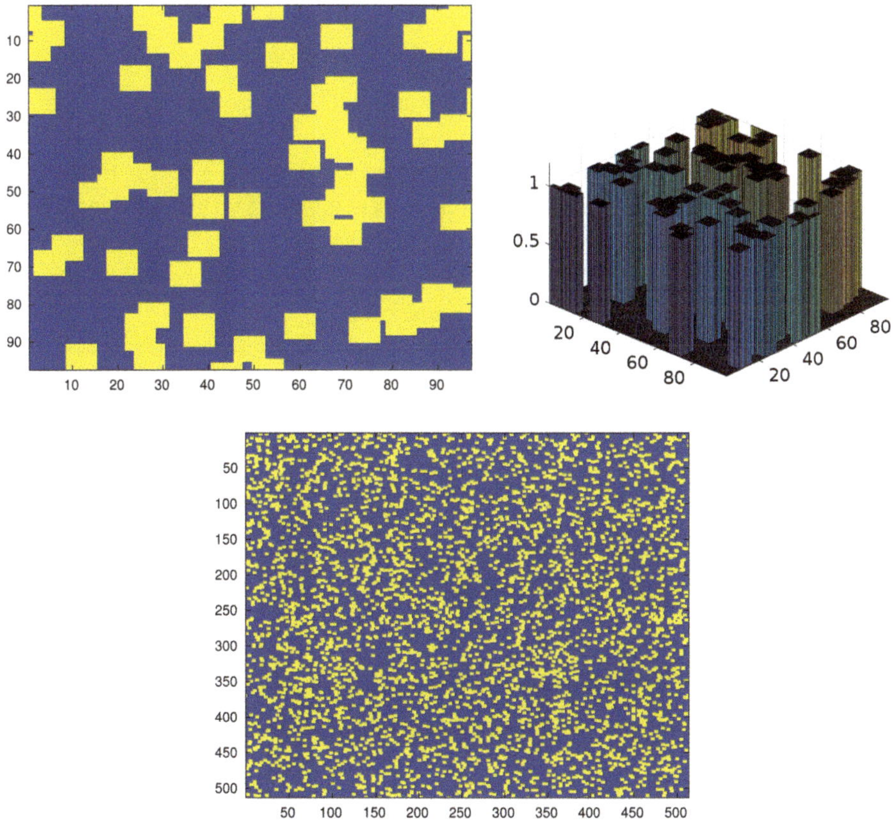

Fig. 5.42: Coefficients for the problem with $L = 8$, $\lambda = 0.1$, $\alpha = 1/4$ (upper left, upper right); coefficients for the problem with $L = 64$, $n_1 = 513$ (bottom).

First, we introduce the uniform $n_s \times n_s$ rectangular grid Ω_{h_s} in Ω with the grid size $h_s = \frac{1}{n_s-1}$, such that $n_s = n_0L + 1$, $n_0 = 2^{p_0}$, i.e., $h_s = \frac{1}{n_0L}$. We assume that the unit cell G_s, $s = 1, \ldots L^2$ of size $\frac{2\alpha}{L} \times \frac{2\alpha}{L}$ adjusts the square grid Ω_{h_s}, such that the center c_s of G_s belongs to the set of grid points in $\Omega_s = \Omega_{n_s}$, while the overlap factor α may take values $\alpha \in \{\frac{1}{n_0}, \frac{2}{n_0}, \ldots \frac{2^{p_0-1}}{n_0}\}$. In this construction the univariate size of the unit cell varies as

$$\frac{2\alpha}{L} = \frac{2\alpha n_0}{n_0 L} = kh_s\,, \quad \text{with} \quad k = 2,\ 4,\ \ldots n_0\,.$$

In the following numerical examples we normally use the overlap constant $\alpha = 1/4$. The size of the unit cell is given by $\frac{1}{L} \times \frac{1}{L}$, which contains $n_0 + 1$ grid points in each spacial direction leading to a $n_1 \times n_1$ rectangular grid with $n_1 = n_0L + 1$.

The FEM discretization of the elliptic PDE in (5.160) can be constructed, in general, on the finer grid Ω_h compared with Ω_s, which serves the resolution of jumping coefficients. To that end, we introduce the $n_1 \times n_1$ rectangular grid Ω_h with the mesh size $h = \frac{1}{n_1-1}$, $n_1 \geq n_s$ that is obtained by dyadic refinement of the grid Ω_s, such that the relation

$$n_1 - 1 = (n_s - 1)2^p\,, \quad \text{with} \quad p = 0, 1, 2, \dots\,, \tag{5.162}$$

holds, implying $h_s = 2^p h$. Now the size of the unit cell G_s on the finer grid Ω_h is given by $(n_0 2^p + 1) \times (n_0 2^p + 1)$.

Given a set of piecewise affine finite elements $\{\psi_i(x)\}$, with $i = 1, \dots, N$, $N = n_1^2$, incorporating periodic boundary conditions, we are looking for the traditional FEM-Galerkin approximation to the exact solution in the form

$$\varphi(x) \approx \sum_{i=1}^{N} u_i \psi_i(x)\,.$$

Introducing the coefficients vector $\mathbf{u} = (u_1, \dots, u_N)^{\mathrm{T}} \in \mathbb{R}^N$, we define the Galerkin FEM discretization of (5.160) as follows:

$$A\mathbf{u} = \mathbf{f}\,, \quad A = [a_{ij}] \in \mathbb{R}^{N \times N}\,, \quad \mathbf{f} = [f_i] \in \mathbb{R}^N\,, \tag{5.163}$$

where

$$a_{ij} = \int_{\Omega} (\lambda \nabla \psi_i \cdot \nabla \psi_j + (1 - \lambda) a(x) \nabla \psi_i \cdot \nabla \psi_j) dx\,, \quad f_i = \int_{\Omega} f(x) \psi_i dx\,. \tag{5.164}$$

We represent the stiffness matrix A in the additive form

$$A = \lambda A_\Delta + (1 - \lambda)\widehat{A}_s\,, \tag{5.165}$$

corresponding to (5.160). Here, A_Δ represents the $N \times N$ FEM Laplacian matrix in a periodic setting that has the standard two-terms Kronecker product form, while the matrix $\widehat{A}_s$ represents the FEM approximation to the 'stochastic part' in the elliptic operator corresponding to (5.161). The latter is determined by the sequence of random coefficients distributions in the course of stochastic realizations, numbered by $m = 1, \dots, M$.

In the case of complicated jumping coefficients the matrix generation in the elliptic FEM usually constitutes the dominating part of the overall solution cost. Our discretization scheme computes all matrix entries at low cost by using local Kronecker product representations. This allows us to store the resultant stiffness matrix in the sparse matrix format. Such a construction only includes the precomputing of tridiagonal matrices representing 1D elliptic operators with jumping coefficients in a periodic setting. In the next sections, we shall describe the efficient construction of the 'stochastic' term A_s.

In the course of stochastic realizations the equation (5.163) has to be solved many hundred or even thousand times, such that every time one has to recompute the stiffness matrix A. Since in the case of complicated jumping coefficients the matrix generation is normally the dominating part in the discretization solution process in the Galerkin FEM, this part of the computations constitutes the main bottleneck in the stochastic simulations.

To enhance the time consuming matrix assembling process (especially in 3D), we apply the FEM-Galerkin discretization (5.164) of equation (5.160) by means of tensor product piecewise linear (affine) finite elements

$$\{\psi_{\mathbf{i}}(x) := \psi_{i_1}(x_1)\cdots\psi_{i_d}(x_d)\}, \quad \mathbf{i} = (i_1, \dots, i_d), \quad i_\ell \in \mathcal{I}_\ell = \{1, \dots, n_\ell\}, \ \ell = 1, \dots, d.$$

Note that the univariate grid size n_ℓ is of the order of $n_\ell = O(1/\epsilon)$, where the small homogenization parameter is given by $\epsilon \approx 1/(n_0 L)$, designating the total problem size

$$N = n_1 n_2 \cdots n_d = O(1/\epsilon^d).$$

The $N \times N$ stiffness matrix is constructed by the standard mapping of the multi-index $\mathbf{i}$ into the N-long univariate index i for the active degrees of freedom in a periodic setting. For instance, we use the so called big-endian convention for $d = 3$ and $d = 2$,

$$\mathbf{i} \mapsto i := i_3 + (i_2 - 1)n_3 + (i_1 - 1)n_2 n_3, \quad \mathbf{i} \mapsto i := i_2 + (i_1 - 1)n_2,$$

respectively.

Taking into account the rectangular structure of the grid, we use the simple finite difference (FD) scheme for the matrix representation of the Laplacian operator $-\Delta$. In this case the scaled discrete Laplacian incorporating periodic boundary conditions takes the form

$$A_\Delta \equiv \Delta_h = \Delta_1 \otimes I_{n_2} + I_{n_1} \otimes \Delta_2, \tag{5.166}$$

where

$$-\Delta_1 = \operatorname{tridiag}\{1, -2, 1\} + P^{(1)} \in \mathbb{R}^{n_1 \times n_1},$$

such that the entries of the ‘periodization’ matrix $P^{(1)} \in \mathbb{R}^{n_1 \times n_1}$ are all zeros except

$$P^{(1)}_{1,n_1} = P^{(1)}_{n_1,1} = 1, \quad \text{and} \quad P^{(1)}_{1,1} = P^{(1)}_{n_1,n_1} = -1.$$

Here $I_{n_1} \in \mathbb{R}^{n_1 \times n_1}$ is the identity matrix, $\Delta_1 = \Delta_2$ is the 1D finite difference Laplacian (endorsed with the Neumann boundary conditions), and $\otimes$ denotes the Kronecker product of matrices. We say that the Kronecker rank of A_Δ in (5.166) equals to 2.

Note that the $n_1 \times n_1$ Laplacian matrices for the Neumann and periodic boundary conditions read as

$$\Delta_N = \begin{bmatrix} -1 & 1 & \dots & 0 & 0 \\ 1 & -2 & \dots & 0 & 0 \\ \vdots & \vdots & \ddots & \vdots & \vdots \\ 0 & 0 & \dots & -2 & 1 \\ 0 & 0 & \dots & 1 & -1 \end{bmatrix} \quad \text{and} \quad \Delta_P = \begin{bmatrix} -2 & 1 & \dots & 0 & 1 \\ 1 & -2 & \dots & 0 & 0 \\ \vdots & \vdots & \ddots & \vdots & \vdots \\ 0 & 0 & \dots & -2 & 1 \\ 1 & 0 & \dots & 1 & -2 \end{bmatrix} \tag{5.167}$$

respectively. In the case $n_1 = 2$ these matrices simplify to

$$\Delta_N = \begin{bmatrix} -1 & 1 \\ 1 & -1 \end{bmatrix} \quad \text{and} \quad \Delta_P = \begin{bmatrix} -1 & 1 \\ 1 & -1 \end{bmatrix}.$$

In the d dimensional problem setting we have the similar Kronecker rank-d representations. For example, if $d = 3$ the 'periodic' Laplacian matrix A_Δ takes a form

$$A_\Delta = A_{1,P} \otimes I_2 \otimes I_3 + I_1 \otimes A_{2,P} \otimes I_3 + I_1 \otimes I_2 \otimes A_{3,P} \,,$$

such that its Kronecker rank equals to 3, and it is similar for the arbitrary $d \geq 3$.

5.7.3 Fast matrix assembling for the stochastic part

The Kronecker form representation of the 'stochastic' term in (5.164) further denoted by A_s is more involved. For given stochastically chosen distribution of overlapping cells G_s, $s = 1, \dots, L^2$, we construct the minimal nonoverlapping decomposition of the full covered grid domain $\widehat{G} = \cup_{s=1}^{L^2} G_s$ colored yellow in Figure 5.42 (we have $a(x) = 1$ for $x \in \widehat{G}$ and $a(x) = \lambda$ for $x \in \Omega \setminus \widehat{G}$) in a form of a union of elementary square cells S_k, $k = 1, \dots, K$, $K \geq L^2$, each of the grid size $\overline{n}_0 \times \overline{n}_0$,

$$\widehat{G} = \cup_{k=1}^{K} S_k \,. \tag{5.168}$$

Here $\overline{n}_0 = 2^p + 1$, and $p = 0, 1, 2, \dots$ is fixed as above by relation $n_1 - 1 = (n_s - 1)2^p$. In this construction, the nonoverlapping elementary cells S_k for different k are allowed to have the only common edges of size $\overline{n}_0$. Note that in the case of nonoverlapping decomposition (5.158) the set of cells $\{S_k\}$ coincides with the initial set $\{G_s\}$, which allows us to maximize the size $\overline{n}_0 \times \overline{n}_0$ of each S_k, $k = 1, \dots, L^2$, to the largest possible, i.e., to $\overline{n}_0 = n_0 2^p + 1$.

To finalize the matrix generation procedure for A_s, we define the local $\overline{n}_0 \times \overline{n}_0$ matrices representing the discrete Laplacian with Neumann boundary conditions,

$$\widehat{Q}_{\overline{n}_0} := \operatorname{tridiag}\{1, -2, 1\} + \operatorname{diag}\{1, 0, \dots, 0, 1\} \in \mathbb{R}^{\overline{n}_0 \times \overline{n}_0} \,,$$

and the diagonal matrix

$$\widehat{I}_{\overline{n}_0} := \operatorname{diag}\{1/2, 1, \dots, 1, 1/2\} \in \mathbb{R}^{\overline{n}_0 \times \overline{n}_0} \,;$$

see the visualization in (5.167). Here, we may select $\overline{n}_0 = 2, 3, 5, \dots$ that correspond to the choices $p = 0, 1, 2, \dots$.

Let the subdomain S_k be supported by the index set $I_k^{(1)} \times I_k^{(2)}$ of size $\overline{n}_0 \times \overline{n}_0$ for $k = 1, \dots, K$. Introduce the block diagonal matrices $\overline{Q}_k \in \mathbb{R}^{n_1 \times n_1}$ and $\overline{I}_k \in \mathbb{R}^{n_1 \times n_1}$ by inserting matrices $\widehat{Q}_{\overline{n}_0}$ and $\widehat{I}_{\overline{n}_0}$ as diagonal blocks into a $n_1 \times n_1$ zero matrix in the positions $I_k^{(1)} \times I_k^{(1)}$ and $I_k^{(2)} \times I_k^{(2)}$, respectively.

Now the stiffness matrix A_s is represented in the form of Kronecker product sum as follows:

$$A_s = \sum_{k=1}^{K} (\overline{Q}_k \otimes \overline{I}_k + \overline{I}_k \otimes \overline{Q}_k) + P^{(2)} \,, \tag{5.169}$$

where

$$P^{(2)} = P^{(1)} \otimes I_{n_1} + I_{n_1} \otimes P^{(1)} \in \mathbb{R}^{N \times N}$$

is the 'periodization' matrix in 2D. In the d dimensional case the representation (5.169) generalizes to a sum of d-term Kronecker products

$$A_s = \sum_{k=1}^{K} (\overline{Q}_k \otimes \overline{I}_k \otimes \cdots \otimes \overline{I}_k + \cdots + \overline{I}_k \otimes \cdots \otimes \overline{I}_k \otimes \overline{Q}_k) + P^{(d)} , \tag{5.170}$$

where $P^{(d)}$ is the 'periodization' matrix in d dimensions, constructed as the d-term Kronecker sum similar to the case $d = 2$.

The Kronecker product representations (5.166) and (5.169) imply the similar form of the total stiffness matrix A. This allows the efficient implementation of the matrix assembly and storage, preserving the Kronecker sparsity. Hence, it proves the storage complexity for the matrix A as follows:

Lemma 5.43 ([193]). *The storage size for the stiffness matrix A is bounded by*

$$\mathrm{Stor}(A) \approx \mathrm{Stor}(A_s) = O(d\overline{n}_0 K + dn_1) .$$

Here, in general, the number K of elementary cells is larger than L^2, and it coincides with L^2 in the case of nonoverlapping decomposition $\widehat{G} = \cup_{s=1}^{L^2} G_s$, where different patches G_s are allowed to have the only joint pieces of boundary.

In the general case $d \geq 2$ and $K \geq L^2$, the Kronecker rank of the matrix A is bounded by

$$\mathrm{rank}_{\mathrm{Kron}}(A) \leq d\,K .$$

The Kronecker rank of the stiffness matrix reduces dramatically in the two cases:
(a) For the case of nonoverlapping cells G_s, $s = 1, \ldots, L^2$, we have

$$\mathrm{rank}_{\mathrm{Kron}}(A) \leq L^d .$$

(b) In the case of cell centered locations of subdomains G_s (special case of geometric homogenization) there holds

$$\mathrm{rank}_{\mathrm{Kron}}(A) \leq d .$$

Let the right hand side in (5.164) satisfy $\langle F, 1\rangle = 0$, then with fixed m, the equation

$$A_m \mathbf{u} = (\lambda A_\Delta + (1 - \lambda) A_{s,m})\mathbf{u} = \mathbf{f} \tag{5.171}$$

has the unique solution. We solve this equation by preconditioned conjugate gradient (PCG) iteration (routine *pcg* in MATLAB library) with the preconditioner

$$B = \frac{1+\lambda}{2} A_\Delta + \delta I = \frac{1+\lambda}{2} \Delta_h + \delta I ,$$

where $\delta > 0$ is the small regularization parameter introduced only for the stability reason (can be ignored in the theory) and I is the $N \times N$ identity matrix.

It can be proven that the condition number of the preconditioned matrix is uniformly bounded in n_1, L, and in $m = 1, \ldots, M$; see [193].

5.7.4 Computational scheme for the homogenized coefficient via stochastic average

In this section, we estimate numerically the mean constant coefficient in the system (5.160) depending on L and other model parameters at the limit of $M \to \infty$; see [115, 116] for the respective problem setting. The numerical realization of the discretized scheme to be considered below was described in full detail in [193].

Consider stochastic realizations specifying the variable part in the 2×2 coefficient matrix $\widehat{\mathbb{A}}_m(x)$, $m = 1, \dots, M$. With fixed coefficient $\widehat{\mathbb{A}}_m(x)$, for $i = 1, 2$ we solve the periodic elliptic problems in Ω,

$$-\lambda\Delta\Phi_i - (1-\lambda)\nabla\cdot\widehat{\mathbb{A}}_m(\cdot)(\nabla\Phi_i + \mathbf{e}_i) = 0\,, \qquad (5.172)$$

where the unit vectors $\mathbf{e}_i$, $i = 1, 2$, are given by $\mathbf{e}_1 = (1,0)^{\mathrm{T}}$ and $\mathbf{e}_2 = (0,1)^{\mathrm{T}}$. The matrix valued equation coefficient is specified by

$$\mathbb{A}_m(x) = \lambda I_{2\times 2} + (1-\lambda)\widehat{\mathbb{A}}_m(x)\,,$$

which corresponds to the diagonal entry in $\mathbb{A}_m(x)$,

$$a_m(x) = \lambda + (1-\lambda)\widehat{a}_m(x)\,. \qquad (5.173)$$

The right hand side in equation (5.172), rewritten in the canonical form (5.160), takes a form

$$f_{i,m} = f_i(x) = (1-\lambda)\nabla\cdot\widehat{\mathbb{A}}_m(x)\mathbf{e}_i\,.$$

Taking into account (5.161), where the diagonal of $\widehat{\mathbb{A}}_m(x)$ is defined in terms of the scalar function $\widehat{a}_m(x)$ in (5.173), we arrive at

$$f_1(x) = (1-\lambda)\frac{\partial\widehat{a}_m(x)}{\partial x_1}\,, \quad f_2(x) = (1-\lambda)\frac{\partial\widehat{a}_m(x)}{\partial x_2}\,.$$

The discretized equation (5.172) takes a form of (5.163),

$$A\mathbf{u}_i = \mathbf{f}_i\,, \quad \text{for} \quad i = 1, 2\,, \qquad (5.174)$$

where the FEM-Galerkin matrix A generated by the equation coefficient $\mathbb{A}_m(x)$ is calculated as described above.

The corresponding vector representation $\mathbf{f}_i \in \mathbb{R}^N$ of the right hand side $f_i(x)$ is computed by multiplication of the discrete upwind gradient matrix ∇_h with a vector $\mathbf{y}_i \in \mathbb{R}^N$. Here the vector $\mathbf{y}_i$ represents the multiple of the vector $\mathbf{e}_i$, $i = 1, 2$, and the equation coefficient $\mathbb{A}_m = \mathbb{A}(x) = [A_{p,q}(x)]$, $p, q = 1, 2$, discretized on the grid Ω_h, i.e., each entry of the 'discretized' matrix coefficient $\mathbb{A}(x) \mapsto A_h$, is given by an N vector array with $N = n^2$, $A_{p,q}(x) \mapsto (A_h)_{p,q} = [A_{p,q}(x_h)]_{x_h\in\Omega_h} \in \mathbb{R}^N$. Hence, we finally arrive at

$$\mathbf{f}_i = (1-\lambda)\nabla_h\cdot\mathbf{y}_i\,, \quad \mathbf{y}_i = [\mathbf{y}_i(x_h)] \in \mathbb{R}^N \text{ with } \mathbf{y}_i(x_h) = A_h(x_h)\mathbf{e}_i\,, \quad x_h \in \Omega_h\,.$$

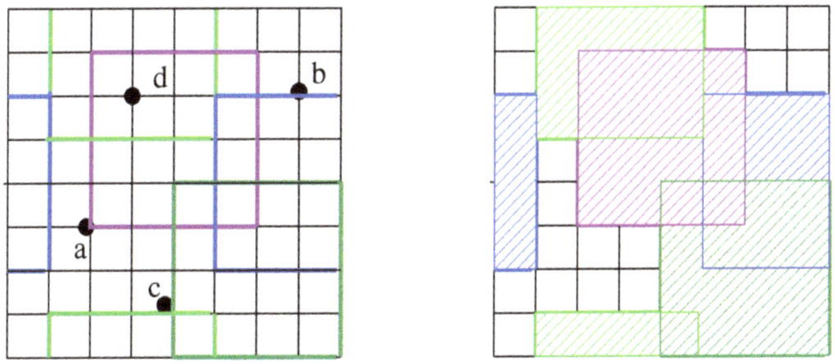

Fig. 5.43: Example of the covering domain $\widehat{G}$ (right) and the typical locations of sampling points for the grid representation of $\mathbb{A}_h(x_h)$ (left).

Specifically, given the grid point $x_h \in \Omega_h$, the corresponding diagonal value of $\mathbb{A}_h(x_h)$ is defined by $a_m(x_h)$; see (5.173). Here the variable part $\widehat{a}_m(x_h)$, describing the jumping coefficient, is assigned by 1 for interior points in $\widehat{G}$, by $1/2$ for interface points (the angle equals to $\pi/2$), by $3/4$ for the 'interior' L-shaped corners (the angle equals to $3\pi/4$), and by $1/4$ for the 'exterior' corner of $\widehat{G}$ (the angle equals to $\pi/4$); see points (d), (b), (c) and (a) in Figure 5.43 respectively. This figure corresponds to $L = 2$, the discretization parameter $n_0 = 4$, and periodic completion of the geometry. One observes the complicated shape of the strongly jumping coefficients and the corresponding right hand sides f_i.

With fixed m, by definition we compute the averaged coefficient matrix $\mathbb{A}_{\text{hom},m} \in \mathbb{R}^{2\times 2}$ with the block constant entries defined by the relation

$$\mathbb{A}_{\text{hom},m}\mathbf{e}_i = \int_\Omega \mathbb{A}_m(x)(\nabla\Phi_i + \mathbf{e}_i)dx\,,$$

which means

$$(A_{\text{hom},m})_{i,j} = \int_\Omega [(\lambda I_{2\times 2} + (1-\lambda)\widehat{\mathbb{A}}_m(x))(\nabla\Phi_i + \mathbf{e}_i)]_j dx\,, \quad i,j = 1,2\,. \tag{5.175}$$

The latter leads to the entrywise representation of the matrix $\mathbb{A}_{\text{hom},m} = [A_{i,j}]$, $i,j = 1,2$,

$$\begin{aligned}
A_{1,1} &= \int_\Omega a_m(x)\left(\frac{\partial\Phi_1}{\partial x_1} + 1\right)dx = \int_\Omega \left(a_m(x) - \frac{\partial a_m(x)}{\partial x_1}\Phi_1\right)dx\,,\\
A_{1,2} &= \int_\Omega a_m(x)\frac{\partial\Phi_1}{\partial x_2}dx = -\int_\Omega \frac{\partial a_m(x)}{\partial x_2}\Phi_1 dx\,,\\
A_{2,1} &= \int_\Omega a_m(x)\frac{\partial\Phi_2}{\partial x_1}dx = -\int_\Omega \frac{\partial a_m(x)}{\partial x_1}\Phi_2 dx\,,\\
A_{2,2} &= \int_\Omega a_m(x)\left(\frac{\partial\Phi_2}{\partial x_2} + 1\right)dx = \int_\Omega \left(a_m(x) - \frac{\partial a_m(x)}{\partial x_2}\Phi_2\right)dx\,.
\end{aligned} \tag{5.176}$$

In numerical implementation, we apply the same variational scheme as for FEM discretization of the right hand side in the initial equation, thus preserving the symmetry of the matrix $\mathbb{A}_{\text{hom},m}$ inherited from the exact variational formulation.

Furthermore, integrals over Ω in (5.175), as they are written for the exact matrix entries $(A_{\text{hom},m})_{i,j}$, $i,j = 1,2$, are calculated (approximately) by the scalar product of the N vector of all-ones with the discrete representation of integrand on the grid Ω_h.

Now we define the stochastic entities of interest that should be analyzed numerically at the limit of large M and L. With fixed L, the homogenized matrix coefficient $\mathbb{A}^L_{\text{hom},M}$ is computed as the sample average of a sequence $\{\mathbb{A}_{\text{hom},m}\}$ (expectation),

$$\mathbb{A}^L_{\text{hom},M} = \frac{1}{M}\sum_{m=1}^{M} \mathbb{A}_{\text{hom},m}\,. \tag{5.177}$$

The respective 2×2 matrix valued deviation (stochastic fluctuation) is denoted by

$$\widehat{\mathbb{A}}^L_{\text{hom},m} = \mathbb{A}_{\text{hom},m} - \mathbb{A}^L_{\text{hom},M}\,, \quad m = 1,\dots,M\,.$$

Finally, we denote

$$\mathbb{A}_{\text{hom},M} = \lim_{L\to\infty} \mathbb{A}^L_{\text{hom},M}\,, \quad \mathbb{A}_{\text{hom}} = \lim_{M\to\infty, L\to\infty} \mathbb{A}^L_{\text{hom},M}\,.$$

The expected asymptotic convergence rate in terms of M and L can be estimated by ([114])

$$\left\langle |\mathbb{A}^L_{\text{hom},M} - \mathbb{A}_{\text{hom}}|^2 \right\rangle^{1/2} \leq C\left(\frac{1}{\sqrt{M}} L^{-d/2} + L^{-d}\ln^d L\right). \tag{5.178}$$

In the description of numerical tests we use the notation $\mathbb{A}_{\text{hom},m} = [A_{ij,m}]$, $i,j = 1,2$. We are interested in calculation of empirical variance V_{12} and V_{diag} of A_{12} and $A_{11} - A_{22}$ respectively, computed by

$$V^L_{12} = \sqrt{\frac{1}{M}\sum_{m=1}^{M} A^2_{12,m}}\,, \quad V^L_{\text{diag}} = \sqrt{\frac{1}{M}\sum_{m=1}^{M} (A_{11,m} - A_{22,m})^2} \tag{5.179}$$

for M realizations, where $A_{ij,m}$ are the matrix entries of the homogenized matrix $\mathbb{A}^L_{\text{hom},m}$.

5.7.5 Empirical variance versus the size of representative volume elements

Here we sketch some numerical results reported in [193].

Due to tensor based construction of the stiffness matrix and usage of sparse representation of matrix entities, in our MATLAB experiments we are able to implement a large number of generated homogenization cells (the size of representative volume elements) in the domain $\Omega = [0,1]^2$ up to $L^2 = 128^2$. This corresponds to the problem sizes $n_s = n_0 L + 1 = 513$ for $n_0 = 4$ and $n_s = 1025$ with $n_0 = 8$.

Tab. 5.15: Variance of A_{12} and $(A_{11} - A_{22})$ versus $L = 2, 4, \ldots, 128$, for $M = 20$ realizations of stochastic processes, with $n_0 = 4$, $\alpha = \frac{1}{4}$, $\lambda = 0.4$, $\varepsilon = 10^{-8}$.

L	V_{12}^L	V_{diag}^L
2	0.002 7017	0.007 0336
4	0.002 5309	0.005 6678
8	0.001 4574	0.003 3991
16	0.000 6413	0.001 5161
32	0.000 3516	0.000 8180
64	0.000 1415	0.000 4544
128	0.000 0676	0.000 2263

First, we check in [193] that our FEM discretization scheme reveals the symmetry in the matrix $A_{\text{hom},m}$ if the discrete system of equations (5.174) is solved accurately enough. This property is the consequence of (5.176).

Table 5.15 represents empirical variance of A_{12} and $(A_{11} - A_{22})$ versus $L = 2, 4, \ldots, 128$, for $M = 20$ realizationsof stochastic processes, with $n_0 = 4$, $\alpha = \frac{1}{4}$, $\lambda = 0.4$, $\varepsilon = 10^{-8}$. We observe the expected decay factor about 2 between the results for the subsequent values of representative volume element, L.

Figure 5.44 presents the empirical variance of A_{12} and $(A_{11} - A_{22})$ versus $L = 2, 4, \ldots, 128$, corresponding to $M = 20$ realization. We observe perfect C/L fitting for a fixed number of realizations $M = 20$, as is shown by the red lines corresponding to the expected C/L fit on the same log-log scale.

Table 5.16 illustrates the behavior of the systematic errors V_{12}^2 and V_{diag}^2 versus $L = 2^p$, $p = 1, \ldots, 7$, computed for $M = 20$ realizations. These data indicate the satisfactory C/L^2 fit, predicted by the theory.

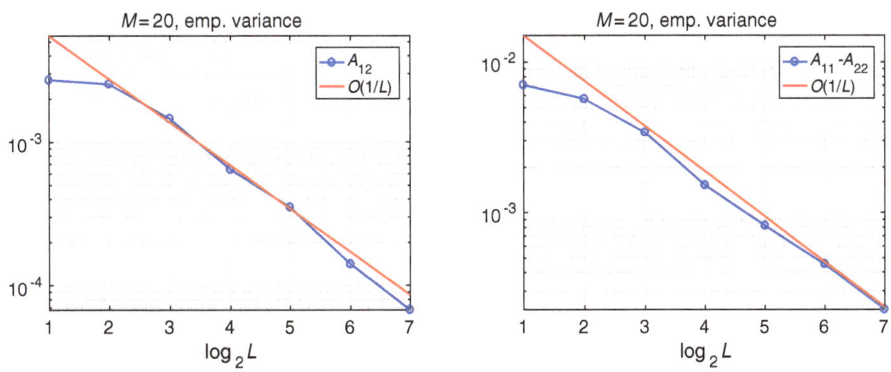

Fig. 5.44: Empirical variance for A_{12} (left) and $A_{11}-A_{22}$ (right) versus L, for $M = 20$, and for sequence $L = 2^p$, $p = 1, 2, \ldots, 7$.

Tab. 5.16: Systematic error of A_{12} and $(A_{11} - A_{22})$ versus $L = 2, 4, \dots, 128$, for $M = 20$ realizations of stochastic processes, with $n_0 = 4$, $\alpha = \frac{1}{4}$, $\lambda = 0.4$, $\varepsilon = 10^{-8}$.

L	V_{12}^2	V_{diag}^2
2	$0.3305 \cdot 10^{-4}$	$0.1834 \cdot 10^{-3}$
4	$0.0601 \cdot 10^{-4}$	$0.1519 \cdot 10^{-3}$
8	$0.0465 \cdot 10^{-4}$	$0.0143 \cdot 10^{-3}$
16	$0.0034 \cdot 10^{-4}$	$0.0019 \cdot 10^{-3}$
32	$0.0014 \cdot 10^{-4}$	$0.0005 \cdot 10^{-3}$
64	$0.0002 \cdot 10^{-4}$	$0.0002 \cdot 10^{-3}$
128	$0.000\,04 \cdot 10^{-4}$	$0.000\,04 \cdot 10^{-3}$

5.7.6 Asymptotic empirical average versus the number of stochastic realizations

Furthermore, the asymptotic of empirical average versus M was analyzed numerically in [193]. The numerical tests for the central limit theorem (CLT) applied to sequences $\mathbb{A}_{\text{hom},m}$ at the limit of the large number of stochastic realizations M have been performed. In particular, the long sequences for $M = 2^{20}$ and for moderate values of L were tested and the results confirmed the CLT asymptotic.

Recall that the empirical average (expected value), S_M, and empirical variance, V_M, for the random sequence $\{X_1, \dots, X_M\}$, are defined by

$$\mu^* = \lim_{M\to\infty} S_M, \quad S_M = \frac{1}{M}(X_1 + \cdots + X_M); \quad V_M = \left(\frac{1}{M}\sum_{m=1}^{M}(X_m - \mu^*)^2\right)^{1/2}, \quad (5.180)$$

so that (5.179) is the particular case of (5.180). We refer to [193], where the asymptotic $O(1/\sqrt{M})$ for the empirical average for A_{12} versus M is confirmed by numerical experiments.

5.8 Sketch of other applications

5.8.1 Operator dependent RS tensor approximation of the Dirac delta

Many radial basis functions, for example the commonly used Matérn functions [271], represent the fundamental solutions of the elliptic operators with constant coefficients. Since the discretized radial basis functions can be represented in the RS tensor format, we may apply the discrete elliptic operator to the RS tensor to obtain the corresponding operator dependent splitting of the Dirac delta. Hence, the latter is constructed by application of the discrete elliptic operator with constant coefficients (stiffness matrix) to the RS tensor representation of the fundamental solution (Green's kernel). In what follows, we sketch the results in [205].

To fix the idea, we consider the simplest case of the single atom with charge 1 located at the origin, such that $u(x) = \frac{1}{\|x\|}$, $x \in \mathbb{R}^3$. Recall that the Newton kernel (5.136) discretized by the R-term sum of Gaussians living on the tensor grid Ω_n, is represented by a sum of the short and long range tensors,

$$\frac{1}{\|x\|} \rightsquigarrow \mathbf{P}_R = \mathbf{P}_{R_s} + \mathbf{P}_{R_l} \,,$$

where

$$\mathbf{P}_{R_s} = \sum_{k=R_l+1}^{R} \mathbf{p}_k^{(1)} \otimes \mathbf{p}_k^{(2)} \otimes \mathbf{p}_k^{(3)} \,, \quad \mathbf{P}_{R_l} = \sum_{k=1}^{R_l} \mathbf{p}_k^{(1)} \otimes \mathbf{p}_k^{(2)} \otimes \mathbf{p}_k^{(3)} \,. \tag{5.181}$$

Let us formally discretize the exact equation for the potential

$$-\Delta \frac{1}{\|x\|} = 4\pi\, \delta(x)$$

by its FD analogy by substitution of $\mathbf{P}_R$ instead of $u(x)$ and FEM Laplacian matrix A_Δ instead of Δ. This leads to the grid representation of the Dirac delta (up to the factor $1/4\pi$)

$$\delta(x) \rightsquigarrow \boldsymbol{\delta}_h := -A_\Delta \mathbf{P}_R \,, \tag{5.182}$$

which is associated with its operator dependent differential representation corresponding to the given elliptic operator with constant coefficients, say with Laplacian.

Recall that in the case $d = 3$ the FEM Laplacian matrix A_Δ takes a form

$$A_\Delta = \Delta_1 \otimes I_2 \otimes I_3 + I_1 \otimes \Delta_2 \otimes I_3 + I_1 \otimes I_2 \otimes \Delta_3 \,, \tag{5.183}$$

where $-\Delta_k = \operatorname{tridiag}\{1, -2, 1\} \in \mathbb{R}^{n_k \times n_k}$ denotes the discrete univariate Laplacian, such that the Kronecker rank of A_Δ equals to 3. Hence, if we apply this to the low rank RS representation of the Green's kernel considered in the previous sections, we arrive at the range separated rank structured decomposition to the discretized Dirac delta.

To that end, we introduce the short and long range splitting of the $\boldsymbol{\delta}_h$ function in the form

$$\boldsymbol{\delta}_h = \boldsymbol{\delta}_s + \boldsymbol{\delta}_l$$

by using the representation (5.182). This leads to the particular expressions generated by the discrete Laplacian,

$$\boldsymbol{\delta}_s := -A_\Delta \mathbf{P}_{R_s} \,, \quad \text{and} \quad \boldsymbol{\delta}_l := -A_\Delta \mathbf{P}_{R_l} \,.$$

Now it is possible to construct the RS tensor based splitting scheme for the solution of the Poisson–Boltzmann equation; see Section 5.5.5. The idea is to modify the right hand side ρ_f in such a way that the short range part in the solution u can be computed independently and the initial equation applies only to its long range part, see [32] for more details. The latter is a smooth function, hence the FEM approximation error can be reduced dramatically even in the case of relatively coarse grids in 3D.

We observe that by definition the short range part vanishes on the interface Γ, hence it satisfies the discrete Poisson equation in Ω_m with the respective right hand side in the form $\boldsymbol{\delta}_s$ and zero boundary conditions on Γ. Then we deduce that this equation can be subtracted from the full discrete linear system, such that the long range component of the solution, $\mathbf{P}_{R_l}$, will satisfy to the same linear system of equations (same interface conditions), but with the right hand side corresponding to the weighted sum of the long range tensors $\boldsymbol{\delta}_l$ only. In our example, we arrive at the particular Poisson equation for the long range part in the considered potential $\mathbf{P}_R$, $\mathbf{u}_l = \mathbf{P}_{R_l}$,

$$-A_\Delta \mathbf{u}_l = \boldsymbol{\delta}_l \,.$$

This scheme can be easily extended to the case of many-atomic systems and to the general class of elliptic differential operators $\mathcal{L}$ with constant coefficients. The corresponding $\mathcal{L}$-generated (operator dependent) splitting to the Dirac delta can be applied for the solution of elliptic equations with jumping coefficients including the governing operator $\mathcal{L}$.

5.8.2 Tensor approach to isogeometric analysis

Isogeometric analysis (IGA) is the modern FEM method for the efficient solution of elliptic equations on complicated geometry. We refer to [5, 8, 24, 76] for the detailed description of the approach. One of the most computationally extensive parts in the numerical implementation of the discretized IGA scheme is the matrix generation on the complicated decomposition of the computational domain composed of many nonoverlapping patches.

In the recent papers [268, 269] the new matrix generation scheme via patchwise low rank approximation to the integrated functions was developed for the solution of the elliptic equation on complicated geometry,

$$\text{Find} \quad u \in H_0^1(\Omega): \qquad -\operatorname{div}(a(x)\nabla u) = f \quad \text{in } \Omega \,, \quad f \in L^2(\Omega) \,.$$

The approach is based on the FEM-Galerkin approximation in $\Omega \in \mathbb{R}^d$ by a patchwise mapping onto reference domain, and fast evaluation of the stiffness matrix on that patch by reducing the d dimensional integration to the univariate operations applied to the low rank representation of the integrand.

Figure 5.45 demonstrates the values of the Jacobian determinant of the selected patch and its image on the parameter domain; see [268] for the details.
Numerical tests demonstrated significant advantages when using the rank structured tensor representations for computation of the stiffness matrix in both 2D and 3D calculations. The details of this approach can be found in [268, 269].

Recently IGA was applied to the space-time discretization of the parabolic problem; see [245]. We refer to [244] concerning the related basic theory of parabolic equations.

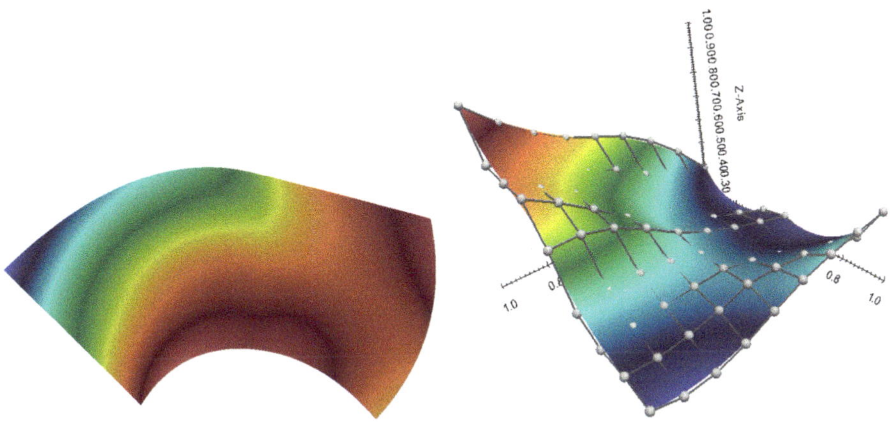

Fig. 5.45: The Jacobian determinant values of the selected patch on the domain (left). The graph and control net in the parameter domain (right).

5.8.3 Quantized-CP approximation of function generated data

Paper [220] introduces the iterative schemes to compute the canonical (CP) approximation of the quantized tensor generated by a function discretized on a large uniform grid in an interval on the real line. The ALS iterative scheme combined with the cross approximation approach was constructed to compute the rank-r CP approximation of the quantized vectors. This requires only $2rL \ll 2^L$ parameters for storage (which is less than $2r^2L$ for QTT format). This representation was called the $\mathrm{Q_{Can}}$ format or the QCP representation. Numerical tests for the ALS algorithm in calculating the QCP approximation applied to various functions demonstrated the exponential error decay in the QCP separation rank.

5.8.4 Superfast QTT wavelet transform

Other applications of the QTT approximation method include the superfast wavelet transform by using the QTT data format [219]. This construction is designed in the spirit of superfast FFT-QTT transform. However, the principal difference is that the wavelet transform operator (matrix) admits the explicit low rank QTT decomposition (which is not the case for the FFT matrix). Hence, only the low rank representation (approximation) to the target vector (signal) is required to perform the QTT wavelet transform. The complexity of the fast QTT wavelet transform scales logarithmically in the size of the input vector. We refer to [291] for the related approach.

Bibliography

[1] P.-A. Absil, R. Mahony and R. Sepulchre, *Optimization Algorithms on Matrix Manifolds*. Princeton University Press, Princeton, NJ, 2008.

[2] E. Acar, D. M. Dunlavy, T. G. Kolda, A scalable optimization approach for fitting canonical tensor decompositions. *Journal of Chemometrics* 25(2) (2011), 67–86.

[3] J. Almlöf, *Direct methods in electronic structure theory*. In D. R. Yarkony: Modern Electronic Structure Theory, Vol. II. World Scientific, Singapore, 1995, 110–151.

[4] A. Ammar, E. Cueto, F. Chinesta, Reduction of the chemical master equation for gene regulatory networks using proper generalized decompositions. *Int. J. Numer. Meth. Biomed. Engng.* 00 (2011), 1–15.

[5] P. Antolin, A. Buffa, F. Calabró, M. Martinelli and G. Sangalli, Efficient matrix computation for tensor-product isogeometric analysis: The use of sum factorization. *Comp. Meth. Appl. Mech. Engrg.* 285 (2015), 817–828.

[6] F. Aquilante, L. Gagliardi, T. B. Pedersen and R. Lindh, Atomic Cholesky decompositions: A root to unbiased auxiliary basis sets for density fitting approximation with tunable accuracy and efficiency. *J. Chem. Phys.* 130 (2009), 154107.

[7] D. Z. Arov and I. P. Gavrilyuk, A method for solving initial value problems for linear differential equations in Hilbert space based on the Cayley transform. *Numer. Funct. Anal. Optimization* 14(5–6) (1993), 456–473.

[8] F. Auricchio, L. Beirão da Veiga, T. J. R. Hughes, A. Reali and G. Sangalli, Isogeometric collocation methods. *Mathematical Models and Methods in Applied Sciences* 20(11) (2010), 2075–2107.

[9] P. Y. Ayala and G. E. Scuseria, Linear scaling second-order Møller–Plesset theory in the atomic orbital basis for large molecular systems. *J. Chem. Phys.* 110(8) (1999), 3660–3671.

[10] E. A. Ayrjan, B. N. Khoromskij and E. P. Zhidkov, Fast relaxation method for solving the difference problem for the Poisson equation on a sequence of grids. *Comp. Phys. Commun.* 29 (1983), 125–130.

[11] I. Babuska, F. Nobile and R. Tempone: A stochastic collocation method for elliptic PDEs with random input data. *SIAM Review* 52(2) (2010), 317–355.

[12] M. Bachmayr and W. Dahmen, Adaptive near-optimal rank tensor approximation for high-dimensional operator equations. *Foundations of Comp. Math.* 15(4) (2015), 839–898.

[13] B. W. Bader and T. G. Kolda: MATLAB Tensor Classes for Fast Algorithm Prototyping. SANDIA Report, SAND2004-5187, Sandia National Laboratories, 2004.

[14] N. S. Bakhvalov and G. Panasenko, *Homogenisation: Averaging Processes In Periodic Media: Mathematical Problems In The Mechanics Of Composite Materials*. Springer, Dordrecht, The Netherlands, 1989.

[15] J. Ballani and L. Grasedyck, Tree adaptive approximation in the hierarchical tensor format. *SIAM J Sci. Comp.* 36(4) (2014), A1415–A1431.

[16] J. Ballani, L. Grasedyck and M. Kluge, Black box approximation of tensors in hierarchical Tucker format. *Linear Alg. Appl.* 428 (2013), 639–657.

[17] M. Barrault, E. Cancés, W. Hager and Le Bris, Multilevel domain decomposition for electronic structure calculations. *J. Comput. Phys.* 222 (2007), 86–109.

[18] H. Bateman and A. Erdelyi, *Higher Transcendental Functions, v. 2*, MC Graw-Hill Book Comp., New York, Toronto, London, 1988.

[19] M. Bebendorf, Adaptive cross approximation of multivariate functions. *J. Constructive approximation* 34(2) (2011), 149–179.

https://doi.org/10.1515/9783110365917-006

[20] M. Bebendorf, *Hierarchical matrices.* Lect. Notes Comput. Sci. Eng., vol. 63, Springer, Berlin, (2008).

[21] M. Bebendorf, S. Rjasanow, Adaptive low-rank approximation of collocation matrices. *Computing* 70(1) (2003), 1–24.

[22] M. Bebendorf, W. Hackbusch, Existence of H-matrix approximants to the inverse FE-matrix of elliptic operators with L8-coefficients. *Numerische Mathematik* 95(1) (2003), 1–28.

[23] N. H. F. Beebe and J. Linderberg, Simplifications in the generation and transformation of two-electron integrals in molecular calculations. *Int. Quantum Chem.* 12(7) (1977), 683–705.

[24] L. Beirão da Veiga, A. Buffa, G. Sangalli and R. Vázquez, Mathematical analysis of variational isogeometric methods. *Acta Numerica* 23(5) (2014), 157–287.

[25] R. E. Bellman, *Dynamic programming.* Princeton University Press, 1957.

[26] P. Benner and T. Breiten, Low rank methods for a class of generalized Lyapunov equations and related issues. *Numerische Mathematik* 124(3) (2013), 441–470.

[27] P. Benner, S. Dolgov, V. Khoromskaia and B. N. Khoromskij, Fast iterative solution of the Bethe–Salpeter eigenvalue problem using low-rank and QTT tensor approximation. *Journal of Computational Physics* 334 (2017), 221–239.

[28] P. Benner, H. Faßbender and M. Stoll, Solving Large-Scale Quadratic Eigenvalue Problems with Hamiltonian Eigenstructure using a Structure-Preserving Krylov Subspace Method. *Electronic Transactions on Numerical Analysis* 29 (2008), 212–229.

[29] P. Benner, V. Khoromskaia and B. N. Khoromskij, A reduced basis approach for calculation of the Bethe–Salpeter excitation energies using low-rank tensor factorizations. *Molecular Physics* 114(7–8) (2016), 1148–1161.

[30] P. Benner, V. Khoromskaia and B. N. Khoromskij, *Range-separated tensor formats for numerical modeling of many-particle interaction potentials.* E-preprint, http://arxiv.org/abs/1606.09218, 2016.

[31] P. Benner, V. Khoromskaia and B. N. Khoromskij, *Range-separated tensor format for many-particle modeling.* SIAM J. Sci. Comput., 40 (2) (2018), A1034–A1062.

[32] P. Benner, V. Khoromskaia, B. N. Khoromskij, C. Kweyu and M. Stein, *Application of the Range-separated Tensor Format in Solution of the Poisson–Boltzmann equation.* Manuscript, 2018.

[33] P. Benner, V. Khoromskaia, B. N. Khoromskij and C. Yang, *Density of states for optical spectra by low-rank and QTT tensor approximation.* ArXiv: 1801.03852, 2018.

[34] P. Benner, A. Onwunta, M. Stoll, *Low-rank solution of unsteady diffusion equations with stochastic coefficients.* SIAM/ASA Journal on Uncertainty Quantification 3 (1), 622–649, 2015.

[35] A. Bensoussan, J.-L. Lions and G. Papanicolaou, *Asymptotic analysis for periodic structures.* Amsterdam: North-Holland, 1978.

[36] S. N. Bernstein, Sur lórdre de la meilleure approximation des fonctions continues par des polynomes de degrȇ donnè. *Mem. Acad. Roy. Belg.* 4(2) (1912), 1–104.

[37] S. N. Bernstein, Sur la limitation des valeurs dún polynomes $P(x)$ de degré n sur tour un segment par ses valeurs en $(n+1)$ points du segment. *Izv. Akad. Nauk SSSR* 7 (1931), 1025–1050.

[38] C. Bertoglio and B. N. Khoromskij, Low-rank quadrature-based tensor approximation of the Galerkin projected Newton/Yukawa kernels. *Comp. Phys. Communications* 183(4) (2012), 904–912.

[39] G. Beylkin, J. Garcke and M. J. Mohlenkamp, Multivariate regression and machine learning with sums of separable functions. *SIAM. J Sci. Comp.* 31(3) (2009), 1840–1857.

[40] G. Beylkin, Ch. Kurcz and L. Monzón: Fast algorithms for Helmholtz Green's function. *Proc. Roy. Soc., Ser. A* 464 (2008), 3301–3326.

[41] G. Beylkin, M. J. Mohlenkamp and F. Pérez, Approximating a wavefunction as an unconstrained sum of Slater determinants. *Journal of Math. Phys.* 49 (2008), 032107.

[42] G. Beylkin and M. J. Mohlenkamp, Numerical operator calculus in higher dimensions. *Proc. Natl. Acad. Sci. USA* 99 (2002), 10246–10251.

[43] M. Bieri, R. Andreev and Ch. Schwab, Sparse Tensor Discretization of Elliptic sPDEs *SIAM J Sci. Comput.* 31(6) (2009), 4281–4304.

[44] M. Bieri and Ch. Schwab: Sparse high order FEM for elliptic sPDEs. *Comp. Meth. Appl. Mech. Engg.* 198 (2009), 1149–1170.

[45] P. Binev, A. Cohen, W. Dahmen, R. DeVore, G. Petrova, P. Wojtaszczyk, Convergence rates for greedy algorithms in reduced basis methods. *SIAM J. Math. Anal.* 43(3) (2011), 1457–1472.

[46] F. A. Bischoff and E. F. Valeev, Low-order tensor approximations for electronic wave functions: Hartree–Fock method with guaranteed precision. *J. of Chem. Phys.* 134 (2011), 104104-1-10.

[47] T. Blesgen, V. Gavini, V. Khoromskaia, Approximation of the electron density of Aluminium clusters in tensor-product format. *J. Comp. Phys.* 231(6) (2012), 2551–2564.

[48] S. F. Boys, G. B. Cook, C. M. Reeves and I. Shavitt, Automatic Fundamental Calculations of Molecular Structure. *Nature* 178 (1956), 1207–1209.

[49] S. Börm, *Efficient numerical methods for non-local operators.* EMS, Zürich (2010).

[50] D. Braess, *Nonlinear approximation theory.* Springer-Verlag, Berlin, 1986.

[51] D. Braess, Asymptotics for the Approximation of Wave Functions by Exponential-Sums. *J. Approx. Theory* 83 (1995), 93–103.

[52] J. Braun and M. Griebel, On a Constructive Proof of Kolmogorov's Superposition Theorem. *Constr. Approx.* 30 (2009), 653–675.

[53] S. Brenner and R. Scott, *The mathematical theory of finite element methods.* Springer, Berlin, 1994.

[54] S. C. Brenner and C. Carstensen, *Finite element methods.* In: Encyclopedia of computational mechanics. John Wiley & Sons, Weinheim, 1(4) (2004), 73–118.

[55] H. J. Bungartz and M. Griebel, Sparse grids. *Acta Numerica* 13 (2004), 1–123.

[56] E. Cancés, A. Deleurence and M. Lewin, A new approach to the modeling of local defects in crystals: The reduced Hartree–Fock case. *Comm. Math. Phys.* 281 (2008), 129–177.

[57] E. Cancés and C. Le Bris, On the convergence of SCF algorithms for the Hartree–Fock equations. *ESAIM: M2AN* 34(4) (2000), 749–774.

[58] E. Cancés, V. Ehrlacher and T. Leliévre, Convergence of a greedy algorithm for high-dimensional convex nonlinear problems. *Mathematical Models and Methods in Applied Sciences.* 21(12) (2011), 2433–2467.

[59] E. Cancés, V. Ehrlacher and T. Leliévre, Greedy algorithms for high-dimensional non-symmetric linear problems. *ESAIM* 2013 (41), 95–131.

[60] E. Cancés, V. Ehrlacher and Y. Maday, Periodic Schrödinger operator with local defects and spectral pollution. *SIAM J. Numer. Anal.* 50(6) (2012), 3016–3035.

[61] M. E. Casida, Time-dependent density-functional response theory for molecules. *In: Recent advances in density functional methods, part I. D. P. Chong, eds., World Scientific, Singapoure* 155 (1995), 1207–1216.

[62] J. D. Carrol and J. Chang: Analysis of individual differences in multidimensional scaling via an *N*-way generalization of 'Eckart-Young' decomposition, *Psychometrika* 35 (1970), 283–319.

[63] J. D. Carrol, S. Pruzansky and J. B. Kruskal, CANDELINC: A general approach to multidimensional anlysis of many-way arrays with linear constraints on parameters, *Psychometrika* 45 (1980), 3–24.

[64] P. L. Chebyshev, Sur les questions de minima qui se rattachent á la representation approximative des fonctions. *Mem. Acad. Sci. Pétersb* 7 (1859), 199–291.

[65] P. G. Ciarlet and C. Le Bris, eds.: *Handbook of Numerical Analysis, v. X, Computational Chemistry.* Elsevier, Amsterdam, 2003.

[66] S. R. Chinnamsetty, M. Espig, W. Hackbusch, B. N. Khoromskij, H-J. Flad, Kronecker tensor product approximation in quantum chemistry. *Journal of Chemical Physics* 127 (2007), 084110.

[67] A. Cichocki and Sh. Amari, *Adaptive Blind Signal and Image Processing: Learning Algorithms and Applications.* Wiley, The Atrium, Chichester, UK, 2002.

[68] A. Cichocki, N. Lee, I. Oseledets, A. H. Pan, Q. Zhao and D. P. Mandic, Tensor networks for dimensionality reduction and large-scale optimization: Part 1 low-rank tensor decompositions. *Foundations and Trends in Machine Learning* 9(4–5) (2016), 249–429.

[69] A. Cichocki, A. H. Phan, Q. Zhao, N. Lee, I. Oseledets, M. Sugiyama, D. P. Mandic, Tensor Networks for Dimensionality Reduction and Large-scale Optimization: Part 2 Applications and Future Perspectives. *Foundations and Trends in Machine Learning* 9(6) (2017), 431–673.

[70] A. Cohen, R. DeVore and Ch. Schwab: *Convergence rates of best N-term Galerkin approximations for a class of elliptic sPDEs*, Report 2009-02, Seminar for Applied Mathematics, ETH Zürich, CH-8092, Zürich 2009.

[71] P. Comon, Tensor decompositions. *Mathematics in Signal Processing* V (2002), 1–24.

[72] P. Comon and C. Jutten, *Handbook of Blind Source Separation: Independent component analysis and applications.* Academic press, 1 (2010), 1–22.

[73] J. W. Cooley, J. W. Tukey, An algorithm for the machine calculation of complex Fourier series. *Math. Comput.* 19 (1965), 297–301. doi:10.2307/2003354.

[74] D. Coppersmith and Sh. Winograd, Matrix multiplication via arithmetic progressions. *Journal of Symbolic Computation* 9(3) (1990), 251.

[75] E. Corona, A. Rahimian, D. Zorin, *A Tensor-Train accelerated solver for integral equations in complex geometries.* Journal of Computational Physics 334 (2017), 145–169.

[76] J. A. Cotrell, T. J. R. Hughes and Y. Bazilevs, *Isogeometric Analysis, Toward Integration of CAD and FEA.* John Wiley & Sons, The Atrium, Chichester, UK, 2009.

[77] W. Dahmen, R. Devore, L. Grasedyck, E. Süli, Tensor-sparsity of solutions to high-dimensional elliptic partial differential equations. *Foundations Comp. Math.* 16(4) (2016), 813–874.

[78] P. J. Davis, *Circulant matrices.* John Wiley & Sons, Inc., NY, 1979.

[79] S. Dolgov, V. Kazeev and B. N. Khoromskij, *Direct tensor product solution of one-dimensional elliptic equations with parameter-dependent coefficients.* Mathematics and Computers in Simulation, 2017; doi:10.1016/j.matcom.2017.10.009. Preprint 51/2012, Max-Planck Institute for Math. in the Sciences, Leipzig 2012.

[80] S. Dolgov, B. N. Khoromskij, A. Litvinenko and H. G. Matthies, Computation of the Response Surface in the Tensor Train data format. *SIAM/ASA J. Uncertainty Quantification* 3 (2015), 1109–1135, E-preprint arXiv:1406.2816, 2014.

[81] S. Dolgov, B. N. Khoromskij, D. Savostyanov and I. Oseledets, Computation of extreme eigenvalues in higher dimensions using block tensor train format. *Comp. Phys. Comm.* 185(4) (2014), 1207–1216.

[82] S. V. Dolgov, B. N. Khoromskij, I. Oseledets and E. E. Tyrtyshnikov, A reciprocal preconditioner for structured matrices arising from elliptic problems with jumping coefficients. *Linear Algebra Appl.* 436(9) (2012), 2980–3007.

[83] S. V. Dolgov, B. N. Khoromskij, I. Oseledets and E. E. Tyrtyshnikov, Low-rank tensor structure of solutions to elliptic problems with jumping coefficients. *J. Comp. Math.* 30(1) (2012), 14–23.

[84] S. V. Dolgov, B. N. Khoromskij and I. Oseledets, Fast solution of multi-dimensional parabolic problems in the TT/QTT formats with initial application to the Fokker–Planck equation. *SIAM J. Sci. Comp.* 34(6) (2012), A3016-A3038.

[85] S. V. Dolgov, B. N. Khoromskij and D. Savostyanov, Superfast Fourier transform using QTT approximation. *J. Fourier Anal. Appl.* 18(5) (2012), 915–953.

[86] S. Dolgov and B. N. Khoromskij, Simultaneous state-time approximation of the chemical master equation using tensor product formats. *Numer. Lin. Algebra Appl.* 22(2) (2015), 197–219.

[87] S. V. Dolgov and B. N. Khoromskij, Two-level Tucker-TT-QTT format for optimized tensor calculus. *SIAM J. on Matr. Anal. Appl.* 34(2) (2013), 593–623.

[88] S. V. Dolgov and I. V. Oseledets, Solution of linear systems and matrix inversion in the TT-format. *SIAM J. Sci. Comp.* 34(5) (2012), A2718–A2739, arxiv:1406.2816.

[89] S. V. Dolgov and D. V. Savostyanov, Alternating minimal energy methods for linear systems in higher dimensions. *SIAM J Sci. Comp.* 36(5) (2014), A2248–A2271.

[90] S. Dolgov and D. V. Savostyanov, Alternating minimal energy methods for linear systems in higher dimensions. Part II: Faster algorithm and application to nonsymmetric systems. E-preprint: arXiv:1304.1222, 2013.

[91] R. Dovesi, R. Orlando, C. Roetti, C. Pisani and V. R. Sauders, The Periodic Hartree–Fock Method and its Implementation in the CRYSTAL Code. *Phys. Stat. Sol. (b)* 217 (2000), 63.

[92] T. H. Dunning Jr, Gaussian basis sets for use in correlated molecular calculations. I. The atoms boron through neon and hydrogen. *J. Chem. Phys.* 90 (1989), 1007–1023.

[93] V. Ehrlacher, C. Ortner and A. V. Shapeev, Analysis of boundary conditions for crystal defect atomistic simulations. *Archive for Rational Mechanics and Analysis* 222(3) (2016), 1217–1268.

[94] M. Espig and W. Hackbusch, A regularized Newton method for the efficient approximation of tensors represented in the canonical tensor format. *Numer. Math.* 122(3) (2012), 489–525.

[95] M. Espig, W. Hackbusch, T. Rohwedder and R. Schneider, Variational calculus with sums of elementary tensors of fixed rank. *Numerische Mathematik* 122(3) (2012), 469–488.

[96] P. P. Ewald, Die Berechnung optische und elektrostatischer Gitterpotentiale. *Ann. Phys.* 64 (1921), 253–287.

[97] M. Fenn, S. Kunis and D. Potts, Fast evaluation of trigonometric polynomials from hyperbolic crosses. *Num. Algorithms* 41 (2006), 339–352.

[98] H.-J. Flad, W. Hackbusch and R. Schneider, Best *N*-term approximation in electronic structure calculations: I. One-electron reduced density matrix. *ESAIM: M2AN* 40 (2006), 49–61.

[99] H.-J. Flad, W. Hackbusch, B. N. Khoromskij and R. Schneider, *Concept of data-sparse tensor-product approximation in many-particle modeling.* In: "Matrix Methods: Theory, Algorithms, Applications", E. Tyrtyshnikov, et. al. eds., World Scientific Publishing, Singapoure, 2010, 313–347.

[100] H.-J. Flad, B. Khoromskij, D. Savostianov and E. Tyrtyshnikov, Verification of the cross 3d algorithm on quantum chemistry data. *Rus. J. Numer. Anal. and Math. Modelling* 4 (2008), 1–16.

[101] L. E. Figueroa and E. Süli, *Greedy Approximation of High-Dimensional Ornstein–Uhlenbeck Operators.* Foundations Comp. Math.,12 (2012), 573–623.

[102] S. Friedland, V. Mehrmann, A. Miedlar and M. Nkengla, Fast low rank approximation of matrices and tensors. *Electron. J. Linear Algebra* 22 (2011), 1031–1048.

[103] Friedman, A. (1976): Partial Differential Equations. R. E. Krieger Pub. Co., Huntington, NY, 1994.

[104] L. Frediani, E. Fossgaard, T. Flå and K. Ruud, Fully adaptive algorithms for multivariate integral equations using the non-standard form and multiwavelets with applications to the Poisson and bound-state Helmholtz kernels in three dimensions. *Molecular Physics* 111 (2013), 9–11.

[105] I. P. Gavrilyuk and V. L. Makarov, Exact and approximate solutions of some operator equations based on the Cayley transform. *Linear Algebra and its Applications* 282 (1998), 97–121.

[106] I. P. Gavrilyuk, Strongly P-positive operators and explicit representations of the solutions of initial value problems for second order differential equations in Banach space. *Journal of Math. Anal. and Appl.* 236 (1999), 327–349.

[107] I. P. Gavrilyuk, Super exponentially convergent approximation to the solution of the Schrödinger equation in abstract setting. *CMAM* 10(4) (2010), 345–358.

[108] I. P. Gavrilyuk, W. Hackbusch and B. N. Khoromskij, Data-Sparse Approximation to Operator-Valued Functions of Elliptic Operator. *Math. Comp.* 73 (2003), 1297–1324.

[109] I. P. Gavrilyuk, W. Hackbusch and B. N. Khoromskij, Data-Sparse Approximation to a Class of Operator-Valued Functions. *Math. Comp.* 74 (2005), 681–708.

[110] I. P. Gavrilyuk, W. Hackbusch and B. N. Khoromskij, Hierarchical Tensor-Product Approximation to the Inverse and Related Operators in High-Dimensional Elliptic Problems. *Computing* 74 (2005), 131–157.

[111] I. P. Gavrilyuk, W. Hackbusch and B. N. Khoromskij, $\mathcal{H}$-Matrix Approximation for the Operator Exponential with Applications. *Numer. Math.* 92 (2002), 83–111.

[112] I. P. Gavrilyuk, W. Hackbusch and B. N. Khoromskij, Tensor-product approximation to elliptic and parabolic solution operators in higher dimensions. *Computing* 74 (2005), 131–157.

[113] I. V. Gavrilyuk and B. N. Khoromskij, Quantized-TT-Cayley transform to compute dynamics and spectrum of high-dimensional Hamiltonians. *Comp. Meth. in Applied Math.* 11(3) (2011), 273–290.

[114] A. Gloria, S. Neukamm and F. Otto, Quantification of ergodicity in stochastic homogenization: optimal bounds via spectral gap on Glauber dynamics. *In: Inventiones mathematicae* 199(2) (2015), 455–515; MIS-Preprint: 91/2013.

[115] A. Gloria and F. Otto, An optimal error estimate in stochastic homogenization of discrete elliptic equations. *In: The annals of applied probability* 22(1) (2012), 1–28.

[116] A. Gloria and F. Otto, Quantitative estimates on the periodic approximation of the corrector in stochastic homogenization. *In: ESAIM / Proceedings* 48 (2015), 80–97. Preprint 12/2015, Max-Planck Institute for Math. in the Sciences, Leipzig, 2015.

[117] R. Glowinski, J.-L. Lions, R. Trémolierés. Analyse numérique des inéquations variationnelles. Dunod, Paris, 1976.

[118] J. A. Goldstein, *Semigroups of Linear Operators and Applications.* Oxford University Press and Clarendon Press, New York, Oxford, 1985.

[119] S. A. Goreinov, E. E. Tyrtyshnikov, N. L. Zamarashkin, A theory of pseudoskeleton approximations. *Linear algebra and its applications* 261(1–3) (1997), 1–21.

[120] L. Grasedyck and W. Hackbusch, Construction and arithmetics of H-matrices. *Computing* 70(4) (2003), 295–334.

[121] L. Grasedyck, W. Hackbusch and B. N. Khoromskij, Solution of large scale algebraic matrix Riccati equations by use of hierarchical matrices. *Computing* 70 (2003), 121–165.

[122] L. Grasedyck, Existence and computation of low Kronecker-rank approximations for large linear systems of tensor product structure. *Computing* 72(3) (2004), 247–265.

[123] L. Grasedyck, Hierarchical Singular Value Decomposition of Tensors. *SIAM. J. Matrix Anal. and Appl.* 31 (2010), 2029.

[124] L. Grasedyck, *Polynomial Approximation in Hierarchical Tucker Format by Vector Tensorization.* Preprint 308, Institut für Geometrie und Praktische Mathematik, RWTH Aachen, 2010.

[125] L. Grasedyck, D. Kressner and C. Tobler, A literature survey of low-rank tensor approximation techniques. *GAMM-Mitteilungen* 36(1) (2013), 53–78.

[126] L. Greengard and V. Rochlin, A fast algorithm for particle simulations. *J. Comp. Phys.* 73 (1987), 325.

[127] W. H. Greub, *Multilinear Algebra.* 2nd edn. Springer, Berlin (1978).

[128] M. Griebel and J. Hamaekers, Sparse grids for the Schroedinger equation. *M2AN* 41 (2007), 215–247.

[129] M. Griebel and D. Oeltz, A Sparse Grid Space-Time Discretization Scheme for Parabolic Problems. *Computing* 81(1) (2007), 1–34.

[130] M. Griebel, D. Oeltz and P. Vassilevski, Space-time approximation with sparse grids. *SIAM J. Sci. Comput.* 28(2) (2005), 701–727.

[131] A. Grothendieck, Produits tensoriels topologiques et espaces nucléaries. *Mem. Amer. Math. Soc.* 16 (1955), 336.

[132] G. H. Golub, C. F. Van Loan, *Matrix Computations.* Johns Hopkins University Press, Baltimore, MD, 1996.

[133] W. Hackbusch, A sparse matrix arithmetic based on H-matrices. Part I. Introduction to H-matrices. *Computing* 62(2) (1999), 89–108.

[134] W. Hackbusch, Convolution of hp-functions on locally refined grids. *IMA J Numer. Anal.* 29 (2009), 960–985.

[135] W. Hackbusch, Efficient convolution with the Newton potential in d dimensions. *Numerische Mathematik* 110(4) (2008), 449–489.

[136] W. Hackbusch, *Elliptic Differential Equations: Theory and Numerical Treatment.* Springer, Berlin, 1992.

[137] W. Hackbusch, *Hierarchische Matrizen.* Springer, Heidelberg, 2009.

[138] W. Hackbusch, *Tensor spaces and numerical tensor calculus.* Springer, Berlin, 2012.

[139] W. Hackbusch and B. N. Khoromskij, A Sparse $\mathcal{H}$-Matrix Arithmetic: General Complexity Estimates. *J. of Comp. and Appl. Math.* 125 (2000), 479–501.

[140] W. Hackbusch and B. N. Khoromskij, A Sparse $\mathcal{H}$-Matrix Arithmetic. Part II: Application to Multi-Dimensional Problems. *Computing* 64 (2000), 21–47.

[141] W. Hackbusch and B. N. Khoromskij, Low-rank Kronecker product approximation to multi-dimensional nonlocal operators. Part I. Separable approximation of multi-variate functions. *Computing* 76 (2006), 177–202.

[142] W. Hackbusch and B. N. Khoromskij, Low-rank Kronecker product approximation to multi-dimensional nonlocal operators. Part II. HKT representations of certain operators. *Computing* 76 (2006), 203–225.

[143] W. Hackbusch and B. N. Khoromskij, Tensor-product approximation to multi-dimensional integral operators and Green's functions. *SIAM J. Matr. Anal. Appl., 30, no.* 3 (2008), 1233–1253.

[144] W. Hackbusch and B. N. Khoromskij, Tensor-product Approximation to Operators and Functions in High Dimension. *Journal of Complexity* 23 (2007), 697–714.

[145] W. Hackbusch and B. N. Khoromskij, Towards $\mathcal{H}$-Matrix Approximation of the Linear Complexity. *Operator Theory: Advances and Applications*, Birkhäuser Verlag, Basel, 121 (2001), 194–220.

[146] W. Hackbusch and S. Kühn, A new scheme for the tensor representation. *Fourier Analysis and Applications* 15 (2009), 706–722.

[147] W. Hackbusch, B. N. Khoromskij and R. Kriemann, Hierarchical Matrices based on a Weak Admissibility Criterion. *Computing* 73 (2004), 207–243.

[148] W. Hackbusch, B. N. Khoromskij and S. Sauter, Adaptive Galerkin Boundary Element Methods with Panel Clustering. *Numer. Math.* 105 (2007), 603–631.

[149] W. Hackbusch, B. N. Khoromskij and S. Sauter, *On $\mathcal{H}^2$-Matrices.* In: Lectures on Applied Mathematics, H.-J. Bungartz, R. Hoppe, C. Zenger, eds., Springer-Verlag, Berlin, 2000, 9–30.

[150] W. Hackbusch, B. N. Khoromskij, S. Sauter and E. E. Tyrtyshnikov, Use of tensor formats in elliptic eigenvalue problems. *Numer. Lin. Alg. Appl.* 19(1) (2012), 133–151.

[151] W. Hackbusch, B. N. Khoromskij and E. E. Tyrtyshnikov, Hierarchical Kronecker tensor-product approximations. *J. Numer. Math.* 13 (2005), 119–156.

[152] W. Hackbusch, B. N. Khoromskij and E. E. Tyrtyshnikov, Approximate iteration for structured matrices. *Numer. Math.* 109 (2008), 365–383.

[153] W. Hackbusch and R. Schneider, *Tensor Spaces and Hierarchical Tensor Representations.* In: Lecture Notes in Computer Science and Engineering, 102, S. Dahlke, W. Dahmen, et al. eds. Springer, Berlin, 2014.

[154] J. Haegeman, Ch. Lubich, I. Oseledets, B. Vandereycken and F. Verstraete, Unifying time evolution and optimization with matrix product states. *Physical Review B* 94(16) (2016), 165116.

[155] N. Hale and L. N. Trefethen, Chebfun and numerical quadrature. *Sci. China Math.* 55(9) (2012), 1749–1760.

[156] H. Harbrecht, M. Peters and R. Schneider, On the low-rank approximation by the pivoted Cholesky decomposition. *App. Numer. Math.* 62(4) (2012), 428–440.

[157] R. J. Harrison, G. I. Fann, T. Yanai, Z. Gan and G. Beylkin, Multiresolution quantum chemistry: Basic theory and initial applications. *J. of Chemical Physics* 121(23) (2004), 11587–11598.

[158] D. R. Hartree, *The Calculation of Atomic Structure*, Wiley, New York, 1957.

[159] M. Head-Gordon, J. A. Pople and M. Frisch, MP2 energy evaluation by direct methods. *Chem. Phys. Letters* 153(6) (1988), 503–506.

[160] L. Hedin, New method for calculating the one-particle Green's function with application to the electron-gas problem. *Phys. Rev.* 139 (1965), A796.

[161] T. Helgaker, P. Jørgensen and J. Olsen, *Molecular Electronic-Structure Theory.* Wiley, New York, 1999.

[162] Hegland M., Burden C., Santoso L. et al., A solver for the stochastic master equation applied to gene regulatory networks. *J Comp. Appl. Math.*, 2007. V. 205, 2, p. 708–724.

[163] N. Higham, *Analysis of the Cholesky decomposition of a semi-definite matrix*, In M. G. Cox and S. J. Hammarling, eds. Reliable Numerical Computations, Oxford University Press, 1990, 161–185.

[164] F. L. Hitchcock, Multiple invariants and generalized rank of a p-way matrix or tensor. *J. Math. Phys.* 7 (1927), 39–79.

[165] F. L. Hitchcock, The expression of a tensor or a polyadic as a sum of products. *J. Math. Phys.* 6 (1927), 164–189.

[166] M. Holst, N. Baker and F. Wang, Adaptiv multilevel finite element solution of the Poisson–Boltzmann equation: algorithms and examples. *J. Comp. Chem.* 21 (2000), 1319–1342.

[167] S. Holtz, T. Rohwedder and R. Schneider, On manifold of tensors of fixed TT-rank. *Numer. Math.* 120(4) (2012), 701–731.

[168] S. Holtz, T. Rohwedder and R. Schneider, The alternating linear scheme for tensor optimization in the tensor train format. *SIAM J Sci. Comp.* 34(2) (2012), A683–A713.

[169] G. C. Hsiao and W. L. Wendland, *Boundary integral equations.* Springer, Berlin (2008).

[170] T. Huckle, K. Waldherr and T. Schulte-Herbrüggen, Computations in quantum tensor networks. *Linear Algebra Appl.* 438 (2013), 750–781, doi:10.1016/j.laa.2011.12.019.

[171] A. Iserles and D. Levin, Asymptotic expansion and quadrature of composite highly oscillatory integrals. *Math. Comp.* 80 (2011), 279–296.

[172] M. Ishteva, P. A. Absil, S. Van Huffel, L. De Lathauwer, Best low multilinear rank approximation of higher-order tensors, based on the Riemannian trust-region scheme. *SIAM J Matrix Anal. Appl.* 32(1) (2011), 115–135.

[173] M. Ishteva, L. De Lathauwer, P. A. Absil, S. Van Huffel, Differential-geometric Newton method for the best rank-(R1, R2, R3) approximation of tensors. *Numerical Algorithms* 51(2) (2009), 179–194.

[174] T. Jahnke and W. Huisinga, *A Dynamical Low-Rank Approach to the Chemical Master Equation.* Bulletin of Mathematical Biology. 2008. V. 70. P. 2283–2302.

[175] V. V. Jikov, S. M. Kozlov and O. A. Oleinik, *Homogenization of differential operators and integral functionals.* Berlin: Springer, 1994.

[176] T. Kailath and A. Sayed, *Fast reliable algorithms for matrices with structure.* SIAM Publication, Philadelphia, 1999.

[177] V. Kazeev and B. N. Khoromskij, Explicit low-rank QTT representation of Laplace operator and its inverse. *SIAM J Matr. Anal. Appl.* 33(3) (2012), 742–758.

[178] V. Kazeev, M. Khammash, M. Nip and Ch. Schwab, *Direct Solution of the Chemical Master Equation Using Quantized Tensor Trains.* PLoS Comput Biol 10(3), 2014.

[179] V. Kazeev, B. N. Khoromskij and E. E. Tyrtyshnikov, Multilevel Toeplitz matrices generated by tensor-structured vectors and convolution with logarithmic complexity. *SIAM J. Sci. Comp.* 35(3) (2013), A1511–A1536.

[180] V. Kazeev, I. Oseledets, M. Rakhuba and Ch. Schwab, QTT-finite-element approximation for multiscale problems. I: Model problem in one dimension. *Advances in Computational Mathematics* 43(2) (2017), 411–442.

[181] V. Kazeev, O. Reichmann and Ch. Schwab, *hp*-DG-QTT solution of high-dimensional degenerate diffusion equations. Techn. Report 2012-11, SAM, ETH Zurich, 2012.

[182] V. Kazeev, O. Reichmann and Ch. Schwab, Low-rank tensor structure of linear diffusion operators in the TT and QTT formats. *Numer. Lin Algebra Appl.* 438(11) (2013), 4204–4221.

[183] V. Khoromskaia, Black-box Hartree–Fock solver by tensor numerical methods. *Comp. Meth. in Applied Math.* 14(1) (2014), 89–111.

[184] V. Khoromskaia, Computation of the Hartree–Fock Exchange in the Tensor-structured Format. *Comp. Meth. in Applied Math.* 10(2) (2010), 204–218.

[185] V. Khoromskaia, *Numerical Solution of the Hartree–Fock Equation by Multilevel Tensor-structured methods.* PhD Dissertation, TU Berlin, 2010. http://opus.kobv.de/tuberlin/volltexte/2011/2948/

[186] V. Khoromskaia, D. Andrae and B. N. Khoromskij, Fast and Accurate 3D Tensor Calculation of the Fock Operator in a General Basis. *Comp. Phys. Comm.* 183 (2012), 2392–2404.

[187] V. Khoromskaia and B. N. Khoromskij, Block circulant and Toeplitz structures in the linearized Hartree–Fock equation on finite lattices: tensor approach. *Comp. Meth. Appl. Math.* 2017; 17 (3):431–455.

[188] V. Khoromskaia and B. N. Khoromskij, *Fast tensor method for summation of long-range potentials on 3D lattices with defects.* Numer. Lin. Algebra Appl., 2016, v. 23: 249–271.

[189] V. Khoromskaia and B. N. Khoromskij, Grid-based lattice summation of electrostatic potentials by assembled rank-structured tensor approximation. *Comp. Phys. Comm.* 185(12) (2014), 3162–3174.

[190] V. Khoromskaia and B. N. Khoromskij, Møller–Plesset (MP2) Energy Correction Using Tensor Factorizations of the Grid-based Two-electron Integrals. *Comp. Phys. Comm.* 185(1) (2014), 2–10.

[191] V. Khoromskaia and B. N. Khoromskij, Tensor numerical methods in quantum chemistry: from Hartree–Fock to excitation energies. *Physical Chemistry Chemical Physics* 17 (2015), 31491–31509.

[192] V. Khoromskaia and B. N. Khoromskij, *Tensor numerical methods in quantum chemistry.* To appear. De Gruyter, Berlin, 2018.

[193] V. Khoromskaia, B. N. Khoromskij and F. Otto, A numerical primer in 2D stochastic homogenization: CLT scaling in the representative volume element. Preprint 47/2017, Max-Planck Institute for Math. in the Sciences, Leipzig 2017.

[194] V. Khoromskaia, B. N. Khoromskij and R. Schneider, QTT Representation of the Hartree and Exchange Operators in Electronic Structure Calculations. *Comp. Meth. Appl. Math.* 11(3) (2011), 327–341.

[195] V. Khoromskaia, B. N. Khoromskij and R. Schneider, Tensor-structured calculation of two-electron integrals in a general basis. *SIAM J. Sci. Comput.* 35(2) (2013), A987–A1010.

[196] B. N. Khoromskij, *O*(*d* log *N*)-Quantics Approximation of *N*-*d* Tensors in High-Dimensional Numerical Modeling. Preprint 55/2009, Max-Planck Institute for Math. in the Sciences, Leipzig 2009. http://www.mis.mpg.de/publications/preprints/2009/prepr2009-55.html.

[197] B. N. Khoromskij, *O*(*d* log *N*)-Quantics Approximation of *N*-*d* Tensors in High-Dimensional Numerical Modeling. *J. Constr. Approx.* 34(2) (2011), 257–289.

[198] B. N. Khoromskij, An Introduction to Structured Tensor-Product Representation of Discrete Nonlocal Operators. *Lecture Notes, Max-Planck Institute for Mathematics in the Sciences, Leipzig* 27 (2005), 1–279.

[199] B. N. Khoromskij, Data-sparse Elliptic Operator Inverse Based on Explicit Approximation to the Green Function. *J. Numer. Math.* 11(2) (2003), 135–162.

[200] B. N. Khoromskij, *Data-Sparse Approximation of Integral Operators.* Lecture notes No. 17, Max-Planck Institute for Mathematics in the Sciences, Leipzig, 2003, 1–61.

[201] B. N. Khoromskij, Fast and Accurate Tensor Approximation of a Multivariate Convolution with Linear Scaling in Dimension. J. Comput. *Appl. Math.* 234 (2010), 3122–3139.

[202] B. N. Khoromskij, Hierarchical Matrix Approximation to Green's Function via Boundary Concentrated FEM. *J. of Numer. Math.* 11(3) (2003), 195–223.

[203] B. N. Khoromskij, Introduction to Tensor Numerical Methods in Scientific Computing. Lecture Notes, University/ETH Zuerich, Preprint 06–2011, Uni. Zuerich 2011, pp. 1–238, http://www.math.uzh.ch/fileadmin/math/preprints/06_11.pdf.

[204] B. N. Khoromskij, On Tensor Approximation of Green Iterations for Kohn–Sham Equations. *Computing and Visualization in Sci.* 11 (2008), 259–271.

[205] B. N. Khoromskij, Operator-dependent approximation of the Dirac delta by using range separated tensor format. Manuscript, 2017.

[206] B. N. Khoromskij, Structured Rank-$(r_1, \dots, r_d)$ Decomposition of Function-related Operators in R^d. *Comp. Meth. Appl. Math.* 6(2) (2006), 194–220.

[207] B. N. Khoromskij, Structured data-sparse approximation to high order tensors arising from the deterministic Boltzmann equation. *Math. Comp.* 76 (2007), 1292–1315.

[208] B. N. Khoromskij, Tensor-Structured Preconditioners and Approximate Inverse of Elliptic Operators in $\mathbb{R}^d$. *J. Constructive Approx.* 30 (2009), 599–620.

[209] B. N. Khoromskij, Tensors-structured Numerical Methods in Scientific Computing: Survey on Recent Advances. *Chemometrics and Intellingent Lab. Systems* 110 (2012), 1–19.

[210] B. N. Khoromskij, Tensor Numerical Methods for High-dimensional PDEs: Basic Theory and Initial Applications. *ESAIM* 48 (2015), 1–28.

[211] B. N. Khoromskij and V. Khoromskaia, Low Rank Tucker-Type Tensor Approximation to Classical Potentials. *Central European J. of Math.* 5(3) (2007), 1–28.

[212] B. N. Khoromskij and V. Khoromskaia, Multigrid accelerated tensor approximation of function related multi-dimensional arrays. *SIAM J. on Sci. Comput.* 31(4) (2009), 3002–3026.

[213] B. N. Khoromskij, V. Khoromskaia, S. R. Chinnamsetty, H.-J. Flad, Tensor Decomposition in Electronic Structure Calculations on 3D Cartesian Grids. *J Comp. Phys.* 228 (2009), 5749–5762.

[214] B. N. Khoromskij, V. Khoromskaia and H.-J. Flad, *Numerical Solution of the Hartree–Fock Equation in Multilevel Tensor-structured Format.* SIAM J Sci. Comp., 33 (1), 45–65 (2011).

[215] B. N. Khoromskij and A. Litvinenko, Data sparse computation of the Karhunen–Loéve expansion. *In: Numerical Anal. Appl. Math., eds. T. Simos, G. Psihoyios, Ch. Tsitouras. AIP Conf. Proc., Melville, New York* 1048 (2008), 311–314.

[216] B. N. Khoromskij, A. Litvinenko and H. G. Matthies, Application of hierarchical matrices for computing the Karhunen–Loéve expansion. *Computing* 84 (2009), 49–67.

[217] B. N. Khoromskij and J. M. Melenk, An Efficient Direct Solver for the Boundary Concentrated FEM in 2D. *Computing* 69 (2002), 91–117.

[218] B. N. Khoromskij and J. M. Melenk, Boundary Concentrated Finite Element Methods. *SIAM J Numer. Anal.* 41(1) (2003), 1–36.

[219] B. N. Khoromskij and S. Miao, Superfast Wavelet Transform Using QTT Approximation. I: Haar Wavelets. *Comp. Meth. Appl. Math.* 14(4) (2014), 537–553.

[220] B. N. Khoromskij, K. K. Naraparaju and J. Schneider, Quantized-CP Approximation and Sparse Tensor Interpolation of Function Generated Data. arXiv preprint, arXiv:1707.04525, 2017.

[221] B. N. Khoromskij and I. Oseledets, DMRG+QTT approach to the computation of ground state for the molecular Schrödinger operator. Preprint 68/2010, Max-Planck Institute for Math. in the Sciences, Leipzig 2010.

[222] B. N. Khoromskij and I. Oseledets, Quantics-TT approximation of elliptic solution operators in higher dimensions. *Russ. J. Numer. Anal. Math. Modelling* 26(3) (2011), 303–322.

[223] B. N. Khoromskij and I. Oseledets, Quantics-TT collocation approximation of parameter-dependent and stochastic elliptic PDEs. *Comp. Meth. Appl. Math.* 10(4) (2010), 34–365.

[224] B. N. Khoromskij and S. Repin, A fast iteration method for solving elliptic problems with quasiperiodic coefficients. *RJNAMM* 30(6) (2015), 329–344.

[225] B. N. Khoromskij and S. Repin, Rank structured approximation method for quasi–periodic elliptic problems. *Computational Methods in Applied Mathematics* 17(3) (2017), 457–477.

[226] B. N. Khoromskij, S. Sauter and A. Veit, Fast Quadrature Techniques for Retarded Potentials Based on TT/QTT Tensor Approximation. *Comp. Meth. Appl. Math.* 11(3) (2011), 342–362.

[227] B. N. Khoromskij and Ch. Schwab, Tensor-Structured Galerkin Approximation of Parametric and Stochastic Elliptic PDEs. *SIAM J. Sci. Comp.* 33(1) (2011), 1–25.

[228] B. N. Khoromskij and A. Veit, Efficient computation of highly oscillatory integrals by using QTT tensor approximation. *Comp. Meth. Appl. Math.* 16(1) (2016), 145–159.

[229] B. N. Khoromskij and G. Wittum, *Numerical Solution of Elliptic Differential Equations by Reduction to the Interface.* Research monograph, LNCSE, No. 36, Springer-Verlag, Berlin, Heidelberg, 2004.

[230] O. Koch, Ch. Lubich., *Dynamical low rank approximation*, SIAM J. Matrix Anal. Appl. 29(2) (2007), 434–454.

[231] T. Kolda, Orthogonal tensor decompositions. *SIAM J. Matrix Anal. Appl.* 23 (2001), 243–255.

[232] T. G. Kolda and B. W. Bader, Tensor Decompositions and Applications. SIAM Review 51(3) (2009), 455–500.

[233] T. G. Kolda, R. M. Lewis and V. Torczon, Optimization by direct search: New perspectives on some classical and modern methods. *SIAM review* 45(3) (2003), 385–482.

[234] A. N. Kolmogorov, On the representation of continuous functionsof many variables by superpositions of continuous functions of one variable and addition. *Dokl. Akad. Nauk USSR* 14(5) (1957), 953–956.

[235] A. N. Kolmogorov and V. M. Tikhomirov, ϵ-entropy and ϵ-capacity of sets in function spaces. *Usp. Mat. Nauk 13(2) (1959), 3–86. English translation: Am. Math. Soc. Transl* 17(2) (1961), 277–364.

[236] V. A. Kotel'nikov, *On the Transmission Capacity of the "Ether" and Wire in Electrocommunications.* First All-Union Conference on Questions of Communication, Izd. Red. Upr. Svyazi RKKA, Moscow, 1933 (Translated on English by V. E. Katsnelson.).

[237] D. Kressner, *Numerical Methods for General and Structured Eigenvalue Problems.* Lecture Notes in Computational Science and Engineering. Springer, Berlin/Heidelberg, vol. 46, 2005.

[238] D. Kressner, M. Steinlechner and A. Uschmajew, Low-rank tensor methods with subspace correction for symmetric eigenvalue problems. *SIAM J Sci. Comp.* 36(5) (2014), A2346–A2368.

[239] D. Kressner and C. Tobler, *Preconditioned low-rank methods for high-dimensional elliptic PDE eigenvalue problems.* Comp. Meth. Appl. Math., 2011. V. 11, 3. P. 363–381.

[240] D. Kressner and C. Tobler, *Krylov Subspace Methods for Linear Systems with Tensor Product Structure.* SIAM J. Matrix Anal. Appl., 31(4), 2010, 1688–1714.
[241] P. M. Kroonenberg and J. De Leeuw, Principal component analysis of three-mode data by means of alternating least squares algorithms. *Psychometrika* 45 (1980), 69–97.
[242] J. B. Kruskal, *Three-way arrays: rank and uniqueness of trilinear decompositions, with applications to arithmetic complexity and statistics.* Linear Algebra Appl., 18, (1977) 95–138.
[243] L. Laaksonen, P. Pyykkö and D. Sundholm, *Fully numerical Hartree–Fock methods for molecules*, Comput. Phys. Rep. 4 (1986), 313–344.
[244] O. A. Ladyzhenskaya, V. A. Solonnikov and N. N. Uraltseva, *Linear and Quasilinear Equations of Parabolic Type.* AMS, Providence, RI, 1968.
[245] U. Langer, S. Moore and M. Neumüller, Space-time isogeometric analysis of parabolic evolution equations. *Comput. Methods Appl. Mech. Engrg.* 306 (2016), 342–363.
[246] J. M. Landsberg, *Tensors: geometry and applications.* Providence, R.I.: American Mathematical Society, 2012.
[247] L. De Lathauwer, B. De Moor, J. Vandewalle, A multilinear singular value decomposition. *SIAM J. Matrix Anal. Appl.* 21 (2000), 1253–1278.
[248] L. De Lathauwer, B. De Moor, J. Vandewalle, On the best rank-1 and rank-$(R_1, \dots, R_N)$ approximation of higher-order tensors. *SIAM J. Matrix Anal. Appl.* 21 (2000), 1324–1342.
[249] C. Le Bris, *Computational chemistry from the perspective of numerical analysis.* Acta Numerica (2005), 363–444.
[250] C. Le Bris, ed(s)., *Handbook of Numerical Analysis, Vol. X, Computational Chemistry*, North-Holland, 2003.
[251] C. Le Bris, T. Leliévre and Y. Maday, *Results and Questions on a Nonlinear Approximation Approach for Solving High-dimensional Partial Differential Equations.* Constr. Approx., 2009. V. 30. P. 621–651.
[252] C. Le Bris and F. Legoll. Examples of computational approaches for elliptic possibly multiscale PDEs with random inputs. *J. Comp. Phys.*, 328 (2017) 455–473.
[253] L. Lin, Y. Saad and C. Yang, Approximating spectral densities of large matrices. *SIAM Review* 58(1) (2016), 34–65.
[254] A. Litvinenko and H. G. Matthies, Sparse data representation of random fields. *Proceedings in Applied Mathematics and Mechanics* 9 (2009), 587–588.
[255] A. Litvinenko, D. Keyes, V. Khoromskaia, B. N. Khoromskij and H. G. Matthies. *Tucker tensor analysis of Matérn functions in spatial statistics.* arXiv:1711.06874, 2017.
[256] S. A. Losilla, D. Sundholm, J. Juselius The direct approach to gravitation and electrostatics method for periodic systems. *J. Chem. Phys.* 132 (2) (2010) 024102.
[257] A. Lozinski and C. Chauviére, *A fast solver for Fokker–Planck equation applied to viscoelastic flows calculations: 2D FENE model.* J. of Comput. Physics. 2003, v. 189 (2), 607–625.
[258] Ch. Lubich, *From quantum to classical molecular dynamics: reduced models and numerical analysis.* Zurich Lectures in Advanced Mathematics, EMS, 2008.
[259] Ch. Lubich, On Variational Approximations in Quantum Molecular Dynamics. *Math. Comp.* 74 (2005), 765–779.
[260] Ch. Lubich, T. Rohwedder, R. Schneider and B. Vandereycken, Dynamical approximation of hierarchical Tucker and tensor-train tensors. *SIAM J Matrix Anal. Appl.* 34(2) (2013), 470–494.
[261] Ch. Lubich and I. V. Oseledets, A projector-splitting integrator for dynamical low-rank approximation. *BIT Numer. Mathematics* 54(1) (2014), 171–188.
[262] Ch. Lubich, I. V. Oseledets and B. Vandereycken, Time integration of tensor trains. *SIAM J Numer. Anal.* 53(2) (2015), 917–941.
[263] J. Lund and K. L. Bowers, *Sinc Methods for Quadrature and Differential Equations.* SIAM, Philadelphia, 1992.

[264] O. Mali, P. Neittaanmaki, S. Repin, *Accuracy verification methods. Theory and algorithms.* Springer, Berlin, Heidelberg, 2014.

[265] S. G. Mallat, *A Wavelet Tour of Signal Processing.* Academic Press, San Diego, 1999.

[266] F. R. Manby, Density fitting in second-order linear-r_{12} Møller–Plesset perturbation theory, *J Chem. Phys.* 119(9) (2003), 4607–4613.

[267] F. R. Manby, P. J. Knowles and A. W. Lloyd, The Poisson equation in density fitting for the Kohn–Sham Coulomb problem. *J. Chem. Phys.* 115 (2001), 9144–9148.

[268] A. Mantzaflaris, B. Jüttler, B. N. Khoromskij and U. Langer, Low Rank Tensor Methods in Galerkin-based Isogeometric Analysis. *Computer Methods in Applied Mechanics and Engineering* 316 (2017), 1062–1085.

[269] A. Mantzaflaris, B. Jüttler, B. N. Khoromskij and U. Langer, *Matrix Generation in Isogeometric Analysis by Low Rank Tensor Approximation.* In: *Lecture Notes in Computer Science.* Boissonnat et. al. ed., LNCS, Springer, Berlin, 2015, p. 321–340.

[270] G. I. Marchuk and V. V. Shaidurov, *Difference methods and their extrapolations.* Applications of Mathematics, New York: Springer, 1983.

[271] B. Matérn, *Spatial Variation.* Vol. 36 of Lecture Notes in Statistics, 2nd edn. Springer-Verlag, Berlin, New York, 1986.

[272] H. Matthies and A. Keese, Galerkin methods for linear and nonlinear elliptic stochastic partial differential equations. *Computer Methods in Applied Mechanics and Engineering* 194 (2005), 1295–1331.

[273] H. Matthies, A. Litvinenko, O. Pajonk, B. V. Rosić and E. Zander, *Parametric and uncertainty computations with tensor product representations in Uncertainty quantification in scientific computing.* A. M. Dienstfrey and R. F. Boisvert, eds., vol. 377 of IFIP Advances in Information and Communication Technology, Springen, Berlin, Heidelberg, 2012, 139–150.

[274] S. A. Matveev, D. A. Zheltkov, E. E. Tyrtyshnikov and A. P. Smirnov, Tensor train versus Monte Carlo for the multicomponent Smoluchowski coagulation equation. *J Comp. Phys.* 316 (2016), 164–179.

[275] V. Mazyja, G. Schmidt, *Approximate Approximations*, Math. Surveys and Monographs, Vol. 141, AMS 2007.

[276] J. Mercer, Functions of positive and negative type and their connection with the theory of integral equations. *Philosophical Transactions of the Royal Society A.* 209(441–458) (1909), 415–446.

[277] H.-D. Meyer, F. Gatti and G. A. Worth, *Multidimensional Quantum Dynamics: MCTDH Theory and Applications.* Willey-VCH, Wienheim, 2009.

[278] M. J. Mohlenkamp and L. Monzón, Trigonometric identities and sums of separable functions. *The Mathematical Intellingencer* 27 (2005), 65–69.

[279] M. J. Mohlenkamp and T. Young, *Convergence of Green Iterations for Schrödinger Equations.* In: Recent Advances in Computational Sciences: Selected Papers from the International Workshop on Computational Sciences and its Education, Beijing, China, 2005.

[280] C. Møller and M. S. Plesset, Note on an Approximation Treatment for many-Electron Systems. *Phys. Rev.* 46 (1934), 618.

[281] B. Munsky and M. Khammash, The finite state projection algorithm for the solution of the chemical master equation. *The Journal of chemical physics* 124 (2006), 044104.

[282] P. Neittaanmaki and S. Repin, *Reliable methods for computer simulation. Error control and a posteriori estimates.* Elsevier, 2004.

[283] A. Nouy, Low-rank methods for high-dimensional approximation and model order reduction. *Model Reduction and Approximation: Theory and Algorithms* 15 (2017), 171.

[284] G. Onida, L. Reining and A. Rubio, Electronic excitations: density-functional versus many-body Green's-function approaches. *Rev. of Modern Physics* 74(2) (2002), 601–659.

[285] Ch. Ortner, L. Zhang, *Atomistic/Continuum Blending with Ghost Force Correction.* SIAM J. Sci. Comp. 38(1) (2016), A346–A375.
[286] I. V. Oseledets, Approximation of $2^d \times 2^d$ matrices using tensor decomposition. *SIAM J. Matrix Anal. Appl.* 31(4) (2010), 2130–2145.
[287] I. V. Oseledets, Approximation of matrices with logarithmic number of parameters. *Doklady Math.* 428(1) (2009), 23–24.
[288] I. V. Oseledets, Constructive representation of functions in low-rank tensor formats. *Constructive Approximation* 37(1) (2013), 1–18.
[289] I. V. Oseledets, Tensor train decomposition. *SIAM J. Sci. Comp., v.* 33(5) (2011), 2295–2317.
[290] I. V. Oseledets et al., Tensor Train Toolbox. https://github.com/oseledets/TT-Toolbox; 2014.
[291] I. V. Oseledets and E. E. Tyrtyshnikov, Algebraic wavelet transform via quantics tensor train decomposition. *SIAM J. Sci. Comput.* 33(3) (2011), 1315–1328.
[292] I. V. Oseledets and E. E. Tyrtyshnikov, Breaking the Curse of Dimensionality, or How to Use SVD in Many Dimensions. *SIAM J. Sci. Comp.* 31 (2009), 3744–3759.
[293] I. V. Oseledets and E. E. Tyrtyshnikov, TT-Cross Approximation for Multidimensional arrays. *Liner Algebra Appl.* 432(1) (2010), 70–88.
[294] I. V. Oseledets, B. N. Khoromskij and R. Schneider, *Efficient time-stepping scheme for dynamics on TT-manifolds.* Preprint 24/2012, Max-Planck Institute for Math. in the Sciences, Leipzig, 2012. http://www.mis.mpg.de/preprints/2012/preprint2012_24.pdf.
[295] I. V. Oseledets, D. V. Savostyanov and E. E. Tyrtyshnikov, Tucker dimensionality reduction of three-dimensional arrays in linear time. *SIMAX* 30(3) (2008), 939–956.
[296] J-M. Papy, L. De Lathauwer and S. Van Huffel, *Exponential Data Fitting Using Multilinear Algebra: the Decimative Case.* Chemometrics, (www.interscience.willey.com) doi:10.1002/cem.1212.
[297] P. Parkkinen, S. A. Losilla, E. Solala, E. A. Toivanen, W.-H. Xu, D. Sundholm. *A generalized grid-based fast multipole method for integrating Helmholtz kernels.* J. Chem. Theory Comp., 13(2), 654–665, 2017.
[298] D. Perez-Garcia, F. Verstraete, M. M. Wolf and J. I. Cirac, Matrix Product State Representations. Quantum Information & Computation 7(5) (2007), 401–430.
[299] T. Petersdorff and C. Schwab, Numerical solution of parabolic equations in high dimensions. *Math Model Numer Anal* 38 (2004), 93–128.
[300] C. Pisani, M. Schütz, S. Casassa, D. Usvyat, L. Maschio, M. Lorenz and A. Erba, *CRYSCOR: a program for the post-Hartree–Fock treatment of periodic systems*, Phys. Chem. Chem. Phys. 14 (2012), 7615–7628.
[301] R. Polly, H.-J. Werner, F. R. Manby and P. J. Knowles. Fast Hartree–Fock theory using density fitting approximations. Mol.Phys. 102 (2004), 2311–2321.
[302] P. Pulay, Improved SCF convergence acceleration. *J. Comput. Chem.* 3 (1982), 556–560.
[303] M. V. Rakhuba, I. V. Oseledets, Fast multidimensional convolution in low-rank formats via cross approximation. *SIAM J Sci. Comp.* 37(2) (2015), A565–A582.
[304] M. V. Rakhuba, I. V. Oseledets, Grid-based electronic structure calculations: The tensor decomposition approach. *J Comp. Phys.* 312 (2016), 19–30.
[305] G. Rauhut, P. Pulay, H.-J. Werner, Integral transformation with low-order scaling for large local second-order Møller–Plesset calculations. *J. Comp. Chem.* 19 (1998), 1241–1254.
[306] H. Rauhut, R. Schneider and Z. Stojanac, Low rank tensor recovery via iterative hard thresholding. *Lin. Algebra Appl.* 523 (2017), 220–262.
[307] E. Rebolini, J. Toulouse and A. Savin, Electronic excitation energies of molecular systems from the Bethe–Salpeter equation: Example of H2 molecule. *In: Concepts and Methods in Modern Theoretical Chemistry (S. Ghosh and P. Chattaraj eds), Electronic Structure and Reactivity* 1 (2013), 367.

[308] M. Reed and B. Simon, *Functional analysis*. Academic Press, Cambridge, Massachusetts 1972.

[309] S. Reine, T. Helgaker and R. Lindh, Multi-electron integrals. *WIREs Comput. Mol. Sci.* 2 (2012), 290–303.

[310] S. Repin, *A Posteriori Estimates for Partial Differential Equations*. Walter de Gruyter, Berlin, 2008.

[311] S. Repin, T. Samrowski and S. Sauter. *A posteriori error majorants of the modeling errors for elliptic homogenization problems*, C. R. Math. Acad. Sci. Paris 351(23–24) (2013), 877–882.

[312] T. Rohwedder and A. Uschmajew, On local convergence of alternating schemes for optimization of convex problems in the tensor train format. *SIAM J Numer. Anal.* 51(2) (2013), 1134–1162.

[313] T. Rohwedder and R. Schneider, Error estimates for the Coupled Cluster method. *ESAIM* 47(6) (2013), 1553–1582.

[314] Y. Saad, J. R. Chelikowsky and S. M. Shontz, Numerical Methods for Electronic Structure Calculations of Materials. *SIAM Review* 52(1) (2010), 3–54.

[315] E. E. Salpeter and H. A. Bethe, A relativistic equation for bound-state problems. *Phys. Review* 82(2) (1951), 309–310.

[316] S. A. Sauter and Ch. Schwab, *Boundary Element Methods*. Springer, Berlin, Heidelberg, 2011.

[317] D. V. Savostyanov and I. V. Oseledets, *Fast adaptive interpolation of multi-dimensional arrays in tensor train format*. Proceedings of 7th International Workshop on Multidimensional Systems (nDS), University of Poitiers, France, 2011, IEEE, doi:10.1109/nDS.2011.6076873,

[318] R. Schatten, *A theory of cross-spaces*. University Press, Princeton (1950).

[319] R. Schneider and A. Uschmajew, Approximation rates for the hierarchical tensor format in periodic Sobolev spaces. *Journal of Complexity* 30(2) (2014), 56–71.

[320] R. Schneider and A. Uschmajew, Convergence results for projected line-search methods on varieties of low-rank matrices via Lojasiewicz inequality. *SIAM J. Optim.* 25(1) (2015), 622–646.

[321] U. Schollwöck, The density-matrix renormalization group. *Reviews of modern physics* 77(1) (2005), 259.

[322] U. Schollwöck, *The density-matrix renormalization group in the age of matrix product states*, Ann. Phys.326(1) (2011), 96–192.

[323] C. Schwab, *p- and hp-Finite Element Methods*. Oxford University Press, Oxford, UK, 1998.

[324] C. Schwab and R.-A. Todor, Karhunen–Loéve approximation of random fields by generalized fast multipole methods. *J. of Comp. Phys.* 217 (2006), 100–122.

[325] C. Schwab and R. Todor, Sparse finite elements for elliptic problems with stochastic loading. *Numer. Math.* 95(4) (2003), 707–713.

[326] C. Schwab and R. Stevenson, *Space-time adaptive wavelet methods for parabolic evolution problems*. Math. Comp. 78(267) (2009), 1293–1318.

[327] E. Schmidt, Zur Theorie der linearen und nichlinearen Integralgleichungen. I. Teil: Entwiklung willkürlichen Functionen nach Systemen vorgeschriebener. *Mathematische Annalen* 63 (1907), 433–476.

[328] V. De Silva and L.-H. Lim, Tensor rank and the ill-posedness of the best low-rank approximation problem. *SIAM J. Matrix Anal. Appl.* 30(3) (2008), 1084–1127.

[329] I. Sloan and H. Wozniakowski, When Are Quasi-Monte Carlo Algorithms Efficient for High Dimensional Integrals. *J. of Complexity* 14(1) (1998), 1–33.

[330] S. A. Smolyak, Quadrature and interpolation formulas for tensor products of certain class of functions. Dokl. Akad. Nauk SSSR 148(5) (1963), 1042–1053. *Transl.: Soviet Math. Dokl.* 4 (1963), 240–243.

[331] P. K. Suetin, *Classical Orthogonal Polynomials (in Russian)*. Nauka, Moscow, 1979.

[332] A. Smilde, R. Bro, P. Geladi, *Multi-way Analysis*. Wiley, Hoboken, New Jersey, USA, 2004.

[333] A. Stegeman and N. D. Sidiropoulos, On Kruskal's uniqueness condition for the Candecomp/Parafac decomposition. *Lin. Alg. Appl.* 420 (2007), 540–552.

[334] F. Stenger, *Numerical methods based on Sinc and analytic functions*. Springer-Verlag, Heidelberg, 1993.

[335] J. Strang and G. J. Fix, *An Analysis of the Finite Element Method*. Prentice-Hall, inc. N. J., 1973.

[336] V. Strassen, *Gaussian Elimination is not Optimal*, Numer. Math. 13 (1969), 354–356.

[337] Stoll, M. and Breiten, T, *A low-rank in time approach to PDE-constrained optimization*. Max-Planck Institute for Mathem. in the Sciences, Magdeburg Preprint, 13(08), 2013.

[338] A. Szabo and N. Ostlund, *Modern Quantum Chemistry*. Dover Publication, New York, 1996.

[339] G. Szegö, *Orthogonal Polynomials*. American Mathematical Society, New York, 1959.

[340] E. Tadmor, The exponential accuracy of Fourier and Chebyshev differencing methods. *SIAM J. Numer. Anal.* 23 (1986), 1–10.

[341] R.-A. Todor and Ch. Schwab, *Convergence rate for sparse chaos approximations of elliptic problems with stochastic coefficients*. IMA J. of Numer. Anal. (2007) 27, 232–261.

[342] L. Tamellini, O. Le Maitre and A. Nouy, Model reduction based on proper generalized decomposition for the stochastic steady incompressible Navier–Stokes equations. *SIAM Journal on Scientific Computing* 36(3) (2014), A1089–A1117.

[343] V. N. Temlyakov, Greedy Algorithms and *M*-Term Approximation with Regard to Redundant Dictionaries. *J. of Approx. Theory* 98 (1999), 117–145.

[344] L. N. Trefethen, *Approximation Theory and Approximation Practice*. SIAM Publ., Philadelphia, 2013.

[345] L. N. Trefethen and D. Bau III, *Numerical linear algebra*. SIAM Publ., Philadelphia, 1997.

[346] E. E. Tyrtyshnikov, Incomplete cross approximation in the mosaic-skeleton method. *Computing* 64 (2000), 367–380.

[347] E. E. Tyrtyshnikov, Kronecker-product approximations for some function-related matrices. *Linear Algebra Appl.* 379 (2004), 423–437.

[348] E. E. Tyrtyshnikov, Mosaic-skeleton approximations. *Calcolo* 33(1) (1996), 47–57.

[349] E. E. Tyrtyshnikov, Tensor approximations of matrices generated by asymptotically smooth functions. *Sbornik: Mathematics* 194(5–6) (2003), 941–954 (translated from *Mat. Sb.* 194(6) (2003), 146–160.

[350] L. R. Tucker, Some mathematical notes on three-mode factor analysis. *Psychometrika* 31 (1966), 279–311.

[351] C. F. Van Loan, The ubiquitous Kronecker product. *J. of Comp. and Applied Math.* 123 (2000), 85–100.

[352] C. F. Van Loan and J. P. Vokt, Approximating Matrices with Multiple Symmetries. *SIAM. J. Matr. Anal. Appl.* 36(3) (2015), 974–993.

[353] J. VandeVondele, M. Krack, F. Mohamed, M. Parinello, Th. Chassaing, J. Hutter, QUICKSTEP: Fast and accurate density functional calculations using a mixed Gaussian and plane waves approach. *Comp. Phys. Comm.* 167 (2005), 103–128.

[354] F. Verstraete, V. Murg and J. I. Cirac, Matrix product states, projected entangled pair states, and variational renormalization group methods for quantum spin systems. *Advances in Physics* 57(2) (2008), 143–224.

[355] F. Verstraete, D. Porras and J. I. Cirac, DMRG and periodic boundary conditions: A quantum information perspective. *Phys. Rev. Lett.* 93(22) (Nov. 2004), 227205.

[356] G. Vidal, Efficient classical simulation of slightly entangled quantum computations. *Phys. Rev. Lett.* 91(14) (2003), 147902-1–147902-4.

[357] V. S. Vladimirov, *Equations of mathematical physics*. 3rd edn. Izdatel'stvo Nauka, Moscow, 1976.

[358] A. Uschmajew, Local Convergence of the Alternating Least Squares Algorithm for Canonical Tensor Approximation. *SIAM. J. Matrix Anal. & Appl.* 33(2) (2012), 639–652.

[359] A. Uschmajew, Well-posedness of convex maximization problems on Stiefel manifolds and orthogonal tensor product approximations. *Numer. Math.* 115(2) (2010), 309–331.

[360] H. Wang and M. Thoss, Multilayer formulation of the multiconfiguration time-dependent Hartree theory. *J. Chem. Phys.* 119 (2003), 1289–1299.

[361] J. H. M. Wedderburn, *Lectures on matrices, colloquium publications*, AMS, NY XVII (1934), 205.

[362] H.-J. Werner, P. J. Knowles, et al., MOLPRO, Version 2002.10, A Package of Ab Initio Programs for Electronic Structure Calculations.

[363] H.-J. Werner, P. J. Knowles, G. Knozia, F. R. Manby and M. Schuetz, Molpro: a general-purpose quantum chemistry program package. *WIREs Comput. Mol. Sci.* 2 (2012), 242–253.

[364] H.-J. Werner, F. R. Manby and P. J. Knowles, Fast linear scaling second order Møller–Plesset perturbation theory (MP2) using local and density fitting approximations. *J. Chem. Phys.* 118 (2003), 8149–8160.

[365] S. R. White, Density-matrix algorithms for quantum renormalization groups. *Phys. Rev. B* 48(14) (1993), 10345–10356.

[366] J. M. Whittacker, *Interpolatory Function Theory.* Cambridge Tracts in Mathematics and Mathematical Physics, Cambridge University Press, London, 33, 1935.

[367] P. Wind, W. Klopper and T. Helgaker, Second order Møller–Plesset perturbation theory with terms linear in interelectronic coordinates and exact evaluation of three-electron integrals. *Theor. Chem. Acc.* 107 (2002), 173–179.

[368] T. Yanai, G. Fann, Z. Gan, R. Harrison and G. Beylkin, *Multiresolution quantum chemistry: Hartree–Fock exchange.* J. Chem. Phys.121(14) (2004), 6680–6688.

[369] Y. Yang, Y. Kurashige, F. R. Manby and G. K. L. Chan, Tensor factorizations of local second-order Møller–Plesset theory. *J. Chem. Phys.* 134 (2011), 044123:1–13.

[370] H. Yserentant, *Regularity and Approximability of Electronic Wave Functions.* Lecture Notes in Math., 2000, Springer, Dordrecht, 2010.

[371] H. Yserentant, The hyperbolic cross space approximation of electronic wavefunctions. *Numer. Math.* 105 (2007), 659–690.

[372] E. Zeidler. *Applied Functional Analysis: Applications to Mathematical Physics.* Springer, Berlin, 1995.

[373] T. Zhang and G. Golub. Rank-One approximation to high order tensors. SIAM J. Matrix Anal. Appl. 23 (2001), 534–550.

Index

https://doi.org/10.1515/9783110365917-007

Radon Series on Computational and Applied Mathematics

Volume 20
Fluid-Structure Interaction. Modeling, Adaptive Discretisations and Solvers
Stefan Frei, Bärbel Holm, Thomas Richter, Thomas Wick, Huidong Yang (Eds.)
ISBN: 978-3-11-049527-0, e-ISBN: 978-3-11-049425-9

Volume 18
Variational Methods. In Imaging and Geometric Control
Maitine Bergounioux, Gabriel Peyré, Christoph Schnörr, Jean-Baptiste Caillau, Thomas Haberkorn (Eds.), 2016
ISBN: 978-3-11-043923-6, e-ISBN: 978-3-11-043039-4

Volume 17
Topological Optimization. Optimal Transport in the Applied Sciences
Maitine Bergounioux, Édouard Oudet, Martin Rumpf, Filippo Santambrogio, Guilaume Carlier, Thierry Champion (Eds.), 2016
ISBN: 978-3-11-043926-7, e-ISBN: 978-3-11-043041-7

Volume 16
Algebraic Curves and Finite Fields. Cryptography and Other Applications
Harald Niederreiter, Alina Ostafe, Daniel Panario, Arne Winterhof (Eds.), 2014
ISBN: 978-3-11-031788-6, e-ISBN: 978-3-11-031791-6

Volume 15
Uniform Distribution and Quasi-Monte Carlo Methods. Discrepancy, Integration and Applications
Peter Kritzer, Harald Niederreiter, Friedrich Pillichshammer, Arne Winterhof (Eds.), 2014
ISBN: 978-3-11-031789-3, e-ISBN: 978-3-11-031793-0

Volume 14
Direct and Inverse Problems in Wave Propagation and Applications
Ivan G. Graham, Ulrich Langer, Jens M. Melenk, Mourad Sini (Eds.), 2013
ISBN: 978-3-11-028223-8, eISBN: 978-3-11-028228-3

www.ingramcontent.com/pod-product-compliance
Ingram Content Group UK Ltd.
Pitfield, Milton Keynes, MK11 3LW, UK
UKHW051022091026
15624UKWH00012B/626

9 783110 370133